ENGINEERING DESIGN
COMMUNICATION

ENGINEERING DESIGN COMMUNICATION

Conveying Design Through Graphics

Second Edition

SHAWNA D. LOCKHART

CINDY M. JOHNSON

Prentice Hall
Boston   Columbus   Indianapolis   New York   San Francisco   Upper Saddle River
Amsterdam   Cape Town   Dubai   London   Madrid   Milan   Munich   Paris   Montreal   Toronto
Delhi   Mexico City   São Paulo   Sydney   Hong Kong   Seoul   Singapore   Taipei   Tokyo

Editorial Director: Vernon R. Anthony
Acquisitions Editor: Sara Eilert
Editorial Assistant: Doug Greive
Director of Marketing: David Gesell
Senior Marketing Manager: Harper Coles
Senior Marketing Coordinator: Alicia Wozniak
Marketing Assistant: Crystal Gonzalez
Project Manager: Maren L. Miller
Senior Managing Editor: JoEllen Gohr
Associate Managing Editor: Alexandrina Benedicto Wolf
Senior Operations Supervisor: Pat Tonneman
Operations Specialist: Deidra Skahill

Senior Art Director: Diane Y. Ernsberger
Cover Director: Jayne Conte
Cover Designer: Suzanne Behnke
Cover Image: Santa Cruz Bicycles and www.E-Cognition.net
AV Project Manager: Janet Portisch
Copyeditor: Barbara Liguori
Full-Service Project Management: Karen Fortgang, bookworks
Composition: S4Carlisle Publishing Services
Printer/Binder: Courier/Kendallville
Cover Printer: Lehigh-Phoenix Color/Hagerstown
Text Font: Times 10/12

Credits and acknowledgments borrowed from other sources and reproduced, with permission, in this textbook appear on the appropriate page within the text. Unless otherwise stated, all artwork has been provided by the author.

Certain images and materials contained in this text were reproduced with the permission of Autodesk, Inc. © 2011. All rights reserved. Autodesk, AutoCAD, DWG, and the DWG logo are registered trademarks of Autodesk, Inc., in the U.S.A. and certain other countries.

SolidWorks® is a registered trademark of DS SolidWorks Corp.

Library of Congress Cataloging-in-Publication Data
Lockhart, Shawna D.
   Engineering design communication : conveying design through graphics / Shawna D. Lockhart,
Cindy M. Johnson. —2nd ed.
       p. cm.
   ISBN–13: 978-0-13-705714–6
   ISBN–10: 0-13-705714–8
   1. Engineering design—Graphic methods.   2. Engineering graphics.
I. Johnson, Cindy M.   II. Title.   III. Title: Conveying design through graphics.
   TA337.L63 2012
   620'.0042—dc22

                                                        2011016875

10 9 8 7 6 5 4 3 2 1

www.pearsonhighered.com

ISBN 10:   0-13-705714-8
ISBN 13: 978-0-13-705714-6

CONTENTS

CHAPTER SIX
CONSTRAINT-BASED MODELING
AND DESIGN   218

CHAPTER SEVEN
MODELING FOR MANUFACTURE
AND ASSEMBLY   266

CHAPTER EIGHT
USING THE MODEL FOR ANALYSIS AND PROTOTYPING   322

CHAPTER NINE
DOCUMENTATION DRAWINGS   368

CHAPTER OPENER SECTIONS

Topics you can expect to learn about are listed here.

A large illustration and an interesting overview give you a real-world context for what this chapter is about.

8 USING THE MODEL FOR ANALYSIS AND PROTOTYPING

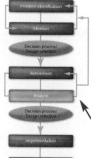

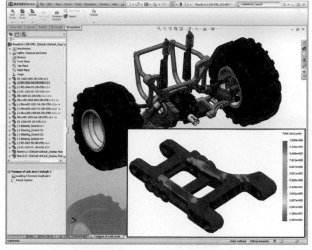

Fatigue analysis uses peak loads to estimate the lifetime of critical components. "Hot" colors in the link arm indicate sections that will fatigue soonest. *(Image courtesy of DS SolidWorks Corp.)*

OBJECTIVES

When you have completed this chapter you should be able to:

1. Extract mass properties data from your CAD models.
2. Evaluate the accuracy of mass properties calculations.
3. Define the file formats used for exporting CAD data.
4. Describe how analysis data can be used to update parametric models.
5. List analysis methods that can use the CAD database.
6. Describe how rapid prototyping systems create physical models from CAD data

The design icon highlights the stage of the design process addressed in the chapter

TESTING, TESTING, TESTING

The refinement phase forms an iterative loop with the analysis phase of design. Test results feed back into model changes, and the revised model is used for further testing. Your 3D solid model database defines the geometry of the design and provides information for engineering analysis, such as determining mass, volume, surface area, and moments of inertia.

You can use tools such as a spreadsheet, equation solver, motion simulator, and finite element package

with model data to check for stress concentrations; determine deflections, shear forces, bending moments, heat transfer properties, and natural frequencies; perform failure analysis and vibration analysis; and make many other calculations.

You can use rapid prototyping equipment to create physical parts as well as "virtual prototypes" that are computer simulations of model behavior. Both kinds of prototypes allow customers and others to interact with and evaluate the design.

323

SPOTLIGHT

Spotlight boxes present background or more in-depth information about a topic.

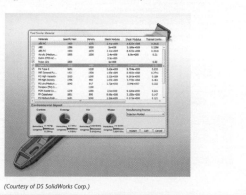

SPOTLIGHT

Getting Easier Being Green

SolidWorks Sustainability is a module that provides life cycle assessment (LCA) within the SolidWorks user interface. It gives you information about environmentally friendly options for your engineering design decisions. This software allows you to conduct life cycle analyses of parts and assemblies. You can use it to see the environmental impact in terms of air, carbon, energy, and water.

To see if you improve your design's environmental impact as you modify it, you can capture the existing baseline, then track how your new design compares. The **Find Similar Material** dialog box shown lets you search the material database for alternative materials based on their mechanical properties. As you choose them you can visually assess the magnitude and direction of their environmental impact with the charts in the panel at the bottom.

(Courtesy of DS SolidWorks Corp.)

STEP BY STEP ACTIVITIES

The Step-by-Step tab identifies these activities.

TWO-POINT PERSPECTIVE

To sketch an object using two vanishing points, follow these steps.

1 Sketch the front corner of the object at its true height. Locate two vanishing points (VPL and VPR) on a horizon line (at eye level). Distance CA may vary—the greater it is, the higher the eye level will be and the more you will be looking down on top of the object. A rule of thumb is to make C–VPL one third to one fourth of C–VPR.

2 Estimate depth and width, and sketch the enclosing box.

3 Block in all details. Note that all parallel lines converge toward the same vanishing point.

4 Darken all final lines. Make the outlines thicker and the inside lines thinner, especially where they are close together.

INDUSTRY CASES

Industry Cases feature engineers talking about the decisions they make and the steps they take when modeling for design.

A gray header with an Industry Case tab identifies these pages.

INDUSTRY CASE

TESTING THE MODEL: VIBRATION ANALYSIS

Lasers are subjected to some intense conditions. Installed in a wind tunnel, the assembly needs to withstand supersonic wind velocities. In a helicopter, the laser needs to function despite vibrations from the engine and rotors. Each of these generates vibrations at different frequencies. Before Quantel USA builds a prototype for testing, it uses the 3D model and SolidWorks Simulation to "shake up" the laser assembly to see how the parts will respond to vibrations at frequencies typical for the application.

8.53 The circuit board attaches to the sheet metal mount, which attaches via two bolts.

The circuit board shown in Figure 8.53 generates a high-voltage pulse to a laser optic. It is mounted to a cantilevered sheet metal part that attaches to the laser assembly with two bolts. When the circuit board and mounting assembly are subjected to vibrations at different frequencies, will the resultant deflection (bending) result in failure? Laine McNeil, Senior Mechanical Engineer, set up a simulation to help find out.

First, he prepared the model for testing by generating a mesh appropriate for each part. Choosing the right kind of mesh and the right density (number of nodes) is a judgment call driven by the nature of the test. For example, the board components had been tested before, so McNeil could focus on the board and the cantilevered sheet metal part. For the circuit board, McNeil specified a 3D mesh to reflect the thickness of the board (see Figure 8.54). For the sheet metal mount, he chose a shell mesh and selected sheet metal as the material whose properties the mesh would use.

Boundary condition markers

8.54 The Mesh Applied to the Model

It is common to simplify the 3D model (by removing fillets, perhaps) to reduce the number of nodes in the mesh. Adding nodes improves accuracy, but it also increases the time needed for analysis. For example, the shell mesh chosen for the mount uses one surface of the part and applies the stiffness properties uniformly over the entire part. This generated fewer nodes, but provided enough complexity to adequately represent the sheet metal part.

McNeil next set up the boundary conditions for the assembly: what are the constraints on how the parts can move? First, how are the parts attached to each other? Parts can be bonded or attached in a way that they cannot move or slide (fixed geometry), hinged, connected via a roller/slider, or supported by a spring. The circuit board was mounted in place with three rubber interfacing pads. The rubber was bonded on one end to the sheet metal mount and to the circuit board at the other for this analysis.

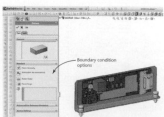

Boundary condition options

8.55 Choosing Boundary Conditions for the Circuit Board

Second, how are the parts connected to the outside world? The sheet metal mount is attached solely by its two bolts, so boundary conditions were set for the two bolt holes (shown in green in Figure 8.55). That is, only the bolt forces and friction are holding the mount; everything else can compress and deflect in the simulation. McNeil could have chosen "on flat surface" for the mount so that the part would not be allowed to deflect downward, only up, but that was not needed for this test.

The model was now analyzed at frequencies expected for its target use. Typical vibration frequencies range from 100 to 2000 hertz. Some clients have minimum standards that must be met, such as the U.S. military's "minimum integrity test." For the circuit board analysis, no deflection was detected in the sheet metal mount until almost 280 hertz. Figure 8.56(c) shows the deflection. The material between the bolt holes is shaded a lighter blue to indicate a small amount of movement, and the upper cantilevered portion of the sheet metal mount is colored red, indicating significantly higher motion. The circuit board had already been tested to conform to the minimum integrity test. Because the circuit board operated reliably despite the first resonant frequency mode shape showing that deformation occurred at as low as 120 hertz, McNeil could be confident that the assembly would be satisfactory as designed.

This result was not always the case. When the circuit board components were initially tested, the cube (representing a power supply) at the top left in Figure 8.56(a) fell off. It had been soldered onto the board, but the solder pads were not strong enough to hold at typical vibrations. Glue was added beneath the power supply to strengthen the bond.

In addition to simple resonant frequency shape generation, analysis can also simulate the energy driving the vibration and the associated displacement. Two examples of vibration-generating devices can readily demonstrate the difference. A cell phone vibrates, but the energy driving the vibration is not enough to shake a room. A hydraulically driven vibrator, used to dislodge material from the bed of a dump truck, has enough energy to shake a building one city block away. For the circuit board assembly, McNeil used the built-in capabilities of SolidWorks, so the energy behind the vibration was assumed to be infinite, and the displacement shown for the circuit board parts was not meaningful. In this instance, this was not the focus of the analysis. If determining actual displacement values is required, the model can be analyzed with a higher-end FEA package, in which a power spectral density curve can be used to define the energy levels causing motion.

At Quantel USA, testing the "virtual prototype" does not replace the need for a physical model. For each design a physical model is built and shaken at the frequencies expected during use. But if the model testing is done properly, the physical model passes the test and only one prototype is needed.

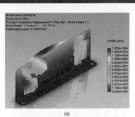

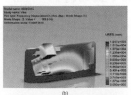

8.56 Results of Testing at (a) 121.70, (b) 183.5, and (c) 274.69 Hertz

(All images courtesy of Quantel USA.)

(continued)

DESIGNER'S NOTEBOOK

Tan pages with a tab on the outer edge identify the Designer's Notebook pages, which show real-life examples of design notebooks and drawings.

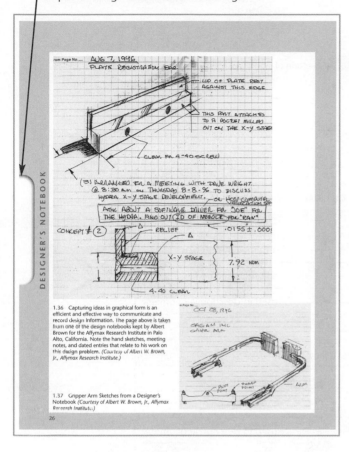

1.36 Capturing ideas in graphical form is an efficient and effective way to communicate and record design information. The page above is taken from one of the design notebooks kept by Albert Brown for the Affymax Research Institute in Palo Alto, California. Note the hand sketches, meeting notes, and dated entries that relate to his work on this design problem. (Courtesy of Albert W. Brown, Jr., Affymax Research Institute.)

1.37 Gripper Arm Sketches from a Designer's Notebook (Courtesy of Albert W. Brown, Jr., Affymax Research Institute.)

26

REVERSE ENGINEERING PROJECT

Reverse Engineering Project pages at the end of each chapter are a continuing project to measure and model a household can opener.

HANDS ON TUTORIALS

A plan for following tutorials that are part of the student edition of SolidWorks is suggested at the end of each chapter. SolidWorks student edition may be purchased as a single price bundle with the text. This is an easy way to provide the software to students. Directions for locating and starting the tutorial are given.

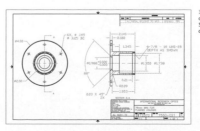

ILLUSTRATIONS

Colored callouts differentiate explanatory text from annotations in technical drawings.
Consistent use of color helps to differentiate the meaning of projection lines, fold lines, and other drawing elements.
A color key is provided for easy reference.

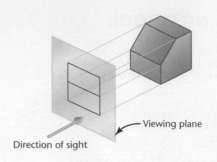

Direction of sight

Viewing plane

Color Key for Instructional Art

Item	In instructional art	In a technical drawing	
Callout arrow	⟶	*	
Dimension line	⟵⟶	⟵⟶	a thin (0.3 mm) black line
Projection line	————	————	a lightly sketched line
Folding line	— — — —	— — —	used in descriptive geometry
Picture plane on edge	————	*	
Plane of projection		*	
Cutting plane on edge	— — — —	↑⌐ — — ⌐↑	(see Chapter 4)
Cutting plane		*	
Reference plane on edge	————	— — —	used in descriptive geometry
Reference plane		*	
Viewing direction arrow	⟹	↑⌐ — — ⌐↑	
Horizon + ground line	————	————	
Rotation arrow	↶	30° ↶	

* Not a typical feature of technical drawings. (Shown in this book for instructional purposes.)

CHAPTER EXERCISES

Exercises 8–11 Use your modeling package to create solid, surface, or 3D wireframe models of the items in Chapter 4, Exercises 4 a–d.

Exercises 12–39 Use your modeling package to create solid, surface, or 3D wireframe models of the objects depicted in Chapter 3, Exercises 7–34. Use the units indicated to model the objects full size.

Exercises 40–45 Use a solid or surface modeling package to revolve the shapes in Chapter 4, Exercises 7–12 about the axes shown.

Exercises 46–49 Use a solid or surface modeling package to extrude the shapes in Chapter 4, Exercises 13–16.

50. For each part shown, list the modeling method you would choose to create it if you were creating the original product definition for mass-producing these items. Give the reasons for your choices.

a. *(Shutterstock.)*

b. *(Shutterstock.)*

c. *(Sergei Devyatkin/Shutterstock.)*

d. *(Shutterstock.)*

e. *(Shutterstock.)*

f. *(Photos.com.)*

g. *(Shutterstock.)*

h. *(Shutterstock.)*

i. *(Shutterstock.)*

Exercises are easy to find. The color stripe on the outer edge of the page corresponds to the chapter so you can flip to them quickly.

PREFACE

APPROACH

Engineering Design Communication is designed to support students who are learning to use graphics as a tool for engineering design, especially those who are learning to model in 3D. We take a more streamlined approach than traditional texts, with special emphasis on three key skills: 2D sketching, reading and visualizing objects from 2D views, and creating 3D models that will function as the design database.

We use the design process to motivate and anchor these skills, explaining why they are important and showing how they are used in the continuum from problem definition through ideation, refinement, analysis, and documentation. Case studies and industry examples illustrate the many ways these skills support practicing engineers in their work.

Unlike tutorial guides and some traditional texts, we strive to include the topics students need to develop models that:

- capture the design intent for a product or system;
- are updated properly when changes are made; and
- serve the many purposes associated with their role as the design database.

We expect that students will be learning in the lab how to create models using a specific software package. Our emphasis throughout is on helping them understand *why* and *when* to use the techniques they are learning. We provide practical tips and step-by-step instructions for sketching and modeling that support the hands-on nature of the course and that can be used with any software package.

Several options are available for bundling student software with this text, including the SolidWorks Student Design Kit and the tutorials it includes. Contact your sales representative for more details.

Our approach is also rooted in the role that the CAD database now plays in concurrent engineering. Engineers today are expected to function in a team environment, communicate with customers as well technical personnel, and prepare models that will be useful for analysis, manufacturing, and presentation purposes. A broad understanding of the many functions of the CAD database prepares students for the workplace and also equips them to evaluate and adapt to new software tools as they are developed. Three optional chapters on the Web present manufacturing issues, drawing management, and presentation graphics to complete this broad overview.

AUDIENCE

This book is suited for introductory graphics courses in which students are learning to sketch and model in 3D; introductory engineering design courses that introduce engineering graphics and modeling as a tool for design; and 3D modeling courses that may support a senior project or build on a 2D drawing/modeling course .

NEW IN THIS EDITION

- Coverage is streamlined. Ten core chapters cover engineering graphics in the design process. Manufacturing issues, animation, and drawing management are enrichment chapters available at www.pearsonhigered.com/lockhart. Additional topics may be added to a custom version of the book.
- Design intent is emphasized. There are more examples and discussion of what it means to capture design intent in a model.
- A step-by-step format is used for specific sketching, modeling, and analysis techniques.
- A reverse engineering project in Chapters 2–10 provides practice with sketching, measuring, modeling, analyzing, and creating drawings for a real product.
- Illustrations use color consistently to help students see key elements and to distinguish annotations from engineering drawing elements.
- The text is fully updated, with a new Chapter 1 case study, contemporary topics such as sustainability, and new cases and examples throughout.
- Exercise sections are enhanced and expanded.
- Suitable SolidWorks tutorials are suggested at chapter end.

HALLMARK FEATURES

Design Framework

Chapter 1 presents the process of design, which is then used as an organizing framework for the rest of the book. Each chapter opens with an introduction that explains how the graphic skills presented in the chapter relate to and are used in engineering practice. The design icon used in each chapter makes it possible to cover chapters earlier in the text without losing the link to the design process.

Industry Cases

Case histories at the end of each chapter reinforce key topics by showing students how a practicing engineer applies them. These case histories present specific instances of general principles presented in the text, showing how engineers use the graphics tools presented in that chapter to solve a design problem.

Designer's Notebook

The role of sketches in planning and visualization is reinforced in each chapter through the Designer's Notebook. Each chapter closes with sketches and notes from a practicing engineer's notebook that relate to the chapter's content. Each excerpt gives students real examples of how they might use their own design notebook. (Chapter 1 encourages them to begin and use such a notebook throughout the course.)

ORGANIZATION/PATHWAYS THROUGH THE BOOK

Sketching

Sketching is the focus of Chapters 2 and 3, which correspond to the ideation and visualization stages of the design process. Many topics traditionally taught in the context of documentation drawings are presented in these chapters so that students learn to use a full range of drawing views and projections in their sketches. Actual sketches are used to illustrate these early chapters so that students may compare their efforts with other sketches, not with instrument or CAD drawings.

Modeling

Modeling—and creating the design database—is the focus of Chapters 4 through 8, in which modeling is linked to the process of design refinement and analysis. These chapters begin with an introduction to geometry that takes students from the 2D sketching environment to the 3D environment they will work with on-screen. After a survey of different modeling methods, separate chapters on constraint-based modeling and assembly modeling introduce modeling considerations unique to 3D solid models. Each general modeling principle is illustrated with real models and examples showing the application of key ideas. Chapter 8, which covers using the model data for analysis and with downstream applications (including rapid prototyping), presents additional considerations for creating a useful design database.

Documentation

Documentation is the focus of Chapters 9 and 10. As in the modeling chapters, documentation and tolerancing techniques are presented in the context of the 3D model that students will use to generate drawings, rendered views, files for numerically controlled machinery, and more.

Web Chapters

Because many students lack an understanding of basic manufacturing processes, Chapter 11, "Implementation," presents key questions to be asked and answered before a design is ready for manufacture. This stand-alone chapter may be covered at any point before the documentation chapter. Because manufacturing issues are addressed in the context of modeling in Chapter 7, some instructors may want to assign Chapter 11 before Chapter 7.

Instructors who wish to ask students to start making drawings from their solid models earlier may cover Chapter 9, "Documentation Drawings," anytime after Chapter 3.

Chapter 12, "Drawing Control and Data Management," may be covered at any point in the text. For courses with more emphasis on rendering and animation, Chapter 13, "Animation and Presentation Graphics," may be covered anytime after Chapter 5.

Web chapters are available at www.pearsonhighered.com/lockhart.

COMPANION WEBSITE

Includes models and drawings for the Reverse Engineering Project; go to www.pearsonhighered.com/lockhart.

POWERPOINT LECTURE SLIDES

Instructors may order complimentary PowerPoint slides. Request ISBN-10: 013-260699-2.

ACKNOWLEDGMENTS

Thank you to the many individuals who provided feedback and suggestions as this book was developed:

R. Glenn Allen, Southern Polytechnic State University
Todd Amundson, Lower Columbia College
Radha Bala, Northern Illinois University
Chuck Bales, Moraine Valley Community College
Kirk Barnes, Ivy Tech Community College
Samir B. Billatos, University of Texas, Brownsville
John Cherng, University of Michigan, Dearborn
Robert A. Chin, East Carolina University
Garrett M. Clayton, Villanova University
Clark Cory, Purdue University
Barry Crittenden, Virginia Polytechnic Institute and State University
Thomas Doyle, McMaster University
Timothy Fox, California State University
Howard Fulmer, Villanova University
Ping Ge, Oregon State University
Frank Gerlitz, Washtenaw College
Larry Goss, University of Southern Indiana
Randy J. Gu, Oakland University
Kevin Henry, Columbia College, Chicago
Robert B. Jerard, University of New Hampshire
Craig Johnson, Central Washington University
Steve Kaminaka, California Polytechnic State University, San Luis Obispo

Hamid A. Khan, East Carolina University
Kathleen Kitto, Western Washington University
Douglas Koch, Southeast Missouri State University
Steve Lamm, San Antonio College
Dee Lauritzen, Eastern Arizona College
Bill Law, Tektronix, Inc.
Dennis Lieu, University of California, Berkeley
Brian Matthews, North Carolina State University
Chris Merritt, CADMAX Consulting
Fritz D. Meyers, Ohio State University
Gregory Mocko, Clemson University
Ahad Nasab, Middle Tennessee State University
John G. Nee, Central Michigan University
Ben Nwoke, Virginia State University
Patrick O'Connor, Texas State Technical College
Richard Patton, Mississippi State University
Ken Perry, Indiana University–Purdue University,
    Fort Wayne
David Planchard, Worcester Ploytechnic Institute
S. Manian Ramkumar, Rochester Institute of Technology
Virgil Seaman, California State University, Los Angeles
Arijit Sengupta, New Jersey Institute of Technology
James Shahan, Iowa State University
Tony Shay, Eastern Michigan University
Randy Shih, Oregon Institute of Technology
William Singhose, Georgia Tech
Pamela Speelman, Eastern Michigan University
Nancy E. Study, Southwest Missouri State University
Vivek Tandon, The University of Texas at El Paso
Roger Toogood, University of Alberta
Lee Waite, Rose-Hulman Institute
Mel Whiteside, Butler Community College
Jianping Yue, Essex Community College

For the many case studies and the models and drawings that illustrate them, we owe a huge debt of gratitude to:

- Joe Graney and Mike Ferrentino, Santa Cruz Bicycles
- Leo Greene, www.E-Cognition.net
- Jeff Salazar and Vanessa Jesperson, LUNAR
- Karen Markus Weil, Cronos Integrated Microsystems, Inc.
- Marty Albini and Bill Clem, Strategix ID

- Kent Svendseid, Marty Albini, and Aki Hirota, Strategix Vision
- William Townsend and Brandon Larocque, Barrett Technology, Inc.
- Stan Mclean, VKI Technologies
- Laine McNeil and Mark Perkins, Quantel USA
- Jae Ellers, Color Printing and Imaging Division, Tektronix, Inc.
- Carl Fehres, IRI International
- Hugo Haselhuhn, C&K Systems
- Robert Reisinger, Mountain Cycle

We sincerely thank all the individuals and companies who shared their experience, expertise, drawings, models, and advice with us in the making of this book. We are grateful to the following individuals for their help with this edition of the book:

Marty Albini; Mark Biasotti, Dassault Systèmes Solid-Works Corp.; David and Caroline Collett; Evan Cosgriff; Joe Evers and Scott Schwartzenberger, Dynojet Research Inc.; Gordon Geheb and Rick Wunderlich, FANUC Robotics America, Inc.; Marla Goodman; Susan Goss, Focus Products Group, LLC; Anthony Gugino, MSC.Software Corporation; Tom Jungst; Mary Kane, Boeing; Dianna Kegel, TowHaul Corp.; Robert Kincaid; Zach Olson, PBC Linear; Richard Patton, Missisippi State University; Mark Perkins; David Pinchefsky, NexGen Ergonomics Inc.; Stephen Sanford, David Yakos, and Bryan Walthall of Salient Technologies, Inc.; Mark Soares, J. E. Soares; Bryan Strobel; Lee Sutherland; Douglas Wintin; and Jeff Zerr.

CONTACTING THE AUTHORS

Our goal is a textbook that presents a contemporary view of the engineering graphics concepts and skills needed for a strong foundation. We welcome your comments and encourage you to join us in continuing to shape a text that will meet the needs of tomorrow's engineers. Please send e-mail to us at edc2@comcast.net.

Shawna D. Lockhart
Cindy M. Johnson

ENGINEERING DESIGN COMMUNICATION

1

GRAPHICS AND THE DESIGN PROCESS

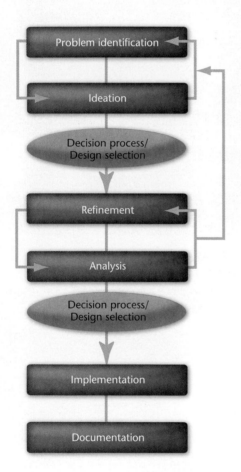

OBJECTIVES

When you have completed this chapter you should be able to:

1. Describe the large and varied role that graphical information plays in design.

2. Visualize a model of the design process.

3. List different types of design and design processes.

4. Identify the role of graphics in and the benefits of concurrent engineering.

5. Explain how communication of all kinds is critical to the team approach to design.

Santa Cruz Bicycles uses Pro/ENGINEER to design racing bikes—and continually improve them. The V10, shown here, took first and second place in the 2009 World Championships and has been ridden to 11 World Cup victories. You will learn how Santa Cruz uses the 3D model throughout the design process to visualize proposed features, simulate the bike's performance, and document the design for manufacture. *(Courtesy of Gary Perkin.)*

COMMUNICATING DESIGN INFORMATION

Why is a picture worth "a thousand words"? Graphical representations convey many kinds of information simultaneously—shape, size, relationships, color—and eliminate the language barriers that hamper written communication. They can be simple sketches for quickly showing an idea to a colleague, or detailed drawings that use well-established conventions to accurately describe 3D objects. They may also be generated from a 3D model of the object, which can be used to simulate the object graphically and physically. In this chapter, you will learn how engineering teams use graphics and 3D models to share and evaluate design ideas with team members and other partners across the globe.

1.1 In this rendered view of the hot-tub vacuum created from the solid model, lights and background were added to add realism to the image. *(Courtesy of SWS Corporation and Brad Wright, Salient Technologies, Inc.)*

1.2 Wireframe views of the model from any angle can be generated automatically from the 3D model. *(Courtesy of SWS Corporation and Brad Wright, Salient Technologies, Inc.)*

1.3 Finite element analysis software uses the solid model to analyze physical properties and display the results graphically. *(Courtesy of SWS Corporation and Brad Wright, Salient Technologies, Inc.)*

Modeling software has transformed the way engineers use graphics. Figures 1.1–1.6 were all derived from a single solid model created in Solid-Works, a constraint-based solid modeling system. When engineers model a part in a solid modeling package, they are creating a database that describes the object in three dimensions as if it were a physical entity. This database makes it possible to generate not only technical drawings but also images that are useful to nonengineers, such as marketing specialists and customers. An accurate database also makes it possible to simulate and test the object's behavior and performance before it is built. The accurate 3D representation of the object can also be used to communicate the design to the manufacturer.

Computer-aided design (CAD) tools are used in many ways throughout the design process. By supporting design tasks and facilitating paperless communication, they enable companies to work in teams more effectively, reduce costs, and get products to market faster—even to change the design process itself.

For example, Boeing's 777 was the first airplane ever produced for which a full-scale model (called a *prototype*) was not built. Because all the parts (except for some fasteners) were designed and modeled electronically, the models functioned as a *virtual prototype* that was used in place of a real one to check the design, fit, and manufacturability of each part. 3D software tools made this transformation possible, but the shift was driven by the company's desire to work smarter. They used software tools to integrate the many teams needed to design an airplane and help them work together efficiently.

Engineers use graphics for design in three basic ways: to *visualize* the design idea, to *communicate* the idea so others can evaluate it, and to *document* the

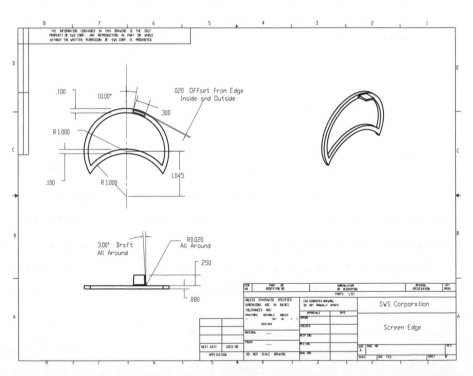

1.4 Detailed dimensioned drawings of each individual part can be created from the information in the solid model. *(Courtesy of SWS Corporation and Brad Wright, Salient Technologies, Inc.)*

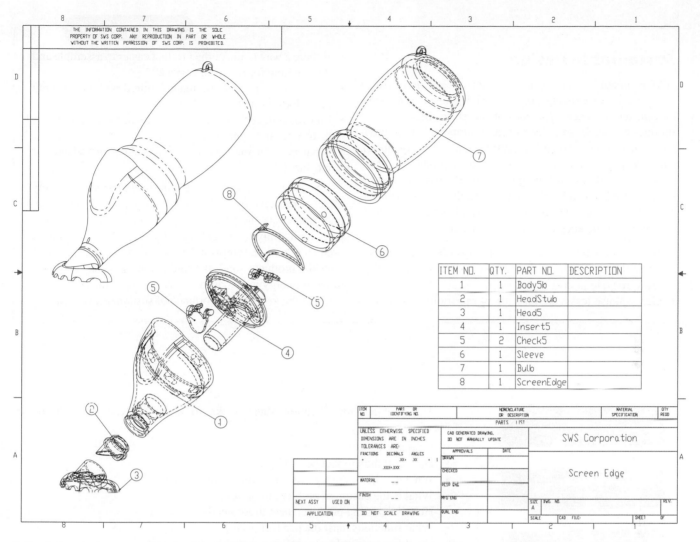

THE INFORMATION CONTAINED IN THIS DRAWING IS THE SOLE PROPERTY OF SWS CORP. ANY REPRODUCTION IN PART OR WHOLE WITHOUT THE WRITTEN PERMISSION OF SWS CORP. IS PROHIBITED.

ITEM NO.	QTY.	PART NO.	DESCRIPTION
1	1	Body5b	
2	1	HeadStub	
3	1	Head5	
4	1	Insert5	
5	2	Check5	
6	1	Sleeve	
7	1	Bulb	
8	1	ScreenEdge	

SWS Corporation

Screen Edge

1.5 Information associated with the individual parts in the model can be shown to document the part for manufacture and assembly. An exploded view generated from the assembled solid model makes it easy to see how the parts fit together. *(Courtesy of SWS Corporation and Brad Wright, Salient Technologies, Inc.)*

design so the product or process can be reliably reproduced and maintained. Understanding the role of graphics in design will help you use these tools more efficiently. It will also help you evaluate new technologies for design as they are developed.

1.1 WHAT IS ENGINEERING DESIGN?

Engineering design is the act of creating the specifications for a product or process that best satisfies the design criteria. Every design task is undertaken to solve a problem: to make a faster race car, to build a larger airplane, to design a more environmentally friendly computer. And every design task ends with specifications (which can include models and drawings) that are used to implement the design.

The process starts with a clear understanding of the problem to be solved. Criteria the design must meet, called *design constraints* limit the possible solutions.

1.6 Exploded views can be wireframe, as in Figure 1.5, or shaded, as above. *(Courtesy of SWS Corporation and Brad Wright, Salient Technologies, Inc.)*

Sustainable Design

"Going green" is adding new design criteria and driving the redesign of existing products. The goal is *sustainable design*, an approach that looks at the entire lifecycle of a product and seeks to reduce negative impacts on the environment (while keeping the product's price competitive). This comprehensive view considers how a product is made, what it is made of, the energy it requires, and the waste it generates. Social and ethical aspects, such as manufacturing conditions in supplier countries, may also be considered. While designing keep the following questions in mind:

- Is there a different material—perhaps one not available when the product was first designed—that would be easier to recycle or cost less to ship?
- Can you use less material without sacrificing function?

- Is there a way to make the product easier to assemble and disassemble for repair or service?
- How can the product be manufactured without creating hazardous waste?
- Can you reduce the energy used to manufacture, package, distribute, and operate the product?
- What parts of the product are reusable, upgradable, or recyclable?

Some groups, such as the Industrial Designers Society of America, have quantified the environmental impact for many materials and processes so designers can easily compare their impact. Legal requirements for reducing a product's "carbon footprint," consumer preferences for sustainable products, and the reduction in costs achieved through energy and material savings are some of the reasons for the growing importance of sustainable design.

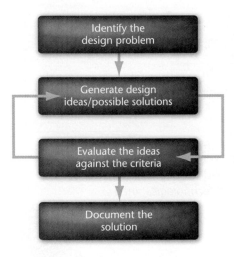

1.7   Simplified Design Model

The problem identification stage seeks to answer questions such as the following:

- Who is the customer/user?
- How will the product be used?
- What are the customer needs that the product must satisfy?
- What is the customer problem to be solved?
- How do existing products meet those needs?
- What functions must the product perform?
- What limits exist for its size and shape?
- What performance goals must be met?
- What profitability goals must be met?
- How is it to be manufactured and what materials will be used?
- What are the environmental impacts to be minimized or avoided?
- How will the product be serviced, repaired, and ultimately, disposed of?

Design problems have many possible solutions. Effective design requires creativity, or the ability to generate many alternative design solutions. A design solution is evaluated by seeing how well it satisfies the design criteria. The heart of the design process is an *iterative loop*: generate a design idea, evaluate it against the criteria, refine the idea, and test it again, until the design idea becomes the design solution.

Once a solution has been determined, the design process ends with documentation for reliably manufacturing or reproducing it.

A simple model of the design process is shown in Figure 1.7.

1.2 THE ROLE OF GRAPHICS

The simple design model illustrates the three key roles that graphics play in design: *visualization*, *communication*, and *documentation*.

Visualization

Learning to draw helps you communicate your design idea and also develops your ability to visualize design solutions—making you a better designer. As you identify the design problem, graphics can help you "see" the problem and possible solutions.

1.8 These sketches depict many alternative concepts to the challenge of designing a valve for a sleep apnea therapy device. (*LUNAR*)

Good designers capture their design ideas as hand-drawn sketches and use these sketches to show the ideas to others for feedback (see Figure 1.8).

For every design problem, there may be several viable solutions. As you generate design ideas, you will want to capture a variety of approaches to the problem before you select one to pursue. Investing too much time in a fully detailed drawing or solid model at this stage may be more than is needed for others to react to the design idea—and it may narrow your vision of the solution prematurely. Learning to sketch your ideas quickly and effectively will allow you to capture all your ideas (before you forget them) so you can share them or return to them in the future.

Few design problems will come to you fully defined. As you generate design ideas and ask others to evaluate them, you will learn more about the design constraints and how different ideas satisfy them. The ability to sketch your ideas by hand is a powerful tool for letting others "see" your ideas so they can understand and evaluate them.

Communication

As you refine your design idea, you will add more and more detail to your drawing of the solution. At the same time, the drawing must remain flexible so you can add and change details without too much effort. Computer-aided design tools make it easy to start with less detail and make design changes easily.

During the evaluation and refinement stages, designers communicate with the members of the product team—as well as customers—about the design idea. In the past, it was difficult to get good feedback from nonengineers without making a physical model of the design or preparing detailed and rendered drawings. Today, software tools—even including 3D printing methods (Figure 1.9)—make it possible to show a design in a variety of ways much earlier in the process.

Earlier in this chapter, in Figures 1.1–1.6, you saw how a single solid model can generate realistic-looking views of the finished product that can be used to evaluate how the customer sees the manufactured piece. Manufacturing personnel can use the same model to check that the design can be made with the equipment and materials specified—and project the cost of manufacturing the product or system.

Despite the finished look of the drawings generated by the solid model, the model itself remains flexible. As you refine your design idea you can incorporate feedback and modify the model to further improve it.

Documentation

Once you've reached the optimal design solution, graphics are a means of permanently recording the solution. **Documentation drawings** are a set of drawings that include all the information necessary to unambiguously describe the part or process so it can be made.

Traditionally, paper drawings were used to document the design. All the information needed was captured on paper and interpreted by the machine operators who manufactured the part. Today, more and more designs are captured as solid models that contain all the information needed and can interface directly with manufacturing software, in the process called **computer-aided manufacturing (CAM)**. Electronic files can be sent via the Internet to suppliers around the world, who use the files to directly control manufacturing processes. Two-dimensional (2D) drawings may be sent with the model to highlight important information, but they are likely to be electronic also, not printed on paper. In all cases, documentation (2D drawing or 3D model) serves as a "contract" with the manufacturer that says, "If you make the parts as described here, we will pay you for them."

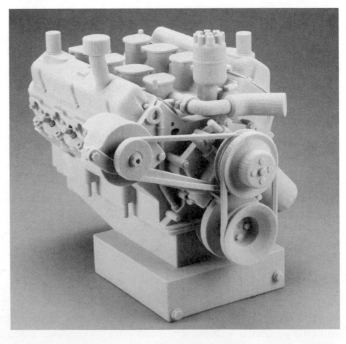

1.9 The 3D printer on the left will make a rapid prototype from a 3D model. The prototype at right is "printed" in layers of material and can be used to simulate a part for evaluation. *(Courtesy of Z Corporation.)*

Learning to document your designs involves using conventions that will be understood by the manufacturer. You will learn how designs are checked after manufacture and how to specify the criteria for determining whether a part is acceptable (Figure 1.10).

Documentation in a CAD/CAM environment also means preparing a model that will be useful for manufacturing. Some models may look good and generate acceptable 2D drawings but may not operate correctly when interpreted by CAM software. Structuring a solid model correctly is part of good documentation (as well as an important facet of being able to easily modify the model).

Documentation also serves legal and archival purposes. Proper documentation captures your design for patent applications and other legal uses. Good file storage habits and an understanding of the documentation requirements in your discipline are part of your design responsibilities. You must learn good habits for backing up files and for documenting changes to existing files.

Engineers and designers use a wide variety of tools to help them visualize, communicate, and document their work. From pencil and paper to modeling software to rendering and animation tools, each technique is used at various points in the design process.

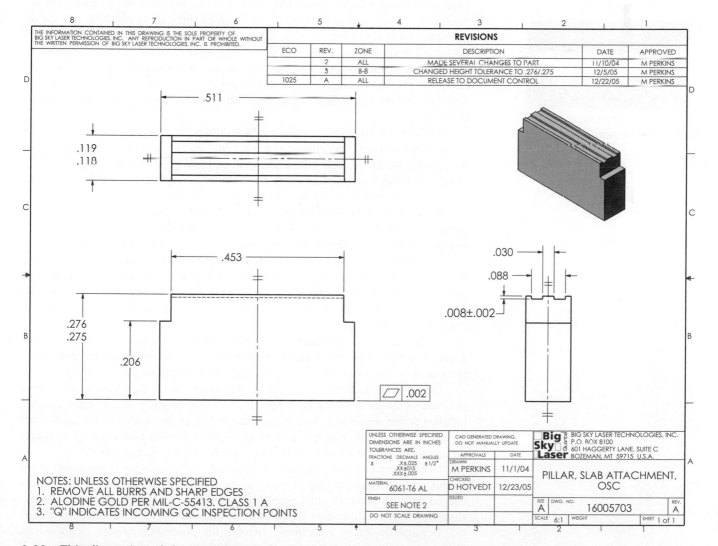

1.10 This dimensioned drawing defines the shape of the part, its dimensions, and tolerances. It also includes other information needed to manufacture the part. *(Courtesy of Quantel USA.)*

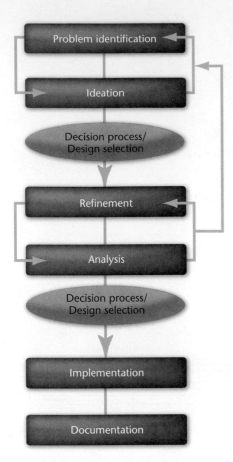

1.11   The Design Process

1.3 THE DESIGN PROCESS IN DETAIL

The simplified design process in Figure 1.7 illustrates the iterative nature of design, but its simplicity doesn't show the complex process of product design.

To be competitive, most firms encourage their designers' creativity and want them to examine a wealth of possible solutions to a design problem. At the same time, the firms must control the cost of refining and developing each of the ideas. One way to balance the importance of generating a wealth of possible solutions with the cost of refining them is to use a design process that breaks development into two stages.

The design process model shown in Figure 1.11 includes three main loops: one for problem identification and ideation, one for refinement and analysis, and a third showing iteration between the previous two. In fact, iteration can occur at all points in the process if circumstances demand it.

In the problem identification/ideation loop, the goal is to define the design problem and identify possible ways the problem may be solved. In this stage, the design team strives to translate customer needs into specifications for the product. A written **needs statement** or problem definition can help keep the project on track. Good design solves the problem that was identified, not some other related problem.

If the customer needs a faster airplane, how fast is fast enough? Detailed design criteria make it possible to evaluate design ideas to see if they can meet specified performance benchmarks and other criteria. Because design problems are frequently loosely defined at the outset, it can be difficult to determine all the requirements ahead of time. A needs statement may be very general at the start. As design ideas are generated and evaluated, they serve to further define the problem. If a concept that gains speed by reducing cargo space is unacceptable, the specification for minimum cargo space should be defined. New ideas are generated to satisfy the problem statement as it becomes better defined until a viable set of criteria is developed.

At this point you must determine which design ideas to pursue. This may be an informal process or it may consist of formal sign-offs from departments responsible for making, marketing, and selling the product. Ideas generated in the first cycle of problem identification and ideation are narrowed to a smaller number that pass muster and proceed to the refinement and analysis stages.

In the refinement/analysis loop, also called the *product development stage,* viable ideas are fleshed out to include more detail. Models are created to analyze and test the design against the engineering specifications. The results of these tests feed back into the refinement process to further improve the design. As detail is added, new features may require the team to revisit concept development for that particular part of the project. Design modifications may be needed to meet the manufacturer's requirements, eliminate assembly difficulties, address product safety, or create a more sustainable design.

Before the product is manufactured, another round of approvals is required. The decision-makers at this stage may differ from those at the first stage, and considerations such as cost, assembly, packaging, and sustainability may be foremost.

By the time a design has reached this stage, the design criteria may have changed. A competitor may have introduced a new product; a new, more energy-efficient power supply may have become available; or the customers for the product may have changed. A company may decide to go back to the ideation stage to generate alternative designs, adding a third iterative loop to the design process.

Finally, any remaining details are clarified, and the design is documented for manufacturing. Changes may be made to the design to incorporate the use of purchased or stock parts. Product assembly is clarified. Allowable tolerances (variations from the specifications) are defined and added to the model or detail drawings.

Because design is a means to an end, the process itself is a flexible one. Different combinations of design teams, software tools, and problem types will generate slightly different versions of this basic model. In the next section, you will learn about the various tools available to accomplish the goals of each stage.

1.4 GRAPHICS TOOLS IN ACTION

Engineers use graphics for visualization, communication, and documentation throughout the design process. A case study from Santa Cruz Bicycles will illustrate the tools used in the redesign of its V10 model, a downhill racing mountain bike, shown in Figure 1.12.

Problem Identification

The impetus for the V10, as for most racing bikes, was higher performance. In 2000, Santa Cruz Bicycles (SCB) purchased three patents for suspension systems that had never been developed. Most bikes have suspension on the front wheel; these patents showed how rear suspension could be manipulated to improve performance and control.

The bicycle's suspension has two roles: absorbing bumps and translating the pulsing motion coming from the human pedaler into forward movement. The balance between how far the wheel can move up and down and maximal forward motion determines bike performance.

The first V10 was ridden in 2002. Its rear suspension allowed the wheel to travel vertically 10 inches. At that time, 8–8.5 inches of travel was common among competitors.

When SCB set out to update the V10 design, better performance was a key criterion. The design team included the product manager (who interfaced with marketing), two design engineers, and racers who rode the V10. The team members had plenty of information for defining the problem: they were familiar with the design drawings, performance specifications, and manufacturing costs of the V10 already being produced. Racing teams provided feedback about how the current design handled and where they envisioned improvement. Racers and engineers viewed race videos to see how the bike performed on courses with more or less rugged terrain. Marketing provided information about competitors' bikes and how the V10 measured up against them.

At the problem identification stage, the team worked to enlarge the criteria for judging the solution. The team identified three primary goals for the redesign:

- Get maximum forward movement from a suspension that also absorbs bumps.
- Improve the handling to give the rider better control.
- Minimize the weight to make a lighter bike.

1.12   The V10 is a generation of downhill racing bikes first sold in 2002 and continually improved based on race team results and rider feedback. The 3D model of the new design is shown here. *(Courtesy of Santa Cruz Bicycles.)*

SPOTLIGHT

Santa Cruz Bicycles

The team at Santa Cruz Bicycles believes that to build a good bike, you have to first *really love* riding bikes. They all ride daily—from the owner to the engineers to the salespeople. This creates a very strong relationship between the design team and their customers—each model is constantly being refined and improved based on feedback from riders who want to—and do—win races with a Santa Cruz bike.

Their engineers are proficient 3D modelers who use the many features of Pro/ENGINEER to visualize, analyze, and test designs before they are built. More important, the 3D model supports the continual improvement that is built into the Santa Cruz culture. They can quickly modify parts, see how the new part affects the movement of the bike, assess the stresses on the new part, and send it to be prototyped for the ultimate test: riding on unpredictable terrain.

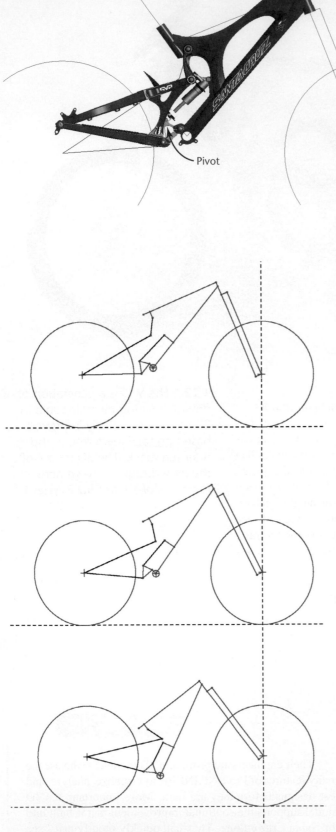

1.13 The V10's rear wheel attaches to a triangular swingarm that connects to the front part of the frame with a pivot. This allows the rear wheel to travel up and down. *(Courtesy of Santa Cruz Bicycles.)*

Ideation

The design team focused first on the geometry of how the various parts of the bike are arranged. Unlike the frame of a conventional bicycle, the V10's frame is in two parts. The back wheel connects to a triangular "swingarm" that attaches to the front frame with a pivot (see Figure 1.13). The length of the parts, the angles between them, and their movement determine how the bike handles.

In the ideation phase, the team captured and evaluated different combinations of handlebar placements, seat, crank, pivots to the wheels and shock absorbers, and more. How much the shock absorbers allow the wheel to travel vertically, the location of the pivots, and the size of the links all affect how the bike will perform.

At the ideation stage, being able to draw a readable freehand sketch is vital to "selling" your idea to your design team colleagues. This stage is called ***universal possibilities*** because the group seeks to consider every possible solution and not limit the design by preconceived notions of what will be best.

For each subsystem of the bike, the team generated and evaluated many concepts. For example, they brainstormed all the ways a pivot assembly might be made. They also employed a kind of "contest" to spur creativity and generate options. One engineer would be given 1.5 days to come up with a combination of parts and materials to make a subsystem, such as a link mechanism. Then, the design tasks would be shuffled among engineers and a different engineer would have 1.5 days to define an alternative approach to the link mechanism. Using a whiteboard, the team would then list the pros and cons of each subsystem design.

Preliminary Design Selection

The design selection process varies by company and by the design challenge. It may be very informal and conducted by a single individual, or it may be a highly formal process in which different teams compete for development dollars for new products. In all cases, design selection narrows the field of options for the final design.

SCB's team needed to evaluate how the parts would *move*. They created 3D skeleton models in CAD. A skeleton model is a simplified representation of the centerlines and other shapes that capture the design geometry. These skeleton models are useful for kinematic analysis (Figure 1.14).

Data generated from each model showed the behavior of the suspension and how the bike handled. Data from the kinematic analysis were captured in a spreadsheet and used to compare the various arrangements of lengths and angles in the designs (Figure 1.15).

Even at this early stage, drawings and models can be used to elicit customer feedback. If the user interface is important to a product, a 3D model or a computer simulation provides a way to

1.14 Skeleton models were used for kinematic analysis of different combinations of the swingarm geometry. The model shows how much the swingarm can pivot. *(Courtesy of Santa Cruz Bicycles.)*

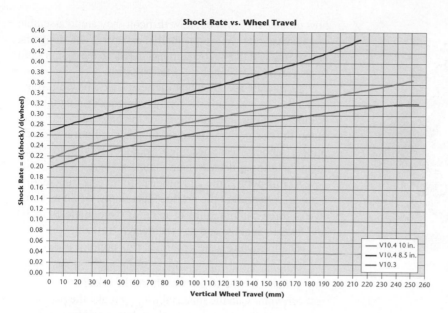

1.15 This graph compares the measure of rear-wheel travel produced from the skeleton model during ideation. *(Courtesy of Santa Cruz Bicycles.)*

get early user feedback. These models may be shown to customers informally, or more formally through focus groups.

At SCB, promising combinations were translated into simple prototypes that could be used to verify the concept. Called "mules," these proto-bicycles were fabricated so that a human rider could evaluate the handling (Figure 1.16).

For a new product, the design team generally prepares a product proposal to show the management group approving new products. Drawings or sketches are used to communicate key features of the device. Rendered models or prototypes convey options for the final look. The proposal may include a product plan, sales estimates, and cost figures for developing the product. It estimates the profit to be generated by the product and describes how it will meet the criteria for a new product, such as helping to achieve company goals or promising the best return on investment. The ability of the design team to communicate effectively through drawings, spreadsheets, written communication, and oral presentation all play a role in the product's acceptance for development.

1.16 This prototype "mule" for the V10 can be ridden and evaluated by a human rider. *(Courtesy of Santa Cruz Bicycles.)*

The preliminary design selection group for the V10 consisted of design engineers, manufacturing engineers, and the product manager. (At some companies, the team might also include members from finance or executive management.)

Function was the primary criterion for this review of the V10's redesign. Spreadsheets with data from the skeleton models, graphs comparing different options, and measurements from the prototypes were used to make the selection. Feedback from the riders who rode the "mules" was an important part of the process as well.

Refinement

When a product is approved for further development, the team moves into the refinement and analysis phases of the project. Because these two phases form another "loop" in the design process, they are often combined into a single stage called the ***development phase***.

During the refinement stage, the product concept is solidified into an accurate plan for making the product. Different options are evaluated in terms of how much they will cost, how long they will take to make, how well they satisfy the customer's requirements,

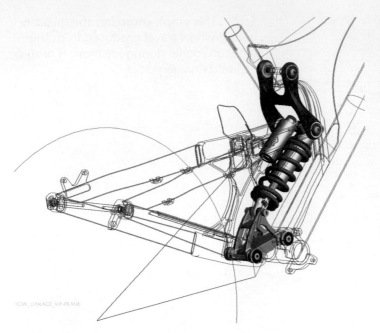

1.17 This subassembly is a combination of features that were evaluated during refinement. *(Courtesy of Santa Cruz Bicycles.)*

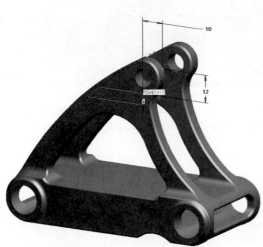

1.18 The rib shown here in red is constrained to be 10 mm back of and 12 mm down from the pivot point. If the pivot location is changed, the rib will be updated to preserve the relationship defined by the constraints. *(Courtesy of Santa Cruz Bicycles.)*

how durable they will be, how difficult they are to assemble and service, how environmentally sustainable the design is, and so forth. At this stage members may join the design team to provide expertise in these different areas.

At Santa Cruz Bicycles the team had chosen a functional design and was now ready to begin defining the part features, styling, and materials for the bike. During refinement, the design is updated as more information is determined about the designed components and their manufacturing processes (see Figure 1.17).

At this point, time and money are invested in creating a model to represent the product accurately. At Santa Cruz, two or three combinations of features were modeled in 3D and used to analyze and test the design further.

The following were criteria at this point:

Weight: How much will a component or choice of material add to the weight of the bike?

Strength: Can the material used to make the part stand up to the stresses generated during a bike race?

Cost: Is a lower-cost option available? Or is the cost justified because of what it adds to performance?

Optimization: Does the choice result in fewer parts to make or better clearance between parts?

The team used an informal concept/criteria matrix to evaluate the options against multiple criteria. For example, manufacturing the frame using carbon fiber instead of aluminum would affect the cost but would decrease the weight of the bike by 15%–17%. The matrix can tally the pluses and minuses of various options to aid in comparing them.

At Santa Cruz, bicycles are modeled in Pro/ENGINEER, a parametric solid modeling package. Parametric models capture *relationships* between part features and the sizes of the features. These relationships capture the **design intent** for the part. They define features in terms of characteristics that must be preserved when the part changes. For example, the key design criteria for a pair of holes for mounting a bracket might be that each be the same distance from the center of the part. Defining the location of the holes to be half the width of the bracket from its midline ensures that when the bracket gets wider, the holes will move accordingly. Similarly, when a part is defined in terms of its relationship to another part, changes in that part will cause any related parts to be updated automatically (see Figure 1.18). This flexibility makes it easy to test many different versions of a design—and is ideal for continually refining and improving racing bikes like the V10.

Analysis

The analysis phase tests the design and feeds back information to refine the design further, forming the second key iteration or "loop" in the design process. Crucial here are the performance criteria that the design must meet.

For the V10, the assembled solid model could be tested like a real bike. By specifying the material for each part, the team assessed the total weight of the bike as designed. Stress analysis (finite element analysis) predicted where parts or the frame might be likely to break (see Figure 1.19). The model was also used to evaluate the bike's fit for riders of different sizes. Human factors software provides a database of different body types and sizes for this purpose. The team can simulate what an Asian female in the 50th percentile for height and weight sees while riding, what she can reach, and the location of her center of mass. The same can be done for a white male in the 50th percentile (see Figure 1.20).

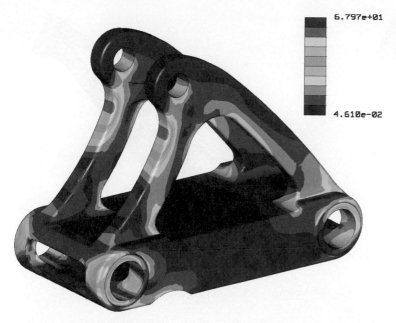

6.797e+01

4.610e-02

1.19 Finite element analysis can be performed within the solid modeling software, or the model can be imported into a separate analysis package. *(Courtesy of Santa Cruz Bicycles.)*

1.20 "Mannequins" of various shapes and sizes can be added to the solid model to assess ergonomic factors. *(Courtesy of Santa Cruz Bicycles.)*

At this point, the model can also be used for manufacturing analysis. At Santa Cruz, manufacturing engineers identify aspects of the design that will be hard to manufacture: a part shape that will require special tooling or be difficult to weld, or parts that must be made to a high degree of precision (tolerance).

Tools that work within the solid modeling packages are available for much of the analysis. Analysis for moving parts, such as the rear wheel suspension, can be done in Pro/ENGINEER (and other modelers) and tested under a wide range of physical conditions—even zero gravity! Human factors and stress testing are two more examples of analysis that can be done in the 3D modeler. For more advanced or specialized testing, a wide range of software can import the 3D model data.

Whether the analysis software recommends a change or simply reports results, it is up to the design engineer to interpret the tests and take steps to refine and test the design further until the desired results are achieved.

Critical Design Review

Once a design is ready to manufacture, another design review may be required. At each stage of the design process, more is invested in the design, and design changes become more costly. At the ideation stage, many ideas are investigated at relatively low cost. In the refinement stage, more time and money are devoted to building and testing models, and the cost of making major design changes increases. After the second design review, companies commission molds and tooling, purchase materials, and commit other resources to manufacturing the product. After this point, major design changes become very costly and can threaten a product's success. The second selection process, often called *critical design review*, is a filter that allows only the products with the greatest chance of success to proceed.

The selection team and review process may be very similar to the one that precedes refinement, or it may involve a completely different set of criteria. Understanding the review criteria helps you make an effective presentation and also helps you make better design decisions in the context of your company's goals.

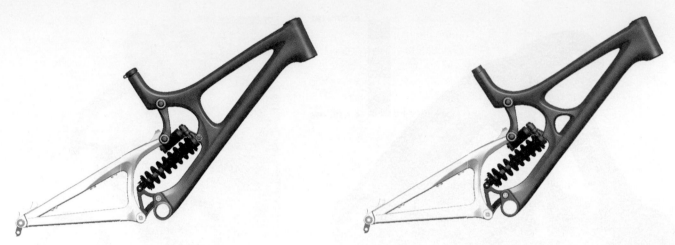

1.21   Multiple options for the look of the frame triangle were considered in the final design review. *(Courtesy of Santa Cruz Bicycles.)*

For the V10, more stakeholders evaluated the design at this point. Representatives from marketing, sales, and graphic design were added to the team, and the aesthetics of the design were considered (see Figure 1.21). How will the logo fit onto the bike as designed? Which look is more marketable? Should a straight or a curved tube be used for the swingarm? What are the implications of the manufacturing cost for the bike's sales potential?

Implementation

During the implementation phase, the design is communicated to those responsible for manufacturing, assembling, and distributing the product. Up to this point, drawings and models served a variety of purposes. In the implementation phase, they become contracts with suppliers. The model and the accompanying drawings (if any) must clearly specify what constitutes an acceptable part.

As in refinement, the 3D solid model serves as the information database. Dimensioned multiview drawings may be generated directly from the solid model. Increasingly often, manufacturing processes are directed from the model itself, with no need for an intermediate drawing. In these cases, tolerances may be noted in the model, or critical dimensions and tolerances marked on a drawing may be sufficient.

Tooling for a bicycle's design can cost as much as $120K, so care is taken to be sure that the tools are correct. SCB uses a detailed checklist to ensure that they are ready to have the tools made for manufacturing the bicycle. Because the 3D model is used to communicate the design to the toolmakers, it is sent to them for review and analysis before the final design is documented.

Sample parts are requested so that the physical parts can be measured against the model and tested to see how they fit together. Mechanical testing, ride testing, and destructive testing are done on the sample parts while there is still time to make a change to alleviate problems (see Figure 1.22). Once engineering is satisfied with parts as manufactured, the models for the design are approved.

1.22   This sample part for the swingarm failed during testing. *(Courtesy of Santa Cruz Bicycles.)*

Although the design has been finalized, there still may be changes during the implementation phase (as well as later when the product is in production). With rapid prototyping and design simulations using the solid model, it is easier and more cost effective to anticipate these issues in the refinement stage, but it is not uncommon for fine-tuning to occur during implementation. Unavailable parts or materials can force a substitution, a supplier may recommend a cheaper alternative, standard parts may be substituted for newly designed ones, or unanticipated problems may crop up during assembly. Sustainability issues can affect a design throughout the lifetime of the product.

Assembly drawings may be generated to document the parts included in an assembly and to illustrate how they fit together. The parts list, or bill of materials, is an important facet of inventory planning and assembly. Parts from the solid model may be catalogued in a company-wide parts database. Parts in the database can be reused or shared by other products, and standard parts inventories can be tracked and monitored.

Documentation

The documentation phase of the design process captures the final design and freezes it. Each company has its own procedures for managing models and controlling changes to the model database, but it must create permanent records for archival purposes, patent applications, and other legal uses (see Figure 1.23).

At Santa Cruz Bicycles, the 3D model is archived at each stage of the process. The model is sufficiently documented so that no 2D drawings are needed. Occasionally, a 2D drawing generated from the model is used to call a toolmaker's attention to a key dimension or tolerance.

Sometimes, after a model is sent to the manufacturer, dimensioned drawings are prepared to document the final product as manufactured and assembled. Measurements are taken from the actual parts, and the solid model is modified to match them, then dimensioned drawings from the final model are stored in the archive. (This is the

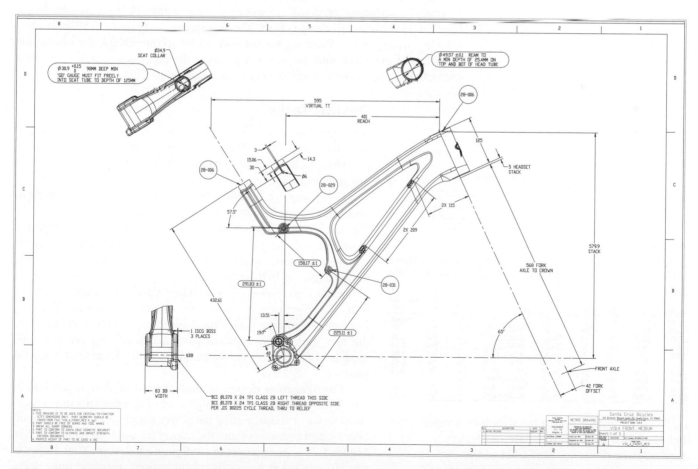

1.23 This drawing generated from the solid model includes notes that call attention to the fit with other parts that the V10's frame must achieve. *(Courtesy of Santa Cruz Bicycles.)*

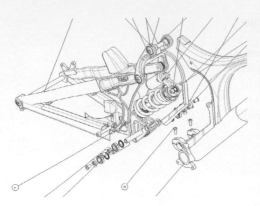

1.24 Exploded views such as this one can be used to illustrate assembly or service manuals for the V10. *(Courtesy of Santa Cruz Bicycles.)*

reverse of the traditional model, in which the documentation drawings are prepared first to transmit the design to manufacturing.)

The 3D design database (solid model) may be archived as the permanent record. Fully dimensioned engineering drawings also may be used to provide a permanent record of the design at some point. Companies that use the model for documentation have well-defined procedures for archiving the final model. Because files stored in electronic format can be altered so that they no longer represent the actual product design, companies must control access to the information in the solid model—or capture it in another medium—to create a permanent record. Once a model has been archived, only authorized personnel may make changes to it.

The solid model can also be used to generate a wide range of graphics used to document a product or process. Views of the model are imported into step-by-step illustrated guides for assembly line personnel (Figure 1.24), and the same guides are sent to service personnel who need to disassemble and reassemble the products after sale.

1.5 TYPES OF DESIGN PROCESSES

The Santa Cruz Bicycles V10 case study illustrates the many ways graphics supports the design process, but it is just one example of the way design tasks can be organized and carried out. The tools the team used and the process they followed were designed to fit the needs of their organization and product.

In large companies, the design process may be well defined and coordinated. In smaller companies, or if you are working alone, it may be up to you to decide how best to organize the design effort. The first step in designing a product is planning the process itself. Planning for the tasks to be accomplished, the interaction among team members, and the decision points in the process will help you select the right tools and use them effectively throughout the process.

Design Tasks

Not every task outlined in the design process model may be required for a design problem. There are several different kinds of design problems, each with a different set of tasks and objectives. Complex design processes may include several different design problems.

For example, a *selection design* problem could be as simple as choosing the appropriate fastener from a catalog (see Figure 1.25). This design problem would be well served by a more simplified model: identify the problem and engineering specifications for the part, select a part, analyze it to see if it works, and document the design decision. In this simple example you would need the ability to read and interpret existing drawings (and written specifications) to be sure the fastener satisfies the criteria for the part. You may also need to edit the design database to incorporate a reference to the new part. Many standard parts manufacturers provide solid models of their parts that can be added to the 3D assembly for the product.

Configuration design is another kind of design problem in which the challenge is to arrange a set of components into a redefined space or shape. Fitting the power supply, LCD panel, circuit boards, and connectors into the casing for the handheld device in Figure 1.26 is an example of configuration design. Your ability to visualize and sketch various configurations, then use more accurate computer models (either 2D or 3D) of the various components

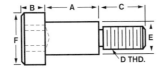

Jig and Fixture Type

Many uses in the construction of tool, jig and fixture work. Steel is 4140L, heat treated with black oxide finish. Stainless is type 303.

Steel

Cat. No.	A	B	C	D	E	F	Key Size	Price Each 1-11
SS-00	.250	3/16	3/8	10-24	.2480/.2460	3/8	1/8	$1.68
SS-7	.250	7/32	7/16	1/4-20	.3105/.3085	7/16	5/32	1.87
SS-0	.312	7/32	7/16	1/4-20	.3105/.3085	7/16	5/32	1.98
SS-1	.250	3/16	3/8	1/4-20	.3730/.3710	5/8	3/16	2.13
SS-2	.312	3/16	3/8	1/4-20	.3730/.3710	5/8	3/16	2.08
SS-3	.375	1/4	1/2	5/16-18	.4980/.4960	3/4	1/4	2.97
SS-4	.500	1/4	5/8	3/8-16	.4980/.4960	3/4	1/4	3.33
SS-5	.625	1/4	5/8	3/8-16	.4980/.4960	3/4	1/4	3.38
SS-6	.750	1/4	5/8	3/8-16	.4980/.4960	3/4	1/4	3.42
SS-8	1.000	3/8	3/4	1/2-13	.6230/.6210	7/8	5/16	4.34

1.25 This catalog listing for shoulder screws includes a 3D view (in this case, a photo), a dimensioned drawing, and a list of the variable dimensions for each part number. *(Courtesy of Reid Tool Supply Company.)*

to test the configuration will help you solve this type of design problem.

Redesign involves making improvements to an existing part or product. This type of design problem can range from redesigning a component to completely redesigning an entire product. Often, these design challenges address problems that come up after a product is released. Customer complaints, service problems, assembly difficulties, and sustainability are all reasons why a part might be redesigned. For example, the printer part shown in Figure 1.27(a) was originally designed as a steel assembly in six parts plus a fastener. The redesigned part, shown in Figure 1.27(c), is a molded plastic piece that requires no assembly yet provides the strength and flexibility required. Redesign requires you to be able to work with models and drawings for parts that have already been designed and produced. Reading documentation drawings and building a computer model from existing specifications are two skills common to this type of design. Your knowledge of manufacturing processes—or the ability to gather this information from manufacturing engineers—also can be very important to your success with redesign.

Reverse engineering is the process of deconstructing a design and analyzing its operation and function. This is usually done by taking apart an existing product and making measurements and observations. Sophisticated 3D laser scanning technology may be used to gather shape data from which CAD models can quickly be created. Coordinate measuring machines or hand measurement tools, such as digital calipers, are also used. Reverse engineering is generally a lawful method to gain product information. It can be used as a starting point for redesign, as a detailed competition review, or as a way to ensure that a new product fits with existing technology.

Original or new product design encompasses the full range of graphics and communications skills, including sketching to support ideation, 3D modeling, design simulation, building prototypes for user input, making presentations to users and managers, and preparing documentation and assembly drawings.

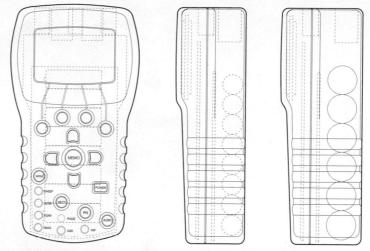

1.26 The dotted lines in this concept drawing represent the size and shape of the components to be contained in the plastic case. The two side views illustrate how a shift from size AA to size C batteries affects the size of the case. *(Copyright 1998 Tektronix, Inc. All rights reserved. Reproduced by permission.)*

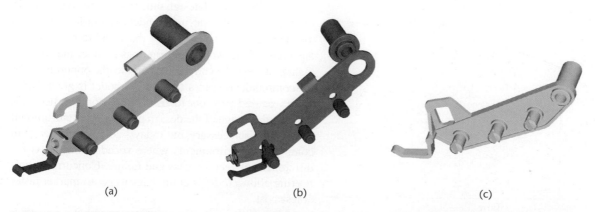

(a)   (b)   (c)

1.27 The exploded view in (b) shows the different pieces that had to be assembled for this part before it was redesigned into a single plastic part, shown in (c). *(Copyright 1998 Tektronix, Inc. All rights reserved. Reproduced by permission.)*

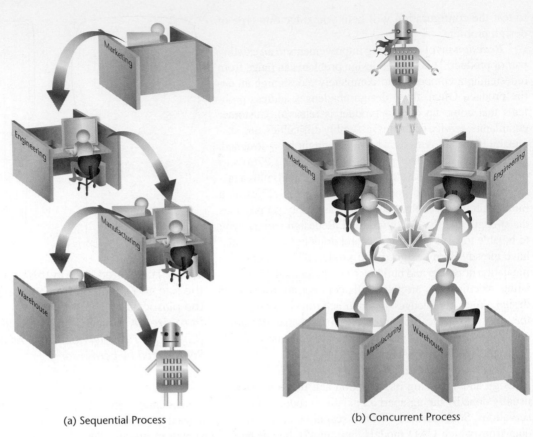

(a) Sequential Process   (b) Concurrent Process

1.28   Traditional versus Concurrent Engineering Design

1.6 CONCURRENT DESIGN

The design process used at Santa Cruz Bicycles is a contemporary one—not just because of the graphics tools the company uses but because it practices ***concurrent design*** (see Figure 1.28). Concurrent design applies the knowledge of different departments early in the design process.

In the traditional design model, the stages correspond to groups that operate independently. Marketing would toss the design problem "over the wall" to engineering, and engineering would, in turn, toss it "over the wall" to manufacturing. As you might imagine, the opportunity for miscommunication in a traditional model is great. Difficulties caused projects to be canceled or returned to a much earlier stage of the design process. A certain amount of iteration is necessary, but today's companies operate in competitive environments where speed to market is the difference between success and failure. Concurrent engineering allows today's companies to get to market faster at less cost.

Most of the costs of producing, assembling, and shipping a product are determined by decisions made early in the design process (see Figure 1.29). Once a product has been released to manufacturing, there are fewer options for controlling costs. When problems are not anticipated, making changes causes delays and adds costs in reworking the design. For complex products, any change to a subassembly or system can ripple through several other systems.

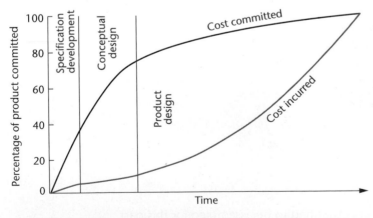

1.29   Early in the design process a large percentage of the life cycle cost of the product is committed. Late in the process, when costs are actually incurred, it may be impossible to reduce costs without major changes to the design. *(Reprinted with the permission of The McGraw-Hill Companies from* The Mechanical Design Process, *Second Edition, by David G. Ullman, 1997.)*

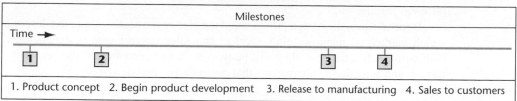

Typical Product Introduction Process

Milestones

Time ➡

1   2   3   4

1. Product concept   2. Begin product development   3. Release to manufacturing   4. Sales to customers

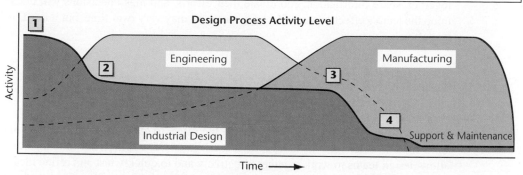

Design Process Activity Level

Engineering

Manufacturing

Industrial Design

Support & Maintenance

Activity

Time ➡

1.30 Industrial design, engineering, and manufacturing are involved throughout the design process in the MBD division at Tektronix, but their activity levels vary over time. The numbered milestones correspond to key events in the typical product introduction process.

1.7 DESIGN TEAMS

Practicing concurrent engineering means building **design teams**. To use teams effectively, it is important to recognize that different functions will have varying levels of involvement. It would not be a good use of anyone's time to build teams that required members from every functional area to participate at the same level throughout the design process (see Figure 1.30). Effective teams match involvement levels to the design process to use each member's time wisely. Deciding *who* should be involved and *when* is a key part of setting up the design process for a given project.

Concurrent engineering recognizes the knowledge base of different individuals and puts it to work. As the design team works to clarify the design problem and draft specifications for a project, it may be working with a vague description of the customer's needs. Marketing's involvement can ensure that the team understands the customer's desires—and help identify additional criteria. Sketches, rendered views, and 3D models play an important role in communicating design ideas to nonengineers—for the simple reason that "a picture is worth a thousand words." Concurrent engineering adds to the design criteria by considering all facets of the product's manufacture, assembly, distribution, repair, and disposal in the problem definition.

Service engineers help design teams anticipate and avoid choices that make produce service and repair difficult. For the motion sensor shown in Figure 1.31, service personnel were on the team to steer the design away from using glue to join the two halves of the molded case. Although glue would be an effective way to seal the case, it would make it difficult to service the components inside and increase the risk that cases would be damaged during service. Planning for a different type of closure helped avoid after-sale replacement costs.

The manufacturing engineer provides expertise about what *can* be built and the options for building it. Just because you can draw a design—or even create a 3D model of it—doesn't mean that it is affordable to make with today's technology. Concurrent engineering involves manufacturing engineers early in the design process.

Similarly, assembly and distribution processes can dramatically affect the cost of getting a product to the customer. For example, IKEA packages all its furniture in flat boxes that stack efficiently to reduce the number of delivery trucks needed.

Assembly engineers evaluate designs to be sure they can be assembled. A fastener in a difficult-to-reach position adds to assembly costs and may require special

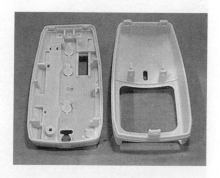

1.31 This prototype for a motion sensor shows the tabs used instead of glue or screws to seal the case. *(Used with the permission of C&K Systems.)*

tools. Identifying it early allows for redesign that can eliminate it or replace it with a different feature.

Depending on the design problem and the size of the organization, the degree to which marketing, manufacturing, service, and sales interact with designers varies considerably. In each case, they will participate in and contribute to the generation, testing, and selection of alternatives that result in the final design.

Once the teams are defined, they need structure. *How* team members work together, share information, coordinate their efforts, and make decisions will determine the team's effectiveness. The team structure may vary over time, but it needs to be defined.

1.8 CONCURRENT DESIGN AND THE CAD DATABASE

The evolution of 3D solid modelers fueled the move to concurrent engineering. 3D solid models offer four advantages over traditional 2D engineering drawings and allow design teams to work together effectively and to quickly test and refine ideas:

- 3D models offer a wide range of ways to view the model, such as animated and rendered views, in addition to traditional 2D views.
- The completeness and accuracy of the design definition in the CAD database make the models suitable for design simulation, such as motion studies and analysis of the object's physical properties.
- The CAD database accurately conveys the model to prototyping and manufacturing software.
- A constraint-based model can automatically update related parts of the model when changes are made.
- More than one individual can view and work with parts of the CAD model at the same time, even before the design is complete.

Creating effective 3D models requires an understanding of the design intent of each component. This means anticipating the range of possibilities for a part and considering key attributes and relationships. For example, a connecting rod may need to vary in its width and thickness, but it will always serve as a connection between two points. By choosing the rod's length as its base feature, and defining the width and thickness of the rod in relation to it, changing the definition of the rod's length will allow the model to be updated. Because the rod's width and thickness are likely to change as the model is refined, you would not want to use them as the base feature. As you learn to use the graphics tools described in this text you will also learn how to prepare models that are easy to modify and reuse.

Teamwork and the CAD Database

Modeling tools and the CAD database are a vital part of how design teams collaborate. For small projects, one person may be responsible for modeling the entire assembly. Weekly team meetings could be a viable mechanism for reviewing progress and providing input to the person building the model. But the complexity of a product such as a bridge or jet airplane—with its many subsystems and integrated mechanical/electronic functions—requires that the effort of many individuals be coordinated into a cohesive design process (see Figure 1.32). Today's modeling tools facilitate this coordination.

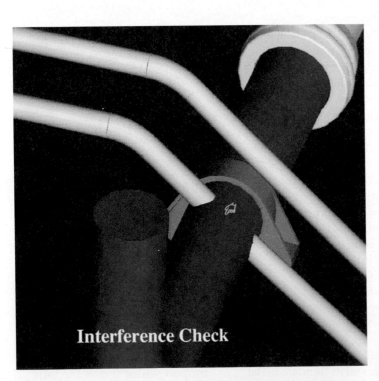

Interference Check

1.32   This close-up view of the solid model assembly shows part interference the design teams had to resolve. *(Copyright © Boeing.)*

Who Is on a Design Team?

Tektronix employs individuals in a variety of positions that may serve on a product team. Their teams consist of representatives from all functional areas involved in a project, including engineering, marketing, manufacturing, service, documentation, industrial design, quality and safety, and purchasing. This "core team" is formed at the product concept stage and carries the product through completion. Member responsibilities include the following:

Program Manager

The program manager has overall program responsibility from conception through introduction and is multidisciplined—usually with a strong technical background and experience in engineering, manufacturing, and marketing.

Mechanical Engineers

Mechanical engineers are responsible for the mechanical and electromechanical product development. They work closely with electrical engineers, industrial designers, manufacturing, and marketing to satisfy the hardware requirements (cooling, ruggedness, aesthetics, component locations, controls, safety, etc.), the look/feel desired by marketing and industrial design, and the manufacturability goals. The mechanical engineer will work closely with purchasing and component vendors as well, and seek early involvement with the component vendor and tool designers. On large projects, mechanical design efforts are divided up in logical groupings. For example, one engineer may be responsible for the exterior plastic package, another for interior sheet metal, another for front panels and graphics.

Electrical Engineers

Electrical engineers are responsible for the electronic functions in a product. This may include front panel controls, power supply, processor system, acquisition or signal conditioning, the calibration/diagnostics algorithm, and integrated circuit design. Responsibilities are divided up functionally among the team. Electrical engineers are responsible for ensuring the product meets required safety and regulatory requirements, and they work closely with the mechanical engineers to ensure the product meets ruggedness requirements of the customer. This includes areas such as operating temperature range, shake, shock, and vibration requirements.

Software Engineers

Almost all products today contain some amount of embedded software. The software engineer works closely with the electrical engineers to develop the code to control the hardware. This may include user interface, power-on diagnostics, internal control, and drivers for remote interfaces.

Quality and Safety Engineers

Responsible for testing the product against applicable safety requirements and product quality expectations, safety engineers are versed in the safety requirements of countries around the world and ensure that products are certified to these standards. Quality engineers work with the design team to predict product reliability based on components used, operating environment, and the like. They develop test plans to validate the product's reliability.

Industrial Design Engineers

The industrial designers work with the design team to ensure the product's "look and feel" meets customer needs and is consistent across similar products. The industrial design engineer is involved with everything from color choices to knob spacing and text font and size.

Industrial Engineers

Industrial engineers are responsible for the overall design of a manufacturing line to support a new product. They will work with the design team to integrate the product into an existing manufacturing line, modify an existing manufacturing line, or develop a new manufacturing line as appropriate. They will develop time estimates for the entire manufacturing process, recommend purchasing strategies, develop models for manufacturing capacity requirements, and develop new test and assembly fixtures as needed.

Manufacturing Test Engineers

Test engineers develop the automated tests necessary to calibrate and test products in manufacturing. Often, they will work with the service group to modify or develop test procedures for periodic calibration or repair of products.

Documentation

Technical writers work with the design team to compose the manuals and other customer documentation for a product. This can include installation/setup information, operator information, use examples, calibration/performance verification, diagnostics, repair, and a programmer's manual for remote operation and control.

CONCURRENT DESIGN AT BOEING

Boeing's shift to a concurrent "design/build" model was adopted in part to reduce the amount of rework required when the many components of a jet airplane come together to be produced. The Boeing 777, the first airplane designed and built under this concurrent model, experienced significant cost reductions owing to design changes, as illustrated in Figure 1.33.

Boeing's concurrent model, shown in Figure 1.34, is one way to illustrate the involvement of different groups in the development phase of new product design. Note the functional groups included in the design/build process circle.

Boeing developed its own proprietary system to enable groups to work in parallel yet communicate effectively as decisions were made. Communication among the many different engineering groups had to be structured so that it did not take more time than what would be saved by working in parallel. (The Commercial Airplane Group had 9700 CATIA workstations in operation!) Boeing used the models themselves to integrate the project teams. Engineers were responsible for uploading their models to Boeing's solids database. This database placed each model into position in a virtual prototype—a "digital preassembly"—of the airplane. Each design team was responsible for using the prototype to identify problems. If, for example, an air duct was positioned in the same space as wiring conduit, as was shown in Figure 1.32, the engineers responsible would work together to resolve the conflict. If no interference was found, no contact would be required.

Today, 3D modeling packages include many of the features developed by Boeing and work with sharing software to integrate team members who may be across the building or across the globe. The scale of the Boeing example illustrates the importance of structuring teamwork so that communication is effective and cost-effective.

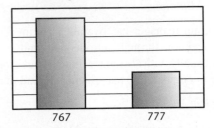

Engineering Change Activity
(through aircraft certification)

767   777

1.33 Boeing tracked and compared the costs incurred for different kinds of changes during the development of the 767 and 777 aircraft and graphed them in the chart shown here. Notice the reduction in change activity with the shift to concurrent engineering and solid modeling. *(Copyright © Boeing.)*

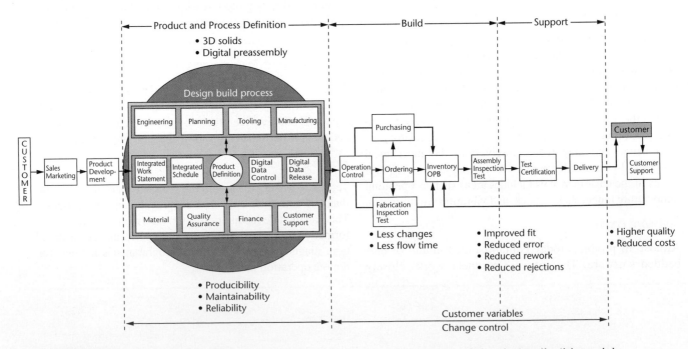

1.34 Boeing's Preferred Business Process illustrates the concurrent nature of the design/build model at the heart of its design process. *(Copyright © Boeing.)*

Communication for Design

This book is devoted to graphical means of communicating. In addition to communicating through graphics, you will need to communicate effectively as a designer and team member. Sometimes, this process will be informal and accomplished through regular meetings, but there are several reasons for formally documenting your work and team decisions. Wherever spoken communication can be misheard, misunderstood, or forgotten, creating a written record can capture decisions more clearly and communicate them to a wider audience.

At the same time, you need to be able to work with tools that allow you to manage and update information shared by the team. Dated memoranda, emails, or instant messages give team members the up-to-date information. This information can also be stored on an intranet, Wiki, or Web page where any team member can find the most current version and use sophisticated search tools. Shared databases facilitate access to sketches, models, performance data, finite element analyses, and team contact information developed over the course of the project. There are several options for "publishing" your information that can make your design team more effective.

In addition to communicating effectively, you also must document your work for legal purposes. Product liability is a large concern and your documentation shows whether products have been analyzed properly with customer safety and environmental impact in mind. Each company will have its own documentation requirements to which you will be expected to adhere.

The Designer's Notebook

Most companies require their designers to keep a notebook as a record of their work, like a personal "diary" of design ideas, project information, and things that need to be done. Legally, it is a record of design ideas generated in the course of an engineer's employment that others may not claim as their own. It also documents procedures and decisions that may come into question at a later date.

The sketches in Figures 1.35–1.37 were taken from a design notebook kept by a practicing engineer. As you work through this book, try keeping your own designer's notebook (see Figure 1.38). Any size bound notebook (not loose-leaf) with grid paper is suitable. Date each entry as you use this notebook to record your design ideas, assignments, sketches, and more. For team projects, use the notebook to capture notes from your team meetings. Use the notebook for all your sketching: at the end of the term, you will have a record of your course and your personal growth.

Writing well, interpreting numerical data, understanding statistical data, reading and creating engineering drawings, sketching, listening, and making presentations are all communications skills that will add to success in your design career.

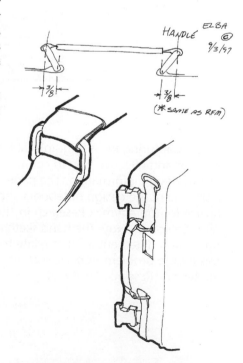

1.35 Notebook Sketches for a Carrying Case *(Courtesy of Scott Ketterer, Tektronix, Inc.)*

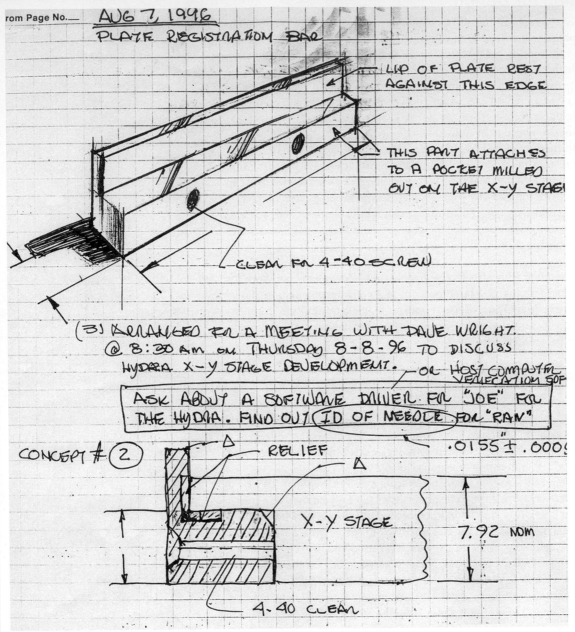

1.36 Capturing ideas in graphical form is an efficient and effective way to communicate and record design information. The page above is taken from one of the design notebooks kept by Albert Brown for the Affymax Research Institute in Palo Alto, California. Note the hand sketches, meeting notes, and dated entries that relate to his work on this design problem. *(Courtesy of Albert W. Brown, Jr., Affymax Research Institute.)*

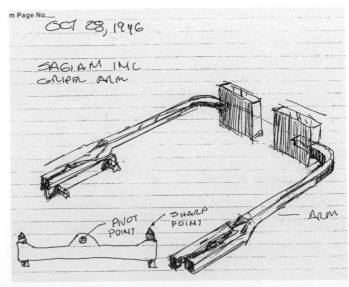

1.37 Gripper Arm Sketches from a Designer's Notebook *(Courtesy of Albert W. Brown, Jr., Affymax Research Institute.)*

Instructions

1. The primary purpose of this notebook is to protect your and the Company's Patent-Rights by keeping records of all original work in a form acceptable as evidence if any legal conflict arises.

2. When starting a page, enter the title, project number, and book number.
   - Use ink for permanence—avoid pencil.
   - Record your work as you progress, including any spur-of-the-moment ideas which may be developed later.
   - Avoid making notes on loose paper to be recopied.
   - Record your work in such a manner that a co-worker can continue from where you stop. You might become ill and to protect your priority it could be urgent that the work continue while you are absent.

3. Give a complete account of your experiments and the results, both positive and negative, including your observations.
   - Record all diagrams, layouts, plans, procedures, new ideas, or anything pertinent to your work including the details of any discussions with suppliers, or other people outside the Company.
   - Do not try to erase any incorrect entries; draw lines deleting them, note the corrections, sign and date the changes. This extra care is worthwhile because of the necessity of original data to prove priority of new discoveries.

4. After entering your data, sign and date the entries.
   - Explain your work to at least two witnesses who are not co-inventors, and have them sign and date the pages in the place provided.
   - Record the names of operators and witnesses present during any demonstration and have at least two witnesses sign the page. If no witnesses are present during an experiment of importance, repeat it in the presence of two witnesses.

5. Since computer programs can be patented these instructions apply to the development of computer software. In this case a description of structure and operation of the program should be recorded in the notebook together with a basic flow diagram which illustrates the essential features of the program. In the course of developing the code, the number of lines of code written each day should be recorded in the notebook together with a statement of the portion of the flow diagram to which the section of code is directed.

This notebook and its contents are the exclusive property of the Company. It is confidential and the contents are not to be disclosed to anyone unless authorized by the Company. You must return it when completed, upon request, or upon termination of employment. It should be kept in a protected place. If loss occurs, notify your supervisor immediately, and make a written report describing the circumstance of the loss.

1.38 Many companies spell out the rationale for, legal issues surrounding, and requirements for an engineer's notebook. This excerpt is taken from the printed instructions for keeping a design notebook used by Affymax Research Institute. *(Reprinted with permission from Harold Gallup, Scientific Notebook Co.)*

KEY WORDS

Communication

Computer-Aided Design (CAD)

Computer-Aided Manufacturing (CAM)

Concurrent Design

Configuration Design

Design Constraints

Design Intent

Design Teams

Development Phase

Documentation

Documentation Drawings

Engineering Design

Foreshortened

Iterative Loop

Needs Statement

Original or New Product Design

Prototype

Redesign

Reverse Engineering

Selection Design

Universal Possibilities

Virtual Prototype

Visualization

EXERCISES

1. Think about a vacation spot with which you are familiar. Write a needs statement for a resort to be located in that area.

2. Consider the recent developments in pocket-size disposable digital cameras. Put yourself in the place of the original designers of these cameras and re-create the written statement of need that might have guided the project. What sustainability issues come to mind?

3. Ergonomics, or human factors engineering, is an important consideration when designing control handles or other interfaces between a tool and the human user. It has been shown that one's productivity improves when more functional interfaces are present. With this in mind, write a needs statement for the ergonomic redesign of the following items:
   a. Toothbrush
   b. Bicycle seat
   c. Fishing rod
   d. Ski pole
   e. Kitchen knife

4. Design the process you would use to design the items for which you have written needs statements (or use your partner's needs statements). Consider the team of people needed, when each of them would be involved, and how and when to present ideas to the team for its approval and input.

5. List five attributes of a successful product design team. What skills are important to its success?

6. You are a member of a three-person design team chosen to develop a pocket-size flashlight that will be sold through high-quality sporting goods/camping equipment catalogs. The basic idea passed on to your team is that a small, flattened, palm-size flashlight would be desirable.
   a. Write a needs statement for this product.
   b. Outline the design process that will be used to go from the first step of identifying a need to production of the part.
   c. Create a simple sketch of three versions of this device to communicate your initial ideas to the other members of your team.

7. Imagine the original design of your cell phone. What if customer feedback presented a need for double the battery life? How would you address this need if it arose early in the design process? What if it arose after the product was introduced? Consider the impact of the change at both stages and list and explain the items that would have to be changed to meet the customers' needs.

8. Software plays an important part in product design. List at least four ways that a design team developing a new product uses computer-generated information. What types of software do you think a mechanical design engineer should be skilled with in this environment?

9. Design Solutions Corporation provides engineering design and analysis consulting services for large industrial customers; company policy is to strictly follow the steps of the design process as shown schematically in Figure 1.11. Explain how a customer benefits from this philosophy, and give three examples of what might happen if steps were skipped.

10. Interview a design engineer who works for a company in your area to better understand their process and the graphics tools used. Compare and contrast the process with that shown in Figure 1.11.

11. Write a set of specifications for packaging that will protect an egg when it is dropped from 30 feet. Develop a process for designing, testing, and refining the device that does not use more than two eggs. Follow your procedure to design the device, and use your notebook to record all your notes, sketches, ideas, and results.

12. Visit the websites for Parametric Technology Corporation (www.ptc.com) and Dassault Systèmes (www.3ds.com), and locate the product names they market for the following stages of the design process:
   a. Problem identification/ideation
   b. Refinement/analysis
   c. Implementation/documentation

13. Visit a website for a product you have used and imagine the needs statement that drove its design. Write the needs statement you think the product addressed.

14. Write a needs statement for a competitor to a product you have used. What would make a better product? List the design criteria for evaluating the design. Outline a process for testing whether the design satisfies those criteria.

15. Identify sustainability issues with a product you use daily. Use the Web to investigate its manufacture, the materials it is made from, its average lifetime, and how it is disposed of. Which attributes of the product would you rate most important to redesign to improve sustainability?

16. Review the flowchart and key personnel listing for Realgud Robots, Inc., at right, and answer the following questions:
   a. At what point in the design process might Lisa Chang meet with a client?
   b. What types of drawing or model information does Tyrell Jones need in order to estimate raw material cost?
   c. Should Gabriel Martinez become involved in the design process? If so, when and why?
   d. Which team members are likely to be in an instant messaging group?
   e. When would Kate Johnson meet formally with the design team?
   f. When would a prototype be shown to Neil Blanco?
   g. What might happen if Lisa Chang wants last-minute design changes?

Gabriel Martinez, CEO
Manages a $84 million per year robotics manufacturing company. He relies on advice from managers and team members to lead the company and make company-wide decisions.

Lisa Chang, Product Line Manager
Steers the product line to garner market share. Manages customer support for the product line.

Tyrell Jones, Purchasing Agent
Prices and sources raw goods, piece parts. Determines vendors and manages vendor accounts and relations.

Neil Blanco, Sales Manager
Manages the 43 person sales staff. Directly manages top five client accounts.

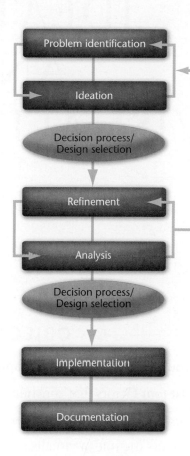

Problem identification

Ideation

Decision process/ Design selection

Refinement

Analysis

Decision process/ Design selection

Implementation

Documentation

Kate Johnson, Production Manager
Manager of in-house assembly and prototyping. Analyzes production costs. Provides input into off-shore contracting.

Pat Winegold, Manufacturing Engineer
Monitors advances in material science and manufacturing processes. Charged with maintaining low product costs while ensuring product manufacturability and performance. Designs tooling and production processes.

Doren Layton, Lead Design Engineer
Directs seven engineers and technicians to develop and test new products and to update existing ones.

2

IDEATION AND VISUALIZATION

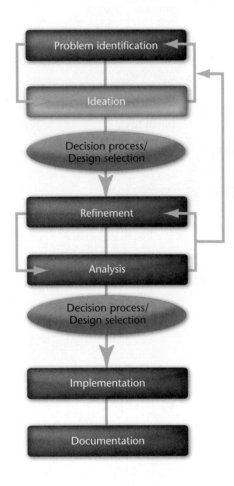

OBJECTIVES

When you have completed this chapter you should be able to:

1. List four factors that constrain designs.
2. Describe techniques for generating ideas.
3. Discuss the role of visualization in ideation and problem identification.
4. Sketch a block diagram of a system or device.
5. Create freehand contour sketches.
6. Create freehand isometric pictorial sketches.
7. Create freehand oblique pictorial sketches.
8. Create simple freehand perspective sketches.
9. Neatly letter notes and labels.

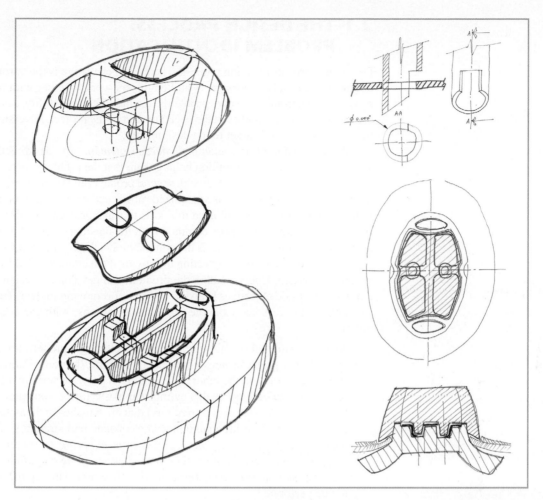

These freehand sketches illustrate the interior portions of a valve in a sleep apnea therapy device. *(LUNAR)*

SKETCHING YOUR IDEAS

The ability to sketch your ideas is key to your ability to capture and evaluate many different design options during the ideation phase. It is also a skill that you use throughout the design process to communicate quickly and effectively.

When you learn to "see" objects as they actually appear, you can better capture their appearance on paper. For any given object, an unlimited number of views can be generated to represent the object. Some views show a more complete view of the object—and some convey its size and shape better. Others might be difficult to interpret and provide little information to the reader of the sketch. The "best" sketch depends on your particular goals for the drawing. As you are developing your sketching skills, keep in mind that sketches are used to communicate your design ideas to others. They should be neatly drawn and clearly show the information you are trying to convey.

2.1 THE DESIGN PROCESS: PROBLEM IDENTIFICATION

Design is a problem-solving process. The design challenge is the "problem" that the engineer must "solve." Designers and engineers work together with team members of varied backgrounds to solve problems like the safety, reliability, aesthetics, ethics, economic feasibility, and even the social impact of the project. Diversity among team members is a valuable asset to the design team.

A problem definition containing a *needs statement* provides direction for the design team. The problem definition helps ensure that the problem solved is the one the design team set out to solve. Although several products and scientific discoveries were the result of accidents, such as vulcanized rubber and penicillin, to be competitive in today's marketplace you will need the skills to solve problems while under time and financial pressure. Much more common than the announcement of a wonderful new accidental discovery or product is the complaint that the designers added features and gadgetry not desired while neglecting to provide the features needed for a marketable product or system. Understanding the problem and developing design specifications that meet the basic objectives keep the design effort moving in the right direction.

Following are some example problem definitions, with the needs statements italicized:

- An environmental science company is developing a process for decontaminating polluted groundwater on-site using bacteria (bioremediation). To demonstrate the process in the laboratory, scientists grow bacteria in a corrosive soup in five pressurized and sealed cylindrical glass, Teflon, and stainless steel containers (called *chemostats* or *reactors*) that are 6 inches in diameter by 18 inches tall. The chemostats are maintained at a constant temperature in a water bath. *A means is needed to keep the bacteria from adhering to the interior surfaces (e.g., the walls) and from settling to the bottom of the container. It is also necessary to stir the soup so it stays completely mixed.* The top of the reactor may not be perforated.
- Removing and disposing of large amounts of snow from parking lots is a major expense for the city of West Yellowstone. Currently, collected snow must be transported several hundred miles to a location where it can be dumped. *A local facility is needed to allow disposal of excess snow removed from streets, parking lots, and other areas.* Research and design an energy-efficient and cost-effective snow-holding and disposal system.
- A therapy center has a client who is an avid hunter and uses a powered wheelchair full-time. *The client needs an adapted wheelchair for hunting.* He finds it easiest to operate devices that are located directly in front of him in the midline position. He would like to be able to store the gun on the wheelchair when he is not shooting. The mount must be accessible to him. It may be possible to use the wheelchair's battery system to power the mount.

If the criteria are too rigidly defined during the initial stages of the design process, they may limit the designer's creativity. If the criteria do not reflect the actual need or problem to be solved, then the design may be wonderfully creative, but not for the problem at hand. The needs statement may be general at first, then modified with additional design constraints or criteria as preliminary research is performed.

Design Constraints

Design constraints limit the range of options that are acceptable. Design constraints can be functional, aesthetic, economic, legal, time-related, or all the above.

Functional design constraints relate to how the product or process will work. What are the limits within which the product or process must operate? A designer needs to know how much load a robot arm must be able to bear, or how much fluid needs to flow through a system. If the design must fit into an

2.1 This cryostat was designed to keep the infrared instruments of the SIRTF space telescope at temperatures as low as 1.4 K. An outer vacuum shell, which has been removed to show detail, shields the cryostat from external temperature extremes. *(Courtesy of NASA and Ball Aerospace & Technologies Corp.)*

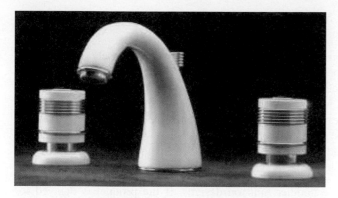

2.2 These faucet handles look nice, but when your hands are wet and soapy, it is hard to grip the smooth, round handle. (For more examples of bad design, visit the website www.baddesigns.com.)

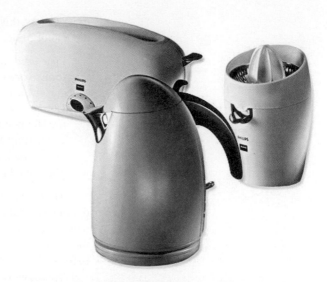

2.3 The Philips-Alessi line of kitchen products is an example of design that combines aesthetics and function. *(Project developed and created by Philips Design.)*

existing space or attach to other equipment, the size may be limited to a certain range. Under what conditions will the product be used? Will it need to operate at extreme temperatures or pressures? (See Figure 2.1.) The Space Shuttle *Challenger* disaster was caused by the failure of an O-ring seal to expand quickly enough in cold weather to make a tight seal under extreme pressure.

Ergonomic constraints concern how well a product fits the human body (see Figure 2.2) and how a user will interact with it. A designer needs to consider who will use the product or process. Visibility may be an issue if an operator needs to see portions of the equipment or beyond it to operate it properly. Designing a handle shape to fit most comfortably in the human hand requires ergonomic testing—and a decision whether to design for left-handed people, too.

Aesthetic constraints express taste and appeal in a certain market (see Figure 2.3). What "look" will speak to the consumer or customer? The changes in product design illustrate how the public's sense of what is contemporary and fashionable changes: "retro" is now "in."

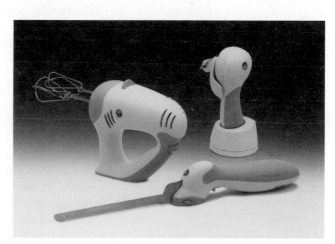

2.4 The Black & Decker Ergo line reflects a greater sensitivity to ergonomics. *(Courtesy of Household Products, Inc., exclusive licensees of Black & Decker household products.)*

Black & Decker revitalized its small-appliance line in the early 1990s to increase aesthetic and ergonomic appeal to the products' consumers. Several of the products from the Black & Decker Ergo line are shown in Figure 2.4.

Economic constraints are also very important. How much can be spent to produce a product? How much should it cost to operate it? Finding a less expensive material that satisfies other design goals can make a product marketable. At the same time, a complex design that requires plant retooling or that cannot be manufactured easily may make the product too expensive to be viable.

Legal constraints are another restriction. Governmental safety and environmental regulations constrain certain design options. Many industries are regulated through codes and standards that must be followed. For example, the American Society of Mechanical Engineers publishes the *ASME Boiler and Pressure Vessel Code*, a U.S. national standard. All commercial, institutional, and governmental design and manufacture of pressure vessels in the United States must follow this code. Section VIII, Part UG, Item UG-5 of this code describes the use of plate. It states:

> Plate used in the construction of pressure parts of pressure vessels shall conform to one of the specifications in Section II for which allowable stress values are given in Subsection C, except as otherwise provided in (b) and UG-4, UG-10, and UG-11. Material not identified in accordance with UG-93 may be used for nonpressure parts such as skirts, baffles, and supporting lugs. A double-welded butt-joint test plate shall be made from each lot of all such material over 3/8 in. thick that is to be welded. Guided bend specimens made from the test plates shall pass the tests specified in QW-451 of Section IX. . . .

Time constraints are another limit on designs. If the product is delayed for 6 months, what other products that fill this need will already be on the market? What new materials or processes will be available that could make this design obsolete? Is the aesthetic appeal rapidly changing in this area, so that the product will look dated before it is ever released?

Today, worldwide competition makes it difficult to succeed if all these major areas are not reflected in the design process. Although engineers may be most comfortable in the realm of functional design, they will undoubtedly work in an environment where customer feedback and manufacturing concerns are integrated into the design process. At the same time, high-speed global communication makes time to market an equally important factor in the ultimate success of the design. Although a certain amount of iteration is necessary (and desirable) in the design process, going back to the drawing board too many times can add costs and allow a competitor to solve the problem before you do.

A *specification* is a measurable statement of the objectives the design should achieve. Developing written specifications for the design provides feedback so that you know when the target design has been achieved. As new considerations come into play in the design, various trade-offs may have to be considered. For example, a slightly higher target cost may have to be considered to meet the schedule, or one of the performance specifications may have to be adjusted to meet the target price.

The early stages of the engineering design process incorporate a wide range of design considerations. As an engineer, you will be required to work with other engineers, but also with representatives from marketing, sales, manufacturing, and other departments who will be involved with the product. In addition, you may be asked to interact directly with key customers. As companies strive to better meet the needs of the market, you as an engineer need the skills to communicate effectively with a variety of individuals, not just to present your work, but to find the information you need to design an acceptable (superior) solution.

Luckily, technology today makes it possible to communicate better and faster and in a wider variety of ways than ever before. Distance communication methods (e-mail, video conferencing, Internet, and intranet sites) present opportunities for working as a team that transcend the limits of traditional face-to-face meetings. Simulation and analysis tools (spreadsheets, finite element analysis software, circuit modeling software, and so on) make it possible to give team members the information they need to evaluate a design. Visualization tools (3D models, animations, and so on) make it easier to address aesthetic and functional concerns and to create more realistic representations of the design necessary for interaction with customers or other nontechnical members of the team.

2.2 IDEATION

Once the problem has been defined, the design team generates ideas for solving it. Have you ever thought of a great idea for a new product only to find out later that someone else is already making it? When you are designing a device or system, the first thing to check is what designs are already in use. A thorough literature search for similar products and background information is a good starting place. It keeps you from "reinventing the wheel" and provides inspiration that improves the quality of the ideas you generate.

In the ideation phase of design, the emphasis is on creativity and new ways of looking at the problem. Look at the problem from different angles. Don't lock into one approach too quickly. Consider many alternatives to investigate further.

As ideas are generated, questions may come up that were overlooked in the problem identification phase. When they do, the design team adds the information to the problem definition, briefly returning to the first step, before continuing to generate ideas.

After the design solutions have been considered, one may rise to the top as the best. More frequently, a smaller group of design ideas fit many criteria, and the information gained is used to generate a second round of options. These options are then refined and evaluated until a satisfactory solution is achieved.

Where Do You Get Your Ideas?

Creativity is a much discussed notion in design. Where do original ideas come from? Many designs and patents are adaptations of other existing devices or ideas. Even ideas that may seem revolutionary started from somewhere. Swiss inventor George de Mestral came up with the revolutionary idea for Velcro while walking his dog. The combination of examining the cockleburs stuck in the dog's fur and trying to solve his wife's problem with a stuck zipper inspired him to invent the new product. But it took years of work before the idea became a marketable product (see Figure 2.5).

Literature Search

Many people claim that all creative problem solving starts with an in-depth knowledge of the subject area and that it requires only ordinary types of thinking skills. Others claim that creative ideas are just new combinations of existing information or other ideas. This is a good reason for starting out your search for inspiration by

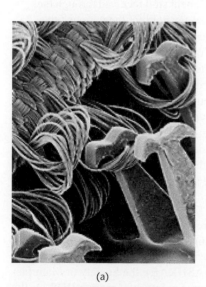

(a)   (b)

2.5 The tenaciousness of a cocklebur inspired the invention of Velcro. The design of the Velcro gripper spines shown in (a) mimics the shape of the hooked ends of the cocklebur, shown in (b). The hooks attach to the loops on the fuzzy side of a Velcro strip in the same way that cockleburs hook onto fabric or hair. *(Cocklebur courtesy of the Brousseau Project, http://elib.cs.berkeley.edu/flowers.)*

researching and collecting data about the subject of your design. The following are some of the places to look for information:

- patent searches
- the World Wide Web
- consumer surveys
- reviews of the competition

An easy way to search the U.S. patent office is on their website: www .uspto.gov. You can search for keywords or browse through different categories. A search on the keywords *eyeglasses* and *retainer*, for example, turned up these results: The word *eyeglasses* occurred 331 times in 163 patents; the word *retainer* occurred 2881 times in 1121 patents; and both *eyeglasses* and *retainer* occurred together in 11 patents. Selecting one of the patents resulted in the following information, depicted in Figure 2.6.

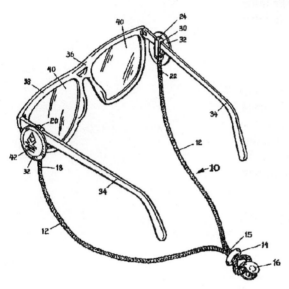

2.6 This figure is typical of the drawings used to document patent applications. Compare the ease of visualizing the product from this drawing with the prose description. *(United States Patent 5,654,787.)*

United States Patent 5,654,787
Eyewear and information holder
Inventors: Barison; Joseph I
(P.O. Box 9787, Denver, CO 80209).
Appl. No.: 549,127
Filed: Oct. 27, 1995
Primary Examiner: Dang; Hung X.
Attorney, Agent or Firm: Crabtree, Edwin H.; Pizarro, Ramon L.; Margolis, Donald W.

Abstract: Eyewear and information holder for holding eyewear and information next to a user of the holder. The holder includes an elongated strap having a first end and a second end. The first end is attached to one end of a first flexible retainer. The second end is attached to one end of a second flexible retainer. The first and second retainers include a stud opening and a temple member opening. When the holder is used for holding eyewear, the temple member openings are adapted for receiving a first and a second temple member therethrough of a pair of eyeglasses. A first and a second advertising piece with an attachment stud on the back of each piece are secured next to the sides of the retainers with the attachment stud received through the stud openings. When the holder is used for holding the various types of information, the temple member openings are adapted for receiving a first and a second engagement arm of a connector device. The connector device includes a foldable attachment arm with tube member thereon and a post extending outwardly from the connector device. The post is adapted for insertion through an opening in a frame used for holding and displaying different types of information. The attachment arm is folded over and the tube member is inserted around the post securing the frame to the connector device. The strap of the holder is then placed around the neck and both the information and the advertising pieces are displayed next to the user of the holder.
12 Claims, 9 Drawing Figures

In addition to the U.S. patent office, there are many other excellent resources for engineering information on the Web. Engineering organizations such as the American Society of Mechanical Engineers (ASME); the American Society of Heating, Refrigeration, and Air Conditioning Engineers (ASHRAE); the Society of Manufacturing Engineers (SME); the Industrial Designers Society of America (IDSA); the American Society for Testing and Materials (ASTM); the Institute of Electrical and Electronics Engineers (IEEE); the American Institute of Chemical Engineers (AICE); and the American Society of Civil Engineers (ASCE) all maintain websites providing guidelines, useful engineering information, and contacts. Many of these organizations also publish engineering references and handbooks that are invaluable in engineering design work. Vendors and manufacturers are another great place to go for information. They often have design guidelines for the products and processes they sell. Appendix 2 contains a list of many useful organizations for engineering and their Web addresses and contact information.

Sites such as GlobalSpec (globalspec.com) and the Thomas Register (www .thomasnet.com) maintain searchable lists of manufacturing companies. You can search them for information about companies, general products, brand names, CAD drawings, and other useful information. For example, if you were considering a design solution that used extruded aluminum, you could search the Thomas Register for companies producing aluminum extrusions, as shown in Figure 2.7. You could then contact specific companies to find out more particulars about pricing, specifications, and manufacturing times. Many companies have websites that you can link to directly from the Thomas Register site. Companies often provide general design guidelines and manufacturing tolerances, material specifications, and other useful information related to their products.

Consumer Surveys

Another place to get information is from the consumer. Marketing surveys and consumer questionnaires provide important feedback. Think about all those little product registration cards that you have filled out. Why does the manufacturer need all that information? Probably to get some idea of who buys the product, how to price it, what other features the consumer would pay for, and what problems could be solved or what improvements could be made. However, you should take consumer surveys with a grain of salt. Sometimes consumers will not state their preferences accurately or will give incomplete information. For example, the customer may rate a particular feature of utmost importance but be willing to pay only a few cents for it. If you are designing your own survey to test customer opinion, you should review the principles of good survey design, taught in most statistics courses, to be sure the data you get back are useful to you.

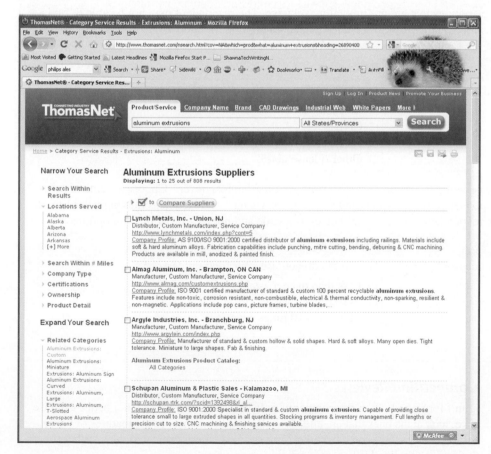

2.7 A search of ThomasNet for aluminum extrusions generated this list of companies that produce them. *(Courtesy of ThomasNet.com.)*

Good Survey Design

A survey is a "conversation" with the consumer. Although your questions are going out on paper, over the Web, or via a professional researcher, they will be answered by a person with time constraints, other interests, and, often, less commitment to the topic. To avoid surveys that are too long or too difficult, put yourself in the place of the respondent and consider the many ways questions could be answered. Some common pitfalls to avoid are as follows:

Leading questions that subtly tell the respondent what you think about a topic.

Poor	Better
Do you think the quality of engineering drawing is as high today as it was 10 years ago? ☐ Yes ☐ No ☐ Undecided	Which of the following best describes the way you feel about the quality of engineering drawing today? ☐ Higher than it was 10 years ago ☐ No change in quality over 10 years
Do you think the quality of engineering drawing is higher today than it was 10 years ago? ☐ Yes ☐ No ☐ Undecided	☐ Lower than it was 10 years ago ☐ No opinion

Double-barreled questions that pose two questions but present only one set of responses. The respondent doesn't know which question to answer. In the following question, a reader who has a degree in engineering technology or civil engineering has to decide whether you really want to know when the degree was earned or if you want to know only about mechanical engineering degrees.

Poor	Better
When did you get your degree in mechanical engineering?	What year did you get your engineering degree? In which engineering discipline?

Questions with incomplete options or options that are not mutually exclusive. The respondent may not be able to find a response that fits and then has to decide whether to skip the question, add an option, or choose one that is close. In interpreting the results, you will have the same problem—deciding how the respondent handled it.

Poor	Better
In which engineering discipline did you get your degree? ☐ Mechanical ☐ Civil ☐ Electrical	In which engineering discipline did you get your degree? ☐ Mechanical ☐ Civil ☐ Electrical ☐ Other (please note): _____

Inconsistent frame of reference: one person's "occasionally" is another person's "almost never." Choose a frame of reference that allows you to compare responses across questions and respondents. Using a scale (such as ranking from 1 to 5) or objective criteria ensures that the respondent will see your choices as a scale and not as a set of random options.

Poor	Better
How often do you use your food processor? ☐ Frequently ☐ Sometimes ☐ Almost never ☐ Never	How often do you use your food processor? ☐ Less than once a week ☐ 2–10 times a week ☐ More than 10 times a week

Competition Reviews

Don't forget to look at the competition. If you are going to design a better exercise bike, look at the ones that are already on the market. What are the features of the top-selling exercise bikes on the market? Are certain features critical to sales? What about their price? Styling? What areas can you improve on?

Sometimes seemingly unrelated products can yield design ideas, too. What other products are commonly used by the same group of people or businesses that will be using yours? Which features are consumers accustomed to and likely to expect from your product, too?

Reverse Engineering

Going beyond merely looking at other products on the market, consider reverse engineering the product. Taking apart an existing design and capturing all the information needed to manufacture it can be a good starting point.

Companies also use this approach when the information for manufacturing an existing product is lost or perhaps never existed in the first place. Whether you are using an existing product as the start of a redesign or capturing the information needed to remake the same product entirely, you can learn a lot by dissecting an existing product. Think about how the parts were manufactured for the existing design. Can you make it more cheaply? Better? More user friendly?

Brainstorming

Another popular method for generating ideas is **brainstorming,** an approach defined and popularized by Alex Osborne. Brainstorming works in two ways: First, it gets as many ideas out of the group as possible. Second, these ideas spark more ideas and generate a larger pool of ideas to work from.

The key to success in using brainstorming is what Osborne calls **deferred judgment.** Deferring judgment—not being critical of the ideas at the time they are presented—may help generate more ideas. One reason is that each person is free to list wild ideas that may not be workable but may spark an idea that is less wild and just might work. Deferring judgment lets you use the creative side of your mind freely by postponing the need to be analytical or saying which ideas may not work. It also may allow people to feel comfortable in presenting their ideas to the group because they won't be criticized. This lessens the chance that good ideas will not be brought up because of an individual's fear of criticism or of appearing silly.

Brainstorming works best with groups of six to ten people working for about an hour. Including representatives from marketing, manufacturing, and consumers can increase the variety of ideas generated. The process is usually headed by a moderator who writes down the ideas, encourages the participants, and keeps any critical analysis of the ideas shut down.

Individual Brainstorming

In his book *Planning and Creating Successful Engineering Designs*, Sydney F. Love describes a brainstorming technique called **individual brainstorming,** which was developed by William J. Osborne and George Muller. In this method, participants are given 120 seconds to silently record key words relating their ideas. The moderator reminds them of the time remaining in 30- to 15-second intervals. (Pressure is supposed

---

Rules for Brainstorming

Alex Osborne gives four rules for brainstorming sessions:

1. *No criticism.* Wait until you have finished generating ideas before you rule any out or begin to criticize. This way all the ideas can be brought out.
2. *Wild ideas are welcome.* Don't limit yourself to ideas that may work. Thinking outside the sometimes closed seeming statement of the problem to come up with novel solutions is encouraged. If the wild idea doesn't work, it might spark another idea that will work.
3. *More is better.* Keep at it until you have generated a long list of ideas. These can be winnowed to form a shorter list after you have finished brainstorming. This is basi-

cally a repetition of the adage that two heads are better than one. If you have a lot of ideas to pick from, some of them should be good ones. It is a common tendency of people to quit brainstorming before they have come up with really creative solutions. Often, the first fifty solutions listed are common and not creative solutions. After that, people have to start stretching their minds for creative solutions.

4. *Work off other ideas.* Each suggestion does not have to be entirely original. If your idea is sparked by another one that was given, don't worry. Your special twist might be just the thing that is needed. Feel free to combine other ideas into a new idea or otherwise change the previous suggestions.

to help the flow of ideas.) Once the first list is generated, the moderator calls on participants one at a time to rapidly present his or her list of ideas. The other participants cross off the duplicated ideas from their list and start a second list of ideas triggered by hearing the speaker. The next participant reads only the ideas that were not crossed off his or her first list. The process continues one more time, with the participants reading the ideas from their second list and generating new ideas in a third list. This process is supposed to help generate ideas because the second and third lists are new ideas and not ones that easily come to mind. Also, the pressure from having to cross off duplicated ideas forces people to think more originally. Or, if the ideas were not duplicated, the originator may be encouraged and produce even more ideas for the second and third lists.

In both kinds of brainstorming sessions, how the question to be solved is posed is also important. A good needs statement can provide needed direction but should not be too narrow. Sometimes a broader statement will generate more creative solutions.

Listing

Another method for generating ideas, sometimes called **listing** or **morphological analysis,** is to break down the system or device into its individual components, then consider every possibility for that individual component. The individual components are combined to come up with new possibilities that may not have been considered. For example, divide a chair into the categories of seat, support base, back, and arms. For each of these components, list every possibility that comes to mind, then combine the different options. For example, one design may mount to the floor with a padded seat and a fabric sling back and arms. Figure 2.8 shows sketches of some of the listed concepts.

Seat	Support	Back	Arms
padded	integral	padded	no arms
wood	star-type	wood	padded arms
plastic	casters	metal	plastic arms
woven	legs	contoured	separate arms
cane	floor mount	straight	attached arms
metal	pedestal	open	metal arms
		fabric sling	sling support

The Decision Process

After a number of possible solutions have been generated, one or more ideas are selected to develop further. There is often a certain amount of subjectivity in the selection of the "best" design concept to pursue further.

The decision can be based on economic considerations by comparing the potential cost-to-profit ratio for each design. Marketing surveys or focus groups may be used to help select among design alternatives. Often, a **decision matrix** is used to help rank the alternatives in a systematic way.

There are two steps in producing a decision matrix. The first is to identify all the design constraints and rank them in order of importance. The ranking may be done by assigning a certain number of points to each design consideration. Often, the weighting number is based on 10 or 100 because most people have some sense of how they would rank based on this number of points. The criteria can also be divided into categories of must have (needs) or would like to have (wants). This makes it easier to determine trade-offs in the wants category. Any design that will be considered for further evaluation will meet all the needs criteria but can vary among the wants criteria.

Once the criteria are fully identified and ranked, the designs are rated according to each of the criteria. The results are typically listed in a chart or decision matrix

2.8 Even a relatively short list of alternatives can be combined into a large number of different designs, as illustrated by these chair design sketches.

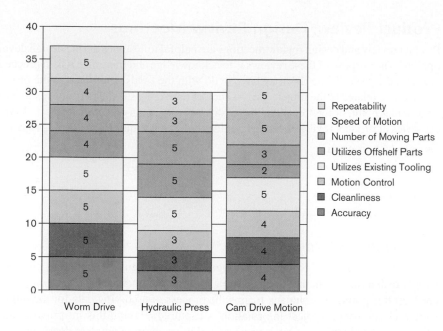

Repeatability
Speed of Motion
Number of Moving Parts
Utilizes Offshelf Parts
Utilizes Existing Tooling
Motion Control
Cleanliness
Accuracy

2.9 A decision matrix was used to evaluate design alternatives for a press brake. The worm drive, hydraulic press, and cam drive motion were rated from 1 to 5 according to the criteria on the right. The ratings were then summed (as shown in the stacked bar chart for each alternative) to determine the option with the highest overall rating. *(Courtesy of Trent Wetherbee, Andrew Beddoe, and Ezra Miller.)*

showing each criterion in one direction and each design alternative in the other direction (see Figure 2.9). The team as a whole can discuss each item and rank each design and record the values. If the team cannot meet together to produce the rankings, the individual members can rank the items, and the results can be averaged. This averaging process, however, can tend to make the alternatives rank toward the middle of the range if individuals perceive the value of the designs differently.

For a weighted decision matrix, the weight for each criterion is multiplied by the rank before the values are summed. The criteria in Figure 2.9 were weighted and a spreadsheet was used to calculate the ranking for each design alternative, as shown in Table 2.1.

For designs that will be divided into need and want categories, the chart can be divided into two portions, the upper portion for the constraints (or needs), and the lower portion for the design criteria (or wants). Often, the need constraint items do not need to be ranked, just checked off. If an item does not meet all the constraints, it does not have to be evaluated further.

Table 2.1 Weighted Decision Matrix for a Press Brake.

Criterion	Weight	Score/Weighted Score					
		Worm Drive		Hydraulic Press		Cam Drive Motion	
Repeatability	10	5	50	3	30	5	50
Speed of motion	8	4	32	3	24	5	40
Number of moving parts	8	4	32	5	40	3	24
Utilizes off-shelf parts	8	4	32	5	40	2	16
Utilizes existing tooling	7	5	35	5	35	5	35
Motion control	7	5	35	3	21	4	28
Cleanliness	7	5	35	3	21	4	28
Accuracy	9	5	45	3	27	4	36
Total score raw/weighted		**37**	**296**	**30**	**238**	**32**	**257**

Product Review/Design Review Meetings

Product review and design review meetings are helpful milestones during design development. The purpose of these reviews is to acknowledge and minimize risk in proceeding with the project. The risk may have to do with the funds committed to the project or it may be related to the project itself—such as a design issue that must be resolved at a particular point, or the threat that the customer will call an end to the project. A typical series of product review meetings might come at these milestones in the project:

- product proposal
- design proposal
- development plan
- engineering release
- product release

Product Review Stages

The review at the ***product proposal*** stage may be driven by marketing and used largely to determine whether there is demand for the product or to identify what users need. At this point, the company begins the process of archiving data for the project so that all the useful project information is captured. Having this information may help in the development of other projects and in providing complete documentation for the current one.

At the ***design proposal*** review, a plan to solve a problem or meet a need (usually one that will produce revenue for the company) is presented. The purpose of the review is to get a commitment to go ahead with the proposal as defined at that stage. A considerable amount of money and people are usually required to move ahead from this point, so the proposal includes enough detail to allow for the review committee to make an informed decision. The proposal should indicate that the product definition, specifications, conceptual design, and identification of the consumer interface are sufficiently developed. It should also show that the product's market potential has been evaluated and present the results of that evaluation. A plan for the timely completion of the project and an estimate of the manpower it will require should be part of the proposal, as should plans for future design review. The design proposal meeting may also include a review of the product's architecture (how the component systems will fit and work together); fastening, joining, and assembly plans; industrial design review; and the business plan for the project.

After the commitment has been made to go ahead with the project as defined in the proposal, many questions still have to be answered. For example, What power supply will we use? How can we cool the device if a certain power requirement is necessary? What is the preferred placement for the user controls? Answers to these questions should be in hand before the ***development plan*** review. The development plan includes all the groups involved with the product. The participants in this level of review usually include representatives from engineering, marketing, manufacturing, service, customer groups, and management. Involving all the groups helps ensure that key concerns are not overlooked in the plan. For example, including members of the service department gives them the opportunity to voice concerns about parts that have historically been problematic or difficult to service (Figure 2.10).

The next phase of the project is engineering and detailed design. In this phase, the design of each part is completed, CAD geometry is finalized, and prototypes of the design are created. The ***engineering release*** is a milestone ascertaining that satisfactory design reviews have been completed for all components and modules and that all outstanding questions have been answered. Questions such as, Can service provide field support? Has the product passed a review for safety? Are all the project CAD data created and archived? may arise at this meeting. After the engineering release of the project, it makes a transition to being manufactured.

During the manufacturing phase, each part produced or acquired is tested to ensure that it will function as designed and produced. This process is called ***qualifying the production.*** The final product goes through a similar series of qualifying tests. When everything is tested and in place, the next milestone is the

2.10   Designers and management gather around a set of drawings to review a design. *(© StockLite/Shutterstock.)*

product release. Approval at this stage means that the product is now ready for use by the customer.

Design Review

As an entry-level designer, you will probably have a more active role in design-review meetings than in product-level reviews. A design review may be a ***product review*** (meaning that it is for the entire system or device) or a ***part review*** for single parts or systems that function at the lowest level of the whole.

Typically at this level, the review is by individuals with a background in the discipline you are working in—mechanical engineers reviewing the work of mechanical part designers, for example. These reviews tend to include more technical details and are often led by the team project leader. For this type of meeting, you need to document major design decisions and the justification for your choices (such as costs, company standard parts, industry codes). You may have to present calculations for cooling data, dynamic analysis, loads, or other information showing how the part as you have designed it will meet required performance specifications. You may use charts and graphs to show thermal curves with accepted engineering margins indicated. You may present sketches or drawings of the part or show a prototype model.

The sketches, graphs, drawings, and CAD models you produce are not just for your use but for the company as a whole. Make your documentation neat and legible. As an engineer you may want to initiate a design review yourself to help you judge risk and design issues for your own work, especially when it is still in a conceptual phase. Getting the input of more experienced engineers or of a cross-disciplinary team can improve your designs, save time later in the process, and contribute to your learning about pertinent processes.

The Role of Visualization in Ideation

What happens when you get an idea? Do you picture something in your head? Many people, when they have ideas about new designs, problem solutions, or other creative things, say that they "see" the answer. They form a mental picture of the solution. There is a lot of debate about whether there can be any thought without language, but many creative inventors say that they see pictures in their mind even

2.11 Words conjure mental images.

before they can express these ideas in words. They communicate these picture ideas to others through sketches. Albert Einstein described this phenomenon in a letter to Jacques Hadamard:

> The words or the language, as they are written or spoken, do not seem to play any role in my mechanism of thought. The psychical entities which seem to serve as elements in thought are certain signs and more or less clear images which can be "voluntarily" reproduced and combined—this combinatory play seems to be the essential feature in productive thought before there is any connection with logical construction in words or other kinds of signs which can be communicated to others.

Certainly the opposite process is true: when we hear expressive language we form a mental image. For example, when you see the bumper sticker "Visualize Whirled Peas," you form a much different mental image than when you see "Visualize World Peace" (Figure 2.11).

The process of visualization is important in forming and expressing your ideas. As an engineer, designer, scientist or technician you will need the ability to visualize things. You will need to visualize how things work, how problems might be solved, what a device drawn on paper would look like if it were real, mental models for complex systems, how parts will fit together, the magnitude of various objects from given dimensions, and many other sorts of things. The ability to apply visualization to solving problems and generating ideas is a useful addition to the verbal and mathematical problem-solving skills that you will be learning throughout your career.

Some people seem to be born with excellent visual abilities; others must learn it. Chances are you enrolled in this course because you are pretty good at visualization, but practice and understanding basic principles will make you better at it.

You will also need to be able to communicate your ideas to members of the design team and others. One of the most direct means of communicating your visual ideas is sketching. Unlike a verbal description, hand sketches do not require verbal processing of the image but seem to flow directly from the mind to the hand. This basic method is one way of "thinking out loud" about mental images. As you sketch you capture the visual ideas for your own future reference as well as for others. Sketching your ideas also develops your ability to visualize.

Making a visual representation of information can help you incorporate it into your thinking about the problem and can help others see the relationships, too. Relationships and comparisons are often best communicated through a visual image. For example, the graph shown in Figure 2.12 makes it easy to quickly see the positive linear relationship between amperage and the dynamic head for the pump. Can you see the relationship as clearly from the data in Table 2.2?

Words and mathematical symbols are important in communicating about design, too, but their pathways through the processing portions of the brain take a less direct route than a visual image. The reader must interpret your verbal or mathematical description and rebuild the visual image to see your design. It is not just coincidence that we say, "I see" when we mean, "I understand." However, written notes to accompany a drawing have an important role in making it easier for someone to understand how to manufacture the device or system you are designing.

The power of visuals is perhaps best exemplified in computer-generated graphics (see Figure 2.13). Think back to the last computer game you played. How were the graphics? Would you have kept on playing it if it were a text-based game? Were the graphics in 3D? If they were not, did you wish they were?

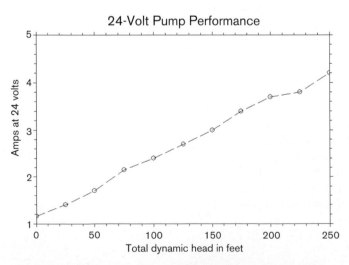

2.12 This plot shows the relationship between the Y-variable (amps at 24 volts) and the X-variable (total dynamic head in feet) by plotting the data points from Table 2.2.

Shaded and rendered 3D computer models are a powerful way to present a very realistic view of your designs. Computer models enable designers to create "virtual worlds" that let the consumer interact with the model as if it were real and responded to events and actions. Using a virtual reality interface with 3D computer models of the design, you can even "touch" the design as well as "see" it as though it were right in front of you. Using as many of your senses as possible can add to your abilities to solve problems.

2.3 BUILDING VISUAL ABILITY

To communicate visually you will need a set of skills. Some of these are physical skills, like sketching clear crisp lines that others can interpret. Because sketch paper is only a 2D medium, you will also need a system to show 3D objects on a flat sheet of paper. In addition, you will need to read drawings someone else has created and build a mental picture of the object. You can build your visual abilities through practice.

Table 2.2 24-Volt Pump Performance.

Total Dynamic Head in Feet	Amps at 24 Volts
0	1.15
25	1.4
50	1.7
75	2.15
100	2.4
125	2.7
150	3.0
175	3.4
200	3.7
225	3.8
250	4.2

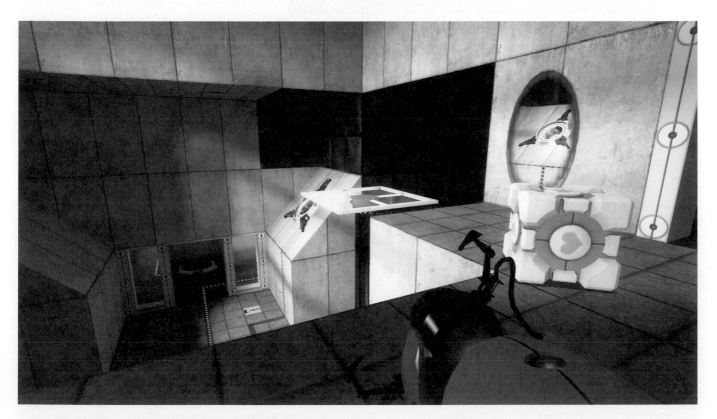

2.13 Role-playing games use 3D modeling to create environments that a player can explore and interact with. Notice how the textures applied to the walls and floor contribute to this scene from the game Portal. *(Courtesy of VALVE.)*

2.14 Grid paper of various types helps you sketch shapes and graphs more accurately.

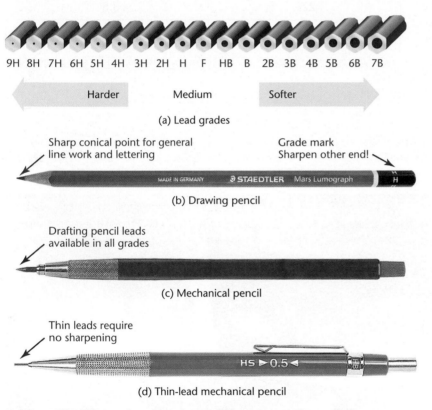

9H 8H 7H 6H 5H 4H 3H 2H H F HB B 2B 3B 4B 5B 6B 7B

Harder   Medium   Softer

(a) Lead grades

Sharp conical point for general line work and lettering

Grade mark
Sharpen other end!

MADE IN GERMANY ✪ STAEDTLER Mars Lumograph H

(b) Drawing pencil

Drafting pencil leads available in all grades

(c) Mechanical pencil

Thin leads require no sharpening

HS ▶ 0.5 ◀ HB

(d) Thin-lead mechanical pencil

2.15 Different grades of lead and different pencils can help you differentiate lines in your sketches.

Tools

The tools you need to begin sketching are simple—just paper, pencil, and eraser (see Figures 2.14–2.15). Later you may want to add engineering markers to your sketching tools, as engineers are often required to sketch in pen or marker to provide a permanent record of their designs. Until you become more proficient at sketching, you will probably prefer to sketch in pencil so you can erase some lines. Modeling clay is also an invaluable tool, especially as you are beginning to visualize and sketch objects. Figures 2.14 and 2.16 show examples of typical sketching supplies.

Pencils are available in different grades of lead (see Figure 2.15). The pencil grade indicates the pencil's hardness or softness, which relates to how light or dark a line it will draw. Hard leads make light lines; soft leads make darker lines. Pencil grades range from the hard leads at 9H through 4H; to medium leads 3H, 2H, H, F, HB, B; then to soft leads 2B through 7B. For most engineering sketches you will need only an HB pencil, or one near that portion of the range. You may also want to use a hard pencil such as 6H to make light lines for blocking in your sketches. When you are more practiced, you can probably switch to using only HB lead, remembering to block in shapes very lightly at first.

It is helpful to sketch on grid paper with divisions of 5 mm, 10 mm, 1/8 inch, or 1/4 inch (see Figure 2.14). The grid lines can help you develop your ability to sketch straight lines, but spend some of your practice time drawing on plain white paper to develop your ability to sketch quickly and not be dependent on the grid lines.

Bound engineering notebooks are frequently used to provide a record of the design. A bound notebook makes it readily apparent if pages have been removed. Also, it is convenient because the sketches, ideas, phone numbers, notes, contact names, and other information for the design are all available in one place.

Practice

Keeping a design notebook is a good way to start practicing. Many people feel awkward jotting down their ideas and making sketches. The tendency to be overly critical of their ability to sketch can prevent some from wanting to try it at all, but practice is necessary to develop the ability in the first place. Start using your notebook to record ideas as sketches. Make sketches of interesting things you see and do not be too critical of your drawing ability. As you sketch it will improve and become a skill you can use to communicate your ideas to others.

2.16 Drawing Templates *(Courtesy of Chartpak.)*

Sketching Basic Shapes

One way to begin freehand sketching is to draw an object's **contour** or outline. An example is shown in Figure 2.17. Contour drawing is really the process of noting the relationship between the object you are trying to draw and the space surrounding it. You can think of it as similar to graphing the shape of the object on a sheet of paper. To draw successfully, you need to represent the object's spatial relationships accurately. Which is the highest point of the object from your point of view? Which is the lowest point? How big are the various features?

Forget the Symbols You Have Learned

As children we learn a way of drawing common objects that does not necessarily describe the object's shape accurately. For example, when you draw a picture of a chair, you may draw a flat seat with legs, as shown in Figure 2.18(a), even though a more realistic view of the chair would show more of the seat and the four legs. These drawings serve more as "symbols" for objects than as accurate representations of their appearance. You can probably easily sketch several of the symbols that appear in elementary school art to depict a cat, pig, house, tree, and so on.

To draw an object's shape as it actually appears, examine it without any preexisting notion of what that object should look like. Two exercises that can help you see the object, and not your mind's representation of what the object looks like, are drawing an object upside down, and drawing the space around the object (not the object itself).

Optical Illusions

Many optical illusions derive from the things the mind "expects" to see. Did you ever wonder why the rising moon on the horizon appears so much larger than the same moon higher in the sky? Because we are more used to seeing the moon overhead (which the mind interprets as closer because there are no other visual references between us and the moon overhead), it appears smaller overhead than when it is on the horizon (which seems to be farther away because we have the context of a large number of objects between us and the moon on the horizon). Actually, the moon occupies the same amount of our field of vision whether it is overhead or on the horizon. Try holding a quarter at arm's length when the moon is on the horizon and appears large. It should just about block out the moon. Then, wait until it has risen and appears smaller. Try the same thing: Judge the size of the moon overhead using a quarter. You should see that the moon is the same size in relationship to the quarter.

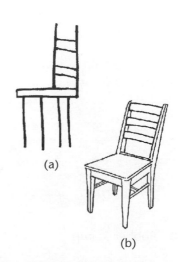

2.17 This freehand contour sketch of a baseball glove shows its shape and features using only lines.

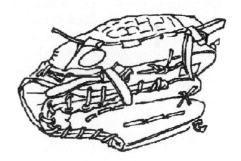

(a)

(b)

2.18 Compare the chair shown as a symbol in (a) with the more realistic contour sketch of the same chair shown in (b).

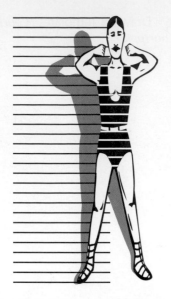

2.19 The appearance of the strong man figure in this illustration shifts as your viewing angle does. Does his face appear to be as thin when you lay the book flat on the desk as it does when you hold it in a vertical position? Do the horizontal lines appear closer together?

Examine the optical illusion shown in Figure 2.19. Lay your book flat on the table and look at the figure. Then, tip the book so that the figure is vertical in space and look again. Do the lines appear closer together when it is lying down or when it is standing up? You should see that the lines appear closer when the figure is lying flat.

The same illusory effect can happen when you are sketching on a sheet of paper lying flat on your desk, leading you to overcompensate in your drawing. It is sometimes helpful to hold your paper vertically and get a different view of your sketch from time to time.

Try creating a contour drawing of the rubber stamp shown in Figure 2.20. Use a sheet of plain white paper attached to a clipboard, if possible, so that you can hold the paper up and look at the sketch straight on. When the paper is laid horizontally on a flat surface like your desktop, the lines you are sketching can appear foreshortened, causing you to make a distorted sketch.

Estimating Proportions

Start out your sketch by noting the relationships of the stamp's features. The overall shape, including the handle, is slightly wider than it is tall. The center handle sticks up above the back edge of the base about half the height. The lowest point is in the lower right. The highest point is on the handle. Imagine tracing the outline of the stamp in the air with your pencil. Next, very lightly sketch the outline of the stamp on your paper. Try to keep it in proportion as shown in Figure 2.21. After you have sketched the basic outline of the stamp, add the details representing the edges of the object. Add the curved contour for the bottom of the handle where it joins the stamp base, and the flat top surface of the handle.

Showing the features of an object in correct proportion to one another is key when sketching. To help you determine the proportions of the object you are trying to sketch, use your pencil or light marks on the edge of your paper to "measure" the object's features. You have probably seen artists on TV or in cartoons holding their paintbrush in front of the object they are painting (see Figure 2.22). Try it yourself by holding your pencil up at arm's length and measuring how many pencil lengths (or what portion of the pencil) make up the vertical height of a door. Holding your arm at the same distance, measure the width. Is the height about three times the

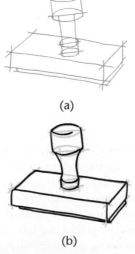

(a)

(b)

2.20 Rubber Stamp

2.21 (a) The blocked-in shape of the rubber stamp shows the overall width, height, and height of the handle over the back edge. (b) The darkened final sketch of the rubber stamp is drawn over the lightly sketched shapes.

2.22 This artist is using his paintbrush to measure the proportions of the castle.

width? You can use this to make rudimentary measurements to keep your sketch in proportion.

Another technique is to hold your paper vertically in front of the object and use the edge of the paper to help you determine whether the object's edges are horizontal, vertical, or at some other angle. Visually judge the size of the angle in relationship to the edge of the paper.

Viewpoint

Keep in mind that as you are sketching objects you want to maintain a consistent viewpoint. This can be easy when you are sketching a picture from a book because you cannot move around the object. But when you can move around an object, you will see a different view of it depending on where you stand. Sometimes people have difficulty sketching because they show parts of the object that cannot really be seen from the viewpoint. For example, knowing that the handle of the rubber stamp appears circular from the top, you may want to show it round in other views, even though it appears elliptical.

Shading

The edges of an object are visible because one surface reflects more light than another. The angle of the light illuminating an object determines how different surfaces appear. Adding shading to your sketch can give it a more realistic appearance because it represents the way the actual object would reflect the light.

Hatch lines, shown in Figure 2.23, and stippling, shown in Figure 2.24, are two different ways to add shading to your sketch. These methods are used rather than continuous pencil shading because they reproduce better on copies. Add hatching lines or stippling to add shading to the stamp's surfaces where they are darker. After you do, darken the outline of the stamp more so that it defines the shape clearly and boldly. Remember that when you are communicating by using a sketch, its subject should be clear, just as it should be when you are writing a sentence. To make the subject—in this case, the stamp—clear, make it stand out with thick bold lines.

Positive and Negative Space

Another way to think about freehand contour sketching is the contrast between the positive and negative space. Figure 2.25(a) shows a drawing of a mixer. Does Figure 2.25(b), showing the negative space, help you see the shape of the contour better? What about breaking it down further, as shown in Figure 2.25(c)?

Some people sketch more accurately when they try to draw the negative space that surrounds the object. An 8.5 × 11-inch sheet of Plexiglas (available at most glass stores) is an excellent tool for developing your sketching ability. Hold the Plexiglas up in front of an object, and using a dry-erase marker, sketch the object on the Plexiglas. The outline should match the object's outline exactly if you do not move. Lower the Plexiglas and look at the orientation of the lines. Are they what you expected? Try looking at the object and drawing the sketch with the Plexiglas lying on your desktop or knees. Then, raise it and see whether your drawing matches the object. Everyday objects work great for developing sketching ability—try your toaster, printer, lamp, as well as exterior and interior views of buildings and equipment.

2.23 Contour Sketch with Hatching

2.24 Contour Sketch with Stippling

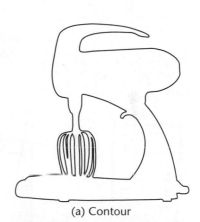

(a) Contour

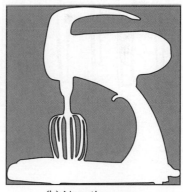

(b) Negative space

(c) Shapes shown

2.25 Negative Space

┌─ **TIP** ─

Practice Drawing Contours

Try sketching the negative spaces that define the shape of a chair. Look at each space as an individual shape. What is the shape of the space between the legs? What is the shape of the space between the rungs and the seat?

Make a sketch of a chair, paying careful attention to sketching the negative spaces of the chair as they really appear. The positive and negative spaces should add up to define the chair.

If you have difficulty, make corrections to your sketch by defining the positive shapes and then check to see if the negative shapes match.

Use an 8.5 × 11-inch sheet of Plexiglas (available at most glass stores) as a tool for developing your sketching ability. Hold the Plexiglas up in front of an object, and using a dry-erase marker, trace the contours of the object on the Plexiglas. If you do not move, the outline should match the object's outline exactly. Lower the Plexiglas and look at the orientation of the lines. Are they what you expected?

Next, try looking at the object and sketching with the Plexiglas lying on your desktop or knees. Then, raise it and see if your drawing matches the object.

Use the same technique to practice drawing the contours of a wide range of everyday objects, especially familiar objects that you learned to draw as symbols as a child.

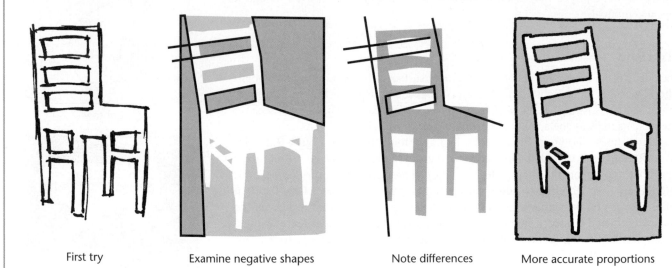

| First try | Examine negative shapes | Note differences | More accurate proportions |

Visual Perception

Optical illusions and magic tricks are both based on the fact that what you see is actually your mind's interpretation of the data coming in through the optic nerve. The more you know about visual effects, the better you will be at re-creating them for another viewer. Learning to sketch is not only an effective means of capturing your design ideas, it can also change the way you "see" the world around you.

One way to practicing "seeing" is to look hard at an image for a few seconds, then close your eyes and try to visualize the scene. Try to re-create as much of the detail as possible, then compare the image you built with the actual scene. Notice what you could "see" in your mind's eye and what you could not.

People's interests and experiences cause them to remember different aspects of a given scene. Did you notice colors? The relative size of items? Texture? You can build your visual ability by practicing different ways of "seeing" and noticing how lighting, movement, and distance alter how your eye perceives different objects.

Imagining how objects and scenes look from different viewpoints is another way to build your visualization skills. One technique that Albert Brown, design engineer at Affymax Research Institute, uses is to select a bird on a rooftop or telephone pole and consider what the buildings, cars, and objects in the area would look like from that viewpoint. What would be visible and what would be obscured by other objects? What shape would a building or parking lot have from a view from the top? The ability to see objects from different vantage points is a key skill in creating and using engineering drawings.

2.4 CONSTRUCTION LINES/SKETCHING AIDS

Contour sketches help you develop your ability to "see" an object as it actually appears so you can re-create that appearance on paper. Another technique is to sketch the basic shape of the object and use that as a guide for the object outline you want to create. ***Construction lines*** are guides that can help you preserve the basic shape and overall dimensions as you add the details of the object.

Before you begin to draw the outline for an object, think about its overall shape and the relationships between its parts. The drawings in Figure 2.26 show how the remote control sketch started with construction lines for the basic shapes. These lines helped preserve the overall shape as the rounded corners of the outline were added. Additional construction lines locating the buttons were added to make it easy to add the details.

Try this technique with objects that are generally square, round, or elliptical. Use the overall shape of the object to help you determine the relative size and placement of features on the object. Think about breaking down more complex objects into their simpler geometric shapes, as shown in Figure 2.27. Sketch the shapes showing their relationships to one another accurately.

Look for the essential shapes of objects. If you were to make a clay model of an object, what would you start with? A ball? A box? Squint your eyes and look at familiar objects. Do you see their basic shapes better?

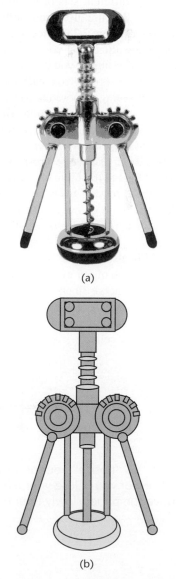

(a)

(b)

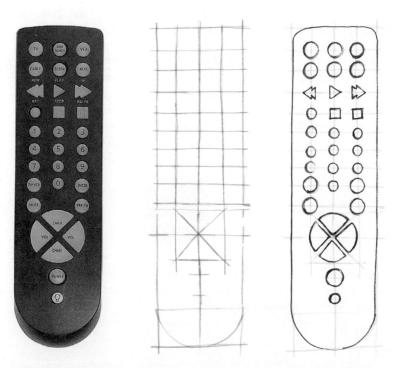

2.26 Construction lines were used to block in the lines for the shape of the remote control. Details were then added, keeping the sketch in proportion.

2.27 The essential shapes of the corkscrew in (a) are sketched in (b).

The Physiology and Psychology of Seeing

What we see is actually an interpretation of nerve impulses sent from the retina to the brain. Light reflected from an object enters the eye through the lens and is focused on the retina at the back of the eye. The retina contains light-sensitive receptor cells called rods and cones. Each eye contains about 125 million rods, which are responsible for recognizing light and dark, and about 7 million cones, which react to colored light. Three kinds of cones—for red, blue, and green—send impulses that are mixed by the brain to create the wide range of colors we "see." Cones are concentrated in the center of the retina, where vision is sharpest; rods are more numerous around the outside of the retina. As a result, peripheral vision (what you see out of the corner of your eye) is primarily black and white.

Two different sets of impulses are created—one from each eye. The impulses from both eyes are combined by the brain (see Figure 2.28). Close one eye and then the other and notice how the object you are looking at seems to shift side to side. The object you see with both eyes is a composite made of the signals from each eye.

But seeing is more than a simple mapping of impulses onto a "screen" in the mind. Brain research has demonstrated that the impulses are combined and recombined as they pass through different parts of the brain. Some impulses from each eye cross over to the opposite brain hemisphere at the optic chiasma, but some do not. This incomplete crossing over means each half of the brain gets signals from both eyes, which scientists believe is responsible for stereoscopic vision, allowing us to see the depth of objects.

This also explains why we are unaware of the "blind spot" in each of our eyes (see Figure 2.29). The optic nerve, where the nerve fibers from the rods and cones come together to exit the eye, has no cells receptive to light, even though it is located on the back of the eye and surrounded by receptor cells.

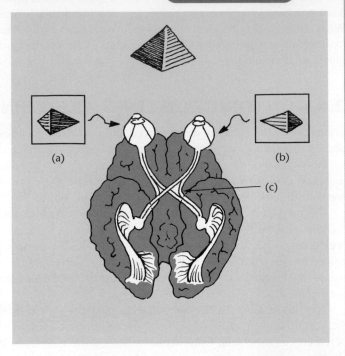

2.28 Light rays from the object strike the retina in the patterns shown in (a) and (b). The impulses cross over the optic chiasma and travel through the brain along the pathways illustrated in (c).

As the visual impulses travel through the brain, different areas of the cerebral cortex affect how the impulses are translated into meaning. Memories, experiences, and visual cues affect how we interpret the impulses—and how we sometimes "see" things that are not there, simply because they were expected. This interpretive seeing is why it can be difficult to "see" an object from a single viewpoint—or to see the actual shapes of objects as they appear from a given viewpoint: The brain adds to the information from the visual stimuli. To improve engineering visualization and drawing skill, it can help to be aware of interpretive seeing and to think about the true geometric shape of objects.

2.29 To "notice" your blind spot, hold this page with the drawing of the + and the dot about 16 inches in front of you. Close your left eye and focus your right eye on the +. Keeping your eye focused on the +, move the paper slowly until the dot moves into your blind spot and disappears.

Standards

ANSI is the American National Standards Institute, a governing body that establishes and publishes standards, or rules, for technical drawings and other engineering practices. The institute publishes standards that govern drawing sheet sizes, linetypes, multiview drawings, and representations for screws and gears, among others.

ISO is the International Organization for Standardization, the governing body for international standards, which is headquartered in Geneva, Switzerland. The group name is always abbreviated ISO regardless of the language it is translated from. Nearly 100 member countries cooperate to develop these international standards.

2.5 SCHEMATIC DRAWINGS

Schematic sketches or drawings use symbols to represent components, constructs, or relationships, instead of showing their actual appearance in the drawing. Think back to the discussion of sketching the chair as a symbol. Children do this without really thinking about the process. For example, a cat has a round head and triangular ears. We can recognize this shape easily and identify it as a cat. Engineering schematic drawings use the same principle. Symbols that are quick to draw and that reproduce well are used to replace the complex shapes of the real objects. For example, electrical and electronic schematics represent electrical power systems and electronic components, such as capacitors and resistors.

Schematic drawings use standard symbols in order to make them easy to interpret. These standard symbols and their sizes are published as *standards* by the American National Standards Institute (ANSI) so they will be used consistently and their meaning will be clear. Standard symbols are published for several different disciplines; some of the most useful ones appear in Appendices 44–48.

Similarly, *freebody diagrams* use vectors, masses, and simplified objects to depict the forces in a mechanical system or device. *Block diagrams* (Figure 2.31, see page 57) show the relationships between equipment used in a system, such as the layout of tanks and piping in a fluid system, or the layout of computer equipment in a network. Figure 2.30 (see page 56) shows examples of freebody, block, and piping schematics.

Schematic drawings do not usually depict the system to any scale. They typically represent the relationships between the features and not their relative sizes. However, sometimes schematic drawings are shown to scale to represent clearly the size and distance relationships between the features, as in the piping diagram shown in Figure 2.30(c).

Sketching a Block Diagram

Block diagrams are used to show the relationship between components, individuals, or groups. Because they can be sketched quickly, you may want to use block diagrams early in the design process to represent how parts in a system will relate. Boxes represent the component items, and lines indicate how they relate to one another. Figure 2.31 (see page 57) shows a block diagram for a corporate structure. Each block indicates a position within the company, and the lines indicate a supervisory relationship between the positions.

2.6 SKETCHING METHODS

As you are developing your sketching skills, keep in mind that sketches are used to communicate your design ideas to others. They should be neatly drawn and clearly show the information you are trying to convey. For concept sketches, it is important to be able to quickly sketch the object to show the relationship between its features. Techniques that use a consistent viewpoint and follow rules for transferring 3D shapes to 2D paper make it easy to create sketches that are easy to interpret. For example, in designing an

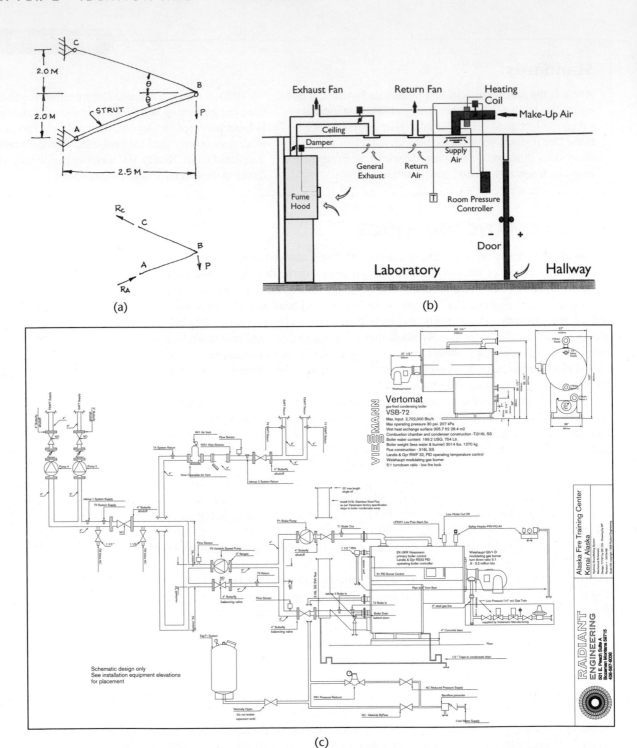

(a)

(b)

(c)

2.30 The freebody diagram in (a), laboratory heating and cooling system block diagram in (b), and piping control diagram in (c) are all examples of schematic drawings that use simple shapes in lieu of realistic images to convey information. *(Part (b) courtesy of Jerry Baker, CTA Architects; part (c) courtesy of Robert Knebel, Radiant Engineering.)*

ergonomic mouse, you may want to convey the appearance of the exterior of the mouse in a sketch such as that shown in Figure 2.32(a).

When designing a piping system, you might create a sketch that shows the relationship and orientation of the piping, valves, and tanks, in an isometric sketch. For a mechanical part like the V-block shown, an oblique sketch may be the best means of conveying the design information quickly. These are all common types of pictorial sketches used in engineering.

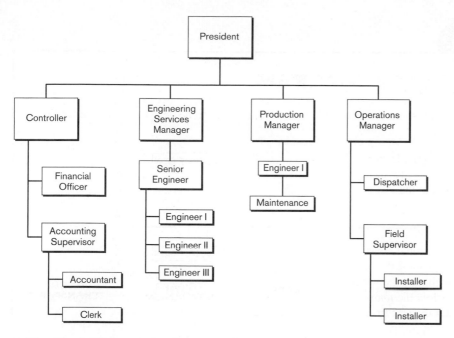

2.31 Block Diagram of a Corporate Structure

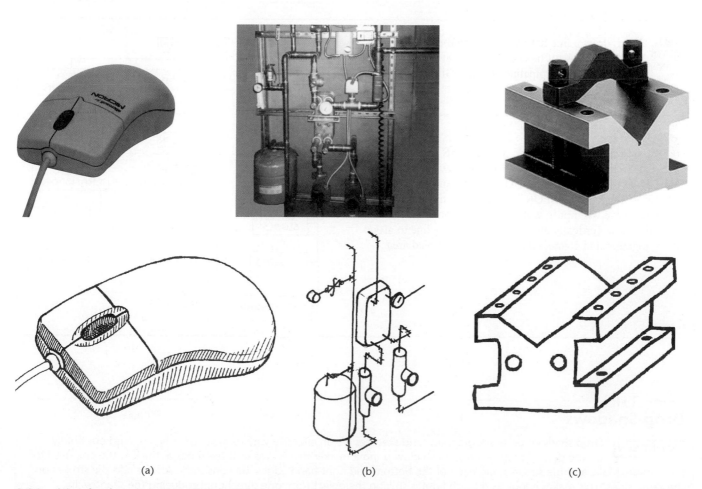

(a)   (b)   (c)

2.32 (a) A freehand sketch of the ergonomic mouse; (b) an isometric sketch of the piping system; (c) an oblique sketch of the V-block. *(Part (c, top) courtesy of Grizzly Industrial, Inc.)*

STEP by STEP

BLOCK DIAGRAM

Follow the steps to sketch a block diagram for a squirt gun. Each block represents a part to be designed. The lines indicate an assembly relationship between parts.

1 Start by identifying the top level of the gun. Letter the text SQUIRT GUN neatly at the top center of the drawing and draw a box around it so that the text is roughly centered in the box.

SQUIRT GUN

2 Next, divide the gun into its major functional areas. This squirt gun has three major subdivisions: the squirt gun body, the trigger mechanism, and the water transport system. Represent each of these below the main squirt gun block. *Tip:* It is easier if you letter the text first and draw the box second.

SQUIRT GUN

BODY   TRIGGER   WATER
            MECHANISM   TRANSPORT

3 Keep this second row of boxes aligned with each other to indicate that they are all at the same level.

SQUIRT GUN

BODY | TRIGGER MECHANISM | WATER TRANSPORT

4 Next, draw a line halfway between the top row and the second row of boxes. Use vertical lines to connect the boxes to this "pipeline" to show that they relate.

SQUIRT GUN

BODY | TRIGGER MECHANISM | WATER TRANSPORT

5 Add the subcategories below each major functional area next. Below BODY, add WATER SUPPLY, HOUSING, and FASTENERS. Below TRIGGER MECHANISM, add SPRING, TRIGGER, and PLUNGER. Below WATER TRANSPORT, add TUBE and SPRAYER. Because these will not fit well if you make another horizontal row, list these vertically and add a vertical pipeline to group them together and connect them to their functional area.

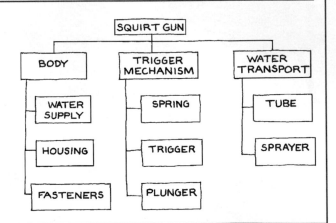

TIP

Drop Shadows

SQUIRT GUN   Drop shadows on block diagrams and sketches are an effective way to make the shapes stand out boldly on the page. Create a dropped shadow using a marker that is two to three times as thick as the one used for the outline. Make a single stroke along each of the bottom and right-hand edges. Be consistent. Always add the strokes on the same sides. This makes it look as though light is striking the object from one direction, producing the shadow. In general, use thick lines to show the important parts of the drawing clearly.

2.7 GEOMETRY OF SOLID OBJECTS

Sketching in engineering is largely used to communicate or record ideas about the shapes of 3D objects. Before you can create accurate engineering sketches, you need to understand just what it is that you are trying to show in your drawings. What is represented in drawing a 3D object?

Solid objects are bounded by the surfaces that contain them. These surfaces can be planar (flat surfaces), single-curved surfaces (a cylinder or a cone, for example), double-curved surfaces (such as a sphere, torus, or ellipsoid), or warped surfaces (like the flowing curved surfaces that you would see on a snowmobile hood, car, or other object). Figure 2.33 shows some common solid shapes and identifies their surfaces.

Many common solids are bounded by a combination of only plane surfaces and single-curved surfaces. For example, the bracket shown in Figure 2.34 is a common mechanical part. All its features are made up of either planar or single-curved surfaces.

The intersection of different surfaces on the object results in an *edge*. Figure 2.36 shows some typical polyhedra (solids that have only plane surfaces as their boundaries) with the edges highlighted. Notice that all the edges on a polyhedron are straight lines, because they represent the intersection of two planar surfaces, but edges can also be curved.

A *vertex* is the location where three or more plane surfaces intersect to form a point. A *face* is another name for a planar surface on an object. A face is bounded by edges where it intersects other surfaces. You can often see the different faces on an object clearly by the way they are lighted. When a face is perpendicular to the height, it is called a *base*. Figure 2.35 shows a shaded solid with its faces and vertices labeled.

When you sketch solid objects, sketch the edges of the object, where surfaces intersect. Each edge begins and ends with a vertex, where three or more plane surfaces come together. Locating the vertices where edges start and stop will help you sketch edges accurately in 3D space. Following are descriptions of the faces and edges of

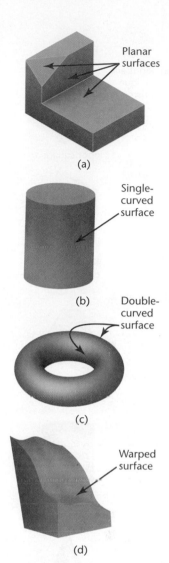

2.33 (a) The U-shaped block is made of planar surfaces only. (b) The cylinder has two planar surfaces (the top and bottom) and a single-curved surface. (c) The torus has just one surface: a double-curved surface. (d) The irregular shape contains a warped surface. The edge of this surface is not a regular geometric shape such as a line or arc.

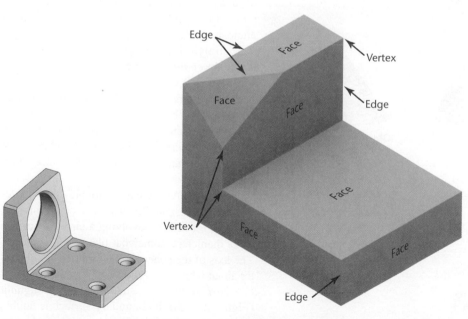

2.34 Bracket

2.35 Faces, Edges, and Vertices of a Solid

Tetrahedron (4 triangles)   Hexahedron (cube)   Octahedron (8 triangles)   Dodecahedron (12 pentagons)   Icosahedron (20 triangles)

2.36 Regular Polyhedra

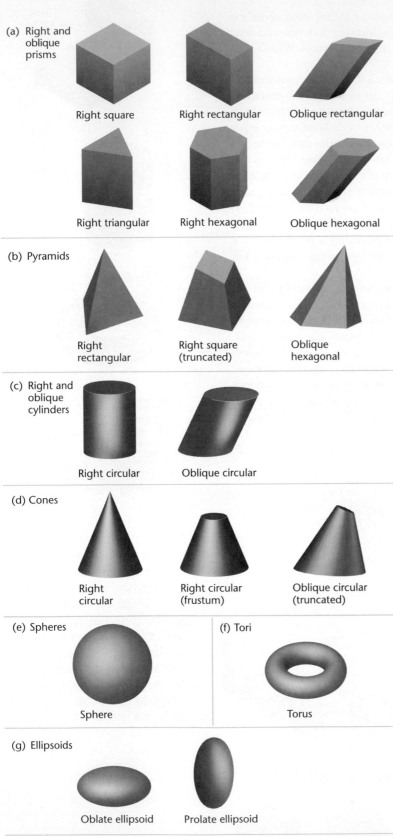

(a) Right and oblique prisms

Right square   Right rectangular   Oblique rectangular

Right triangular   Right hexagonal   Oblique hexagonal

(b) Pyramids

Right rectangular   Right square (truncated)   Oblique hexagonal

(c) Right and oblique cylinders

Right circular   Oblique circular

(d) Cones

Right circular   Right circular (frustum)   Oblique circular (truncated)

(e) Spheres   **(f)** Tori

Sphere   Torus

(g) Ellipsoids

Oblate ellipsoid   Prolate ellipsoid

2.37 Types of Solids

common solids (see Figure 2.37). In the rest of this chapter, you will learn to create sketches by drawing these features of solid objects.

Prisms A prism has two parallel equal polygons as its bases (faces perpendicular to its height). It has three or more additional faces, which are parallelograms (Figure 2.37(a)). A triangular prism has a triangular base, a rectangular prism has rectangular bases, and so on. A right prism has faces and side edges that are perpendicular to the bases; an oblique prism has faces and side edges that are angled to the bases. If one end is cut off to form an end that is not parallel to the bases, the prism is said to be truncated (meaning "shortened by having a part cut off").

Pyramids A pyramid has a polygon for a base and triangular side faces that intersect at a common point called its *vertex* (Figure 2.37(b)). The line from the center of the base to the vertex is called the *axis*. If the axis is perpendicular to the base, the pyramid is a right pyramid; otherwise, it is an oblique pyramid. A triangular pyramid has a triangular base; a square pyramid has a square base; and so on. If a portion near the vertex has been cut off, the pyramid is truncated, or referred to as a frustum.

Cylinders A cylinder has a single-curved exterior surface (Figure 2.37(c)). You can think of a cylinder as being formed by taking a straight line and moving it in a circular path to enclose a volume. Each position of this imaginary straight line in its path around the axis is called an *element* of the cylinder.

Cones A cone has a single-curved exterior surface (Figure 2.37(d)). You can think of it as being formed by moving one end of a straight line around a circle while keeping the other end fixed at the vertex of the cone. An element of the cone is any position of this imaginary straight line.

Spheres A sphere has a double-curved exterior surface (Figure 2.37(e)). You can think of it as being formed by revolving a circle about one of its diameters, somewhat like spinning a coin. The axis of the sphere is the term for the diameter about which it is revolved.

Tori A torus is shaped like a doughnut (Figure 2.37(f)). Its boundary surface is double curved. You can think of it as being formed by revolving a circle (or other curve) around an axis positioned away from the curve.

Ellipsoids An ellipsoid is shaped like an egg (Figure 2.37(g)). It is formed by revolving an ellipse about its minor or major axis.

2.8 PICTORIAL SKETCHING

A pictorial sketch represents a 3D object on a sheet of 2D paper by orienting the object so its width, height, and depth show in a single view.

Pictorial sketches are used frequently during the ideation phase of engineering design to quickly record ideas and communicate them to others. Their similarity to how the object is viewed in the real world makes them easy to understand.

Later in the design process, pictorial drawings are also often used to show how parts fit together in an assembly and in part catalogs and manuals to make it easy to identify the objects.

The following are three common methods used to sketch pictorials:

- isometric
- oblique
- perspective

Figure 2.38 shows perspective, isometric, and oblique views of a stapler. Each of the pictorial methods differs in the way the edges of the object are represented on the paper. A *perspective sketch* is most realistic looking. It shows the object much as it would appear in a photograph—portions of the object that are farther from the viewer appear smaller, and lines recede into the distance. The depth of the sketch in Figure 2.38(c) is exaggerated for emphasis.

Isometric sketches are drawn so that lines do not recede into the distance but remain parallel. This makes isometric views easy to sketch but may reduce the realistic appearance.

Oblique sketches shows the front surface of the object looking straight on. These are easy to create but look the least realistic because the depth of the object seems out of proportion.

Next, you will learn how to create pictorial freehand sketches using each of these methods.

2.9 ISOMETRIC SKETCHES

An isometric sketch shows the object rotated and tipped so that you are looking onto a corner where the major axis directions meet. Each axis is used to show one principal dimension of the object: width, height, and depth. Imagine a cube rotated 45°, so that you are looking onto one of its edges, and then tipped toward you as shown in Figure 2.39. In this isometric view, you are looking along the diagonal of the cube. The bottom edges of the cube are angled at 30° from horizontal.

You can use the edges of this cube to define a Cartesian coordinate system. Figure 2.40 shows an isometric pictorial view of a set of Cartesian coordinate axes. In the isometric view, the angles between the axes represent the 90° angles between the axes in the Cartesian system.

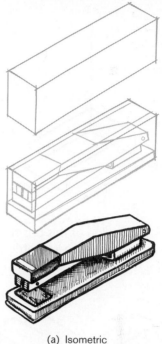

(a) Isometric

(b) Oblique

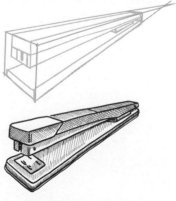

(c) Perspective

2.38 Isometric, Oblique, and Perspective Sketches of a Stapler

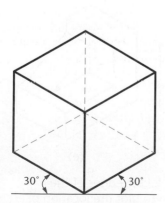

2.39 Isometric View of a Cube

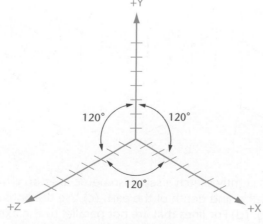

2.40 A Set of Isometric Axes

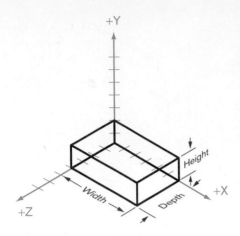

2.41 Each edge of this prism is parallel to one of the isometric axes.

The top, front, and side edges of the cube are tipped away from your view by the same amount. When a line or plane is tipped away from your view it appears *foreshortened* (shown smaller than its actual size) Because the surfaces are all tipped by the same amount, this produces equal foreshortening, the key to an isometric view. In fact, isometric means "equal measure."

Isometric sketches are generally drawn without depicting this foreshortening, so the actual measurements are used. The resulting sketch is larger than a true isometric view that would be produced using CAD or by projection onto a plane, but it is proportionally correct and easier to draw.

To sketch an isometric view, plot points representing corners of the object along the isometric axes. Typically, the X-axis is used to represent the horizontal location of an object relative to the origin or 0,0,0 point; the width of the part can be plotted along the X-axis. The Y-axis is used to represent vertical measurements. In your isometric sketch, use the Y-axis to represent the height of the object. The Z-axis is perpendicular to the X-Y plane and can be used to represent the depth of the part.

The choice of which dimension of the object to plot along which axis line produces different isometric orientations for the part. Overall you should choose to orient the part so that it shows the object clearly. It does not really matter which axis—X, Y, or Z—is labeled with which orientation as long as you are consistent within the sketch. If you know the dimensions of the object, you can plot the width, height, and depth on the isometric axes, as in Figure 2.41.

Orienting Isometric Views

The same object can be sketched with different orientations in isometric views (depending on which diagonal of the box you are looking at). Figure 2.42 shows an object sketched in all the possible isometric orientations. Which one shows the object best?

Select an orientation for your isometric sketch that shows the object clearly. If the object is complex, you may need to show more than one orientation or give a note explaining the depth of a hole. (In general you do not show hidden features in isometric sketches.)

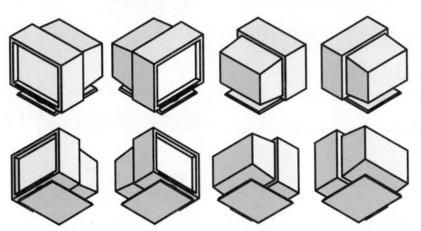

2.42 Eight different isometric orientations of a CRT monitor. Plan your sketch to use an orientation that shows the object clearly.

Sketching Isometrics without Grid Paper

To create an isometric sketch when you are not using grid paper, first sketch light lines at 30° from horizontal, then add a vertical line through their intersection to establish the direction and orientation for the part, as shown in Figure 2.43. Block in

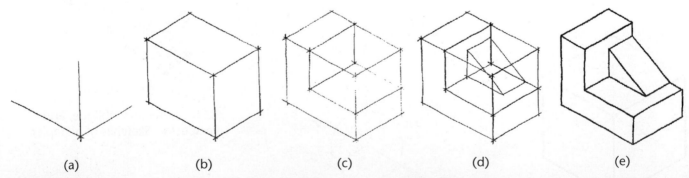

| (a) | (b) | (c) | (d) | (e) |

2.43 Creating an Isometric Sketch: (a) Lightly sketch a set of isometric axes on which to locate edges of the object. (b) Sketch the overall width, height, and depth of the part. (c) Use the lines drawn to plot the location of other surfaces and edges in the drawing. (d) For lines that are not parallel to the axes, locate endpoints in relation to the axes and connect the points to define the surface. (e) Darken the object lines to complete the sketch.

STEP by STEP

CREATING A SKETCH USING ISOMETRIC GRID PAPER

Isometric grid paper makes it easy to plot points and sketch isometric views. An isometric grid has lines angled at 30° from horizontal in two directions. These lines represent two of the axis directions (or the bottom edges of the cube). Vertical lines through points where the 30° lines intersect represent the vertical axis (some isometric grid papers do not show these).

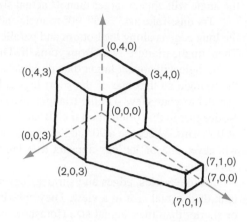

The vertices are labeled for you with their X-, Y-, and Z-coordinates. When you are sketching you will usually start from an idea or a part. When you are drawing from a part, try orienting it like the view you will show.

 Start your isometric sketch by selecting the orientation for your view. Decide which isometric axis will show the width, which will show the height, which will show the depth.

Then, block in the overall dimensions of the object to form a box that will fit the object you will draw. Keep in mind the proportions. If the height is twice the depth, make it appear so in your sketch.

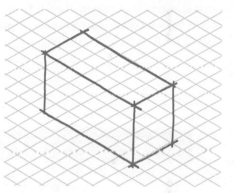

 Next, determine the measurements for features in your sketch. Locate the main corners of the object along the isometric grid lines.

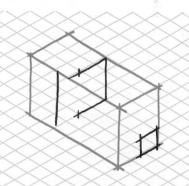

 Locate the remaining corners from previous locations. Connect the vertices to show the object's edges.

Darken in the final lines.

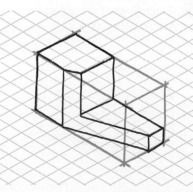

The True Size of an Angle

Angles do not always appear true size in a view. Sometimes the angle will appear larger than its actual size.

Try this: Take a 45°, 45°, 90° triangle and hold it so that the long edge is along the bottom and parallel to your view. Then, tip the triangle away from yourself. The 90° angle on the triangle will appear larger than its actual size. Now, hold the triangle so that one of the short legs is horizontal and parallel to your view. Tip the triangle away from yourself. Notice that in this orientation the 90° angle always appears as 90°. This is because a 90° angle always appears true size whenever one of its legs appears true length (parallel to your view).

Unlike angles, edges and surfaces never appear larger than their actual size in a view. They appear only true size or shorter than their actual size (foreshortened). To visualize this, take a book and tip it away from you. You will find that there is no direction you can tip it so that its edges appear larger than their actual size.

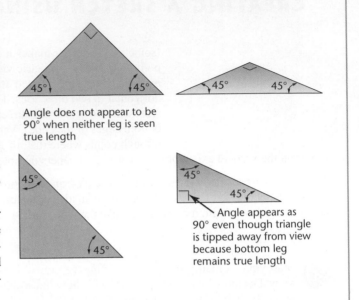

Angle does not appear to be 90° when neither leg is seen true length

Angle appears as 90° even though triangle is tipped away from view because bottom leg remains true length

---

SKETCHING NONISOMETRIC LINES

1 The surface on the front corner is oblique, bounded by three nonisometric lines.

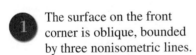

In a full-size isometric sketch, lines that are parallel to the isometric axis lines are sketched to represent their actual lengths on the object (or proportionately smaller if the object is too large to fit on the paper full size). Lines that are not in the same direction as the isometric axis lines will appear other than their actual lengths because they are not foreshortened equally. To draw these nonisometric lines, locate their endpoints by measuring along isometric lines as shown.

3 Use the grid paper or tick marks on the axes to locate the object's vertices along the axis lines.
Locate the vertices for the oblique surface by measuring along the isometric lines on the object.

2 First, block in the overall dimensions of the object on the isometric axes.

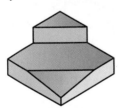

4 Connect the vertices to define the edges of the object.

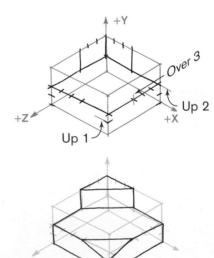

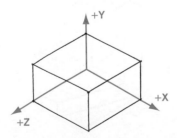

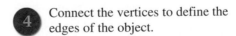

STEP by STEP

DRAWING OBLIQUE SURFACES IN ISOMETRIC

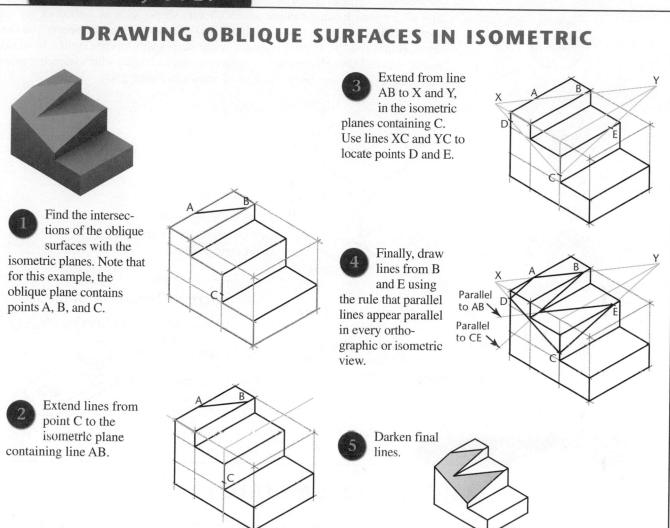

1 Find the intersections of the oblique surfaces with the isometric planes. Note that for this example, the oblique plane contains points A, B, and C.

2 Extend lines from point C to the isometric plane containing line AB.

3 Extend from line AB to X and Y, in the isometric planes containing C. Use lines XC and YC to locate points D and E.

4 Finally, draw lines from B and E using the rule that parallel lines appear parallel in every orthographic or isometric view.

5 Darken final lines.

the overall width, height, and depth of the part along the lines you have drawn and create a box that will fit the entire object. Estimate the sizes and relationships of each feature to keep the sketch in proportion. Keep parallel lines parallel as you box in the object.

Use the edges you have drawn to locate other surfaces and edges on the part. Identify portions of the overall box that would need to be removed from the outer block to form the shape of the object and lightly block in the edges. In an isometric drawing, lines that are parallel on the object will always be parallel to one another in your drawing. For slanting surfaces, locate corners that intersect a line parallel to an axis, or edges that are parallel to the existing lines. Then, sketch the slanting lines between the points you have located.

Isometric Ellipses

As a circular shape is tipped away from you (as in an isometric view) it appears elliptical. Imagine looking straight onto a circular shape, such as the top of the can shown in Figure 2.44. As the can is tipped away from your view it appears as a flattened, elliptical shape. For this reason, circular shapes will appear elliptical in an isometric sketch. These ellipses appear differently depending on which isometric plane they are drawn in—top, front, or side—as shown in Figure 2.45.

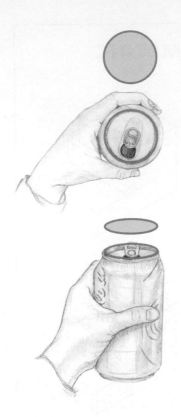

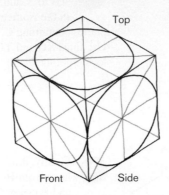

2.45 Circles as ellipses are different in the three isometric planes.

2.44 Starting from an end view of a can and tipping it away from you is a good illustration of how circular shapes appear as ellipses in a foreshortened view.

TIP

Screw Threads in Isometric

Use parallel partial ellipses equally spaced at the symbolic thread pitch to represent only the crests of a screw thread in isometric.

STEP by STEP

SKETCHING AN ISOMETRIC ELLIPSE

To draw an isometric ellipse representing a circle, first locate the center and sketch in the shape of the box that will contain the circular shape. Then, sketch the ellipse tangent to the walls of the box.

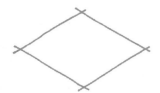

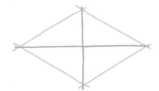

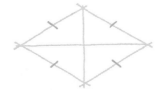

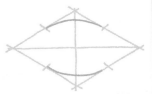

1 Locate the center of the ellipse (the circle's center point) in relation to other features in the sketch, then block in the overall dimensions of the ellipse with lines parallel to the isometric axes for the view.

2 Identify the center of the ellipse with construction lines as shown.

3 Locate the midpoints of the edges; these locate the major and minor axes of the ellipse.

4 Sketch arcs from each pair of midpoints as shown.

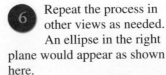

6 Repeat the process in other views as needed. An ellipse in the right plane would appear as shown here.

5 Darken the edges of the final ellipse.

STEP by STEP

DRAWING ISOMETRIC ELLIPSES

Random-Line Method

1 Draw parallel lines spaced at random across the circle.

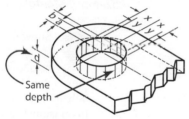

2 Transfer these lines to the isometric drawing. Where the hole exits the bottom of the block, locate points by measuring down a distance equal to the height d of the block from each of the upper points. Draw the ellipse, part of which will be hidden, through these points. Darken the final drawing lines.

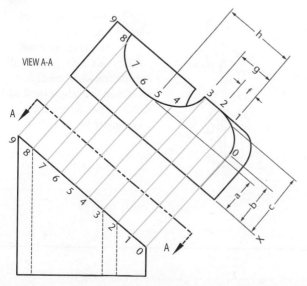

Eight-Point Method

1 Enclose the given circle in a square, and draw diagonals. Draw another square through the points of intersection of the diagonals and the circle as shown.

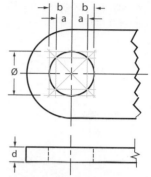

2 Draw this same construction in the isometric, transferring distances a and b. (If more points are desired, add random parallel lines, as above.) The centerlines in the isometric are the conjugate diameters of the ellipse. The 45° diagonals coincide with the major and minor axes of the ellipse. The minor axis is equal in length to the sides of the inscribed square.

When more accuracy is required, divide the circle into 12 equal parts, as shown.

12-point method

Nonisometric Lines

If a curve lies in a nonisometric plane, not all offset measurements can be applied directly. The elliptical face shown in the auxiliary view lies in an inclined nonisometric plane.

1 Draw lines in the orthographic view to locate points.

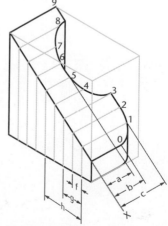

2 Enclose the cylinder in a construction box and draw the box in the isometric drawing. Draw the base using offset measurements and construct the inclined ellipse by locating points and drawing the final curve through them.

Measure distances parallel to an isometric axis (a, b, etc.) in the isometric drawing on each side of the centerline X–X. Project those not parallel to any isometric axis (e, f, etc.) to the front view and down to the base, then measure along the lower edge of the construction box, as shown.

VIEW A-A

STEP by STEP

CREATING A SCALE ISOMETRIC PIPING LAYOUT

Isometric sketches are often used to show the arrangement of pipes, pumps, and valves in process and flow systems, such as the hydronic snow melting system shown here.

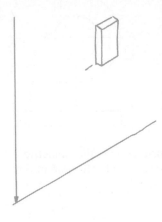

1 Begin by using grid paper and determining the approximate scale you will use. You will not be able to sketch this piping full size on your paper, so let each quarter-inch division on the paper equal 1 foot on the objects. This will give your isometric sketch a scale of 1/4 inch:1 foot. Start by laying out some of the overall dimensions. The height of the back wall (8 feet) and the height of the central heat exchanger (30 inches) are drawn parallel to the isometric Y-axis; the bottom of the wall serves as the Z-axis.

(Courtesy of Radiant Engineering, Bozeman, Montana.)

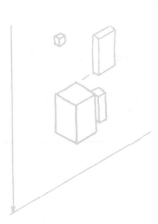

2 Locate the central plate heat exchanger on the wall and sketch a box representing its overall height, width, and depth.

Next, box in the overall dimensions of the pumps, tanks, and boilers in their proper locations and in proportion to the back wall and the heat exchanger.

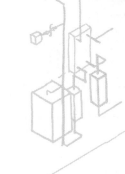

4 Show the valves using standard piping symbols. (You can look these up in Appendix 45.) When the symbols are shown in an isometric view, lines that are horizontal or vertical in the symbol are drawn parallel to the axis directions in the isometric sketch.

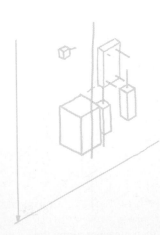

3 Next, draw the centerline of the pipe where it connects to the tank. How many inches does it go straight out before it turns? Represent this proportion of a foot along the isometric axis direction. Draw vertical lines to represent vertical pipe. Use the isometric grid as a reference so that the segments are drawn to the correct length.

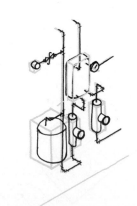

5 Add the shape of the components to the sketch, using the boxes with the overall dimensions as guidelines. Darken the edges (and add shading as desired) to complete the sketch.

2.10 OBLIQUE SKETCHES

Oblique sketches are a type of pictorial drawing. Unlike the isometric sketch, the oblique sketch starts with the front surface of the object, as if you were looking straight onto it (see Figure 2.46). The depth of the object is represented by parallel lines extending from the front surface. You have probably been creating oblique sketches since elementary school, as they are an easy way to add depth to a drawing. They are not very realistic, because you would never be able to look straight onto the front of an object and see its depth clearly at the same time.

Oblique sketches have the advantage that circular shapes in the front view appear circular (not elliptical as in isometric sketches), and the front surface of the object appears true size. For this reason, oblique pictorials can be useful when it is important to provide a true view of an irregular or curved surface that would be difficult to sketch quickly in an isometric view.

Objects shown in oblique sketches are usually oriented so that their most characteristic shape is shown as the front in the sketch. The depth of the object is shown receding from the front view. The angle of the receding lines, showing the object's depth, is arbitrary, but it is often 30° from horizontal because it gives a fairly realistic appearance and is easy to sketch. When the depth is shown full scale (actual length) the drawing is called a *cavalier projection*. When the depth is shown half size it is called a *cabinet projection*. Showing the depth full size adds to the distorted appearance of the object, as you can see in Figure 2.48, so oblique pictorials usually show depth less than full size.

If the back surface of the object is the same size and shape as the front surface, a quick way to create an oblique sketch is to draw the front surface, then sketch it again (to represent the back surface) offset up and to the right. Sketch the receding lines by connecting the corners on the front and back surfaces as shown in Figure 2.47. Darken the visible lines to complete the sketch.

To create an oblique sketch of a more complex object, select the most irregular or characteristic surface of the object as the front. Avoid selecting a surface that will cause the longest dimension of the object to be shown as receding depth lines, as this will make the sketch seem more distorted. Sketch this surface true shape (using grid paper, if desired, or estimating units to keep the drawing in proportion), as shown in Figure 2.48. Next, sketch in the edges that recede from this front surface. Use the same angle for all receding lines that are perpendicular to the front surface so that parallel receding edges appear as parallel lines.

Estimate the depth of the object or the length along the receding lines to the next surface. Draw the next surface that is parallel to the first surface true size, then draw the receding lines from its corners as you did for the first surface.

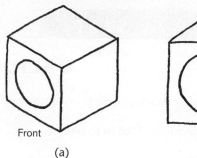
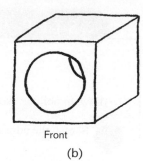

Front   Front

(a)   (b)

2.46 The two cube sketches illustrate one key difference between an isometric and oblique sketch. In an isometric sketch (a), circles appear as ellipses in all views. In an oblique sketch (b), features in the front view are drawn full size, so circles appear round.

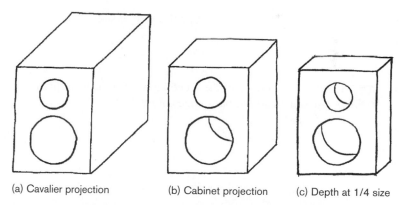

(a) Cavalier projection   (b) Cabinet projection   (c) Depth at 1/4 size

2.47 These three oblique sketches illustrate the effect of changing the scale at which the depth of the object is shown: (a) shows depth at full size (cavalier projection); (b) shows depth at half size (cabinet projection); (c) shows depth at quarter size.

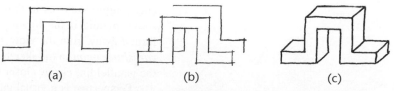

(a)   (b)   (c)

2.48 Three-Step Oblique. When the back surface of the object is the same shape and size as the front: (a) Sketch the front surface true size. (b) Sketch the edges of the front surface again, offsetting the lines by an equal amount up and to the right. (c) Connect the corners to draw the receding edges.

STEP by **STEP**

OBLIQUE SKETCHING

To draw an oblique sketch of a more complex object:

1 Sketch the edges of the front surface.

2 Sketch the edges that recede from the corners of the front surface, keeping parallel edges parallel.

3 Sketch the next surface that is parallel to the front surface at the approximate depth.

4 Draw the receding lines from the second surface, repeating steps 3 and 4 as needed.

5 Connect the corners of the receding lines at the appropriate depth to indicate the back edges of the part.

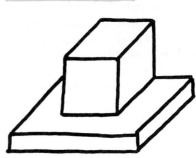

2.11 PERSPECTIVE SKETCHES

Perspective sketches are a third kind of pictorial sketch. They differ from isometric and oblique sketches in that receding lines that are parallel on the object are not shown parallel in the drawing. To create the illusion that part of the object or scene is farther away from the viewer, receding parallel lines on the object converge at the ***vanishing point*** in the sketch. Look at the photograph in Figure 2.49 and notice how the parallel lines grow closer together as they recede from the viewer.

Perspective is a visual cue used to make drawings appear realistic. Leonardo da Vinci was one of the first artists to use perspective in his drawings systematically to convey the appearance of depth. Before the use of perspective, objects in a scene were often sized according to their relative importance to the painter, not according to their distance from the observer. A perspective sketch uses the vanishing point to simulate the way the eye perceives distance, which makes the objects in the sketch more realistic.

The appearance of a perspective sketch depends on your viewpoint in relation to the object. Select some reachable object in the room and move so that you are looking at it from above and really notice its shape. Then, gradually move so that you are looking at it from below. Notice how the change of viewpoint changes the appearance of its surfaces—which ones are visible and their relative size.

The *horizon line* in a perspective sketch is a horizontal line that represents the eye level of the observer. Locating the sketched object below the horizon line produces a view from above (or a bird's-eye view). Locating the sketched object above the horizon line produces a view from below (or a worm's-eye view). Figure 2.50 illustrates the horizon line in a drawing and the effect of placing the object above or below the horizon line. The vanishing point in a perspective sketch is a point on the horizon line where receding lines from the object converge, as illustrated in Figure 2.51.

2.49 The railroad tracks in this photo appear to converge in the distance, even though they remain parallel. When you look at the tracks, you do not "see" them converging because this is a cue that the mind interprets as an effect of increasing distance. Perspective drawings use this technique to trick the viewer into seeing depth in a flat image.

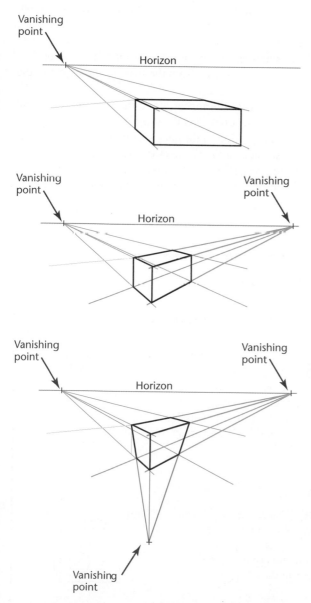

2.51 One-, Two-, and Three-Point Perspective

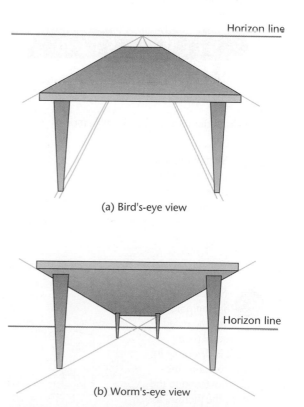

(a) Bird's-eye view

(b) Worm's-eye view

2.50 Placement of the horizon line determines bird's-eye or worm's eye view.

To create the illusion of depth in a sketch, use guidelines, called *projectors,* that run from the corners of the object to the vanishing point. Figure 2.51 illustrates how these guidelines match up to the way the eye perceives distance. In a perspective sketch, the projectors help you keep the back surfaces of the object in proportion. The distance along the projector represents the distance that a feature is from the viewer. Each projector is scaled equally as the object recedes, so the ratios of width to height in the object remain the same.

Perspective sketches can have multiple vanishing points, as shown in Figure 2.51. Two-point perspective sketches have two vanishing points; there are three vanishing points in a three-point perspective. In each case projectors from the object to the vanishing points serve the same purpose as guidelines for drawing receding lines to create the illusion of distance.

Creating a Perspective Sketch

To create a one-point perspective sketch, orient the object you are sketching so that you are looking directly onto the front surface of the object. Sketch this front surface true size on your sheet of paper.

Add the horizon line and vanishing point. This is the location where the projectors converge in the drawing. Draw the projectors from the front of the object to the vanishing point.

Estimate the depth of the object as it will appear along the receding lines and lightly sketch the back surface of the object between the projectors. It should appear smaller than the front surface but in proportion. If the back surface of the object is not the same size or shape as the front surface, use the projectors to calculate its location. That is, if the distance between the projectors is 8 units, and the back surface is 10 units wide, you can use the size of the unit at a given depth to locate the back

---

STEP by STEP

ONE-POINT PERSPECTIVE

To sketch the part in one-point perspective—that is, with one vanishing point—follow these steps.

1 Sketch the true front face of the object, just as in oblique sketching.

2 Select the vanishing point for the receding lines. In many cases it is desirable to place the vanishing point above and to the right of the picture, as shown, although it can be placed anywhere in the sketch. However, if the vanishing point is placed too close to the center, the lines will converge too sharply and the picture will be distorted. Sketch the receding lines toward the vanishing point.

Vanishing Point

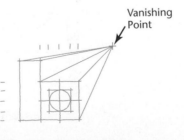

3 Estimate the depth to look good and sketch in the back portion of the object. Note that the back circle and arc will be slightly smaller than the front circle and arc.

Estimate Depth

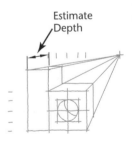

4 Darken all final lines. Note the similarity between the perspective sketch and the oblique sketch earlier in the chapter.

STEP by STEP

TWO-POINT PERSPECTIVE

To sketch an object using two vanishing points, follow these steps.

1 Sketch the front corner of the object at its true height. Locate two vanishing points (VPL and VPR) on a horizon line (at eye level). Distance CA may vary—the greater it is, the higher the eye level will be and the more you will be looking down on top of the object. A rule of thumb is to make C–VPL one third to one fourth of C–VPR.

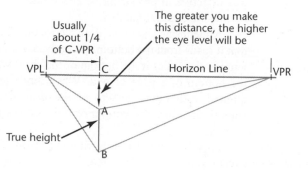

3 Block in all details. Note that all parallel lines converge toward the same vanishing point.

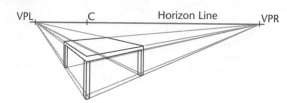

4 Darken all final lines. Make the outlines thicker and the inside lines thinner, especially where they are close together.

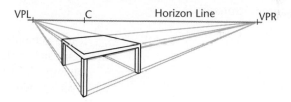

2 Estimate depth and width, and sketch the enclosing box.

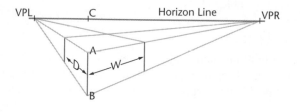

edges. You can also add projectors from the front surface to the vanishing point to locate features along the receding surfaces. Connect the corners of the front and back surfaces to create the edges of the object that recede into the distance. To complete the sketch, darken the visible lines in the final sketch.

You can use these same steps to sketch any object in perspective. To produce a good perspective sketch, you first must consider the orientation for the object so that it will be shown to advantage in the perspective sketch. Imagine a cigar-shaped space vehicle. Orient the long direction of the part in the direction of the vanishing point to call attention to its size.

Although perspective projection produces the most realistic pictorial view, it can be more time consuming to sketch. In perspective, the size of the object shown depends on the distance from the viewer, and the foreshortening in the picture makes it difficult to assess whether objects are the same size or simply farther away.

Perspective views are frequently used in architectural applications, where the large scale of the objects makes the effect of distance so obvious that it must be reflected in the view. For smaller objects and mechanical parts, it is generally as effective to use an isometric or oblique pictorial, which are easier to sketch (although many CAD packages will generate a perspective view automatically).

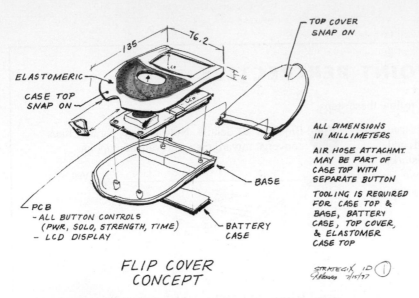

FLIP COVER
CONCEPT

STRATEGIX ID ①

2.52   A specific note requires a leader line to indicate the part of the sketch to which the note applies. Notice the use of dots and arrows with the leader lines in this drawing. *(Courtesy of Robert Mesaros, Strategix ID.)*

2.12 LETTERING

Information needs to be neatly lettered on sketches and drawings. Engineering drawings are usually lettered using 1/8-inch uppercase Gothic-style letters. Gothic style letters have no serifs, which is the name for the ending accent strokes or bars on letters. These letters are easy to make and read. Figure 2.53 shows the letter and number shapes in the Gothic style.

The arrows on the letter and number shapes indicate pencil strokes that will result in well-formed shapes. The strokes used in cursive letter forms are designed for a flowing script and are less effective in producing balanced and evenly rounded letters and numbers.

All lettering should be drawn horizontally so that it reads from the bottom of the sheet with the letters oriented as if the bottom of the sheet were the bottom of a page of text. When lettering is aligned with other objects in the drawing, it should read from the bottom or right side of the drawing.

It is frustrating to read large drawings that have lettering running in different directions, and costly mistakes may result if numbers are misread.

Lettering sizes in a drawing are based on the sheet size of the final drawing. When you are adding text to a CAD drawing, you need to think about the final plotted drawing so that you size your text correctly. Titles and important headings are usually larger than the standard size for notes, dimensions, and other text. Standards for lettering sizes are reproduced in Table 2.3.

General and Specific Notes

General notes are notes that apply to the entire object, design, or process. They are usually placed at the right side of the sheet, preferably near the lower right. Group all the general notes in one place. Specific notes apply just to a particular part or feature.

Table 2.3 Minimum Letter Size Standards.

Use	Minimum Letter Heights	Sheet Sizes
Title block information such as drawing title, drawing size, CAGE code, drawing number, revision number	6 mm (.25")	A0, A1 (larger than 17" × 22")
	3 mm (.125")	A2, A3, A4 (8.5" × 11" up to and including 17" × 22")
Section and view letters	6 mm (.25")	All
Zone letters and numbers in borders	6 mm (.25")	All
Drawing block headings	2.5 mm (.10")	All
All other text	3 mm (.125")	All

To show which part or feature a note refers to, a leader is drawn from the location to the note. The leader should be drawn with a horizontal tail about 1/8 inch long, and an angled line ending at the feature. Usually, an arrowhead is used to point to the edge of the part, and a dot is used if the feature is inside the object's outline. Figure 2.52 shows an example of specific notes and leader use.

Straight-line letters

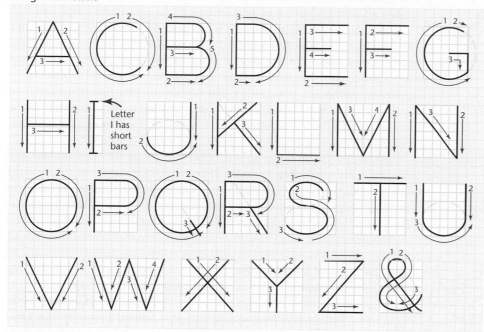

6 units

Letters in
TOM Q. VAXY
are 6 units wide.
All others are 5 units
except I (1 unit)
and W (8 units)

Circular forms:
In addition to the
letter O, the letters
Q, C, G and D are
based on
a true circle

Elliptical forms:
The lower portion
of the letter J and
the letter U is elliptical

The S is based on two
ellipses, like an 8

Numerals

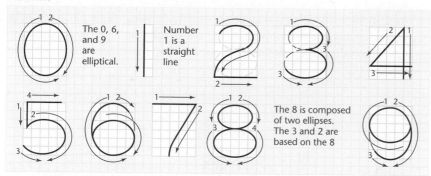

Vertical lowercase letters

2.53   Forming Letters and Numbers in the Gothic Style

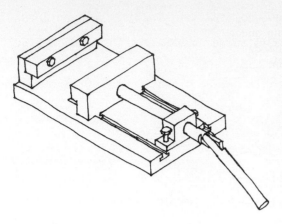

2.54 Isometric Assembly Sketch

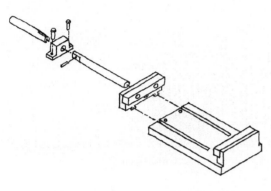

2.55 This isometric assembly sketch has been exploded and centerlines have been added to show where parts line up in the assembly.

2.13 SKETCHING ASSEMBLIES

Assembly drawings are used to show how parts assemble together. Because they do not need to provide all the information to make individual parts, isometric sketches or drawings are often used that show only the exterior view of the assembled parts. You can use assembly sketches as you design to identify areas that must fit together, have common dimensions, or maintain critical distances to ensure that the device will function as expected. Figure 2.54 shows an isometric sketch of an assembly.

Sketched assembly drawings are useful in documenting your ideas in a rougher state. An exploded isometric assembly drawing shows the individual parts moved apart from one another along the isometric axis line directions. A centerline pattern is used to show how the parts relate and fit together. A centerline pattern is a thin dark line that starts with a long line, then has a short dash of about 3 mm (1/8 inch), and ends with a long line. The length of the long portion of the line is adjusted to fit what you are drawing, but it is typically in the range of 40 to 80 mm (1-1/2 to 3 inches)

When creating an exploded assembly sketch, keep the parts lined up with the other parts where they assemble. If possible, move any single parts along only one axis direction. Figure 2.55 shows an exploded isometric assembly in which all the parts are moved along only one axis direction. Notice the centerline pattern showing how the parts align. If you must move a part in two different directions so that it can be seen clearly, be sure to add a centerline showing how the part must move to get to its proper location in the assembly. Figure 2.56 shows an exploded assembly with parts moved in two axis directions. You can see the jog in the centerline indicating how the part has been moved.

Finished assembly drawings that will be given to manufacturing workers to show them how to produce a device also contain a parts list that identifies each part in the assembly, the quantity necessary, the material of each part, and a number identifying which item it is on the drawing. A parts list can also be included on a sketched assembly drawing. Sometimes it is helpful to give a quick assembly sketch to manufacturing along with part drawings, especially if they are parts you need to have manufactured quickly from

2.56 In this isometric assembly, two parts have been exploded in two directions so they do not overlap other parts in the assembly. The centerlines indicate the path the part must follow to assemble correctly. *(Courtesy of Albert W. Brown, Jr.)*

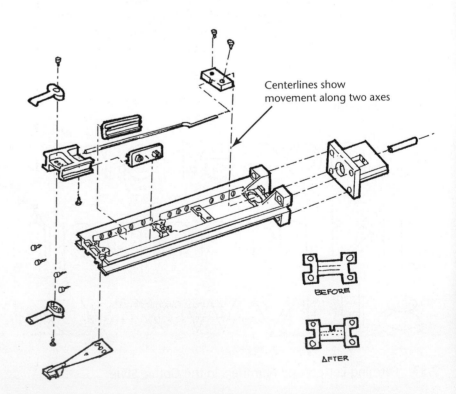

Centerlines show movement along two axes

BEFORE

AFTER

a sketch. Seeing how the parts fit together in assembly will help the manufacturer understand the entire device and identify potential problems.

2.14 SKETCHING IN THE DESIGN PROCESS

The techniques you have learned in this chapter will add to your repertoire of skills used to visualize and communicate your designs. Effective sketching and visualization require practice. To develop these skills, keep a notebook and make drawings showing your ideas and reflecting objects you see around you.

Sketching techniques are generally used early in the design process to generate ideas and show them to other members of the design team. The portability of pencil and paper, however, makes the ability to sketch an important means of capturing your ideas, working out relationships between parts, and recording your work.

As more details of the design are finalized you will use 2D CAD drawings or 3D models to further define your designs, but many of the techniques you learned for sketching 3D shapes will also help you define them in the CAD environment. Typically if you cannot sketch the design, you do not have the details sufficiently worked out to begin modeling in CAD. In the next chapter you will learn about techniques for creating orthographic views to show the true sizes of features accurately in your drawings.

SKETCHING FOR IDEATION: A CROSSACTION BREAKTHROUGH

When Oral-B hired LUNAR to create an innovative approach to brushing (and a new flagship product), they had no preconceived notions of what that would mean. But they knew that they wanted a brush that was ergonomically superior to the traditional "flat stick" design, and they wanted the brush to improve brushing effectiveness. The development of the Oral-B CrossAction Toothbrush is an example of the power of ideation in product design.

Jeff Salazar, lead designer for the project, started first not with sketches but with extensive research into the kinesiology of brushing teeth. A research firm had been hired to do a study of how people hold their toothbrushes as they brush. They found that there are five basic grips. Each person might use more than one depending on the part of the mouth to be brushed. Two or three grips were the ones used by most people.

As the kinesiology study was underway the design team at LUNAR undertook their own study of brushing. Given that the design constraints were focused almost totally on ergonomics, they studied how people brush. They engaged team members and others and watched them brush. They used models of teeth to investigate brushing. They bought "flat stick" toothbrushes, heated them so they could be bent at different angles, and had a variety of people report back on their brushing experience with handles of different shapes. People of various sizes and backgrounds, male and female

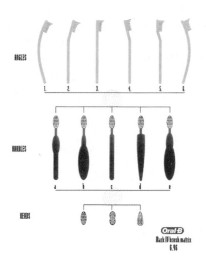

2.57 The Matrix Used to Drive Concept Generation

were asked to brush with the angled handles. This helped them understand the extremes of the angles forward and back that the handle could form.

They also studied the competition. A competitor had come out with what it promised was a superior toothbrush—one that was "bent like a dental instrument." After trying the toothbrush, however, the team concluded that the shape was only superior if someone else (e.g., the dental hygienist) was brushing your teeth. This key insight helped them focus the next stage of ideation on shapes with potential.

After discussing their results, the team selected five to six angle configurations for further study. The ergonomic results of the team's efforts were an instinct that people wanted a toothbrush that would "fill the hand" more than the thin stick. But where should the brush be thicker? The team created a matrix that paired each of the angle configurations with the range of options for the handles' thickness (Figure 2.57). Then, Jeff began sketching all the various combinations (see Figure 2.58).

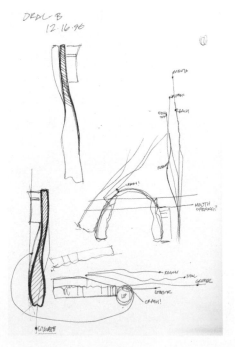
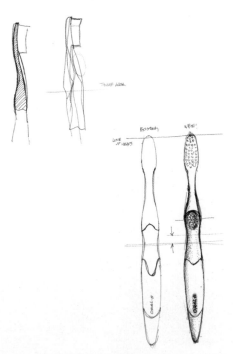
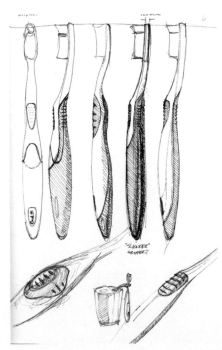

2.58 Sketches and Notes on the Concept as It Was Developed

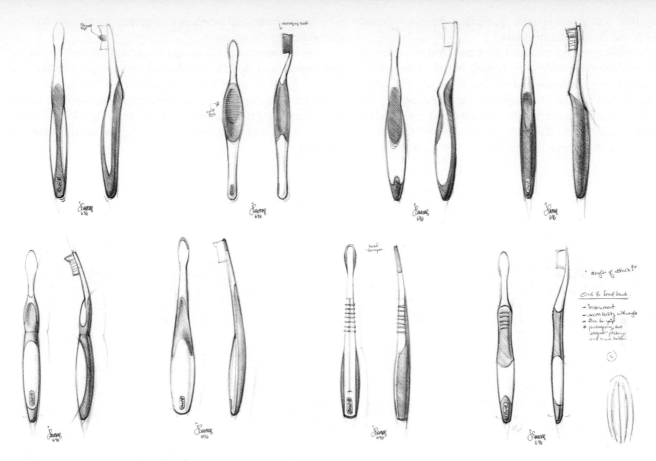

2.59 Full-Size Sketches Used to Create the Foam "3D Sketches"

After about a week of generating pencil sketches, Jeff decided he needed to "sketch" in 3D. Starting with a full-size sketch (Figure 2.59), he used a band saw to cut the shape of the sketch in foam, then shaped the foam into a model that he and others could hold and evaluate for various grips (Figure 2.60). Those that didn't work for one grip or another were eliminated. The Lunar team and their counterparts at Oral-B reviewed the various options and selected four for further development.

At this point, the team created 3D models of the four candidates. Oral-B had prototypes made from the CAD models for the handles (Figure 2.61), added bristles, and embarked on consumer research that would drive the next stage of refinement. Consumers were asked to brush for one month, then attend a focus group to report on their experience. Lunar Design was present to hear consumer reactions.

What they found was that consumers loved the fatter handle. Its shape in the hand gave them more control when moving the brush to different parts of their mouth. The form provided optimal ergonomics for each of the five common grips. The faceted surface on the handle gave users a tactile clue as to where to place their fingers when gripping. The rubberlike material used for the facets was located where users exerted the most finger pressure—and it made it easier to hold the brush when wet. The very positive response indicated that the ergonomic handle could make a significant difference in

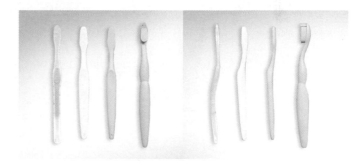

2.60 Foam Models Used as 3D Sketches

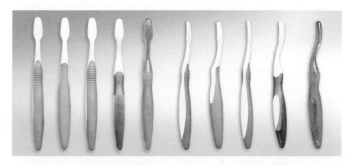

2.61 Prototypes Generated from the 3D Model

(continued)

brushing that consumers would appreciate. People reported that they loved their toothbrush now!

At the same time, Jeff and the Lunar and Oral-B teams had to listen and interpret what they heard. Focus groups are a wealth of different reactions, and it is up to the team to determine what is relevant and what is peripheral. For example, one reaction was that the fatter handle did not fit in the standard toothbrush holder. The team had to assess whether this was an issue that would hamper acceptance of the toothbrush or whether the improved brushing experience would outweigh its importance to the customer.

The refinement of the design proceeded through two more prototypes (and additional ideation around packaging; see Figure 2.62) before the final Oral-B CrossAction was launched. The intensity of the ideation stage—and the many techniques used to generate and evaluate design ideas—were critical to the team's success in achieving this breakthrough in brushing.

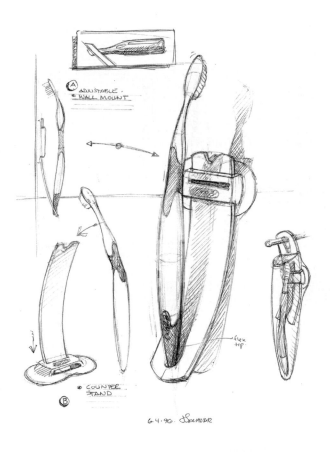

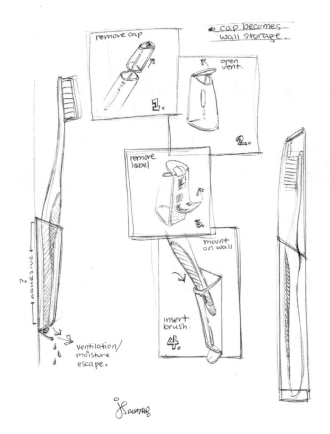

2.62   Ideation Sketches for CrossAction Packaging

(All images courtesy of LUNAR)

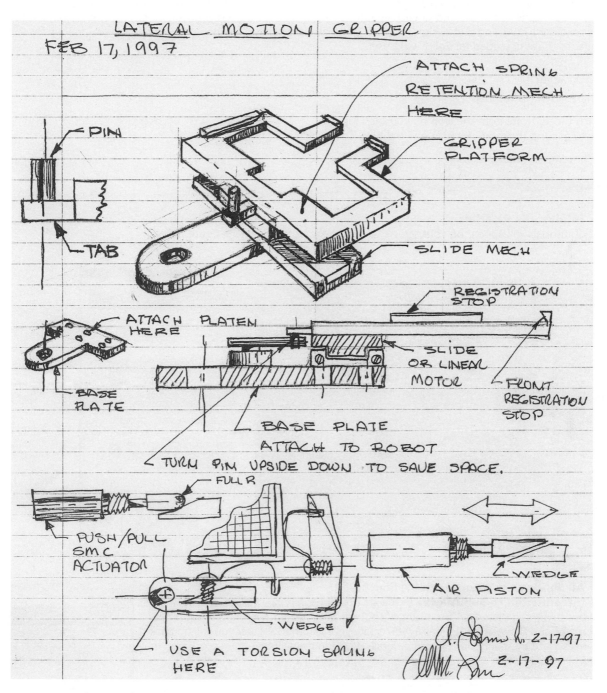

LATERAL MOTION GRIPPER
FEB 17, 1997

PIN

ATTACH SPRING RETENTION MECH HERE

GRIPPER PLATFORM

TAB

SLIDE MECH

REGISTRATION STOP

ATTACH HERE

PLATEN

SLIDE OR LINEAR MOTOR

FRONT REGISTRATION STOP

BASE PLATE

BASE PLATE
ATTACH TO ROBOT
TURN PIN UPSIDE DOWN TO SAVE SPACE.

FULL R

PUSH/PULL SMC ACTUATOR

WEDGE

AIR PISTON

USE A TORSION SPRING HERE

WEDGE

a. Brown 2-17-97
2-17-97

2.63 In designing the lateral motion gripper, Albert Brown used a pictorial (or 3D) sketch to illustrate how the parts of the gripper would appear when assembled. This page from his notebook includes notes about the assembly and how the different parts of the mechanism will interact. *(Courtesy of Albert W. Brown, Jr., Affymax Research Institute.)*

HANDS ON TUTORIALS

SolidWorks Tutorials

The *SolidWorks Student Design Kit* may be bundled with this textbook. (For the correct package ISBN, please contact your Pearson Rep or visit www.pearsonhighered.com.) An excellent set of tutorials is included with the software. You may want to do the tutorials in the order suggested at the end of each chapter. If you have not installed your software, you should install it now.

Other excellent modeling and design software is also available. If you will be using different software, look for software features similar to those described for SolidWorks users.

Getting Started

To find the SolidWorks Tutorials:

- Launch the software from the icon on your desktop.
- Click to expand the topics listed below the help icon, which is a question mark.

- Click **SolidWorks Tutorials.**
- A list of tutorials will appear on your screen, similar to the right side of the figure below.

- Click **All Solidworks Tutorials** (Set 1) from the list.
- Click **Introduction to SolidWorks i**n the list that appears.
- On your own, complete the first tutorial. This will give you an overview of the **Part, Assembly,** and **Drawing** functions as you create your first model. Later you will go back and investigate these functions in more detail. Have fun modeling!

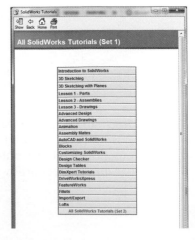

(All images courtesy of SolidWorks Corp.)

REVERSE ENGINEERING PROJECT

The Can Opener Project

In this ongoing project, you will reverse engineer an Amco Swing-A-Way 407WH Portable Can Opener.

This effective and low-cost can opener seems simple in its familiarity, but it is clear when you begin to take one apart that a considerable effort went into designing a product that is inexpensive, reliable, and easy to operate for most people. This can opener is readily available and affordable, so it is recommended that you purchase it, or a similar product, so you can make measurements directly from the product when required in future chapters.

Product Features

- Portable
- Lightweight
- Manually operated
- Comfortable to hold
- Durable construction
- Has a bottle opener
- Colored handles available
- Low maintenance
- 5-year warranty

1. *How many ways?* This is far from the only can opener on the market. Use the Web to research manual can opener designs. Find at least three can opener models that are different from the Amco Swing-A-Way. Make a list of the features of each of the three.
2. *More than meets the eye.* Create a functional block diagram for the can opener. How many distinct parts are used in its manufacture? Which parts can be grouped together into a subassembly?
3. *Building a better can opener.* Sketch a concept for an innovative can opener design.

(Courtesy of Focus Products Group, LLC.)

KEY WORDS

Aesthetic Constraints

Base

Block Diagrams

Brainstorming

Cabinet Projection

Cavalier Projection

Construction Lines

Contour

Decision Matrix

Deferred Judgment

Design Constraints

Design Proposal

Development Plan

Economic Constraints

Edge

Engineering Release

Ergonomic Constraints

Face

Foreshortened

Freebody Diagrams

Functional Design Constraints

Horizon Line

Individual Brainstorming

Isometric Sketches

Legal Constraints

Listing or Morphological Analysis

Oblique Sketches

Part Review

Perspective Sketches

Product Proposal

Product Release

Product Review

Projectors

Qualifying the Production

Schematic

Specification

Standards

Time Constraints

Vanishing Point

Vertex

SKILLS SUMMARY

Being skilled in different methods of communicating about your designs will prepare you to select the most appropriate method for a given task. In addition to being able to represent your ideas so that you can communicate about them with others, you also must be able to interpret information available in drawings and sketches early in the design process. You should now be familiar with one way of conceptualizing the design process, the use of a needs statement, and constraints on designs.

You also should be able to create three types of pictorial drawings used in engineering and the lettering shapes that are used to add notes to engineering drawings. From this point on you should always letter your drawing notes neatly using engineering letter shapes. Practice the letter shapes by adding titles to your sketches in the following exercises. Gain additional practice by using engineering letter shapes when writing out your homework problems for other classes. Practice them while talking on the phone.

As you practice your skill in isometric and oblique sketching create some of your isometric and oblique sketches using grid paper. Do many others on plain white paper to gain experience in quickly capturing the shape of an object or idea. Practice using engineering markers as well as pencil. If you have access to a whiteboard or chalkboard, try this sketching surface also.

EXERCISES

1. List four factors that constrain design problems? Give examples of each.

2. Respond to this statement from an engineer: "My main concern in designing products is their function. If my boss is worried about money she should talk to the accounting department."

3. Many upper-division engineering design courses emphasize student participation in "open-ended" design problems, meaning there may be more than one correct answer to the problem. How many possible solutions are generated depends on the technique used. Describe idea-generation methods from this chapter that would lead to many alternative solutions.

4. You are a member of a team designing an automatic door-opening device for people using wheelchairs to enter and exit buildings on campus. Many design constraints exist, including those associated with the abilities and limitations of the users, the physical environment, security and safety concerns, and so forth. What specific design constraints can you list? Explain five of these constraints, including their origin and their possible impact on your design.

5. Create a survey questionnaire to determine the following:
   a. What computer options are desirable to engineering students?
   b. What options should be included on a 4 × 4 diesel pickup truck?
   c. What features are desired in a new snowboard binding design?
   d. What features are desired in a handheld cordless mixer?

6. Create a decision matrix comparing each of the following:
   a. Three brands of mechanical pencils
   b. Three brands of cereal
   c. Three brands of computers
   d. Three brands of calculators

7. Use Web search tools to find the names and contact information for three companies that manufacture ball-type universal joints. Record the information you find in your notebook. Write a letter to one of them requesting specifications and pricing.

8. Create a simple, proportionally accurate, single-view sketch of a bicycle, from memory. Then, look carefully at a bike, noticing frame geometry, crank length, wheel size. Repeat the sketch, and compare both with the actual bike. How did you do?

9. Imagine that your roommate has asked you to design a car to protect the teenager in the household who is learning to drive. You have probably seen the Nerf foam balls that you can throw in the house without breaking things. The car you design will be entirely made of or coated with a similar material. It must protect the teenager and others from accidents. What will it look like? Will your car start from scratch or be remodeled from the 1976 Dodge Dart that is already sitting in the driveway? What color will it be? Other accessories?

10. Sketch a block diagram for the design of a children's wagon. Divide the design to the furthest part level for your understanding of the components.

11. Sketch a block diagram showing the relationships among your family members for the last three generations.

12. Sketch a set of single-point perspective projectors and try creating a drawing of the houses on your street from the viewpoint of someone standing in the middle of the road and looking down the street.

13. Work with a partner on this exercise. Choose one of the objects in Exercises 20–37. One of you will describe the part while the other sketches it. Try it two ways: one in which the describer cannot see the sketch, the other in which he or she can see it and give feedback while it is being drawn. Then, trade roles. Try it again with a third partner: one of you describes, one sketches, one gives encouragement and feedback.

14. Sketch a freehand perspective view of the oarlock, the thread tap, and the phone receiver shown. Scales are not provided; try to draw all features of each object to the same relative scale.

a.

(Courtesy of Clavey Paddlesports.)

b.

c.

(Shutterstock.)

15. Sketch the objects shown using isometric, oblique, and one- or two-point perspective.

a.

b.

c.

d.

e.

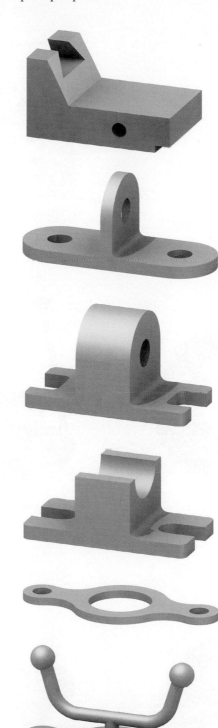

f.

g.

h.

i.

j.

k.

16. Create isometric and oblique sketches of the objects shown. Use plain or grid paper. Draw the objects in proportion.

a. Lightbulb *(Shutterstock.)*

b. Electrical Outlet *(Shutterstock.)*

c. Telephone *(Shutterstock.)*

d. Gear *(Shutterstock.)*

e. Wagon *(Shutterstock.)*

f. Watering Can *(Shutterstock.)*

g. Cup and Saucer *(Shutterstock.)*

h. Bullhorn *(Shutterstock.)*

i. Magnet *(Shutterstock.)*

k. Hat *(Shutterstock.)*

l. Baseball *(Shutterstock.)*

j. Wrench *(Shutterstock.)*

m. Chair *(Shutterstock.)*

n. Birdhouse *(Shutterstock.)*

17. Sketch isometric, oblique, and perspective pictorials of the objects shown in the photographs.

a. Vintage Television *(Shutterstock.)*

b. Male and Female Plugs

18. Create a freehand isometric sketch of the stepper motor shown. Ignore the electrical wiring and connector if you find it easier to visualize that way. Try drawing a perspective sketch of the object. In what ways do the sketches differ?

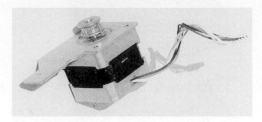

19. The CAD model shown is an isometric view of a tool stop. The model contains (or omits) some information that would usually be omitted (or present) in different drawing representations. Redraw the tool stop as

a. an isometric sketch
b. a perspective sketch

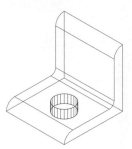

Exercises 20–37: Create isometric and oblique sketches of the objects shown. Use grid paper to draw the objects to size, or scale them consistently. Assume that each grid square equals 1/2 inch, or that each square equals 10 mm.

20.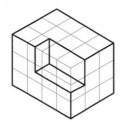

21.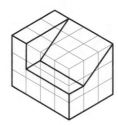

22.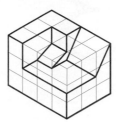

23.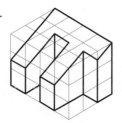

24.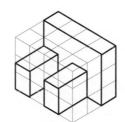

25.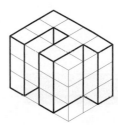

26.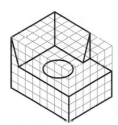

27.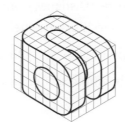

28.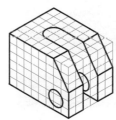

29.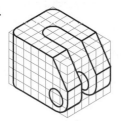

30.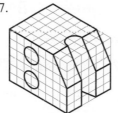

31.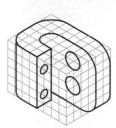

32.

33.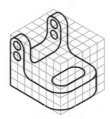

34.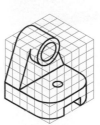

35.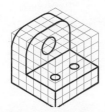

36.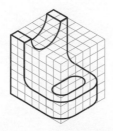

37.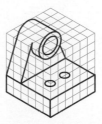

3 MULTIVIEW SKETCHING

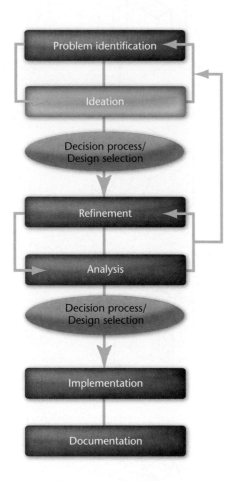

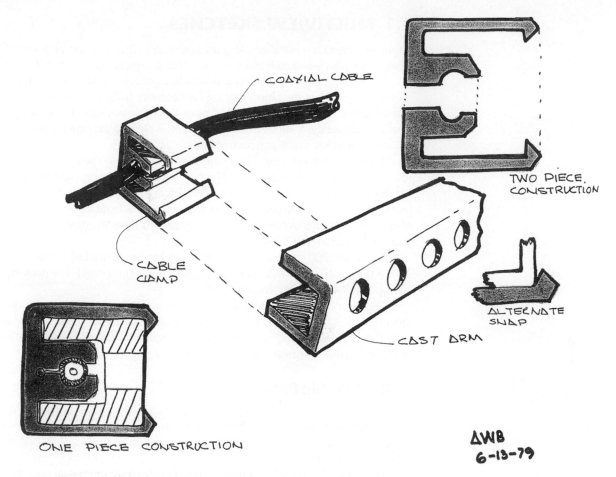

COAXIAL CABLE

TWO PIECE CONSTRUCTION

CABLE CLAMP

ALTERNATE SNAP

CAST ARM

ONE PIECE CONSTRUCTION

AWB
6-13-79

The 3D view of this coaxial cable clamp shows its relationship to the cast arm it will clamp onto, but the sketches of one-piece and two-piece construction, and the alternative shape for the clamp's snap, are orthographic views that show the true shape of the clamp more accurately. (*Courtesy of Albert W. Brown, Jr., Affymax Research Institute.*)

CAPTURING THREE DIMENSIONS ON PAPER

Pictorial sketches are good for getting your ideas down on paper. They also make it easy for other people to visualize the shape of the object. However, pictorial drawings show the object tipped in such a way that the surfaces are foreshortened, making it difficult to interpret the exact sizes of feature. A realistic appearance is often less important than a system that accurately reflects the size and shape of an object and the relationship between its parts. Orthographic projection (or multiview drawing) is a method that has been developed for transferring 3D shapes to 2D paper so that the resulting sketches can show object features true size. The 2D views can accurately define the size of and relationships between features as well as their shape.

3.1 MULTIVIEW SKETCHES

Engineers often communicate using *multiview drawings* that use *orthographic projection*. A multiview sketch and the object that it depicts are shown in Figure 3.1.

The photograph of the V-block shown in Figure 3.1(a) shows the object in perspective. Portions of the object that are farther away from the viewer appear smaller when viewed in perspective. In this case, the back surfaces do not appear much smaller, because the V-block in the photograph is close to the camera, and its surfaces are too short to see much perspective effect.

In engineering drawings, it is important to be able to specify accurately all the information needed to make the object. Perspective drawings are not very useful for this because the back surfaces of the object appear smaller than the front surfaces. In addition, some parts of the object are hidden from sight in the single view. Multiview drawings overcome these limitations to show the object's size and shape accurately (see Figure 3.16(b)).

For the last 200 years, multiview drawings have been used by engineers, whereas 3D CAD is relatively new. Understanding how to interpret multiview drawings is still an important skill for any engineer.

The ability to create orthographic multiview sketches will aid you in visualizing and presenting your designs to others. Many items are easier to sketch orthographically than pictorially. Multiview sketching is another technique you can add to the set of skills you use to convey your engineering designs.

Orthographic Projection

Multiview drawings are made up of views created looking straight onto (perpendicular to) each point on the object from a particular viewing direction. Of course, this is not how you see objects in real life. It is a construct that allows features on the object to be shown true shape.

The drawings are called *multiview drawings* because more than one view is necessary to completely define a 3D object; each view shows only two of the three principal dimensions. Unlike pictorial sketches that show three dimensions in a single 2D view, each 2D orthographic view shows only two of the three dimensions of the object. No single orthographic view can show all three *principal dimensions*: height, width, and depth.

3.1   This V-block clamp is shown here in (a) a catalog photograph and in (b) a standard multiview sketch. *(Part (a) courtesy of Grizzly Industrial, Inc.)*

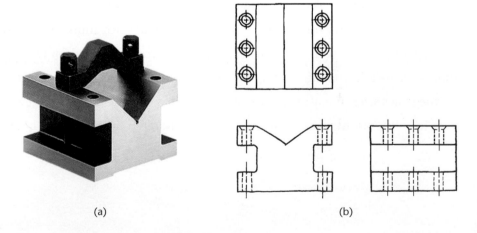

(a)   (b)

Look at the view of the helicopter in Figure 3.2. To see the width of the helicopter, you need another view.

In multiview drawings, the principal dimensions are associated with certain views of the object, as shown in Figure 3.3. The horizontal distance in the front view is always defined as the *width* whether or not it is the widest dimension on the object. The vertical dimension in the front view is defined as the *height*. The *depth* is shown in any view from a viewpoint that is 90° from that of the front view. For example, both the side and top views would show the depth of the object. Associating the dimensions with the views of the object allows the terms *height*, *width*, and *depth* to be used consistently from view to view.

3.2 This view of the helicopter shows only its depth and height. It is not possible to see how long it is from this viewpoint.

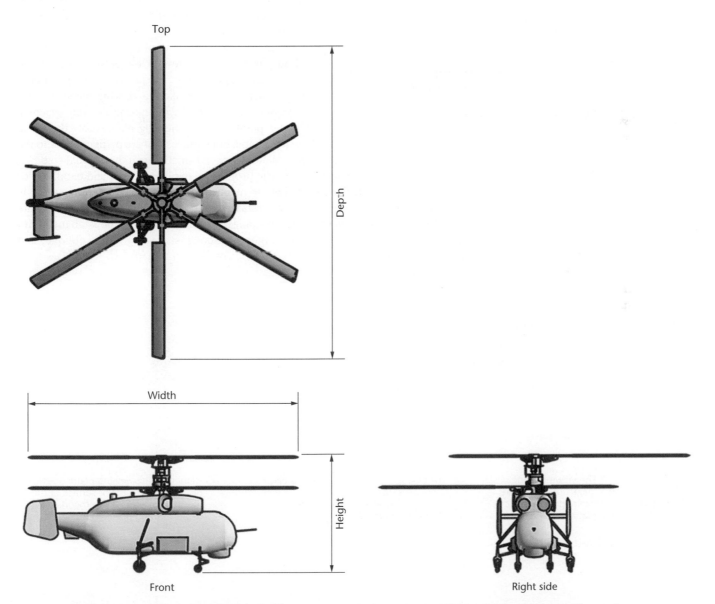

3.3 Each of the three 2D views in a multiview drawing shows only two of the three principal dimensions of the object.

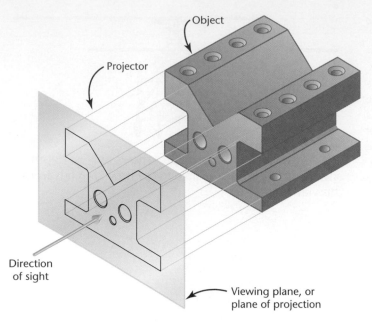

3.4   Orthographic Projection of an Object

Systems of Projection

One way to picture an orthographic view is as a projection onto a 2D viewing plane. Projection systems involve four basic parts: the object, the plane of projection, the viewpoint or direction of sight, and the projectors, as shown in Figure 3.4.

There are two major types of projection: parallel and perspective. You have already seen how perspective projectors converge to a point, whereas parallel projectors remain parallel to one another. The key difference between them is that in perspective projection the distance from the object to the plane of projection affects the size of the object shown. Parallel projection eliminates distance as a variable and represents the true shape of surfaces accurately.

Parallel projection is used to create orthographic, oblique, and isometric views. Oblique sketches are one type of parallel projection: The projectors are parallel to one another, but they strike the viewing plane at some angle other than 90°. In oblique projection systems, surfaces parallel to the viewing plane show their true size and shape, but surfaces that are perpendicular to the viewing plane are foreshortened and are displayed smaller than their actual size.

In both orthographic and isometric projection, the line of sight and parallel projectors are perpendicular to the viewing plane. The key difference between them is that the object in an isometric view is rotated and tipped so that the three coordinate axes appear in the view. In an orthographic view, the viewing plane is aligned with one of the surfaces on the object. Figure 3.5 contrasts orthographic, oblique, and isometric projections.

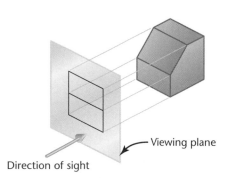

(a) Orthographic: The viewing plane is aligned with a major surface on the object. Projectors are perpendicular to the viewing plane.

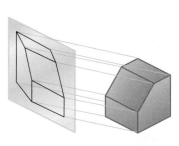

(b) Oblique: The viewing plane is aligned with a surface of the object. Projectors are at an angle other than 90° to the viewing plane.

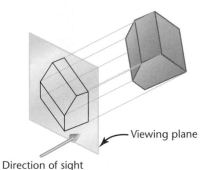

(c) Isometric: The viewing plane does not align with a major surface of the object. Projectors are perpendicular to the viewing plane.

3.5   Projection Methods

The Glass Box

The standard arrangement of orthographic views can be understood by imagining the object projected onto the walls of a glass box. The walls of the glass box are six mutually perpendicular planes, representing the top, front, right-side, left-side, rear, and bottom views of the object. Figure 3.6 shows the projection of an object onto the walls of the glass box. Unfolding the box, as shown in Figure 3.7, produces the

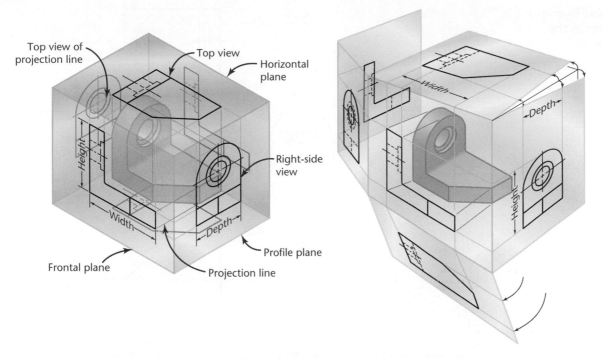

3.6 An orthographic view of the object is projected onto each of the six walls of the "glass box." The walls unfold outward from the front view.

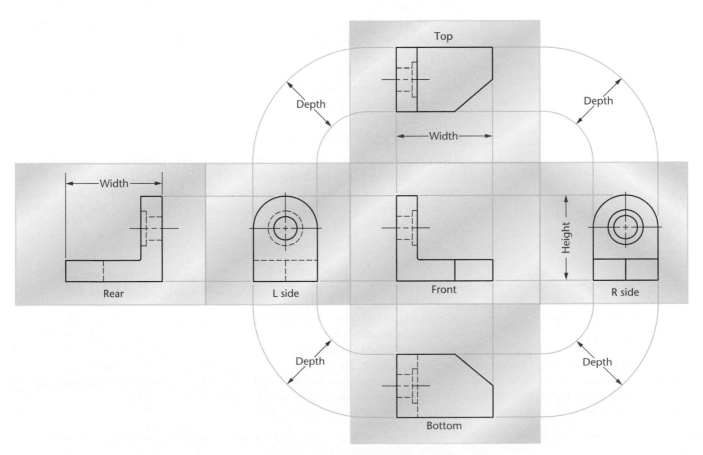

3.7 The box flattens out to produce the standard arrangement of orthographic views.

3.8 Orthographic Drawing
Showing Six Views in the
Standard Third-Angle
Arrangement

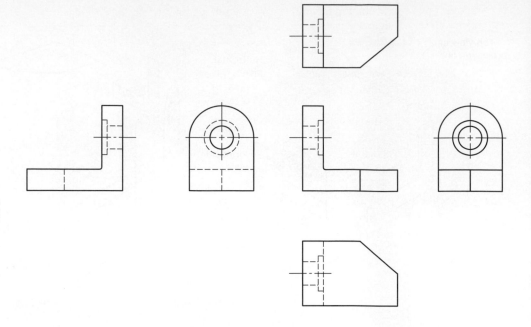

arrangement of views used in the United States (called third-angle projection). The lines representing the edges of the glass box are not shown in engineering drawings. Figure 3.8 shows the final multiview drawing.

Using Projection Lines

Any point on the object shown in the front view must correspond along a straight line, called a ***projection line,*** with that same point on the object in the top view. Any point on the object shown in the side view must correspond along a straight line with that same point on the object in the front view. The unfolded box used to show the arrangement of views in Figure 3.7 is commonly used to show how projection lines (shown in blue) relate the views to the object and to each other in multiview sketches. When you are creating multiview sketches and drawings, develop the views together, and keep them correctly aligned with one another.

Choosing the Necessary Views

Notice that the right- and left-side views are essentially mirror images of each other. The only difference is in the lines that appear hidden. Hidden lines use a dashed line pattern to represent edges of the object that are not directly visible from that direction of sight. Both the right and left views do not need to be shown, so usually the right-side view is drawn.

The Geometry of Projection

Gaspard Monge, born in France in 1746, is considered to be the father of orthographic projection. Prior to his time, there was no systematic approach to engineering drawings. As a professor at the Polytechnic School in France, Monge developed his ideas for orthographic projection. His method for describing a structure or device so it could be built reliably was so important that it was kept a military secret until Monge published *Géométrie descriptive* in 1795. Shortly thereafter, the basic principles of descriptive geometry and the projection of multiple views became a standard part of engineering education.

This is also true of the top and bottom views, and of the front and rear views. The top, front, and right-side views, arranged together, are shown in Figure 3.9. These are called the three *regular views*, because they are the views most frequently used.

A sketch or drawing should contain only the views needed to clearly and completely describe the object. These minimally required views are referred to as the *necessary views*.

Choose the views that have the fewest hidden lines and show essential contours or shapes most clearly. Complicated objects may require more than three views or special views such as partial views.

Many objects need only two views to clearly describe their shape. If an object requires only two views and the left-side and right-side views show the object equally well, use the right-side view. If an object requires only two views and the top and bottom views show the object equally well, choose the top view. If only two views are necessary and the top and right-side views show the object equally well, use the combination that fits best on your paper while remaining aligned in projection.

Transferring Depth Dimensions

The depth dimensions in the top and side views must correspond point-for-point.

The front, top, and right-side views of the object shown in the previous figures are shown in Figure 3.9(a), but instead of a glass box, *folding lines* are shown between the views. These folding lines correspond to the hinge lines of the glass box and can be used as a reference for marking the depth. The H/F folding line, between the top and front views, is the intersection of the horizontal and frontal planes. The F/P folding line, between the front and side views, is the intersection of the frontal and profile planes.

Folding lines are usually left off the drawing. Instead, *reference surfaces* are used for making depth measurements. In Figure 3.9(b) the front surface of the object (A) is used as a reference. To transfer a dimension from one view to a related view (a view that shares that dimension) measure from the edge view of a plane that shows on edge in both views. Note that D1, D2, and all other depth measurements correspond as if folding lines were used.

> **TIP**
> Marking the distances on a scrap of paper and using it like a scale to transfer the distance to another view works well when sketching.

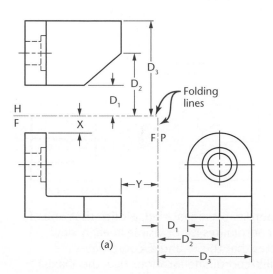

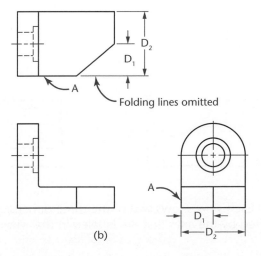

3.9 Orthographic Drawing Showing Three Regular Views (a) with Folding Lines; (b) without Folding Lines

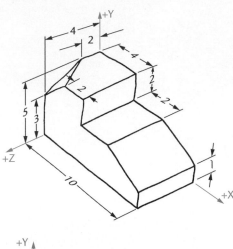

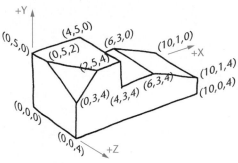

3.10   An Isometric Sketch of an Angled Block

Adjacent and Related Views

When two views sharing a common dimension are placed next to each other so that the dimension is aligned in both views, the views are referred to as ***adjacent views***. The front and side view are adjacent views. They share the height dimension and are aligned so that points with the same height dimension line up along the same projection line.

The side and top views are referred to as ***related views***. Related views are adjacent to the same view (in this case, the front view), and the distance that is measured perpendicular to projection lines between features appearing in both views is the same.

Coordinates and Projection

Points seen in the top and front views have their X-coordinates in common. A point seen in the side and front views aligns because the Y-coordinate of the point is the same in those views. If you think of the object in a "glass box" that has been unfolded to produce the views, it is easy to see why the points line up. A projection line is the set of all the points that share a coordinate value.

Figure 3.10 shows isometric sketches of an angled block with the dimensions labeled and with the coordinates labeled. In Figure 3.11 the orthographic views are shown with the coordinate axes labeled. The orthographic views in Figure 3.12 are labeled with the coordinates for the main corners of the block. Notice that each orthographic view shows only two of the X-, Y-, Z-coordinates, but any point located in two adjacent views defines the location of the object in 3D space (Figures 3.12 and 3.13).

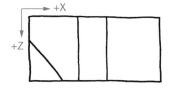

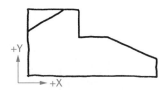

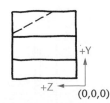

3.11   Each view shows two of the three dimensions, as indicated by the axes that are labeled in that view.

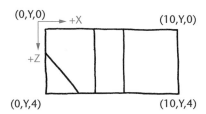

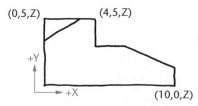

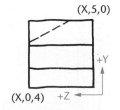

3.12   The coordinates are labeled. Notice that one of the X-, Y-, Z-coordinates is not visible in each view: 10,Y,0 indicates that the X- and Z-coordinates are visible, but all Y-coordinate locations (like the Y-axis) appear as a point.

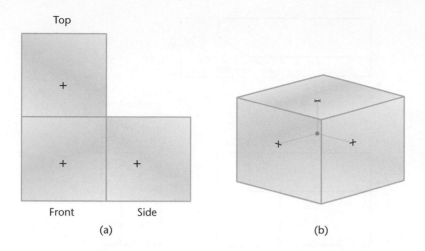

Top

Front   Side

(a)

(b)

3.13 (a) A point as it appears in each view is defined by two of the three dimensions. (b) Any two views can provide the missing third dimension and locate the point in space.

Reference Surfaces (Coordinate System)

Reference surfaces can be based on a Cartesian coordinate system (Figure 3.14). An orthographic view of the object is perpendicular to all the points in the view along the direction of one coordinate system axis. When you look straight down the X-axis, you see the X-Y and X-Z planes on edge and look straight onto the Y-Z plane. When you look straight down the Y-axis, you see the X-Y and Y-Z planes on edge and look straight onto the X-Z plane. When you look straight down the Z-axis, you see the X-Z and Y-Z planes on edge and look straight onto the X-Y plane. Figure 3.14 shows orthographic views of the same Cartesian coordinate system.

Notice that each set of two views shows a common plane of the coordinate system on edge. You can use this edge view of the common plane as a reference surface to transfer measurements from view to view. When you do so, the edge view of the common coordinate plane will have been rotated to preserve its

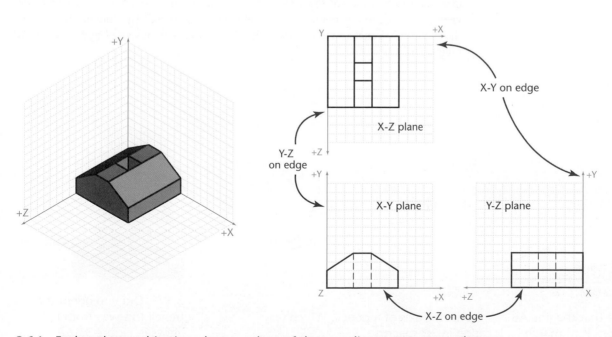

3.14 Each orthographic view shows a plane of the coordinate system on edge.

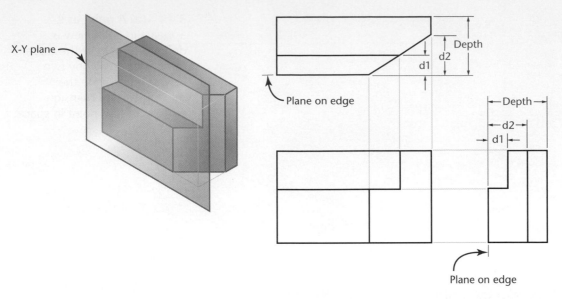

3.15   Measurements can be made from a reference surface.

orientation. In the top and side views, the X-Y coordinate plane is on edge (see Figure 3.15). You can locate the Z-coordinate of a point by measuring from this plane.

Relating Views to the Coordinate System

Figure 3.16 shows three orthographic views of a single line. A set of coordinate axes has been added to each view, with the X-Y plane parallel to the front view. You can interpret the information in the orthographic views to understand the position of points, lines, and planes in a drawing. The height, or location, of endpoint A on the Y-axis is shown in the front and side views. The width information for point A is shown along the X-axis in the front and top views. Point A's depth, or distance back from the front surface of the glass box, is shown on the Z-axis in the top and side views. The location of point B can be understood likewise. The line in 3D space is shown in the isometric view at right. Its orientation is represented by the pencil in Figure 3.17.

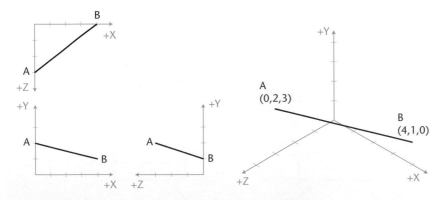

3.16   Imagine that the line AB represents the axis of a pencil. Which way would you orient it to match its location in 3D space?

3.17   Did you orient the pencil this way from the views in Figure 3.16?

STEP by STEP

SKETCHING A MULTIVIEW DRAWING

To begin sketching this angled block, first, decide how you will orient the block to produce the front view.

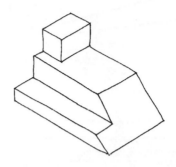

1 Lightly block in the overall dimensions.

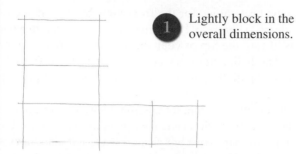

2 Use light lines to locate features in the front view, which should show the overall shape of the object and represent features in proportion.

3 Slightly darken the lines of the object as needed to help you visualize the shape.

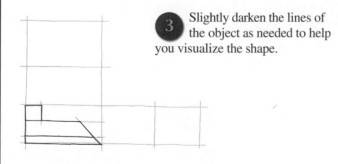

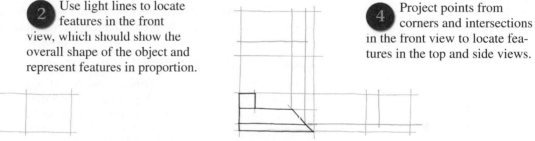

4 Project points from corners and intersections in the front view to locate features in the top and side views.

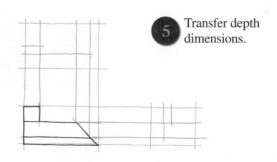

5 Transfer depth dimensions.

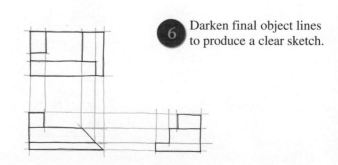

6 Darken final object lines to produce a clear sketch.

Poor choice

Poor choice

Good choice of front view

3.18 Choosing the Best Front View

Choosing the Front View

To sketch orthographic views of the helicopter in Figure 3.18, first, you must decide which view of the object you will orient as the front view. There are some general rules to follow for determining which view of the object should be shown as the front view.

The front view in a drawing should

- show the shape of the object clearly;
- show the object in a usual operating position; and
- have large flat surfaces parallel to the viewing direction.

The side of the helicopter is the best choice for the front view because it shows the distinctive shape most clearly (Figure 3.18). Orient the helicopter in your sketch so that the top rotor is toward the top of your sheet. It would look odd for it to be shown upside down. The helicopter does not have many large flat surfaces, but orient it so that it is essentially parallel to the view and not tipped at some angle.

After you have selected the direction of sight for the front view, it is preferred to show the top view (aligned above the front view) and the right-side view (aligned to the right of the front view), unless other views are clearer. If the left-side view is clearer than the right-side view, you may want to consider reorienting the object in the front view (by 180°) to produce a clear right-side view.

Keep in mind that sketches and drawings do not always have to show three views. If you are drawing an object that has a uniform thickness, you can show a single view and provide the thickness in a drawing note. The time you save by not drawing unnecessary views can be productively spent on other tasks.

Spacing Between Views

Spacing between views is mainly a matter of appearance. Views should be spaced well apart but close enough to appear related to each other (Figure 3.19). You may need to leave space between the views to allow room for notes and dimensions.

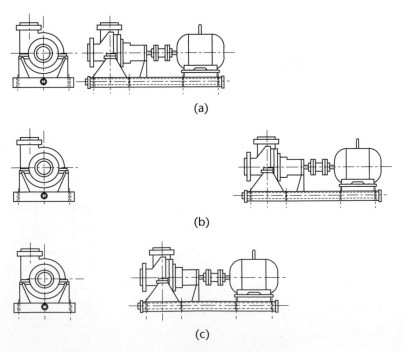

(a)

(b)

(c)

3.19 (a) Views are too close together; (b) views are too far apart; (c) views are spaced well.

Third-Angle versus First-Angle Projection

Third-Angle Projection

The arrangement of views shown below is called *third-angle projection*. It will be used for all multiview drawings in this book. In third-angle projection, the viewing plane for the front view is placed in front of the object, and the front view is projected onto it. The viewing plane for the right-side view is to the right of the object, and the viewing plane for the top view is above the object. The views are then unfolded into the standard arrangement shown below.

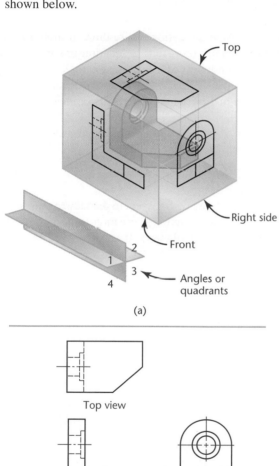

Top

Right side

Front

2
1
3
4

Angles or quadrants

(a)

Top view

Front view

Right-side view

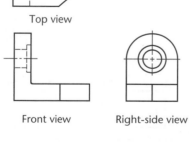

(a) Third-angle projection: Place this symbol on the drawing to indicate the third-angle projection system is being used.

Profile plane

Frontal plane

R side view

Front view

Top view

2
1
3
4

Horizontal plane

Angles or quadrants

2
1
3
4

(b)

Right-side view   Front view

Top view

First-Angle Projection

Many countries draw views arranged using *first-angle projection*. In first-angle projection the front view projects onto a plane behind the front view. The right-side view projects onto a plane to the left of the object, and the top view is projected onto a plane below the object. The views are then unfolded to produce the arrangement shown above.

Both first- and third-angle projection arrange the views around the front view.

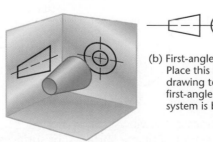

(b) First-angle projection: Place this symbol on the drawing to indicate the first-angle projection system is being used.

Ship arriving too late to save a drowning witch

3.20 Frank Zappa used this single-view illustration on the cover of his album "Ship Arriving Too Late to Save a Drowning Witch." *(Title and cover art used by permission of The Zappa Family Trust. Original illustration by Roger Price 1953, 1981. All rights reserved.)*

3.2 INTERPRETING MULTIVIEW DRAWINGS

Can you interpret the single-view drawing in the cartoon shown in Figure 3.20? If you did not know that the drawing was titled "Ship Arriving Too Late to Save a Drowning Witch," would you be able to "see" the familiar objects and their relationship in 3D? The cartoonist used a single view that is not easily identified, and the humorous title is a surprise. In engineering, less familiar objects are depicted in multiple views to provide more information, but you still need to interpret those views to "see" what the object looks like.

Lines and Points in a Multiview Drawing

A straight visible or hidden *line* in a drawing or sketch has three possible meanings:

- an edge (intersection) between two surfaces
- the edge view of a surface
- the limiting element of a curved surface

Examples are shown in Figure 3.21.

Points in a drawing always represent a location where three or more surfaces join to form a vertex on an object. A point represents a location in space and appears in a drawing as a corner or the endpoint of a line. An example of a point in a drawing and on an object is shown in Figure 3.22. When labeling points in a drawing or sketch, list the closer point first when two points coincide, as in the point view of a line.

Any point on the object shown in a view must align along a projection line with that same point in any other view, as shown in Figure 3.23.

Normal, Inclined, and Oblique Surfaces

A surface can have three possible orientations with respect to the viewing plane: normal, inclined, or oblique (Figure 3.24). Understanding these three typical orientations helps you recognize these surfaces in the standard views. It also helps you understand how to define them in a 3D model.

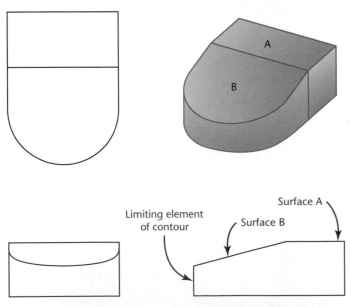

3.21 Most of the lines in the views of this model represent the intersection of planar surfaces, or surfaces viewed edgewise. One line in the side view represents the limiting element of the curved surface.

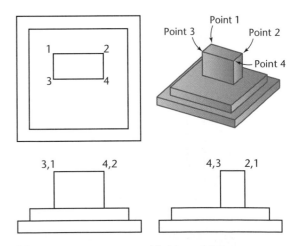

3.22 Points represent vertices and appear as corners.

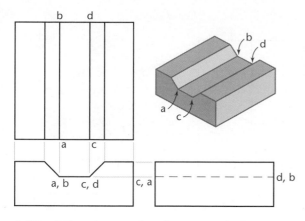

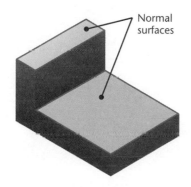

A **normal surface** is parallel to one viewing plane (or perpendicular to the line of sight for the view) and shows true size in that view. It shows on edge in the two other standard views, where it appears as a line (see Figure 3.25).

An **inclined surface** surface is perpendicular to one of the viewing planes but not true size in any view (see Figure 3.26). It appears on edge as an angled line in the view to which it is perpendicular. In the two other standard views, its shape appears foreshortened.

An **oblique surface** surface is not parallel *or* perpendicular to any of the standard views (see Figure 3.27). Its shape is foreshortened in all three standard views; it does not appear on edge in any view.

3.23   Points in any view line up with the same point in any other view.

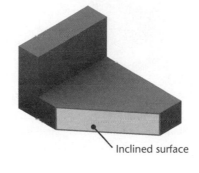

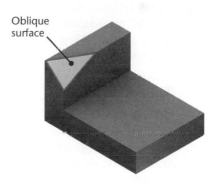

3.24   Surface Orientations

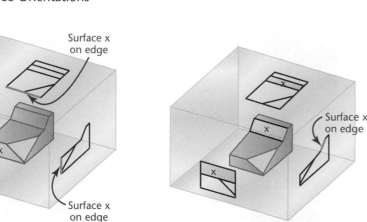

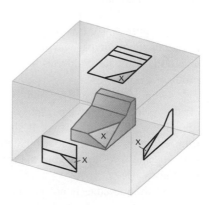

3.25   Surface X is a normal surface. It appears true size in the viewing plane to which it is parallel. It appears on edge as a vertical or horizontal line in the other two views.

3.26   Surface X is an inclined surface. It appears as an angled line in the view to which it is perpendicular. It appears foreshortened in the other two views. It does not appear true size in any standard view.

3.27   Surface X is an oblique surface. It appears foreshortened in each view. It retains its general shape even though foreshortened.

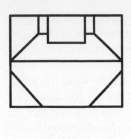

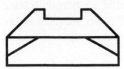

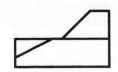

3.28 After you have identified each surface as normal, oblique, or inclined, check the answer on page 110.

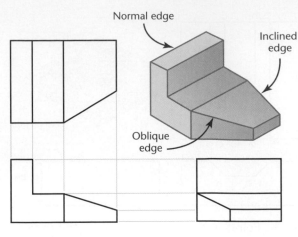

3.29 Normal, Oblique, and Inclined Edges

Figure 3.28 shows normal, inclined, and oblique surfaces in a multiview drawing. Can you identify which is which from the preceding definitions without looking at the key?

Normal, Inclined, and Oblique Edges

The terms *normal*, *inclined*, and *oblique* can also be applied to a single edge on an object. Figure 3.29 shows examples of normal, inclined, and oblique edges.

A normal edge, similar to a normal surface, is parallel to two of the principal views and shows true length in those views (Figure 3.30). It is perpendicular to one of the principal viewing planes, and in that view the endpoints line up to produce a point view.

An inclined edge is parallel to one of the principal viewing planes and tipped toward the other two. It shows true length in the view to which it is parallel, and foreshortened in the other two views.

An oblique edge is not parallel or perpendicular to any of the principal views and appears foreshortened in each of the principal views.

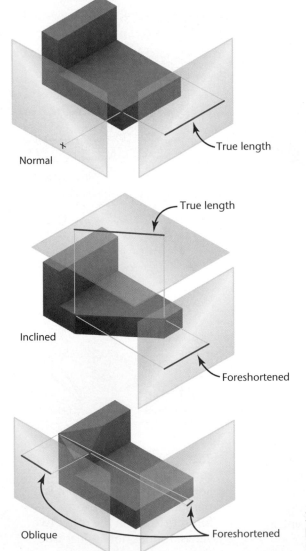

3.30 Projections of Edges in Front and Side Views

Typical Features

Recognizing typical features that are a part of many engineering designs can help you visualize the object (see Figure 3.31). It also allows you to communicate about them by naming them using typical terminology. Many CAD systems have predefined features for these frequently used items to make creating your model even easier.

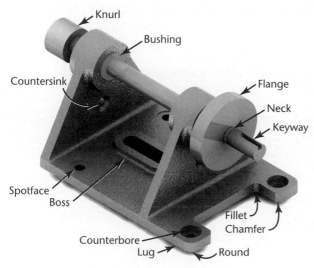

3.31   Commonly Manufactured Features

Feature	Example
Fillet: A rounded interior blend between surfaces; used, for example, to strengthen adjoining surfaces or to allow a part to be removed from a mold	
Round: A rounded exterior blend between surfaces; used to make edges and corners easier to handle, improve strength of castings, and allow for removal from a mold	
Counterbore: A cylindrical recess around a hole, usually to receive a bolt head or nut	
Countersink: A cone-shaped recess around a hole, often used to receive a tapered screw head	
Spotface: A shallow recess like a counterbore, used to provide a good bearing surface for a fastener	
Boss: A short raised protrusion above the surface of a part, often used to provide a strong flat bearing surface	
Lug: A flat or rounded tab protruding from a surface, usually to provide a method for attachment	
Flange: A flattened collar or rim around a cylindrical part to allow for attachment	
Chamfer: An angled surface, used on cylinders to make them easier to start into a hole, or plates to make them easier to handle	
Neck: A small groove cut around the diameter of a cylinder, often where it changes diameter	
Keyway/Keyseat: A shaped depression cut along the axis of a cylinder or hub to receive a key, used to attach hubs, gears, and other parts to a cylinder so they will not turn on it	
Knurl: A pattern form on a surface to provide for better gripping or more surface area for attachment, often used on knobs and tool handles	
Bushing: A hollow cylinder that is often used as a protective sleeve or guide, or as a bearing	

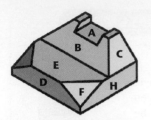

Key to Figure 3.28
Normal: A, E, D, H
Inclined: B, C
Oblique: F

Intersections and Tangencies

Multiview drawings use conventions for depicting the way planar and curved surfaces meet. A plane surface can *intersect* or be *tangent* to a contoured surface, as shown in Figure 3.32. When the surfaces are tangent, no line is drawn to indicate a change between the plane and contoured surface. When the surfaces are intersecting, an edge is formed that is shown as a line in the drawing. Figure 3.33 shows examples of intersecting and tangent surfaces.

You can use projection to draw intersections between surfaces in 2D orthographic drawings and sketches. Figure 3.34 shows some typical intersections between different diameters of pipe. When you are sketching these in orthographic views, identify points on the intersection and project those points into the remaining views as shown on page 112. When you create a 3D model, the software depicts the intersections or tangencies. You can typically turn on or off the display of tangencies.

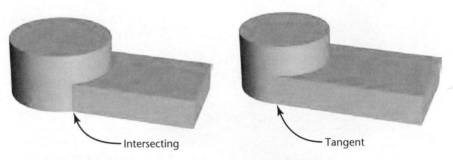

Intersecting   Tangent

3.32   Intersecting and Tangent Planes in Pictorial Views

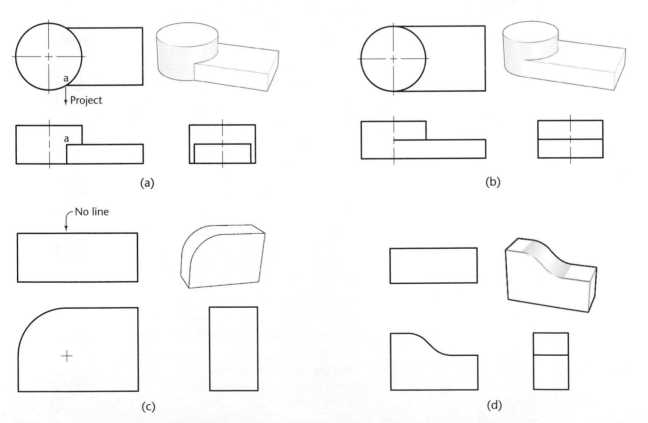

(a)   (b)

(c)   (d)

3.33   Examples of Intersecting and Tangent Planes in Orthographic and Pictorial Views

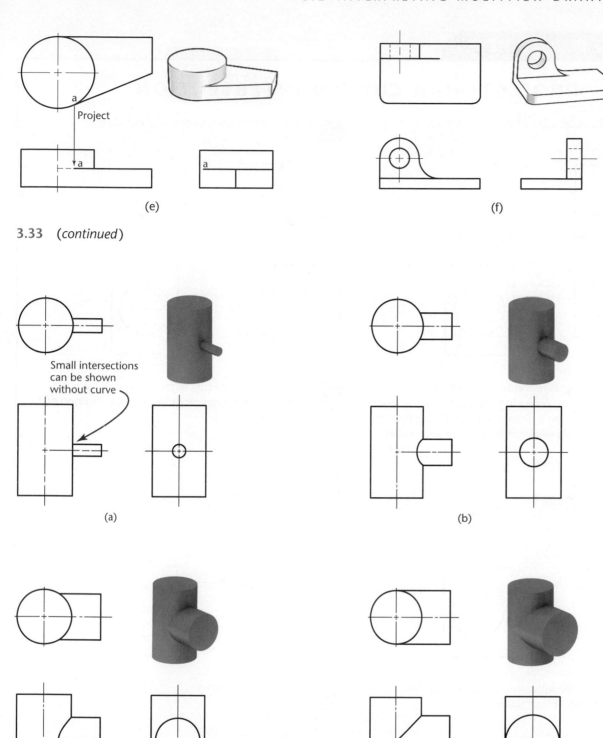

(e)

(f)

3.33 (continued)

Small intersections can be shown without curve

(a)

(b)

(c)

(d)

3.34 Intersections of Cylinders with Different Diameters

STEP by STEP

PROJECTING A CURVED INTERSECTION

1 Project the outer edge of the intersection.

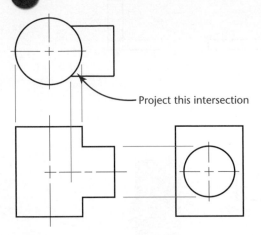

Project this intersection

2 Identify intermediate points on the intersection line. Project these points from the surface to the miter line. (The miter line is just a shortcut for transferring the depth dimensions. You can also measure from a reference surface.)

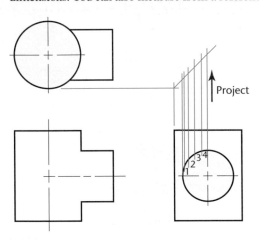

Project

3 Project from the miter line to the related view to find the points of intersection.

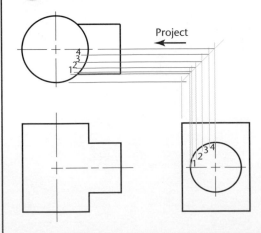

Project

4 Project the points into the adjacent view.

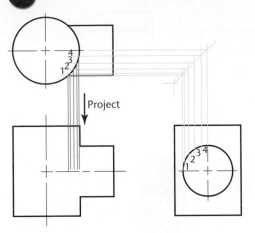

Project

5 Project from the other adjacent view. Where the two projection lines cross must be the location of that point of intersection.

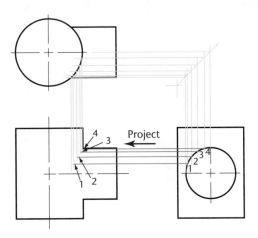

Project

6 Draw the final curve.

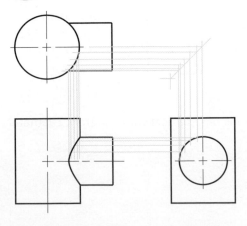

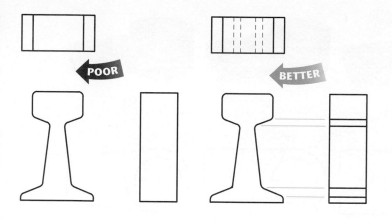

3.35 Conventional lines are used to show changes in direction between two surfaces when leaving lines out might make the view misleading.

Sometimes, a drawing may appear misleading if it contains planar surfaces that change direction through tangency with a contoured surface. Because the surfaces are tangent, no line is drawn where they meet, so the change in direction is not depicted. In such cases, a line may be drawn to indicate the change of direction of the surface even though tangent surfaces do not form a visible edge. Figure 3.35 shows an example where the projection might be misleading without the addition of such lines.

TIP

A 45° miter line can be drawn to aid in transferring distances into the related view. The miter line acts as the diagonal of a square. Because the sides of the square are equal, the vertical distance and horizontal distances will be the same.

STEP by STEP

PROJECTING AN IRREGULAR CURVE

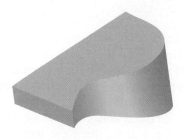

Use projection to locate irregular curves in orthographic views.

1 Mark randomly spaced points along the curve in the view that shows its shape clearly.

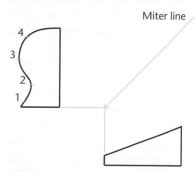

2 Locate the same points where the surface with the curve appears on edge.

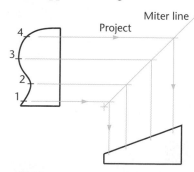

3 Project each point into the view where you need to show the curved shape.

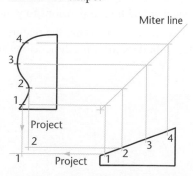

4 Connect the points with a smooth curve. A spline or polyline works well.

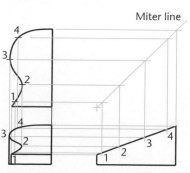

3.36 Examples of Fillets and Rounds in Pictorial Views

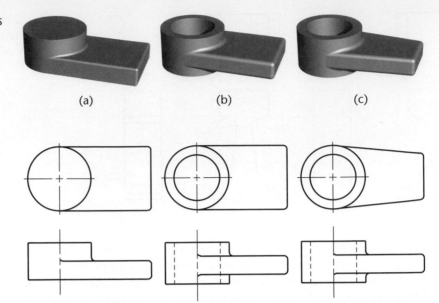

(a)   (b)   (c)

3.37 Drawing Fillets and Rounds in Orthographic Views

(a)   (b)   (c)

Fillets and Rounds

A *fillet* is an interior rounded blend between two surfaces. A *round* is an exterior rounded blend. Both are illustrated in Figure 3.36. You should draw the fillets and rounds on an object in the views where you see their curved shape. In views where you do not see their rounded shapes, ignore the fillets and rounds and draw the views as if there was no fillet or round present, as shown in Figure 3.37.

Scale

Sketched views of parts will not always fit full size on a sheet of paper. A reduced *scale* is used to represent a large part on a small sheet. When sketching you often do not indicate the scale; rather, you show the object so that it fits on the paper and its features are proportioned correctly. You can label the features with sketched dimensions to specify their size.

If you are showing the sketch to a common scale (such as half size or one-tenth size), note the scale for your sketch near the bottom right of the sheet. The scale note gives the relationship between sketched size and real-world size. Some common scales are full scale (1:1), half scale (1:2), one-quarter scale (1:4), one-fifth scale (1:5), one-tenth scale (1:10), double scale (2:1), ten times scale (10:1), and other common multiples. Try to avoid scales such as third scale (1:3), sixth scale (1:6), and other uneven multiples. Figure 3.38 shows a drawing drawn to a smaller scale. Notice the dimensions that provide the sizes of features.

Sketching Dimensions

When you are adding dimensions to your sketches try to follow these guidelines:

- Place dimensions where the feature shows its shape clearly and true size.
- Use thin dark lines for dimensions and their text.

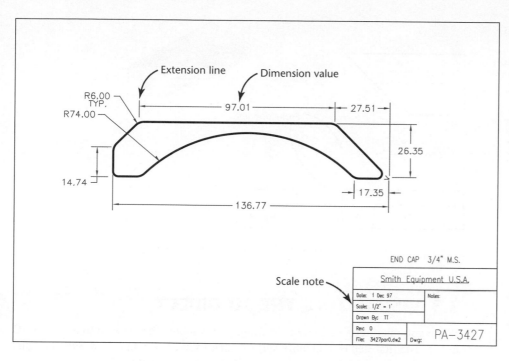

3.38 This large part was drawn at a scale of 1/2 inch = 1 foot. Note the scale indicated in the title block in the lower-right corner of the drawing. *(Courtesy of Smith Equipment USA.)*

- Locate the dimension to the feature using an extension line (see Figure 3.38).
- Place dimensions off the drawing views.
- Do not duplicate dimensions from one view to the next.
- Make arrowheads, text, and other symbols about 3 mm (1/8") tall.
- Space dimensions away from the object by about 9 mm (3/8").
- Group dimensions together to make them easy to read.

Figure 3.39 shows examples of good practice for sketching dimensions.

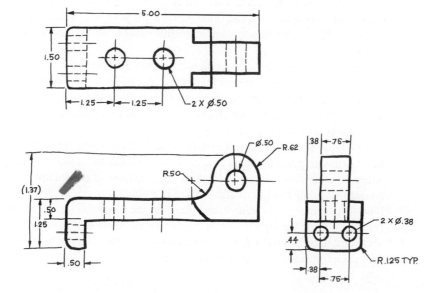

3.39 Sketched Dimensions

3.40 Surface A on the model is the same in each view. The projection lines from one view to the next help you locate the surface in each view.

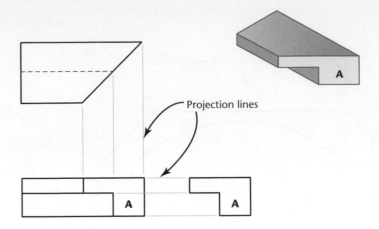

Projection lines

3.3 VISUALIZING THE 3D OBJECT

When you interpret a multiview drawing, you combine the views into a mental 3D image. The first step is to identify surfaces that show clearly in a view. For example, surface A in Figure 3.40 shows its shape in the front and right-side views.

A surface has the same basic shape no matter how it is viewed. It always has the same number of vertices. When a surface is perpendicular to a view, all the vertices line up to produce a straight line, as they do in the top view of the highlighted surface in Figure 3.40. Using projection lines is one way to help identify the same surface in more than one view.

Interpreting Views by Removed Portions

Another way to visualize an object from 2D views is to start with a rectangular block (prism) just large enough to contain the object and then examine the views looking for areas that are cut off from this block.

Figure 3.41 shows three views of an object. Start with a block of clay, foam, or some other material with the same general proportions as the views, or sketch an isometric block that has the same overall height, width, and depth, as shown in Figure 3.42(a).

Now, examine the views to identify inclined or normal surfaces on edge in one of the views. Can you imagine material removed from the block where the normal surface shows on edge in the side view? Mark the ends of the surfaces in your isometric sketch, then lightly block in the portion that is removed. Figure 3.42(b) shows the isometric view that results.

Oblique and inclined surfaces are usually the hardest for people to visualize. It helps to assign point numbers to each vertex and use the number to keep track of the vertex in each view. Figure 3.41 shows the vertices of the inclined surface labeled. Notice that point 1 is left of and in front of point 2. Point 3 is directly below point 2. Find this surface by locating each vertex on the block (as shown in Figure 3.42(c)). Imagine cutting through the object with a broad knife that passes through each vertex. Mark these points by locating them relative to the edges in your isometric sketch, then sketch straight lines between the points.

3.41 To interpret the multiview drawing, first visualize the overall shape of the object—in this case a rectangular block.

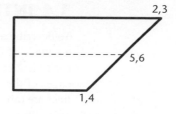

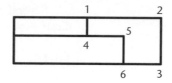

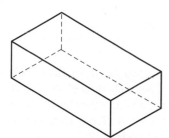

(a) Sketch a block with the same overall proportions as the object views.

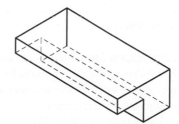

(b) Identify places where material is removed from the block to form the shape. Mark it on the sketch.

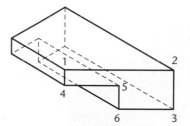

(c) Use numbers to match up corners in each view. Connect the vertices to locate the surfaces and edges.

3.42 Visualizing the Drawing

STEP by STEP

MAKING A CLAY MODEL

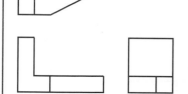

1 Shape a block of clay into the overall width, height, and depth of the views. Use a paring knife to make crisp cuts in the clay to define the surface edges.

2 Lightly scribe the lines of the drawing views on the appropriate surfaces of the clay block.

3 Look at the outlines of the views to identify portions of the block that are missing from each view. Cut these away first.

4 Identify the remaining surfaces on the object. Turn the block to see how the surfaces are defined in the views. Make the final cuts to finish the model.

3.4 INTERPRETING LINETYPES

To describe an object more fully in a 2D drawing, engineering drawings use different kinds of lines, or *linetypes*, to add information to the depiction. Remember that each view of the object shows the entire object *as viewed from that direction*. The edges of the object are depicted with thick lines to clearly outline the object. These lines are referred to as *visible lines*, as they are what you would see when viewing the object from that direction.

Several other kinds of lines are used to represent features on the object or to indicate how views were constructed.

Hidden Lines

When edges or other features are not directly visible in a view, use the *hidden* linetype to represent them. *Hidden lines* allow you to see edges on the back, bottom, left, or interior of the object in an orthographic view. Hidden lines are thin lines made of short dashes.

To visualize hidden lines, imagine an object made of clear plastic or glass, such as the one shown in Figure 3.43. From the front, you can look right through the object to see the edges where surfaces join together inside and in the back of the object. These edges would appear as hidden lines in the drawing as shown in Figure 3.44. Notice how hidden edges in the front view in Figure 3.44 (and in Figure 3.45) are differentiated from the thick visible lines in the same view. The standard linetypes are shown in Figure 3.46.

Standard hidden lines conform to drawing standards to eliminate ambiguity for the reader, as illustrated in Figure 3.44. Hidden lines should do the following:

- Form neat intersections with the edge of the object and other hidden lines.
- Jump across lines (edges) that they do not intersect on the object itself. (This is a convention observed in sketches but may not be implemented in CAD drawings.)
- Not extend another line of a different linetype. Any two lines of different linetypes should never join to form a single straight line. Leave a visible gap in your sketch (about 1/16 inch) whenever two different lines would join to form a straight line.

3.43 A transparent model makes it easy to see edges inside and on the back of the model that would appear as hidden lines in a sketch or drawing.

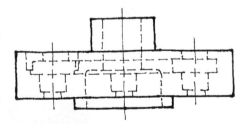

3.45 Hidden Lines in a Sketch

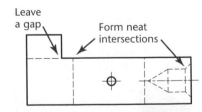

3.44 Good hidden-line technique makes hidden lines unambiguous to the reader.

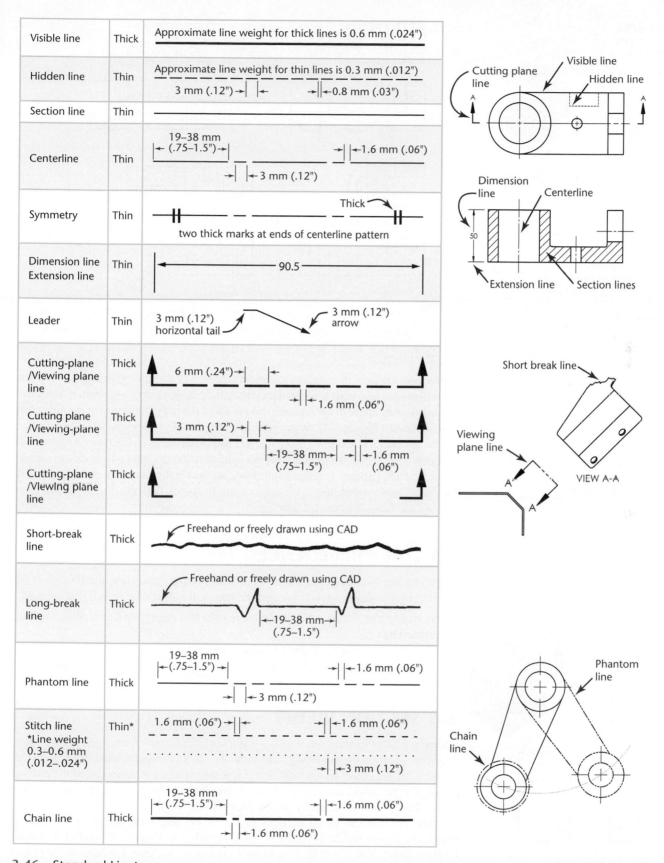

Visible line	Thick	Approximate line weight for thick lines is 0.6 mm (.024")
Hidden line	Thin	Approximate line weight for thin lines is 0.3 mm (.012") 3 mm (.12") → | ← → |← 0.8 mm (.03")
Section line	Thin	
Centerline	Thin	19–38 mm |← (.75–1.5") →| → |←1.6 mm (.06") →| |← 3 mm (.12")
Symmetry	Thin	Thick two thick marks at ends of centerline pattern
Dimension line Extension line	Thin	←————— 90.5 —————→
Leader	Thin	3 mm (.12") horizontal tail 3 mm (.12") arrow
Cutting-plane /Viewing plane line	Thick	6 mm (.24") →| ← →| ← 1.6 mm (.06")
Cutting plane /Viewing-plane line	Thick	3 mm (.12") →| ← |←19–38 mm→| →|←1.6 mm (.75–1.5") (.06")
Cutting-plane /Viewing plane line	Thick	
Short-break line	Thick	Freehand or freely drawn using CAD
Long-break line	Thick	Freehand or freely drawn using CAD |←19–38 mm→| (.75–1.5")
Phantom line	Thick	19–38 mm |←(.75–1.5") →| →|←1.6 mm (.06") →| |← 3 mm (.12")
Stitch line *Line weight 0.3–0.6 mm (.012–.024")	Thin*	1.6 mm (.06") →|| ← →|←1.6 mm (.06") →|←3 mm (.12")
Chain line	Thick	19–38 mm |← (.75–1.5") →| →|←1.6 mm (.06") →| |←1.6 mm (.06")

3.46   Standard Linetypes

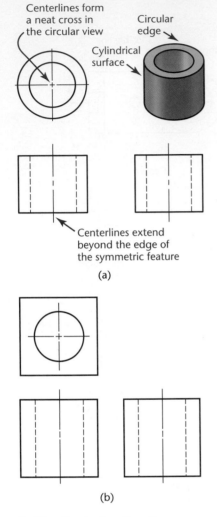

Centerlines form a neat cross in the circular view

Circular edge

Cylindrical surface

Centerlines extend beyond the edge of the symmetric feature

(a)

(b)

3.47   Centerline Practices

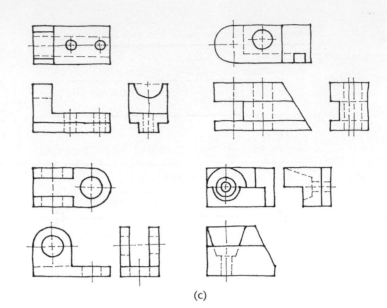

(c)

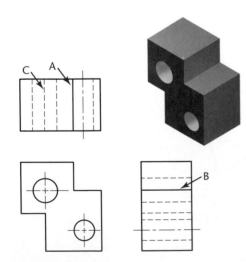

C   A

B

3.48   Line Precedence

Centerlines

Centerlines are used in drawings to indicate the axis of symmetry for a part or feature. Centerlines are also used to show the symmetrical alignment of a pattern of holes (often called a *bolt circle* or *bolt pattern*) and to show the path of motion for moving parts in an assembly. Figure 3.47 shows several examples of correct centerline practices.

Note that thickness as well as pattern is used to distinguish linetypes. Object lines or visible lines are drawn thick so that the subject of the drawing will stand out clearly in your sketch. Hidden lines and centerlines are drawn thin. Practice making vertical, horizontal, and angled lines with each of the linetypes illustrated in Figure 3.46 so you can use them in your sketches.

3.5 LINE PRECEDENCE

When one line is directly behind another in an orthographic view, the following order of line precedence is used to determine which line should be shown (Figure 3.48):

• A visible line always takes precedence over and covers up a centerline (B) or a hidden line (A).
• A hidden line takes precedence over a centerline (C).
• A cutting-plane line for a section view takes precedence over the centerline.

3.6 SECTION VIEWS

Showing a lot of interior detail makes a drawing hard to interpret, even to an experienced eye. Section views allow you to show the object's interior detail. A section view shows what the object would look like if it were cut along a plane, called the *cutting plane*. The portion of the object in front of the plane is removed, and the section view shows the surfaces revealed by the cut (see Figure 3.49).

The cut surfaces (solid portions of the object that were sliced by the cutting plane) are shown *hatched* (Figure 3.50). (*Hatching* is a series of parallel thin lines (section lines) drawn to indicate a surface created by the cut.) Portions of the object that were previously hidden may now show as visible lines.

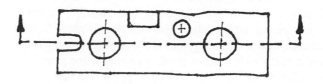

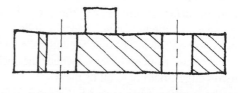

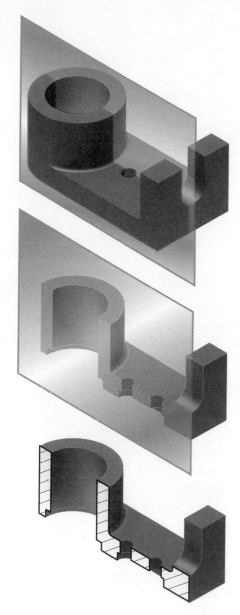

3.49 Creating a Section

3.50 Hidden lines are not used for the hole not revealed by the section view, because it is fully described in other views.

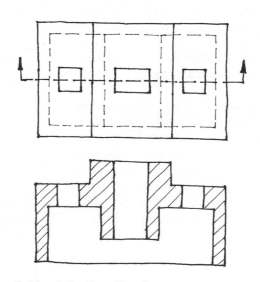

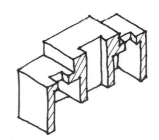

3.51 A Section Sketch

The cutting-plane linetype is used in a single view to show the edge view of the cutting plane indicating where the object is cut. The arrows on the cutting-plane line indicate the line of sight; the section view depicts the object as you would see it when facing in the direction of the arrows.

Sections rarely show hidden lines because their purpose is to make interior details clear without the confusion of hidden lines (Figure 3.51). Section views often replace one of the standard drawing views.

3.7 AUXILIARY VIEWS

To fully describe an inclined or oblique surface, engineering drawings will often include a nonstandard view called an *auxiliary view* (see Figure 3.52). Auxiliary views are created in the same way as other orthographic views, but they depict a view from some direction other than top, front, right side, left side, bottom, or rear. Auxiliary views are most commonly used to show a surface that is not parallel to any of the principal viewing planes (and therefore does not appear true size in any of the views); auxiliary views are often constructed to show the true size of a surface.

For an inclined surface, think back to the example of the glass box. Imagine a new viewing plane added to it that is parallel to the surface to be shown. The new viewing plane is perpendicular to one of the principal views (where the inclined surface appears on edge), but not to all of them.

Notice that the auxiliary view shown in Figure 3.53 is projected from the front view and has the same depth as the top view. The auxiliary viewing plane is hinged so that it is perpendicular to the front view. When the views are unfolded as shown in Figure 3.53, projection lines can be drawn from the front view to the auxiliary view to transfer the height locations of the object's vertices. The depth locations can be transferred by measuring from a reference plane parallel to the front viewing plane.

For oblique surfaces, a second auxiliary view is necessary, as the oblique surface is not perpendicular to any of the principal views.

Auxiliary views are also frequently used to show the true angle between two planes, called the *dihedral angle*. A dihedral angle is shown true size in a view where the line of intersection between the two planes is shown as a point. In the past, the true dihedral angle would be calculated or projected in an auxiliary view and then measured. Today, complex shapes that have many inclined and oblique surfaces often can be created fairly easily using 3D CAD, allowing measurements to be made directly from the model.

Selecting Views

Even if 3D CAD is being used, it is important to understand which views show surfaces, lines, and angles true size. You should understand how to show a line true length, how to show a line as a point view, how to show a surface on edge, and how to show a surface true size. These four types of views can be used to solve many other engineering problems, such as finding the shortest distance between two angled pipes.

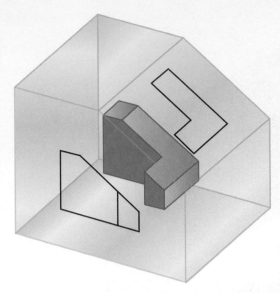

3.52 An auxiliary view of an inclined surface is parallel to the surface to be shown and perpendicular to the principal view where the inclined surface appears on edge.

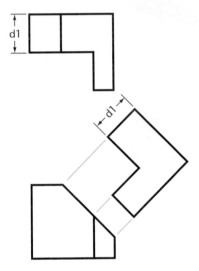

3.53 An auxiliary view of an inclined surface is projected from the front view.

Understanding how to show geometry true size will be useful in orienting your view or work plane when using 3D CAD.

- To show a line true size, create a view so that you are looking perpendicular to the line (see Figure 3.54).
- To show a line as a point view, create a view so that you are looking parallel to the line (perpendicular to the view where you see it true length, as shown in Figure 3.55).
- To show a surface on edge, create a view to show any line in the surface as a point view (see Figure 3.56).
- To show a surface true size, create a view so that you are looking perpendicular to the surface (or perpendicular to the edge view of the surface, as shown in Figure 3.57).

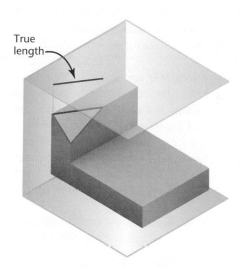

3.54 True-Length View of Line

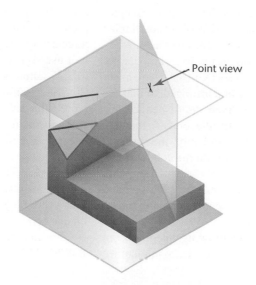

3.55 A Line Viewed as a Point

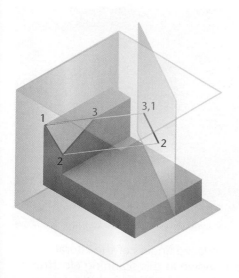

3.56 A Surface on Edge

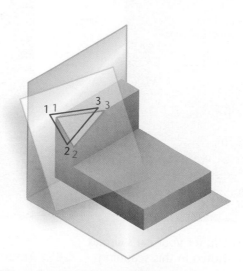

3.57 A Surface Shown True Size

MEMS VISUALIZATION: FROM 2D TO 3D AND BACK AGAIN

Designing microelectromechanical structures (MEMS) presents a visualization challenge. "MEMS use a technology similar to that used to create integrated circuits (IC) (computer chips)," explained Karen W. Markus, VP and Chief Technical Officer of Cronos Integrated Microsystems, Inc. "But MEMS use that manufacturing process to make machines—switches, relays, optical devices, sensors, accelerometers, and other devices that have moving parts—that are built at the scale of a computer chip."

Motion is at the heart of a MEMS device. For example, the airbag in your car is triggered by a MEMS sensor similar to the lateral resonator shown in Figure 3.58. The sensor is basically two sets of comb fingers—one set is fixed and one set is suspended so it can move. The fingers themselves are made of conductive material (polysilicon) and arranged so that the differential capacitance between the fixed and suspended sets can be measured. "When you slam on the brakes, the suspended set will move (through the force of inertia) and change the differential capacitance between the comb fingers from the neutral position," explained Markus. "The system logic looks at this change in capacitance, and if the size and duration of the shift is large enough, it sends a signal to inflate the airbag."

MEMS designs reflect both the growing integration of electrical and mechanical systems and the desire to miniaturize existing functions. Before MEMS, the airbag accelerometer was a ball-and-tube device. A conductive ball was positioned at one end of a metal tube. If the forces were large enough, the conductive ball would roll down the tube and close a switch at the other end, triggering the airbag. The MEMS device provides the same function but is smaller, weighs less, and offers more intelligence. For weight- and size-sensitive applications, such as satellite design, MEMS devices offer new possibilities—a satellite the size of a can of tennis balls, for example. Because they integrate electrical and mechanical properties into their designs, MEMS devices are also much more easily integrated into a system of logic components that will use signals from the MEMS sensor or send signals to a MEMS actuator device.

Like computer chips, MEMS are manufactured by a process that lays down successive layers of material—conductive and nonconductive—on a wafer. For each layer, a film of the material is applied to the entire surface of the wafer. Then, the desired shape of the structure is masked, and unwanted material is removed. The next layer is then applied. For each layer, the designer creates a 2D pattern for the material in that layer. This pattern is essentially a top view of the shapes to be formed on that layer.

For integrated circuits, the 2D representation is sufficient. But MEMS designers need to visualize how the material

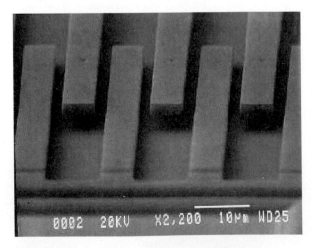

3.58 The MEMS sensor used to trigger your car's airbag is similar to the lateral resonator shown in this scanning electron micrograph. The larger squares are pads anchoring the fixed sets of fingers. The central shuttle is designed to flex and allow the shuttle fingers to move.

3.59 The interlaced fingers of the lateral resonator are shown in this scanning electron micrograph. The air space beneath the fingers was created the removal of a layer of oxide.

deposited in each layer will result in a 3D structure. Section views from the front or side of the chip are necessary to see how shapes on different layers come together to form a 3D device.

The manufacturing process starts with a layer of silicon nitride as an insulating material. Each layer of polysilicon, the conductive material, is preceded by a layer of silicon dioxide. The oxide layers are sacrificial—they will be washed away at the end of the process to create air pockets and gaps between the structures. An orthographic view is made for each layer to describe the structures formed by that layer. The outline for each layer of material is modeled on a separate layer, yet kept properly aligned on top of one another.

"In making the lateral comb resonator, the design goal was two sets of polysilicon fingers—onc fixed and one suspended set—separated by an air space across which the capacitance can be measured. The fixed set must be anchored firmly to the chip, while the suspended set must be designed to allow movement. The center structure of the device shown in Figure 3.58 is a shuttle of fingers with a folded beam suspension at either end of it. This suspension structure is attached to the substrate by anchor points, but the arrangement of the legs allows the shuttle to move when force is applied to it. Each comb structure has fixed (anchored) fingers and suspended fingers (attached to the shuttle) to provide the ability to sense and drive the resonator."

To create the resonator, both the shuttle and the fixed fingers need to be in the same plane, but an air space is needed below the shuttle so it can move (see Figure 3.59). The structures on this layer, however, also need to be attached by anchors. The section views in Figure 3.60 illustrate how the resonator is built up from the different layers. The first layer of polysilicon, Poly 0, forms pads that can be used as electrodes and can also be where the shuttle and fixed fingers attach. The plane of the fingers is formed by the second layer of polysilicon, Poly 1. The air gap under the plane of the fingers is formed by the oxide layer, which is removed after manufacture.

The challenge of designing MEMS devices is anticipating and visualizing the topographical effects of the manufacturing process and planning the layers accordingly. That is, each layer is a film that covers the entire surface of the wafer, and each

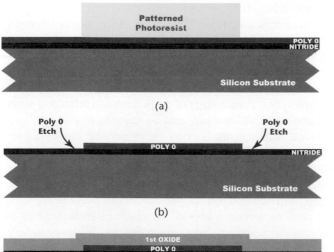

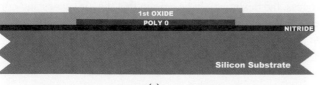

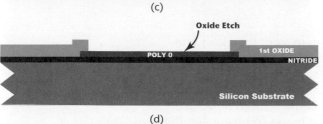

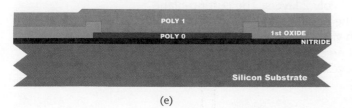

3.60 The basic elements of the MEMS surface micromachining manufacturing process are illustrated in this sequence of section views.

(a) A layer of nitride has been applied, followed by the first layer of polysilicon.

(b) The desired portions of the Poly 0 layer have been masked and the unwanted portions removed.

(c) The next layer is the first sacrificial oxide layer; it is applied to the surface of the chip and conforms to the shape of the structures on the previous layers.

(d) The oxide is masked and etched to expose Poly 0 so the next polysilicon layer will attach to it. The rest of the oxide remains to form a gap between the nitride and the structures in the next polysilicon layer.

(e) The next polysilicon layer is added. The masking and etching process is repeated for subsequent layers.

(continued)

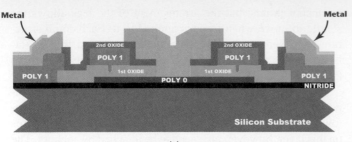

(a)

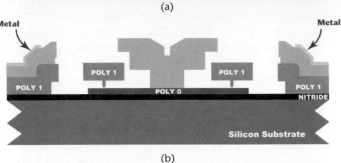

(b)

3.61 The section view in (a) shows the layers and how each conforms to the topography of the structures formed before it. In (b), the cross section shows the device after the oxide has been removed.

subsequent layer flows over whatever structures are in place on the wafer (see Figure 3.61). Each "flat" layer takes on the surface shape, or topography, of the wafer at that stage of the process. Successive section views are vital to visualizing the effect of each layer (see Figure 3.62).

Structures are formed by a process that removes unwanted material and protects the structures to be kept. For example, after the Poly 0 layer covers the wafer, the 2D shape of the pads that will remain on that layer is masked using ultraviolet light and a photosensitive polymer called *photoresist*. Any polysilicon outside the masked areas is then removed in an etching process. This creates islands of polysilicon, as shown in Figure 3.60(b). The next layer is oxide, which will form the air pockets. Notice in the section view in Figure 3.60(c) that this layer covers the entire chip and flows over and around the Poly 0 structures. Because the next layer of polysilicon must attach to the Poly 0 layer, the oxide is removed from the places that will serve as anchor points, as shown in Figure 3.60(d). The remaining oxide separates the next layer of polysilicon, Poly 1, from the substrate. The Poly 1 layer is then deposited and patterned, and excess material is removed to form the interlaced fingers.

To document a MEMS device, only top orthographic views made of basic geometric shapes are needed to convey the layout. In planning a MEMS device, however, section views and knowledge of the manufacturing process are needed to anticipate the geometry of the layers as they are applied. Multiple section views, at each stage of the process and reflecting different cutting-plane locations, may be required to visualize some devices. Although the section views may be hand sketches, consistent use of hatch patterns and careful selection of cutting-plane location are important. The hatch patterns for different layers are vital to differentiating between the oxide layers, which will be removed, and the conductive polysilicon layers. By creating section views of the effect of each stage of the process, the designer can be certain that the air spaces and connections needed for movable structures are drawn correctly in the plan view (see Figure 3.61).

3.62 The interplay of top views and section views allows the designer to visualize the structures formed by the layers and ensure that all design rules are met. The outlines in the bottom figure are layouts for the masks to be applied to the Poly 1 and Poly 2 layers. The cutting-plane lines on the outlines indicate the locations of section views a-a and b-b, which show the resulting structures up to that point of the process.

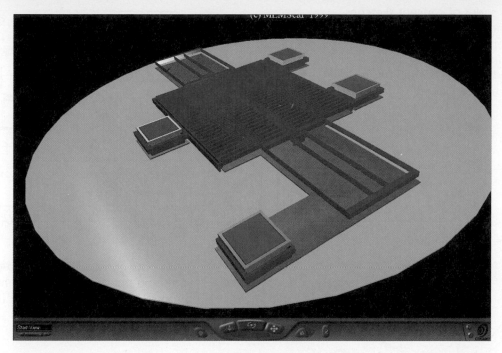

3.63 The lateral resonator layout was translated into a solid model of the MEMs device by Kanaga, a MEMs design environment from MEMScaP. The solid model is "fabricated" from the layout layer by layer so that structures retain the color of the layer from which they were formed. MEMScaP and other design environments use solid modeling tools or VRML to generate 3D representations of the process specified in the layouts for each layer. The model shown here is VRML. *(Courtesy of MEMScaP S.A.)*

Design tools for MEMS devices are being developed to address the hybrid nature of MEMS design. Layout editors used for integrated circuit design are being modified to incorporate the 3D visualization tools needed for MEMS design. At the same time, these tools aim to bridge the gap to the kinds of mechanical analysis needed for movable structures—for stress, fluid flow, thermal properties, and others, in addition to the electrical simulation commonly performed on IC chips. The model in Figure 3.63 was generated by a design environment for MEMS that includes 3D visualization and integrated analysis functions.

For a virtual tour of the MEMS fabrication process and more about MEMS design, visit the MEMS website at www.memsrus.com

(Figures 3.58 through 3.62 are used with the permission of Cronos Integrated Microsystems, Inc.)

3.64 In this series of concept sketches for an array holder, the shape of the attachment end of the holder appears most clearly in the side view. Three variations on this aspect of the design are sketched as side views.

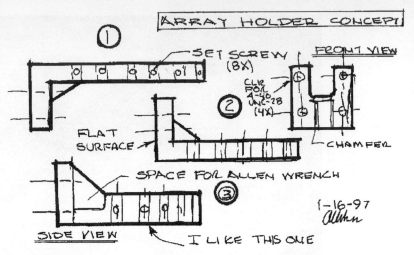

3.65 The construction lines to the vanishing points (labeled VP1 and VP2 in the sketch) serve as a guide for the 3D sketch of the array holder.

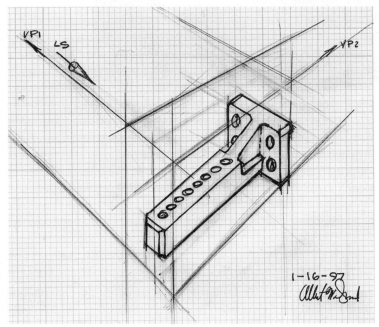

3.66 Notice the arrow in the top left of the final concept sketch that indicates the direction of the light source for the drawing. This arrow is an aid to visualizing the way light will hit the object and which surfaces will be in shadow.

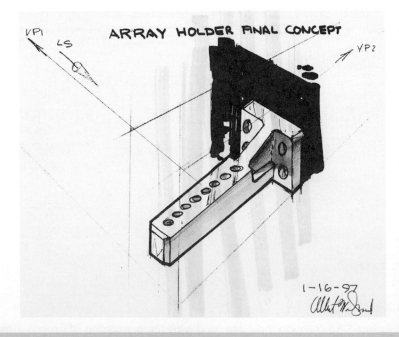

(Courtesy of Albert W. Brown, Jr., Affymax Research Institute.)

SolidWorks Tutorials

Starting from an Imported AutoCAD Drawing

In this tutorial, you will see an example of how files can be exchanged between modelers and will explore the similarities and differences between SolidWorks and AutoCAD.

- Launch SolidWorks from the icon on your desktop.
- Click to expand the topics listed below the "**?**" help icon.
- Click: **SolidWorks Tutorials**.
- Click: **All SolidWorks Tutorials (Set 1)** from the list.
- Click: **AutoCAD and SolidWorks** from the list that appears.

Part A

On your own, complete the **Importing an AutoCAD File** tutorial. A strength of CAD software is the ability to import and use drawings and models from other systems.

- *Caution:* Some steps may not match the software version perfectly. If they don't, use the tooltips and other navigation aids to help you. Try some things on your own. Try double-clicking outside the drawing border. Try double-clicking on drawing lines and circles. What happens, if anything?

Part B

Complete the **Converting an AutoCAD Drawing to 3D** tutorial. This can be handy when you have a shape already created in AutoCAD drawing format. The drawing shown is the one provided for you with the tutorials.

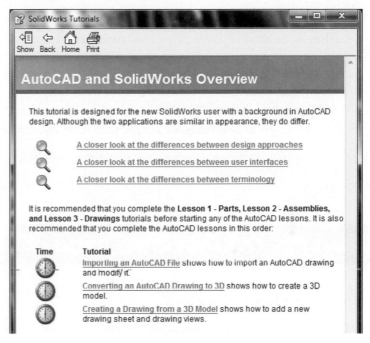

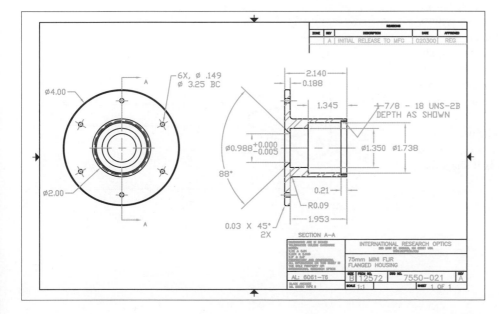

3.67 This is the AutoCAD drawing you will import into SolidWorks. How many errors can you spot?

(continued)

HANDS ON TUTORIALS

Part C

Complete the **Creating a Drawing from a 3D Model** tutorial. This will give you a feel for the interplay between 2D views and the 3D model.

Part D

Choose **A closer look at the differences between design approaches** in the window where you started parts A and B.

- Read through the comparison of SolidWorks and AutoCAD.
- Navigate to the AutoCAD product page at http://usa.autodesk .com/autocad/compare/. Look at the list of features for AutoCAD 2011 and earlier versions.

   Which features in SolidWorks are not found in AutoCAD, if any?

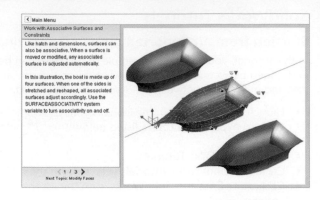

3.68 AutoCAD's Associative Surface Feature *(Autodesk screen shots reprinted with the permission of Autodesk, Inc.)*

REVERSE ENGINEERING PROJECT

The Can Opener Project

You should have the Amco Swing-A-Way 407 WH can opener for this project (see Chapter 2).

1. *Sketch a part.* Create a multiview sketch of the lower handle for the can opener similar to that shown here for the upper handle.
2. *Name the dimensions.* Add dimensions to your orthographic sketch of the lower handle. Do not worry about measurements for now. Give names to the dimensions, such as lower handle length, lower handle height, and hole diameter.
3. *Which are critical?* Make a list of the dimensions that must match dimensions on other parts for the can opener to function. Which dimensions will not be very important to the can opener's function?
4. *Form and function.* The lower handle has a slight bend. From examining your can opener, what is the function of this feature? Why is it shaped as it is?

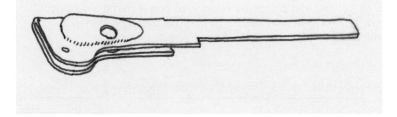

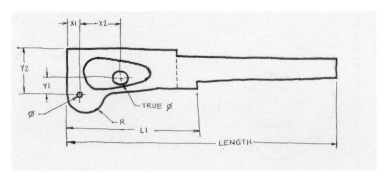

KEY WORDS

Adjacent Views

Auxiliary View

Centerlines

Cutting Plane

Dihedral Angle

Fillet

First-Angle Projection

Folding Lines

Hatching

Hidden Lines

Inclined Surface

Intersect

Line

Linetypes

Multiview Drawings

Necessary Views

Normal Surface

Oblique Surface

Orthographic Projection

Points

Principal Dimensions

Projection Line

Reference Surfaces

Regular Views

Related Views

Round

Scale

Tangent

Third-Angle Projection

Visible Lines

SKILLS SUMMARY

In this chapter you learned how to create multiview orthographic sketches and how to visualize the parts shown in an orthographic drawing. The exercises will give you practice in multiview sketching, using correct line patterns and making sure to show views in alignment. It is a serious error not to line up adjacent drawing views in projection. It is important to make your sketches neat and legible.

You will also practice visualizing a shape from a multiview drawing by creating pictorial sketches from multiview drawings. You will practice your visualization skills further by matching orthographic views with the correct pictorial view. Finally, you will practice determining and drawing necessary views and making section and auxiliary views.

EXERCISES

1. Create your own cartoon that shows a single view of some event in a way that makes it visually simple and comical.

2. Create a three-view orthographic projection sketch showing front, side, and top views of a computer monitor or television that uses a cathode-ray tube (CRT). Do not be too concerned with the control buttons, knobs, or cables. Use a scale of 1:4.

3. New developments in flat-screen video monitor technology permit redesign of computer screens so that depth from screen face to the rear of the case is about 4 inches. Sketch a three-view orthographic projection of this new configuration. Use a scale of 1:4.

4. Sketch a three-view orthographic projection of your telephone. Create an isometric sketch of the same phone. What advantages or disadvantages does each sketch have in communicating ideas?

5. Sketch a freehand three-view orthographic projection of the oarlock, the thread tap, and the phone receiver shown below. Scales are not provided; try to draw all features of each object to the same relative scale.

   a. Oarlock *(Courtesy of Clavey Paddlesports.)*

   b. Thread tap

   c. Phone receiver *(Shutterstock.)*

6. This CAD model is an isometric view of a tool stop. The model contains (or omits) some information that would usually be omitted (or present) in different drawing representations. Redraw the tool stop as a three-view orthographic projection drawing.

   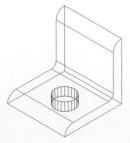

Exercises 7–12 Multiview Sketching Problems. Sketch the orthographic projection views necessary to fully define the object. How many views are needed in each case? Assume that each grid square equals either 1/2 inch or 10 mm.

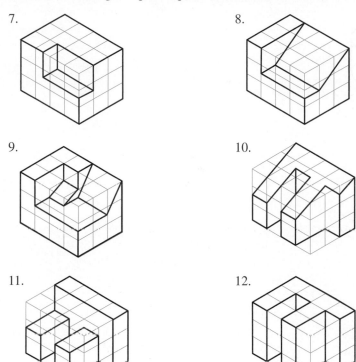

7.

8.

9.

10.

11.

12.

Exercises 13–18 Multiview Sketching Problems. Sketch the orthographic projection views necessary to fully define the object. Assume that each grid square equals either 1/4 inch or 5 mm.

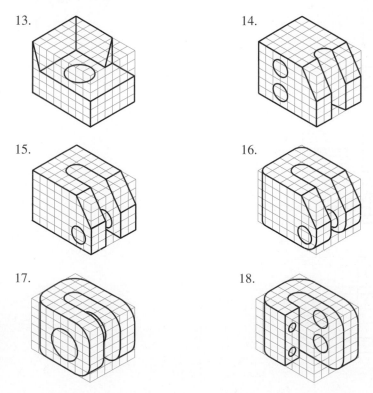

13.

14.

15.

16.

17.

18.

Exercises 19–24 Multiview Sketching Problems. Sketch the orthographic projection views necessary to fully define the object. Assume that each grid square equals either 1/4 inch or 5 mm.

19.
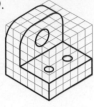

20.
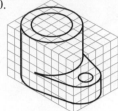

21.
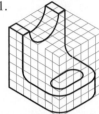

22.
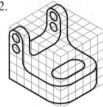

23.
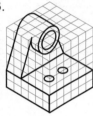

24.
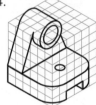

Exercises 25–34 Isometric Sketching. Given the following orthographic views, create an isometric sketch of each object. Assume that each grid square equals either 1/2 inch or 10 mm.

25.
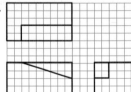

26.
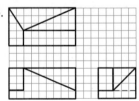

27.
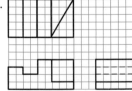

28.
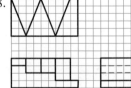

29.
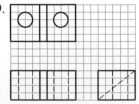

30.
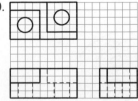

31.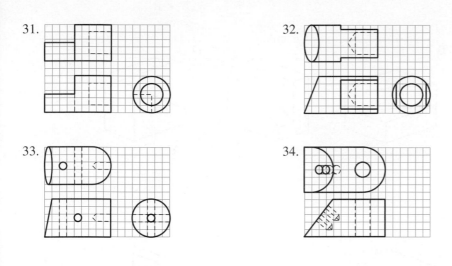

32.

33.

34.

Exercises 35–46 Match each set of orthographic views with the pictorial view that it could represent.

35.

A   B   C   D

36.

A   B   C   D

37.

A   B   C   D

38.

A   B   C   D

39.

A   B   C   D

40.

A   B   C   D

41.

A   B   C   D

42.

A   B   C   D

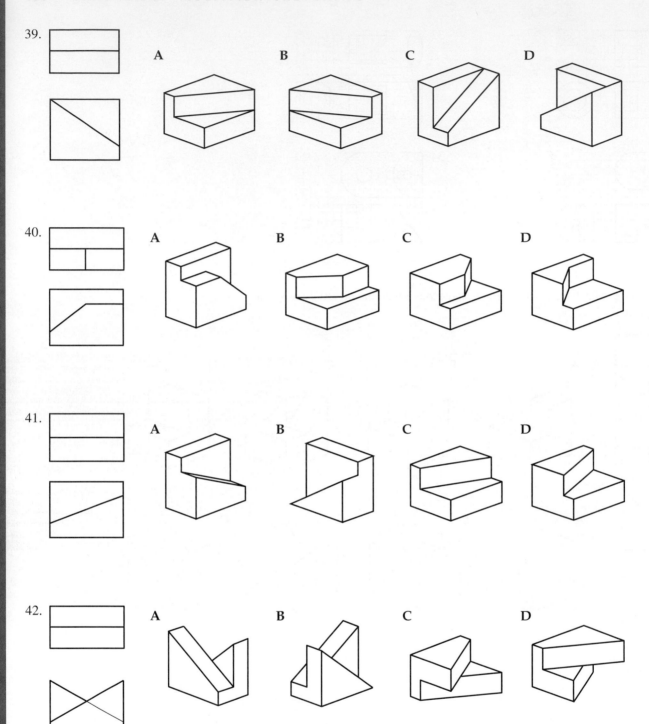

43.

A   B   C   D

44.

A   B   C   D

45.

A   B   C   D

46.

A   B   C   D

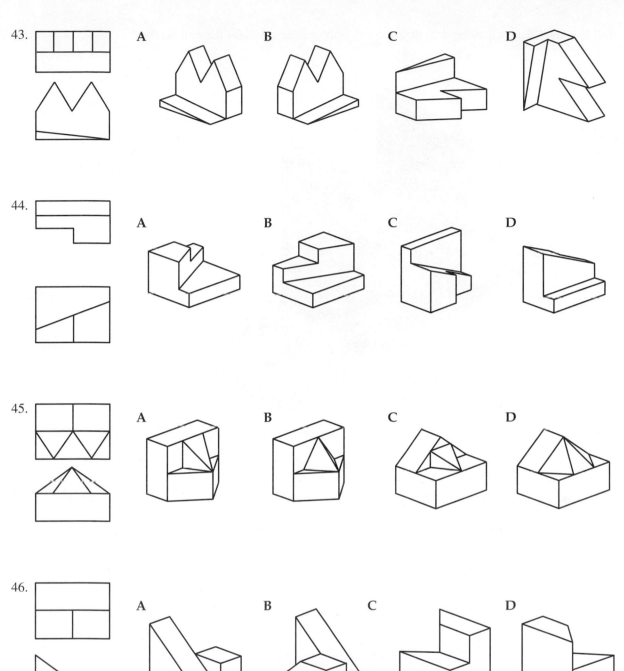

Exercise 47 Wall Bracket. Create a drawing with the necessary orthographic views for the wall bracket.

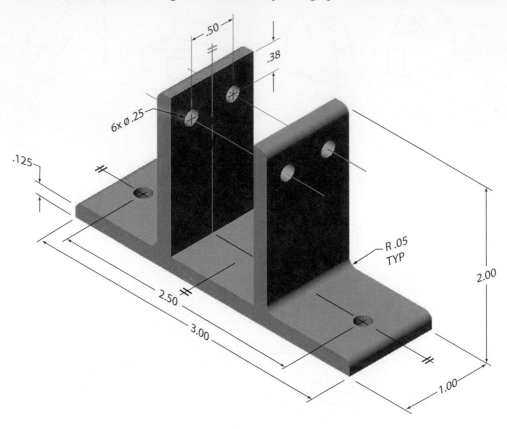

Exercise 48 Sheet Metal Bracket. Create a drawing of the necessary orthographic views for the sheet metal bracket.

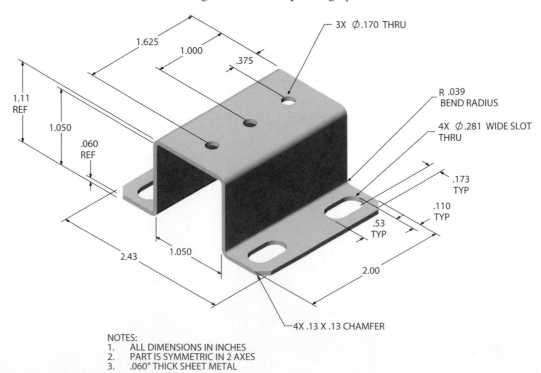

NOTES:
1. ALL DIMENSIONS IN INCHES
2. PART IS SYMMETRIC IN 2 AXES
3. .060" THICK SHEET METAL

Exercise 49 Electronics Mount. Create a drawing with the necessary orthographic views for the sheet metal electronics mount.

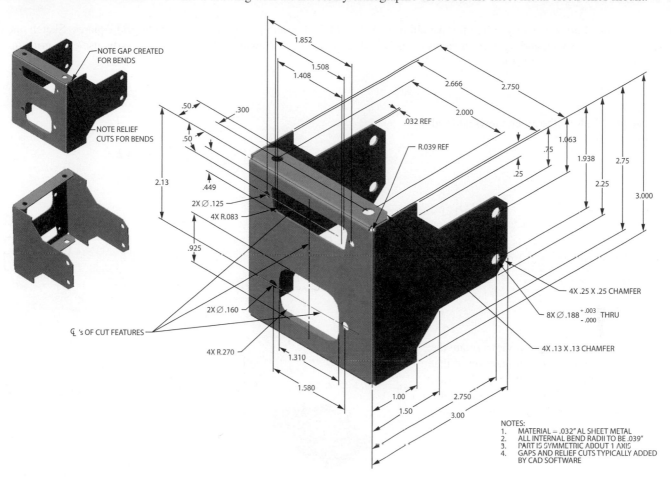

NOTE GAP CREATED FOR BENDS

NOTE RELIEF CUTS FOR BENDS

.032 REF

R.039 REF

2X Ø.125
4X R.083
2X Ø.160
4X R.270

℄ 's OF CUT FEATURES

4X .25 X .25 CHAMFER

8X Ø.188 +.003 -.000 THRU

4X .13 X .13 CHAMFER

1.852
1.508
1.408
2.666
2.750
2.000
1.063
.75
1.938
.25
2.25
2.75
3.000

.50
.300
.50
2.13
.449
.925
1.310
1.580
1.00
1.50
2.750
3.00

NOTES:
1.   MATERIAL = .032" AL SHEET METAL
2.   ALL INTERNAL BEND RADII TO BE .039"
3.   PART IS SYMMETRIC ABOUT 1 AXIS
4.   GAPS AND RELIEF CUTS TYPICALLY ADDED BY CAD SOFTWARE

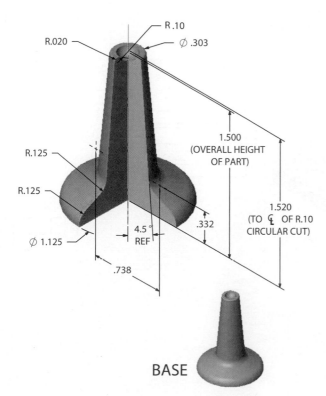

R.10
Ø.303
R.020
R.125
R.125
Ø 1.125
4.5° REF
.332
.738
1.500 (OVERALL HEIGHT OF PART)
1.520 (TO ℄ OF R.10 CIRCULAR CUT)

BASE

Exercise 50 Gyroscope Base. Create a drawing with the necessary orthographic views.

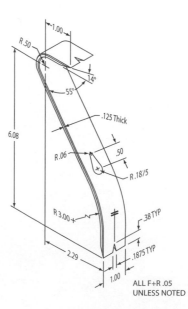

1.00
R.50
14°
55°
.125 Thick
6.08
R.06
.50
R.18/5
R 3.00
.38 TYP
2.29
.1875 TYP
1.00
ALL F+R .05 UNLESS NOTED

Exercise 51 Pry Bar. Create a drawing with the necessary orthographic views for the pry bar.

FREEHAND SECTIONING PROBLEMS

Exercises 52–54 are especially suited for sketching on 8.5 × 11-inch graph paper with appropriate grid squares. Sketch one or two problems per sheet, adding section views as indicated. To make your drawings fit on the paper easily, use each grid square as equal to either 5 mm or 1/4 inch.

Exercise 52 Redraw the given views and add the front section view.

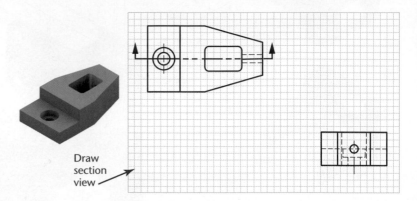

Exercise 53 Redraw the top view, rotate the side view, and move it into a position so that you can project the front view in section. Add the front section view. Each grid square equals 6 mm or 1/4 inch. Rotate the side view into position.

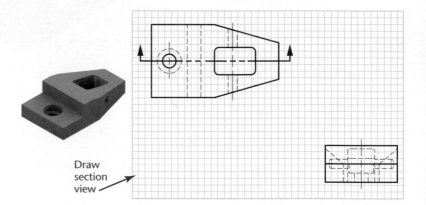

Exercise 54 Use the same directions as for Exercise 53.

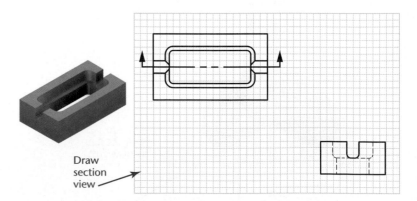

AUXILIARY VIEW EXERCISES

Draw the orthographic views of the part shown. Add an auxiliary view showing the inclined or oblique surface true size. Let each mark equal 5 mm or 1/4 inch.

Exercise 55

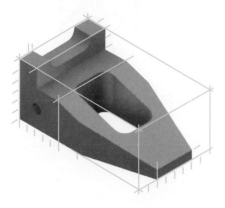

Exercise 56

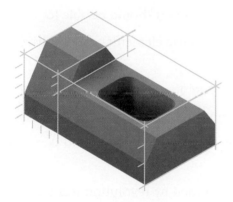

Exercise 57

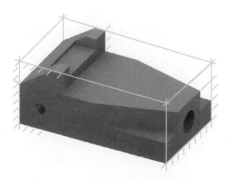

4 USING GEOMETRY FOR MODELING AND DESIGN

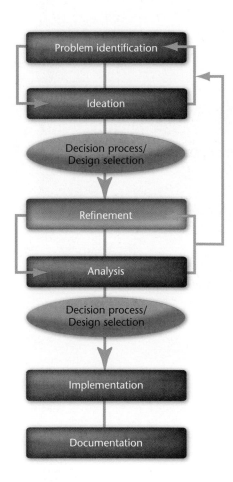

OBJECTIVES

When you have completed this chapter you should be able to:

1. Identify and specify basic geometric elements and primitive shapes.

2. Select a 2D profile that best describes the shape of an object.

3. Identify mirrored shapes and sketch their lines of symmetry.

4. Identify shapes that can be formed by extrusion and sketch their cross sections.

5. Identify shapes that can be formed by revolution techniques and sketch their profiles.

6. Define Boolean operations.

7. Specify the Boolean operations to combine primitive shapes into a complex shape.

8. Work with Cartesian coordinates and user coordinate systems in a CAD system.

9. Identify the transformations common to CAD systems.

Many different geometric shapes were used to model this jetboard. The wireframe view of the top cover reveals several regular geometric shapes used to model the interior components. The graceful lines of the outer hull are defined by the irregular curves used to model it. *(Courtesy of Leo Greene, www.e-Cognition.net.)*

DRAWING GEOMETRY

Engineering drawings, whether sketched or created using CAD, combine basic geometric shapes and relationships to define complex objects. 2D drawings are composed of simple entities such as points, lines, arcs, and circles, as well as more complex entities such as ellipses and curves. Reviewing the basic geometry of these elements helps you define and combine these elements in your drawings and CAD models.

When you draw by hand, you don't need to "tell" your pencil the beginning and ending points that will make up the line. The freehand line is an approximation of the geometric definition of a straight line. Lines drawn using a CAD system are highly accurate definitions—much greater than you can see on a computer monitor. Understanding the geometric properties of these entities is a first step in creating useful CAD models.

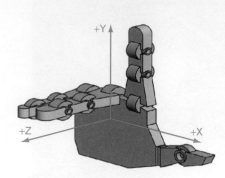

4.1   Right-Hand Rule

4.2   In systems that use the right-hand rule, the positive Z-axis points toward you when the face of the monitor is parallel to the X-Y plane.

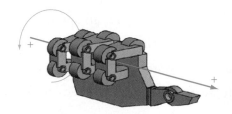

4.3   The curl of the fingers indicates the positive direction along the axis of rotation.

4.1 COORDINATES FOR 3D CAD MODELING

2D and 3D CAD drawing entities are stored in relationship to a Cartesian coordinate system. No matter what CAD software system you will be using, it is helpful to understand some basic similarities of coordinate systems.

Most CAD systems use the ***right-hand rule*** for coordinate systems; if you point the thumb of your right hand in the positive direction for the X-axis and your index finger in the positive direction for the Y-axis, your remaining fingers will curl in the positive direction for the Z-axis (shown in Figure 4.1). When the face of your monitor is the X-Y plane, the Z-axis is pointing toward you (see Figure 4.2).

The right-hand rule is also used to determine the direction of rotation. For rotation using the right-hand rule, point your thumb in the positive direction along the axis of rotation. Your fingers will curl in the positive direction for the rotation, as shown in Figure 4.3.

Though rare, some CAD systems use a left-hand rule. In this case, the curl of the fingers on your *left* hand gives you the positive direction for the Z-axis. In this case, when the face of your computer monitor is the X-Y plane, the positive direction for the Z-axis extends into your computer monitor, not toward you.

A 2D CAD system uses only the X- and Y-coordinates of the Cartesian coordinate system. 3D CAD systems use X, Y, and Z. To represent 2D in a 3D CAD system, the view is straight down the Z-axis. Figure 4.4 shows a drawing created using only the X- and Y- values, leaving the Z-coordinates set to 0, to produce a 2D drawing. Recall that each orthographic view shows only two of the three coordinate directions because the view is straight down one axis. 2D CAD drawings are the same: They show only the X- and Y-coordinates because you are looking straight down the Z-axis.

When the X-Y plane is aligned with the screen in a CAD system, the Z-axis is oriented horizontally. In machining and many other applications, the Z-axis is considered to be the vertical axis. In all cases, the coordinate axes are mutually perpendicular and oriented according to the right-hand or left-hand rule. Because the view can be rotated to be straight down any axis or any other direction, understanding how to use coordinates in the model is more important than visualizing the direction of the default axes and planes.

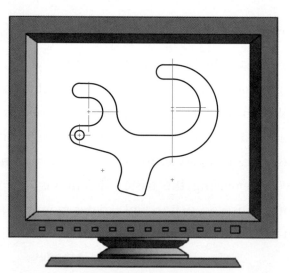

4.4   This drawing was created on the X-Y plane in the CAD system. It appears true shape because the viewing direction is perpendicular to the X-Y plane—straight down the Z-axis.

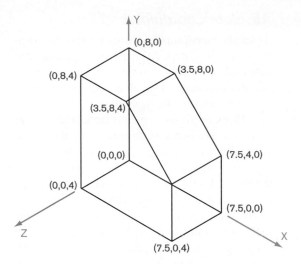

4.5 3D coordinates locate vertices in the CAD system.

The vertices of the 3D block shown in Figure 4.5 are identified by their coordinates. Often, it is useful when modeling parts to locate the origin of the coordinate system at the lower left of the part, as shown in Figure 4.5. This location for the (0,0,0) point on a part is useful when the part is being machined, as it then makes all coordinates on the part positive. Some older numerically controlled machinery will not interpret a file correctly if it has negative lengths or coordinates. CAD models are often exported to other systems for manufacturing parts, so try to create them in a common and useful way.

Specifying Location

Even though the model is ultimately stored in a single Cartesian coordinate system, you may usually specify the location of features using other location methods as well. The most typical of these are relative, polar, cylindrical, and spherical coordinates. These coordinate formats are useful for specifying locations to define your CAD drawing geometry.

SPOTLIGHT

The First Coordinate System

René Descartes (1596–1650) was the French philosopher and mathematician for whom the Cartesian coordinate system is named. Descartes linked algebra and geometry to classify curves by the equations that describe them. His coordinate system remains the most commonly used coordinate system today for identifying points. A 2D coordinate system consists of a pair of lines, called the X- and Y-axes, drawn on a plane so that they intersect at right angles. The point of intersection is called the *origin*. A 3D coordinate system adds a third axis, referred to as the Z-axis, that is perpendicular to the two other axes. Each point in space can be described by numbers, called *coordinates*, that represent its distance from this set of axes. The Cartesian coordinate system made it possible to represent geometric entities by numerical and algebraic expressions. For example, a straight line is represented by a linear equation in the form $ax + by + c = 0$, where the x- and y-variables represent the X- and Y-coordinates for each point on the line. Descartes' work laid the foundation for the problem-solving methods of analytic geometry and was the first significant advance in geometry since those of the ancient Greeks.

4.6   Absolute coordinates define a location in terms of distance from the origin (0,0,0), shown here as a star. These directions are useful because they do not change unless the origin changes.

4.7   Relative coordinates describe the location in terms of distance from a starting point. Relative coordinates to the same location differ according to the starting location.

4.8   Polar coordinates describe the location using an angle and distance from the origin (absolute) or starting point (relative).

Absolute Coordinates

Absolute coordinates are used to store the locations of points in a CAD database. These coordinates specify location in terms of distance from the origin in each of the three axis directions of the Cartesian coordinate system.

Think of giving someone directions to your house (or to a house in an area where the streets are laid out in rectangular blocks). One way to describe how to get to your house would be to tell the person how many blocks over and how many blocks up it is from two main streets (and how many floors up in the building, for 3D). The two main streets are like the X- and Y-axes of the Cartesian coordinate system, with the intersection as the origin. Figure 4.6 shows how you might locate a house with this type of absolute coordinate system.

Relative Coordinates

Instead of having to specify each location from the origin, you can use *relative coordinates* to specify a location by giving the number of units from a previous location. In other words, the location is defined relative to your previous location.

To understand relative coordinates, think about giving someone directions from his or her current position, not from two main streets. Figure 4.7 shows the same map again, but this time with the location of the house relative to the location of the person receiving directions.

Polar Coordinates

Polar coordinates are used to locate an object by giving an angle (from the X-axis) and a distance. Polar coordinates can either be absolute, giving the angle and distance from the origin, or relative, giving the angle and distance from the current location.

Picture the same situation of having to give directions. You could tell the person to walk at a specified angle from the crossing of the two main streets, and how far to walk. Figure 4.8 shows the angle and direction for the shortcut across the empty lot using absolute polar coordinates. You could also give directions as an angle and distance relative to a starting point.

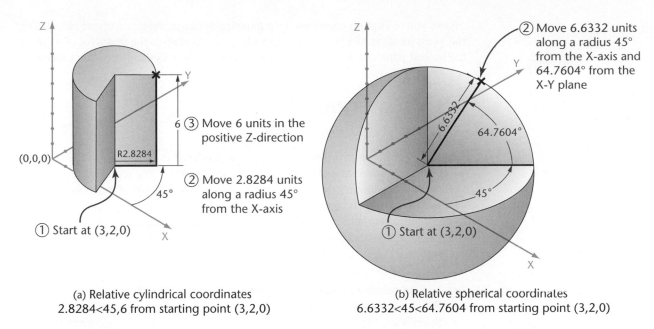

(a) Relative cylindrical coordinates
2.8284<45,6 from starting point (3,2,0)

(b) Relative spherical coordinates
6.6332<45<64.7604 from starting point (3,2,0)

4.9   The target points in (a) and (b) are described by relative coordinates from the starting point (3,2,0). Although the paths to the point differ, the resulting endpoint is the same.

Cylindrical and Spherical Coordinates

Cylindrical and spherical coordinates are similar to polar coordinates except that a 3D location is specified instead of one on a single flat plane (such as a map).

Cylindrical coordinates specify a 3D location based on a radius, angle, and distance (usually in the Z-axis direction). This gives a location as though it were on the edge of a cylinder. The radius tells how far the point is from the center (or origin); the angle is the angle from the X-axis along which the point is located; and the distance provides the height where the point is located on the cylinder. Cylindrical coordinates are similar to polar coordinates, but they add distance in the Z-direction. Figure 4.9(a) depicts cylindrical coordinates used to specify a location.

Spherical coordinates specify a 3D location by the radius, an angle from the X-axis, and the angle from the X-Y plane. These coordinates locate a point on a sphere, where the origin of the coordinate system is at the center of the sphere. The radius gives the size of the sphere; the angle from the X-axis locates a place on the equator. The second angle gives the location from the plane of the equator to a point on the sphere in line with the location specified on the equator.

Even though you may use these different systems to enter information into your 3D drawings, the end result is stored using one set of Cartesian coordinates.

Using Existing Geometry to Specify Location

Most CAD packages offer a means of specifying location by specifying a the relationship of a point to existing objects in the model or drawing. For example, AutoCAD's "object snap" feature lets you enter a location by "snapping" to the endpoint of a line, the center of a circle, the intersection of two lines, and so on (Figure 4.10). Using existing geometry to locate new entities is faster than entering coordinates. This feature also

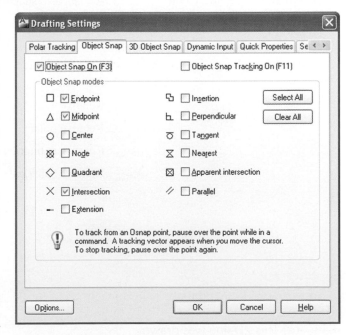

4.10   Object snaps are aids for selecting locations on existing CAD drawing geometry. *(Autodesk screen shots reprinted with the permission of Autodesk, Inc.)*

allows you to capture geometric relationships between objects without calculating the exact location of a point. For example, you can snap to the midpoint of a line or the nearest point of tangency on a circle. The software calculates the exact location.

4.2 GEOMETRIC ENTITIES

Points

Points are geometric constructs. Points are considered to have no width, height, or depth. They are used to indicate locations in space. When you represent a point in a sketch, the convention is to make a small cross, or a bar if it is along a line, to indicate the location of the point (see Figure 4.11). In CAD drawings, a point is located by its coordinates and usually shown with some sort of marker like a cross, circle, or other representation. Many CAD systems allow you to choose the style and size of the mark that is used to represent points.

Most CAD systems offer three ways to specify a point:

- Type in the coordinates (of any kind) for the point.
- Pick a point from the screen with a pointing device (mouse or tablet).
- Specify the location of a point by its relationship to existing geometry (e.g., an endpoint of a line, an intersection of two lines, or a center point; see Figure 4.12).

Picking a point from the screen is a quick way to enter points when the exact location is not important, but the accuracy of the CAD database makes it impossible to enter a location accurately in this way.

Lines

A straight line is defined as the shortest distance between two points. Geometrically, a line has length but no other dimension such as width or thickness. Lines are used in drawings to represent the edge view of a surface, the limiting element of a contoured surface, or the edge formed where two surfaces on an object join. In a CAD database, lines are typically stored by the coordinates of their endpoints.

For the lines shown in Figure 4.13 the table shows how you can specify the second endpoint for a particular type of coordinate entry. (For either or both endpoints, you can also snap to existing geometry without entering any coordinates.)

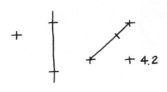

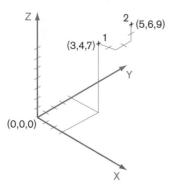

4.11 The crosses and bars represent points as they would appear in a sketch.

4.12 Point 1 was added to the drawing by typing the absolute coordinates 3,4,7. Point 2 was added relative to Point 1 with the relative coordinates @2,2,2.

	(a) Second Endpoint for 2D Line	**(b) Second Endpoint for 3D Line**
Absolute	6,6	5,4,6
Relative	@3,4	@2,2,6
Relative polar	@5<53.13	n/a
Relative cylindrical	n/a	@2.8284<45,6
Relative spherical	n/a	@6.6332<45<64.7606

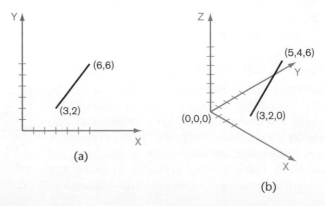

4.13 (a) This 2D line was drawn from endpoint (3, 2) to (6, 6). (b) This 3D line was drawn from endpoint (3, 2, 0) to (5, 4, 6).

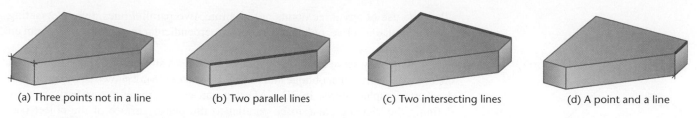

(a) Three points not in a line　(b) Two parallel lines　(c) Two intersecting lines　(d) A point and a line

4.14　The highlighted entities in each image define a plane.

Planes

Planes are defined by any of the following (see Figure 4.14):

- three points not lying in a straight line
- two parallel lines
- two intersecting lines
- a point and a line

The last three ways to define a plane are all special cases of the more general case—three points not in a straight line. Knowing what can determine a plane can help you understand the geometry of solid objects and use the geometry as you model in CAD.

For example, a face on an object is a plane that extends between the vertices and edges of the surface. Most CAD programs allow you to align new entities with an existing plane. You can use any face on the object—whether it is normal, inclined, or oblique—to define a plane for aligning a new entity. The plane can be specified using the geometry of the object as defined in the preceding list

Defining planes on the object or in 3D space is an important skill for working in 3D CAD. CAD software provides tools for defining new planes (see Figure 4.15). The options for these tools are based on the geometry of planes. Typical choices

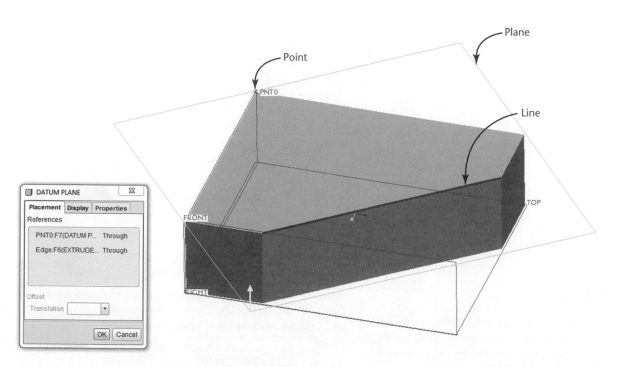

4.15　A point and a line (the edge between two surfaces in this case) were used to define a plane in this Pro/ENGINEER model.

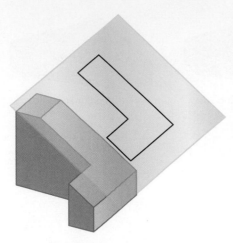

4.16 The viewing plane is aligned with the inclined surface and shows it true size.

allow the use of any three points not in a line, two parallel lines, two intersecting lines, a point and a line, or being parallel to, perpendicular to, or at an angle from an existing plane.

A plane may serve as an orientation that shows a surface true shape. In the last chapter you learned that oblique and inclined surfaces are not shown true size in standard orthographic views. In 3D CAD you can easily create an auxiliary view by defining the viewing plane to be parallel to the planar surface of the object (see Figure 4.16).

It is also handy to define a plane where features are shown true shape when you are modeling those features. You will learn more about orienting work planes to take advantage of the object's geometry later in this chapter.

Circles

A circle is a set of points that are equidistant from a center point. The distance from the center to one of the points is the *radius*. The distance across the center to any two points on opposite sides is the *diameter*. The circumference of a circle contains 360° of arc. In a CAD file, a circle is often stored as a center point and radius.

Most CAD systems allow you to define circles by specifying any one of the following:

- the center and a diameter
- the center and a radius
- two points on the diameter
- three points on the circle
- a radius and two entities to which the circle is tangent
- three entities to which the circle is tangent.

These methods are illustrated in Figure 4.17.

4.17 Ways to Define a Circle

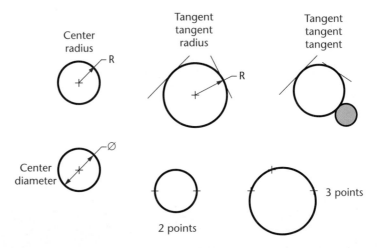

Circles in SolidWorks

The options for defining a circle in SolidWorks are to define the center and a point on the perimeter, or to select three points on the perimeter. These seem to be surprisingly few choices, but all the other ways to define a circle are possible through the use of geometric and dimensional constraints. For example, you can add three constraints requiring the circle to be tangent to entities in your drawing. This will fully define the circle, so specifying a dimension for its radius or diameter is unnecessary.

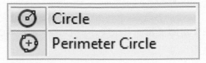

Arcs

An arc is a portion of a circle. An arc can be defined by specifying any one of the following (see Figure 4.18):

- a center, radius, and angle measure (sometimes called the *included angle* or *delta angle*)
- a center, radius, and chord length
- a center, radius, and arc length
- the endpoints and a radius
- the endpoints and a chord length
- the endpoints and arc length
- the endpoints and one other point on the arc (3 points)

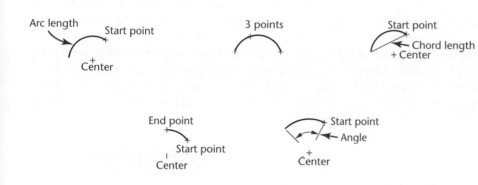

4.18 Arcs can be defined many different ways. Like circles, arcs may be located from a center point or an endpoint, making it easy to locate them relative to other entities in the model.

Formulas for Circles and Arcs

r = radius
C = circumference
π = pi $\cong 3.14159$
a = arc length
A = area
L = chord length
θ (theta) = included angle
rad (radian) = the included angle of an arc length such that the arc length is equal to the radius
$C = 2\pi r$, the curved distance around a circle
$A = \pi r^2$, the area of a circle
$a = 2\pi r \times \theta/360$, so the arc length = $0.01745r\theta$ when you know its radius, r, and the included angle, θ, in degrees
$a = r \times \theta$ (when the included angle is measured in radians)

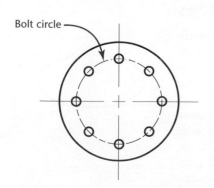

Bolt-Hole Circle Chord Lengths

To determine the distance between centers for equally spaced holes on a bolt-hole circle:

n = 180 / number of holes in pattern
L = sin n × bolt-hole circle diameter
Example: 8-hole pattern on a 10.00-diameter circle:
180/8 = 22.5
sin of 22.5 is .383
.383 × 10 = 3.83 (the chord)

For more useful formulas, see Appendix 49.

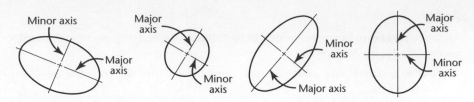

4.19   Major and Minor Axes of Some Ellipses

Ellipses

An ellipse can be defined by its major and minor axis distances. The major axis is the longer axis of the ellipse; the minor axis is the shorter axis. Some ellipses are shown and labeled in Figure 4.19.

An ellipse is created by a point moving along a path where the sum of its distances from two points, each called a ***focus of an ellipse*** (*foci* is the plural form), is equal to the major diameter. As an aid in understanding the shape of an ellipse, imagine pinning the ends of a string in the locations of the foci, then sliding a pencil along inside the string, keeping it tightly stretched, as in Figure 4.20. You would not use this technique when sketching (you would just block in the major and minor axes and the enclosing parallelogram, then sketch the ellipse tangent to it), but it serves as a good illustration of the definition of an ellipse.

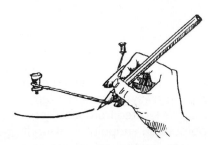

4.20   When an ellipse is created with the pencil-and-string method, the length of the string between the foci is equal to the length of the major axis of the ellipse. Any point that can be reached by a pencil inside the string when it is pulled taut meets the condition that its distances from the two foci sum to the length of the major diameter.

SPOTLIGHT

Locating the Foci of an Ellipse

To locate the foci of an ellipse, draw arcs with their centers at the ends of the minor axis and their radii equal to half the major axis. The intersection of each pair of arcs is a focus of the ellipse. Most CAD packages allow you to specify an ellipse by specifying either the major and minor axes or the foci.

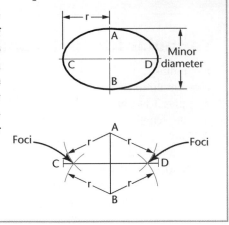

SPOTLIGHT

The Perimeter of an Ellipse

The perimeter of an ellipse is a set of points defined by their distance from the two foci. The sum of the distances from any point on the ellipse to the two foci must be equal to the length of the major diameter. The perimeter of an ellipse may be approximated in different ways. Many CAD packages use infinite series to most closely approximate the perimeter. The mathematical relationship of each point on the ellipse to the major and minor axes may be seen in the following approximation:

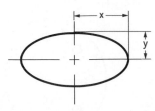

$$P = 2\pi \sqrt{\frac{x^2 + y^2}{2}}$$

Spline Curves

Splines are used to describe complex, or *freeform*, curves. Many surfaces cannot be easily defined using simple curves such as circles, arcs, or ellipses. For example, the flowing curves used in automobile design blend many different curves into a smooth surface (Figure 4.21). Creating lifelike shapes and aerodynamic forms may require spline curves.

The word *spline* originally described a flexible piece of plastic or rubber used to draw irregular curves between points. Mathematical methods generate the points on the curve for CAD applications.

4.21 Spline curves were used for the front end of the Jaguar's body to create shapes that cannot be defined by regular curves such as arcs or ellipses. *(Courtesy of ICEM Technologies.)*

One way to create an irregular curve is to draw curves between each set of points. The points and the tangencies at each point are used in a polynomial equation that determines the shape of the curve. This type of curve is useful in the design of a ship's hull or an aircraft wing. Because this kind of irregular curve passes through all the points used to define the curve, it is sometimes called an *interpolated spline* or a *cubic spline*. An example and its vertices are shown in Figure 4.22.

Other spline curves are approximated: they are defined by a set of vertices. The resulting curve does not pass through all the vertices. Instead, the vertices "pull" the curve in the direction of the vertex. Complex curves can be created with relatively few vertices using approximation methods. Figure 4.23 shows a 3D approximated spline curve and its vertices.

The mathematical definition for this type of spline curve uses the X- and Y- (and Z- for a 3D shape) coordinates and a parameter, generally referred to as u. A polynomial equation is used to generate functions in u for each point used to specify the curve. The resulting functions are then blended to generate a curve that is influenced by each point specified but not necessarily coincident with any of them.

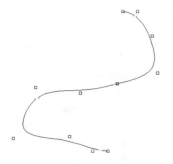

4.23 Except for the beginning and endpoints, the fit points for the spline curve stored in the database do not always lie on the curve. They are used to derive the curve mathematically.

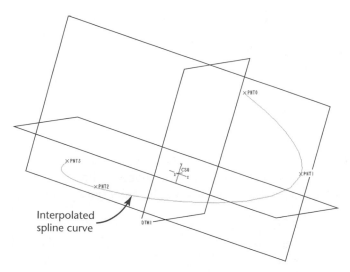

4.22 An interpolated spline curve passes through all the points used to define the curve.

B-Splines

The **Bezier curve** was one of the first methods to use spline approximation to create flowing curves in CAD applications. The first and last vertices are on the curve, but the rest of the vertices contribute to a blended curve between them. The Bezier method uses a polynomial curve to approximate the shape of a polygon formed by the specified vertices. The order of the polynomial is 1 degree less than the number of vertices in the polygon (see Figure 4.24).

The Bezier method is named for Pierre Bezier, a pioneer in computer-generated surface modeling at Renault, the French automobile manufacturer. Bezier sought an easier way of controlling complex curves, such as those defined in automobile surfaces. His technique allowed designers to shape natural-looking curves more easily than they could by specifying points that had to lie on the resulting curve, yet the technique also provided control over the shape of the curve. Changing the slope of each line segment defined by a set of vertices adjusts the slope of the resulting curve (see Figure 4.25). One disadvantage of the Bezier formula is that the polynomial curve is defined by the combined influence of every vertex: a change to any vertex redraws the entire curve between the start point and endpoint.

A **B-spline** approximation is a special case of the Bezier curve that is more commonly used in engineering to give the designer more control when editing the curve. A B-spline is a blended piecewise polynomial curve passing near a set of control points. The spline is referred to as **piecewise** because the blending functions used to combine the polynomial curves can vary over the different segments of the curve. Thus, when a control point changes, only the piece of the curve defined by the new point and the vertices near it change, not the whole curve (see Figure 4.26). B-splines may or may not pass through the first and last points in the vertex set. Another difference is that for the B-spline the order of the polynomial can be set independently of the number of vertices or control points defining the curve.

In addition to being able to locally modify the curve, many modelers allow sets of vertices to be weighted differently. The weighting, sometimes called *tolerance*, determines how closely the curve should fit the set of vertices. Curves can range from fitting all the points to being loosely controlled by the vertices. This type of curve is called a nonuniform rational B-spline, or **NURBS** curve. A rational curve (or surface) is one that has a weight associated with each control point.

4.24   A Bezier curve passes through the first and last vertex but uses the other vertices as control points to generate a blended curve.

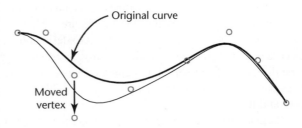

4.25   Every vertex contributes to the shape of a Bezier curve. Changing the location of a single vertex redraws the entire curve.

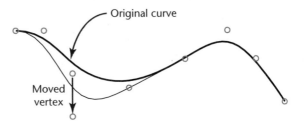

4.26   The B-spline is constructed piecewise, so changing a vertex affects the shape of the curve near only that vertex and its neighbors.

Splines are drawn in CAD systems based on the mathematical relationships defining their geometry. Figure 4.27 shows an approximated spline drawn using AutoCAD. Figure 4.28 shows an interpolated spline drawn using SolidWorks. Both curves are drawn with a spline command, and both provide a dialog box that allows you to change properties defining the curve; however, the properties that are controlled vary by the type of spline being created by the software package. You should be familiar with the terms used by your modeling software for creating different types of spline curves.

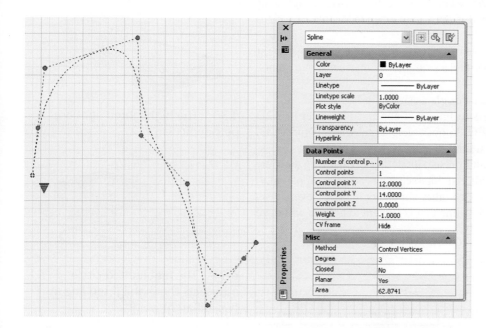

4.27 This spline drawn in AutoCAD is pulled toward the defined control points. The **Properties** dialog box at the right allows you to change the weighting factor for each control point. *(Autodesk screen shots reprinted with the permission of Autodesk, Inc.)*

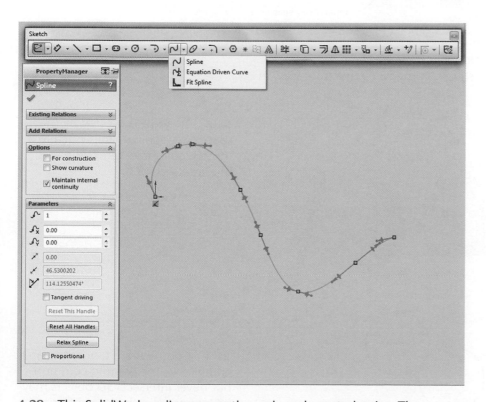

4.28 This SolidWorks spline passes through each control point. The **Property Manager** allows you to control spline properties. *(Image courtesy of DS SolidWorks Corp.)*

4.3 GEOMETRIC RELATIONSHIPS

When you are sketching, you often imply a relationship, such as being parallel or perpendicular, by the appearance of the lines or through notes or dimensions. When you are creating a CAD model you use drawing aids to specify these relationships between geometric entities.

Two lines or planes are *parallel* when they are an equal distance apart at every point. Parallel entities never intersect, even if extended to infinity. Figure 4.29 shows an example of parallel lines.

Two lines or planes are *perpendicular* when they intersect at right angles (or when the intersection that would be formed if they were extended would be a right angle), as in Figure 4.30.

Two entities *intersect* if they have at least one point in common. Two straight lines intersect at only a single point. A circle and a straight line intersect at two points, as shown in Figure 4.31.

When two lines intersect, they define an *angle* as shown in Figure 4.32.

The term *apparent intersection* refers to lines that appear to intersect in a 2D view or on a computer monitor but actually do not touch, as shown in Figure 4.33. When you look at a wireframe view of a model, the 2D view may show lines crossing each other when, in fact, the lines do not intersect in 3D space. Changing the view of the model can help you determine whether an intersection is actual or apparent.

Two entities are *tangent* if they touch each other but do not intersect, even if extended to infinity, as shown in Figure 4.34. A line that is tangent to a circle will have only one point in common with the circle.

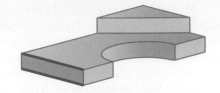

4.29 The highlighted lines are parallel.

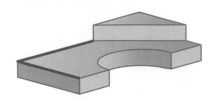

4.30 The highlighted lines are perpendicular.

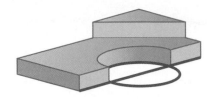

4.31 The highlighted circle intersects the highlighted line at two different points.

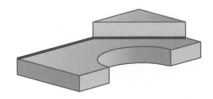

4.32 An angle is defined by the space between two lines (such as those highlighted here) or planes that intersect.

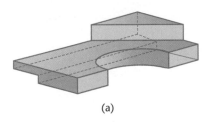
(a)

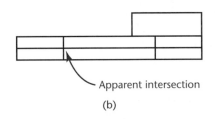
(b)

Apparent intersection

4.33 From the shaded view of this model in (a), it is clear that the back lines do not intersect the half-circular shape. In the wireframe front view in (b), the lines appear to intersect.

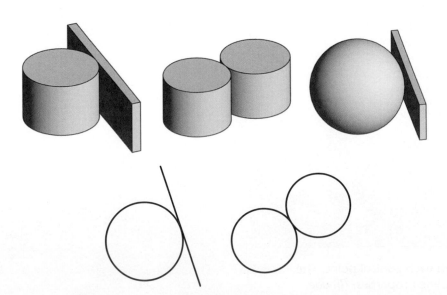

4.34 Lines that are tangent to an entity have one point in common but never intersect. 3D objects may be tangent at a single point or along a line.

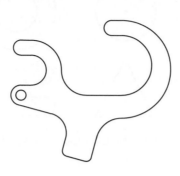

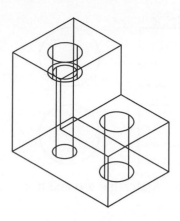

4.35 A radial line from the point where a line is tangent to a circle will always be perpendicular to that line.

4.36 A 2D Drawing Made of Only Lines, Circles, and Arcs

4.37 A 3D Model Made of Only Lines, Circles, and Arcs

When a line is tangent to a circle, a radial line from the center of the circle is perpendicular at the point of tangency, as shown in Figure 4.35. Knowing this can be useful in creating sketches and models.

The regular geometry of points, lines, circles, arcs, and ellipses is the foundation for many CAD drawings that are created from these types of entities alone. Figure 4.36 shows a 2D CAD drawing that uses only lines, circles, and arcs to create the shapes shown. Figure 4.37 shows a 3D wireframe model that is also made entirely of lines, circles, and arcs. Many complex-looking 2D and 3D images are made solely from combinations of these shapes. Recognizing these shapes and understanding the many ways you can specify them in the CAD environment are key modeling skills.

4.4 SOLID PRIMITIVES

Many 3D objects can be visualized, sketched, and modeled in a CAD system by combining simple 3D shapes or *primitives*. They are the building blocks for many solid objects. You should become familiar with these common shapes and their geometry. The same primitives useful when sketching objects are also used to create 3D models of those objects.

A common set of primitive solids used to build more complex objects is shown in Figure 4.38. Which of these objects are polyhedra? Which are bounded

4.38 The most common solid primitives are (a) box, (b) sphere, (c) cylinder, (d) cone, (e) torus, (f) wedge, and (g) pyramid.

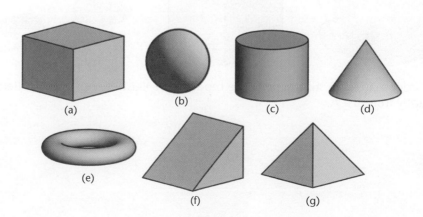

4.39 Match the top and front views shown here with the primitives shown in Figure 4.38.

by single-curved surfaces? Which are bounded by double-curved surfaces? How many vertices do you see on the cone? How many on the wedge? How many edges do you see on the box? Familiarity with the appearance of these primitive shapes when shown in orthographic views can help you in interpreting drawings and in recognizing features that make up objects. Figure 4.39 shows the primitives in two orthographic views. Review the orthographic views and match each to the isometric of the same primitive shown in Figure 4.38.

Look around and identify some solid primitives that make up the shapes you see. The ability to identify the primitive shapes can help you model features of the objects using a CAD system (see Figure 4.40). Also, knowing how primitive shapes appear in orthographic views can help you sketch these features correctly and read drawings that others have created.

Making Complex Shapes with Boolean Operations

Boolean operations, common to most 3D modelers, allow you to join, subtract, and intersect solids. Boolean operations are named for the English mathematician George Boole, who developed them to describe how sets can be combined. Applied to solid modeling, Boolean operations describe how volumes can be combined to create new solids.

4.40 The 3D solid primitives in this illustration show basic shapes that make up a telephone handset. *(Shutterstock.)*

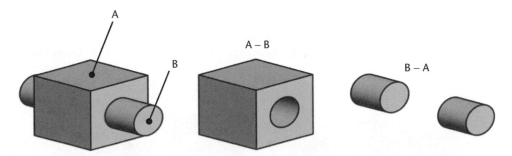

4.41 Order matters in subtraction. The models here illustrate how A – B differs significantly from B – A.

Table 4.1 Boolean Operations.

Name	Definition	Venn Diagram
Union (join/add)	The volume in both sets is combined or added. Overlap is eliminated. Order does not matter: A union B is the same as B union A.	
Difference (subtract)	The volume from one set is subtracted or eliminated from the volume in another set. The eliminated set is completely eliminated—even the portion that does not overlap the other volume. The order of the sets selected when using difference *does* matter (see Figure 4.41). A subtract B is not the same as B subtract A.	
Intersection	The volume common to both sets is retained. Order does not matter: B intersect A is the same as A intersect B.	

The three Boolean operations, defined in Table 4.1, are

- union (join/add)
- difference (subtract)
- intersection

Figure 4.42 illustrates the result of the Boolean operations in pictorial views of two solid models. Look at some everyday objects around you and make a list of the Boolean operations and primitive solid shapes needed to make them.

Figure 4.43 shows a bookend and a list of the primitives available in the CAD system used to create it, along with the Boolean operations used to make the part.

4.43 This diagram shows how basic shapes were combined to make a bookend. The box and cylinder at the top were unioned, then the resulting end piece and another box were unioned. To form the cutout in the end piece, another cylinder and box were unioned, then the resulting shape was subtracted from the end piece.

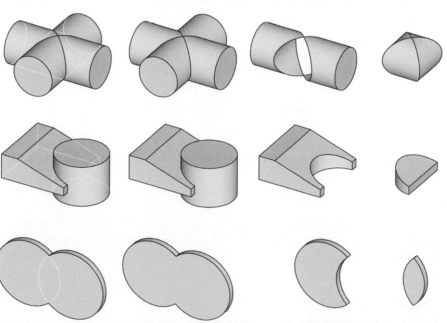

(a) Union   (b) Difference   (c) Intersection

4.42 The three sets of models at left produce the results shown at right when the two solids are (a) unioned, (b) subtracted, and (c) intersected.

4.44 Symmetrical parts can have symmetry about a line or point.

4.5 RECOGNIZING SYMMETRY

An object is *symmetrical* when it has the same exact shape on opposite sides of a dividing line (or plane) or about a center or axis. Recognizing the symmetry of objects can help you in your design work and when you are sketching or using CAD to represent an object. Figure 4.44 shows a shape that is symmetrical about several axes of symmetry (of which two are shown) as well as about the center point of the circle.

Mirrored shapes have symmetry where points on opposite sides of the dividing line (or mirror line) are the same distance away from the mirror line. For a 2D mirrored shape, the axis of symmetry is the mirror line. For a 3D mirrored shape, the symmetry is about a plane. Examples of 3D mirrored shapes are shown in Figure 4.45.

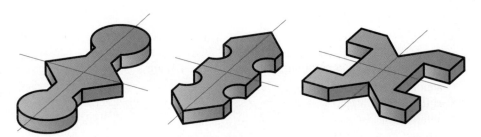

4.45 Each of these symmetrical shapes has two mirror lines, indicated by the thin axis lines. To create one of these parts, you could model one quarter of it, mirror it across one of the mirror lines, then mirror the resulting half across the perpendicular mirror line.

To simplify sketching, you need to show only half the object if it is symmetrical (Figure 4.46). A centerline line pattern provides a visual reference for the mirror line on the part.

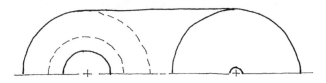

4.46 Orthographic sketches of symmetrical parts may show only half of the object.

Most CAD systems have a command available to mirror existing features to create new features. You can save a lot of modeling time by noticing the symmetry of the object and copying or mirroring the existing geometry to create new features.

Right- and Left-Hand Parts

Many parts function in pairs for the right and left sides of a device. A brake handle for the left side of a mountain bike is a mirror image of the brake handle for the right side of the bike (Figure 4.47). Using CAD, you can create the part for the left side by mirroring the entire part. On sketches you can indicate a note such as RIGHT-HAND PART IS SHOWN. LEFT-HAND PART IS OPPOSITE. Right-hand and left-hand are often abbreviated as RH and LH in drawing notes.

> ── TIP ──
> Using symmetry when you model can be important when the design requires it. When the design calls for symmetrical features to be the same, mirroring the feature ensures that the two resulting features *will be* the same.

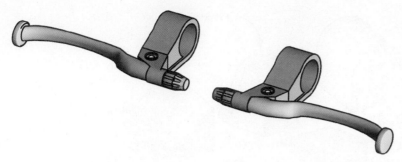

4.47 These brake levers are an example of right- and left-hand parts.

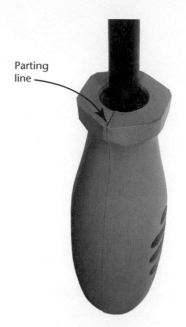

Parting-Line Symmetry

Molded symmetrical parts are often made using a mold with two halves, one on each side of the axis of symmetry. The axis or line where two mold parts join is called a *parting line*. When items are removed from a mold, sometimes a small ridge of material is left on the object. See if you can notice a parting line on a molded object such as your toothbrush or a screwdriver handle like the one shown in Figure 4.48. Does the parting line define a plane about which the object is symmetrical? Can you determine why that plane was chosen? Does it make it easier to remove the part from the mold? As you are developing your sketching and modeling skills think about the axis of symmetry for parts and how it could affect their manufacture.

4.48 The parting line on a molded part is often visible as a ridge of material.

4.6 EXTRUDED FORMS

Extrusion is the manufacturing process of forcing material through a shaped opening (Figure 4.49). Extrusion in CAD modeling creates a 3D shape in a way similar to the extrusion manufacturing process. This modeling method is common even when the part will not be manufactured as an extrusion.

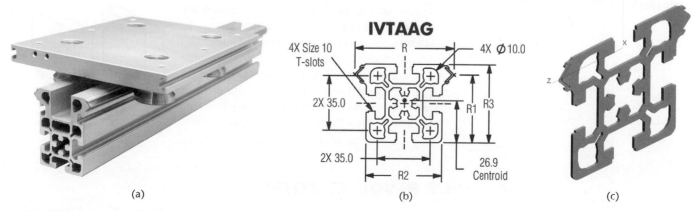

(a)   (b)   (c)

4.49 Symmetry and several common geometric shapes were used to create this linear guide system. The rail in (a) was created by forcing aluminum through an opening with the shape of its cross section. The extruded length was then cut to the required length. The solid model in (c) was created by defining the 2D cross-sectional shape (b) and specifying a length for the extrusion. *(Integrated configuration of Integral V™ linear guides courtesy of PBCLinear.)*

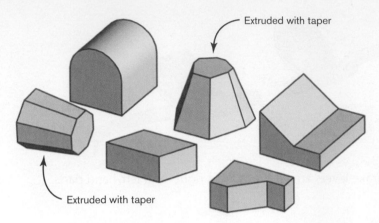

4.50   These CAD models were formed by extruding a 2D outline. Two of the models were extruded with a taper.

To create as shape by extrusion, sketch the 2D outline of the basic shape of the object (usually called a *profile*), and then specify the length for the extrusion. Most 3D CAD systems provide an **Extrude** command. Some CAD systems allow a taper (or draft) angle to be specified to narrow the shape over its length (Figure 4.50).

Swept Shapes

A swept form is a special case of an extruded form. *Sweeping* describes extruding a shape along a curved path. To sweep a shape in CAD, create the 2D profile and a 2D or 3D curve to serve as the path. Some swept shapes are shown in Figure 4.51.

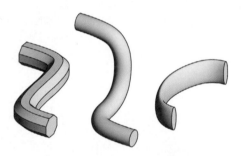

4.51   These shapes started as an octagon, a circle, and an ellipse, then were swept along a curved path.

Sketching Extruded Shapes

Shapes that can be created using extrusion are often easily sketched as oblique projections. To sketch extruded shapes, show the shape (or profile) that will be extruded parallel to the front viewing plane in the sketch. Copy this same shape over and up in the sketch based on the angle and distance you want to use to represent the depth. Then, sketch in the lines for the receding edges.

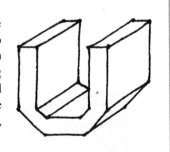

4.7 REVOLVED FORMS

Revolution creates 3D forms from basic shapes by revolving a 2D profile around an axis to create a closed solid object. To create a revolved solid, create the 2D shape to be revolved, specify an axis about which to revolve it, then indicate the number of degrees of revolution. Figure 4.52 shows some shapes created by revolution.

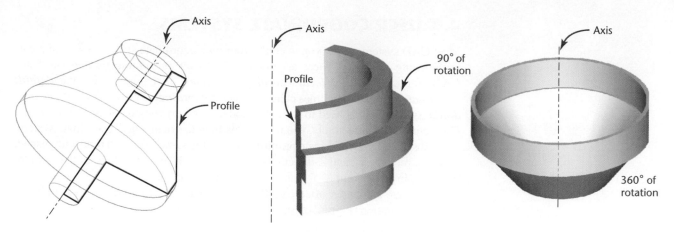

4.52   Each of the solids shown here was created by revolving a 2D shape around the axis.

Often, a 2D sketch is used to create 3D CAD models. Look at the examples shown in Figure 4.53 and match them to the 2D profile used to create the part. For each part, decide whether extrusion, revolution, or sweeping was used to create it.

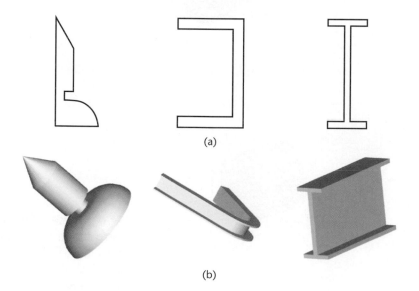

4.53   What operation would you choose to transform the profiles shown in (a) into the models in (b)?

4.8 IRREGULAR SURFACES

Not every object can be modeled using the basic geometric shapes explored in this chapter. Irregular surfaces are those that cannot be unfolded or unrolled to lie in a flat plane. Solids that have irregular or warped surfaces cannot be created merely by extrusion or revolution. These irregular surfaces are created using surface modeling techniques. Spline curves are frequently the building blocks of the irregular surfaces found on car and snowmobile bodies, molded exterior parts, aircraft, and other (usually exterior) surfaces of common objects, such as an ergonomic mouse. An example of an irregular surface is shown in Figure 4.54. You will learn more about modeling irregular surfaces in Chapter 5.

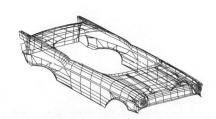

4.54   Irregular Surfaces

4.9 USER COORDINATE SYSTEMS

Most CAD systems allow you to create your own coordinate systems to aid in creating drawing geometry. These are often termed *user coordinate systems* (in Auto-CAD, for example) or *local coordinate systems*, in contrast with the *default coordinate system* (sometimes called the *world coordinate system* or *absolute coordinate system*) that is used to store the model in the drawing database. To use many CAD commands effectively, you must know how to orient a user coordinate system.

Most CAD systems create primitive shapes the same way each time with respect to the current X-, Y-, and Z-directions. For example the circular shape of the cylinder is always in the current X-Y plane, as shown in Figure 4.55.

To create a cylinder oriented differently, create a user coordinate system in the desired orientation (Figure 4.56).

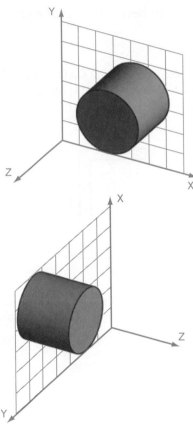

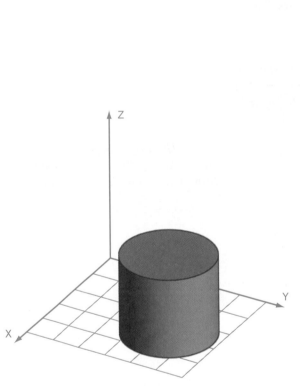

4.55   The cylinder is oriented with the circular base on the X-Y plane and the height in Z.

4.56   These cylinders were created after the *X-Y* plane of the coordinate systems was reoriented.

To create the hole perpendicular to the oblique surface shown in Figure 4.57, create a new local coordinate system aligned with the inclined surface. After you have specified the location of the hole using the more convenient local coordinate system, the CAD software translates the location of the hole to the world (default) coordinate system.

Many CAD systems have a command to define the plane for a user coordinate system by specifying three points. This is often an easy way to orient a new coordinate system—especially when it needs to align with an oblique or inclined surface. Other solid modeling systems allow the user to select an existing part surface on which to draw the new shape. This is analogous to setting the X-Y plane of the user coordinate system to coincide with the selected surface. With constraint-based modelers a "sketch plane" often is selected on which a basic shape is drawn that will be used to form a part feature. This defines a coordinate system for the sketch plane.

A user or local coordinate system is useful for creating geometry in a model. Changing the local coordinate system does not change the default coordinate system where the model data are stored.

TIP

All CAD systems have a symbol that indicates the location of the coordinate axes—both the global one used to store the model and any user-defined one that is active. Explore your modeler so you are familiar with the way it indicates each.

4.57 A new coordinate system is defined relative to the slanted surface to make it easy to create the hole.

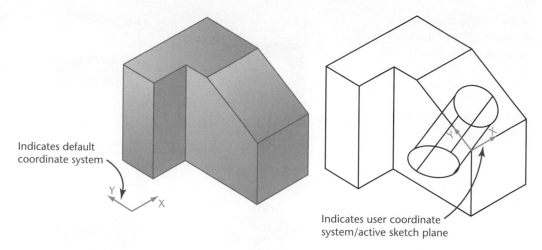

Indicates default coordinate system

Indicates user coordinate system/active sketch plane

4.10 TRANSFORMATIONS

A 3D CAD package uses the default Cartesian coordinate system to store information about the model. One way it may be stored is as a matrix (rows and columns of numbers) representing the vertices of the object. Once the object is defined, the software uses mathematical methods to transform the matrix (and the object) in various ways. There are two basic kinds of transformations: those that transform the model itself (called *geometric* transformations) and those that merely change the view of the model (called *viewing* transformations).

Geometric Transformations

The model stored in the computer is changed using three basic **transformations** (or changes): moving entities (sometimes called **translation**), rotating entities, and scaling entities. When you select a CAD command that uses one of these transformations, the CAD data stored in your model are converted mathematically to produce the result. Commands such as **Move** (or **Translate**), **Rotate**, or **Scale** transform the object on the coordinate system and change the coordinates stored in the 3D model database.

Figure 4.58 shows a part after translation. The model was moved over 2 units in the X-direction and 3 units in the Y-direction. The corner of the object is no longer located at the origin of the coordinate system.

Figure 4.59 illustrates the effect of rotation. The rotated object is situated at a different location in the coordinate system. Figure 4.60 shows the effect of scaling. The scaled object is larger dimensionally than the previous object.

> **TIP**
>
> The following command names are typically used when transforming geometry:
> - **Move**
> - **Rotate**
> - **Scale**

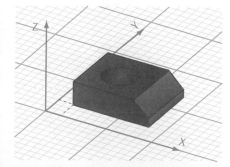

4.58 This model has been moved 2 units in the X-direction and 3 units in the Y-direction.

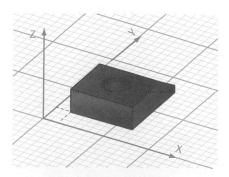

4.59 This model has been rotated in the X-Y plane.

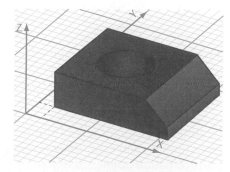

4.60 This model has been scaled to 1.5 times its previous size.

4.61 Note that the location of the model relative to the coordinate axes does not change in any of the different views. Changing the view does not transform the model itself.

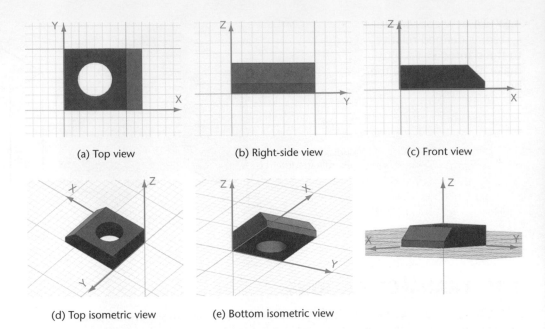

(a) Top view

(b) Right-side view

(c) Front view

(d) Top isometric view

(e) Bottom isometric view

Viewing Transformations

A viewing transformation does not change the coordinate system or the location of the model on the coordinate system; it simply changes your view of the model. The model's vertices are stored in the computer at the same coordinate locations no matter the direction from which the model is viewed on the monitor (Figure 4.61).

Although the model's coordinates do not change when the view does, the software does mathematically transform the model database to produce the new appearance of the model on the screen. This viewing transformation is stored as a separate

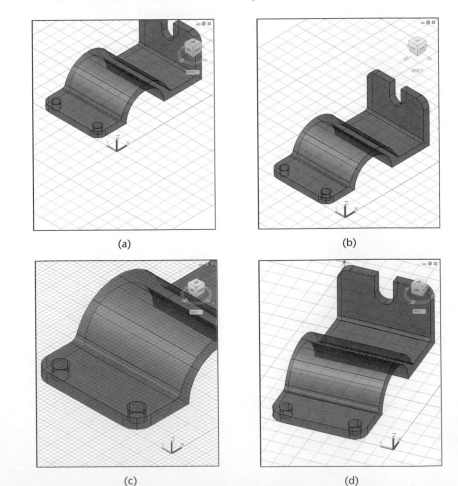

(a)

(b)

(c)

(d)

4.62 Panning moved the view of the objects in (a) to expose a different portion of the part in (b). In (c), the view is enlarged to show more detail. In (d), the view is rotated to a different line of sight. Note that in each case, the viewing transformation applies to all the objects in the view and does not affect the location of the objects on the coordinate system. (Notice that the position relative to the coordinate system icon does not change.)

part of the model file (or a separate file) and does not affect the coordinates of the stored model. Viewing transformations change the view on the screen but do not change the model relative to the coordinate system.

Common viewing transformations are illustrated in Figure 4.62. Panning moves the location of the view on the screen. If the monitor were a hole through which you were viewing a piece of paper, panning would be analogous to sliding the piece of paper to expose a different portion of it through the hole. Zooming enlarges or reduces the view of the objects and operates similar to a telephoto lens on a camera. A view rotation is actually a change of viewpoint; the object appears to be rotated, but it is your point of view that is changing. The object itself remains in the same location on the coordinate system.

Viewing controls transform only the viewing transformation file, changing just your view. Commands to scale the object on the coordinate system transform the object's coordinates in the database.

Examine the six models and their coordinates in Figure 4.63. Which are views that look different because of changes in viewing controls? Which look different because the objects were rotated, moved, or scaled on the coordinate system?

You will use the basic geometric shapes and concepts outlined in this chapter to build CAD models and create accurate freehand sketches. The ability to visualize geometric entities on the Cartesian coordinate system will help you manipulate the coordinate system when modeling in CAD.

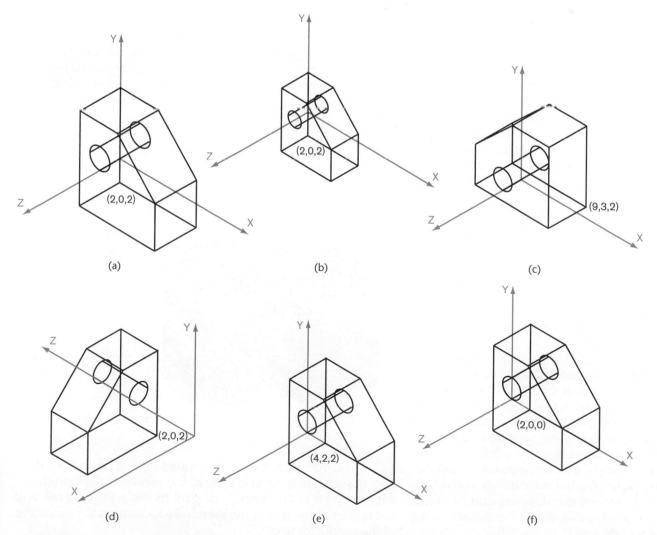

4.63 Three of these models are the same, but the viewing location, zoom, or rotation has changed. Three have been transformed to different locations on the coordinate system.

THE GEOMETRY OF 3D MODELING: USE THE SYMMETRY

Strategix ID used magnets to create a clean, quiet, zero-maintenance brake for the exercise bike it designed for Park City Entertainment. When copper rings on the bike's iron flywheel spin past four rare-earth magnets, they create current in a circular flow (an *eddy current*) that sets up a magnetic field. This opposing magnetic field dissipates power and slows the wheel. Moving the magnets onto and off the copper rings varies the amount of resistance delivered. When Marty Albini, Senior Mechanical Engineer, modeled the plastic magnet carrier for the brake, he started with the magnets and their behavior as the carrier moved them onto and off the copper rings (see Figure 4.64). "There is no one way to think about modeling a part," Albini said. "The key is to design for the use of the part and the process that will be used to manufacture it." To make the magnet carrier symmetrical, Albini started by modeling half of it.

The magnet carrier was designed as a part in the larger flywheel assembly, parts of which were already completed. Each pair of magnets was attached to a backing bar that kept them a fixed distance apart. To begin, Albini started with the geometry he was sure of: the diameter of the magnets, the space between them, and the geometry of the conductor ring. He sketched an arc sized to form a pocket around one of the magnets so that its center point would be located on the centerline of the conductor ring (see Figure 4.65). He then sketched another similar arc but with its center point positioned to match the distance between the centers of the two magnets. He connected the two arcs with parallel lines to complete the sketch of the inside of the carrier. This outline was offset to the outside by the thickness of the wall of the holder. (Because this is an injection-molded plastic part, a uniform wall thickness was used throughout.) One final constraint was added to position the carrier against the rail on the elliptical tube along which it would slide: the outside of the inner arc is tangent to this rail. With the sketch geometry fully defined, Albini extruded the sketch up to the top of the guide tube and down to the running clearance from the copper ring.

To add a lid to the holder, Albini used the SolidWorks **Offset** command to trace the outline of the holder. First, he clicked on the top of the holder to make its surface the active sketch plane. This is equivalent to changing the user coordinate system in other packages: it signals to SolidWorks that points picked from the screen lie on this plane. He then selected the top edges of the holder and used the **Offset** command with a 0 offset to "trace" the outline as a new sketch. To form the lid, he extruded the sketch up (in the positive Z-direction) the distance of the uniform wall thickness.

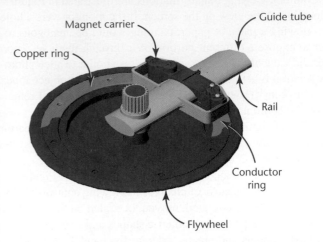

4.64 The magnet carrier for the brake was designed to move onto and off the conductor ring by sliding along an elliptical guide tube, pulled by a cable attached to the small tab in the middle of the carrier.

SolidWorks joined this lid to the magnet holder automatically because both features are in the same part and have surfaces that are coincident. This built-in operation is similar to a Boolean join in that the two shapes are combined to be one.

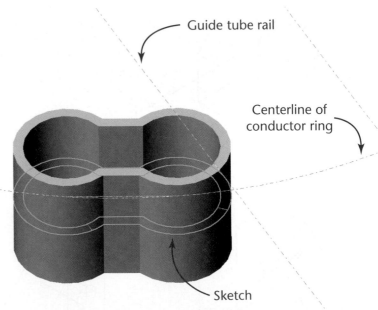

4.65 The magnet carrier was extruded up and down from the sketch, shown here as an outline in the middle of the extruded part. Notice that the sketch is tangent to the guide tube rail, and the centers of the arcs in the sketch are located on the centerline of the conductor ring.

For the next feature, Albini created a "shelf" at the height of the rail on which the holder will slide. Using **Offset** again, he traced the outline of the holder on the sketch plane, then added parallel and perpendicular lines to sketch the outline of the bottom of the shelf. The outline was then extruded up by the wall thickness. The distance from the outside of the magnet holder to the edge of the shelf created a surface that would sit on the rail (see Figure 4.66).

Two walls were added by offsetting the edge of the shelf toward the magnet holder by the wall thickness, then offsetting the edge again by 0. Lines were added to connect the endpoints into an enclosed shape to be extruded. (In Solid-Works, an extrusion can be specified to extend in one or both directions, and to extend to a vertex, a known distance, the next surface, or the last surface encountered.) For the walls, Albini extruded them to the top surface of the magnet holder "lid."

The connecting web between the magnet holders needed to match the shape of the elliptical tube in the flywheel assembly (see Figure 4.67). To make it, Albini sketched an ellipse on the newly created wall. An ellipse is a sketching primitive that can be specified by entering the length of the major and minor axes. Albini used the dimensions from the tube for the first ellipse sketch, then drew a second one with the same center point but with longer axes so that a gap equal to the wall thickness between them would be formed. The two ellipses were trimmed off at the bottom surface of the shelf and at the midpoint, and lines were drawn to make a closed outline. The finished sketch was extruded to the outside surface of the opposite wall.

More walls were sketched and extruded from the bottom surface of the shelf. Then, the wall over the connecting web was sketched and extruded down to the web.

The next step was to add the rounded edges for the top of the magnet holder. Albini invoked the **Fillet** command and selected to round all the edges of the top surface at

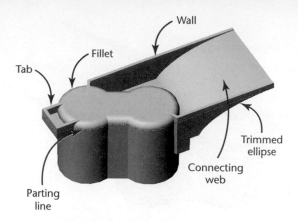

4.67 This view of the magnet carrier shows the elliptical shape of the connecting web and the rectangular shape of the tab. The parting line for the part, shown here as a dotted white line, is located at the edge of the fillet on the top of the magnet chambers.

once. As it created the fillet, SolidWorks maintained the relationship between the wall surfaces that intersected the top edge of the holder and extended them to the new location of the edge.

Next, Albini created a tab at the end of the part that would rest on the plastic collar in the assembly that went all the way around the magnet carrier. He first extruded a rectangular shape up from the top of the collar to form the "floor" of the tab. The walls of the tab required two additional extrusions.

The fillet at the top of the magnet holder provided the location for the parting line—the line where the two halves of the mold would come apart and release the part. Albini added a parting plane and used the built-in **Draft** option to add taper to the part so it would come out of the mold. After selecting all the surfaces below the parting plane, he specified a draft angle, and SolidWorks adjusted all the surfaces. This feature of SolidWorks makes it easy to add the draft angle after a part is finished. When draft is added, the geometry of the part becomes more complex and harder to work with. A cylinder with draft added becomes a truncated cone, for example, and the angles at which its edges intersect other edges vary along its length.

The next step was to add the bosses at the top of the magnet chambers that would support the bolts controlling the depth of the magnets. As it was a design goal to make the top of the chamber as stiff as possible to limit flex caused by the attraction of the magnets to the flywheel, the bosses were placed as far apart as possible, and ribs were added for rigidity. The bosses were sketched as circles on the top surface of the magnet holder with their centers concentric with the holes in the bar connecting the magnets below. Both bosses were extruded up in the same operation.

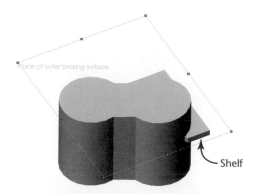

4.66 The surface of the rail was used as the sketch plane for the "shelf" on which the magnet carrier will slide.

(continued)

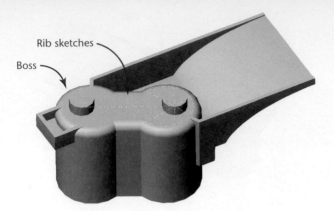

Rib sketches

Boss

4.68 Sketched circles were extruded to form the bosses on the top of the magnet chamber. The dotted lines shown here on the top of the chamber pass through the center point of the bosses and were used to locate the center rib and radial ribs.

Ribs in SolidWorks are built-in features. To create a rib, you simply draw a line and specify a width, and SolidWorks creates the rib and ends it at the first surface it encounters. To create the center rib, Albini sketched a line on the plane at the top of the bosses and specified a width (ribs on a plastic part are usually two thirds of the thickness of the walls). The rib was formed down to the top surface of the holder lid.

For the ribs around the bosses, Albini did as Obi Wan Kenobi might have advised: "Use the symmetry, Luke." He sketched the lines for ribs radially from the center points of the bosses (see Figure 4.68). To create the ribs, Albini created four of them on one boss, then mirrored them once to complete the set for one boss, then mirrored all the ribs from one boss to the other boss. Once all the ribs were formed, he cut the tops off the ribs and bosses to achieve the shape shown in Figure 4.69.

The result was a stiffer rib and a shape that could not be achieved with a single rib operation.

To complete the part, circles were drawn concentric to the bosses and extruded to form holes that go through the part (see Figure 4.70). Draft was added to the ribs and walls to make the part release from the mold easily. Fillets were added to round all the edges, reducing stresses and eliminating hot spots in the mold. Then, the part was mirrored to create the other half. The center rib and tab for attaching the cable were added and more edges filleted. Draft was added to the inside of the holder, and the part was complete.

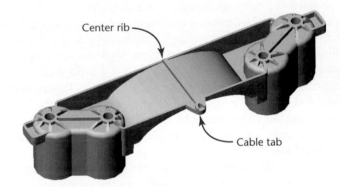

Center rib

Cable tab

4.70 Circles concentric with the bosses were extruded to form the holes shown here in the finished magnet carrier.

4.69 This view of the magnet carrier shows the symmetry of the ribs and the shape that resulted from "slicing off" the top of the bosses after the ribs were formed.

MAR 21, 1997

1). ATTEMPTED TO ALIGN GRIPPER WITH
DROP-JET ARM. GRIPPER NOT ALIGNED WITH
ROBOT. RADII DO NOT MATCH
POSITION A-18 ON ROUND WELL PLATE

$$T = 1 2 3 3 2$$
$$R = 6815$$
$$Z = 171$$

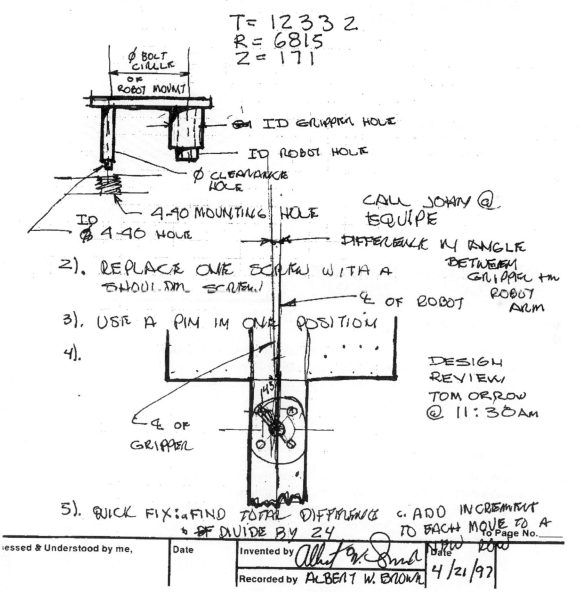

∅ BOLT CIRCLE OF ROBOT MOUNT

ON ID GRIPPER HOLE

ID ROBOT HOLE

∅ CLEARANCE HOLE

4-40 MOUNTING HOLE

ID ∅ 4-40 HOLE

CALL JOAN @ EQUIPE

DIFFERENCE IN ANGLE BETWEEN GRIPPER the ROBOT ARM

2). REPLACE ONE SCREW WITH A SHOULDER SCREW

3). USE A PIN IN ONE POSITION

4).

₵ OF ROBOT

₵ OF GRIPPER

DESIGN REVIEW TOM ORROW @ 11:30AM

5). QUICK FIX: a FIND TOTAL DIFFERENCE c. ADD INCREMENT
b. ÷ DIVIDE BY 24 TO EACH MOVE TO A
NEW ROW

essed & Understood by me,	Date	Invented by *Albert W. Brown*	Date
		Recorded by ALBERT W. BROWN	4/21/97

4.71 Simple geometric shapes are often an important basis for modeling and design. These sketches of a gripper alignment tool from Albert Brown's notebook show a circular pattern of mounting holes. Note the abbreviation ID for interior diameter (OD is used to indicate outside diameter). *(Courtesy of Albert W. Brown, Jr., Affymax Research Institute.)*

HANDS ON TUTORIALS

SolidWorks Tutorials

3D Sketching

In this tutorial, you will practice basic skills in making 3D sketches as the basis for part models.

- Launch SolidWorks from the icon on your desktop.
- Click to expand the topics listed below the "?" help icon.
- Click: **SolidWorks Tutorials**.
- Click: **All SolidWorks Tutorials (Set 1)** from the list.
- Click: **3D Sketching** from the list that appears.

On your own, complete this tutorial. In it you will create 3D sketches to use as paths for sweeping the shapes for an oven rack.

Questions

1. Is this rack made of a single piece of material? If not, how would the pieces be attached to one another?
2. How do you think an oven rack is manufactured?

3D Sketching

Using SolidWorks, you can create 3D sketches. You use a 3D sketch as a sweep path, as a guide curve for a sweep or loft, as a centerline for a loft, or as one of the key entities in a routing system. A useful application of 3D sketching is designing routing systems.

This lesson introduces you to 3D sketching and describes the following concepts:

- Sketching relative to coordinate systems
- Dimensioning in 3D space
- Mirroring features

(Image courtesy of DS SolidWorks Corp.)

REVERSE ENGINEERING PROJECT

The Can Opener Project

To accurately reverse engineer the can opener, you will need to make measurements for the part features. *Metrology* is the science of making measurements. The digital caliper is one commonly used measurement tool. The accuracy of a measurement is dependent on several factors, including the following:

- the skill of the operator
- the temperature at which the measurements are taken
- how stationary the part is while being measured
- the accuracy of the measurement device.

1. *Ready, set, measure.* Measure the critical dimensions of the lower handle part. Make each measurement five times. Calculate the mean and standard deviation for each set of measurements. Determine what value you will use when modeling that dimension. Label the values on the sketch you drew for the lower handle. What factors influence the accuracy of the value you choose for the dimension?

Digital Calipers *(Courtesy of L. S. Starrett Company.)*

2. *Sketch and model centerlines.* Sketch a 3D wireframe drawing showing the main centerlines of the can opener parts.

Use your 3D modeling software to create a single 3D sketch that shows the centerlines of each main part of the can opener, similar to that shown.

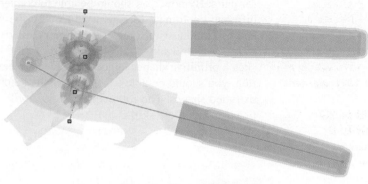

Draw the 3D sketch lines only.
You will model the can opener parts later.

KEY WORDS

Absolute Coordinates

Absolute Coordinate System

Angle

Apparent Intersection

Bezier Curve

B-Spline

Cubic Spline

Cylindrical Coordinates

Default Coordinate System

Diameter

Extrusion

Focus of an Ellipse

Freeform

Interpolated Spline

Local Coordinate System

Mirrored

NURBS Curve

Parallel

Perpendicular

Piecewise

Polar Coordinates

Primitives

Radius

Relative Coordinates

Revolution

Right-Hand Rule

Spherical Coordinates

Spline

Sweeping

Symmetrical

Transformations

Translation

User Coordinate System

World Coordinate System

SKILLS SUMMARY

The skills that you develop from material in this chapter will help you in creating 2D CAD drawings and 3D CAD models for parts. By now, you should be able to convert and interpret different coordinate formats used to describe point locations and be familiar with some of the basic geometry useful in creating CAD drawings. You should also be able to identify and sketch primitive shapes joined by Boolean operations. In addition, you should be able to visualize and sketch revolved and extruded shapes. The following exercises will give you practice using all these skills.

EXERCISES

1. Sketch some objects that you use or would design that have right-hand and left-hand parts, such as a pair of in-line skates or side-mounted stereo speakers for a computer.
2. In solid modeling, simple 3D shapes are often used to create more complex objects. These are called *primitives*. Using an isometric grid, draw seven primitives.
3. What is a Boolean operation? Define two Boolean operations by sketching an example of each in isometric view.
4. Consider primitives and Boolean operations that could be used to create a "rough" model of each of the items shown below. Using the photos as underlays, sketch primitives that could be used to create items a–d.

   a. Handlebar-mount gun rack

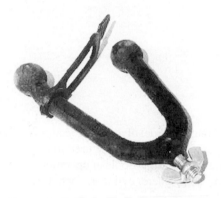

   b. ACME Corporation reduction gear

   c. Ashcroft Model 1305D deadweight pressure tester

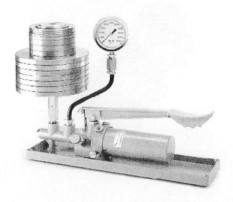

d. Davis Instruments solar-powered digital thermometer

5. Use nothing but solid primitives to create a model of a steam locomotive. Sketch the shapes and note the Boolean operations that would be used to union, difference, or intersect them, or create the model using Boolean operations with your modeling software. Use at least one box, sphere, cylinder, cone, torus, wedge, and pyramid in your design.

6. Identify the solid primitives and Boolean operations you could use to create the following objects.

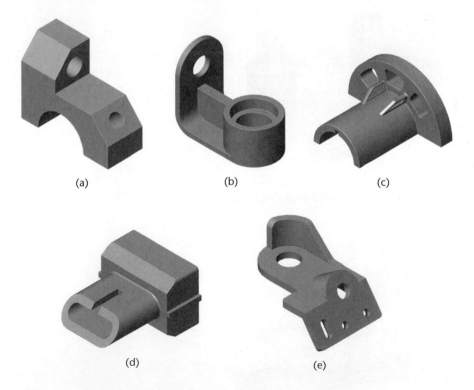

(a)   (b)   (c)

(d)   (e)

Exercises 7–12 Use an isometric grid to help sketch the solids formed by revolving the following shapes about the axis shown. Coordinates are defined by the X-Y-Z icon, with positive X to the right, positive Y up, and positive Z out of the page.

7.

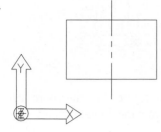

8.

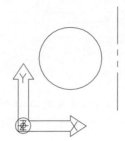

9.

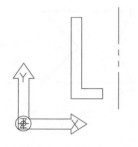

10.

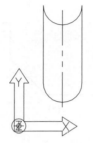

11.

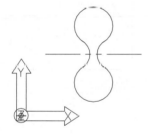

12.

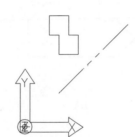

Exercises 13–16 Use an isometric grid to help sketch the solids formed by extruding the following shapes along the axis specified. Coordinates are defined by the X-Y-Z icon, with positive X to the right, positive Y up, and positive Z out of the page.

13. Extrude 6 inches in the positive Z-direction.

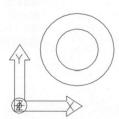

14. Extrude 4 inches in the positive Z-direction.

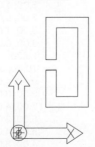

15. Extrude 6 inches in the positive Z-direction.

16. Extrude 4 inches in the positive Z-direction.

 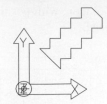

17. Starting at point A in each of the following figures, list the coordinates for each point in order as relative coordinates from the previous point.

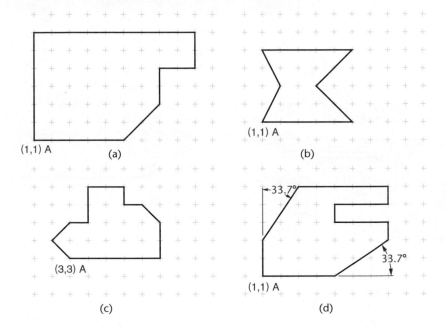

18. Plot the coordinates in each of the following lists on grid paper. Each point represents the endpoint of a line from the previous point, unless otherwise indicated. Relative coordinates are preceded by @.

a. X, Y	b.	c.	d.
1.00, 1.00	0.00, 0.00	0,0	2,2
4.00, 1.00	3.00, 0.00	@2<0	@-1<0
4.00, 2.00	4.00, 1.00	@3<30	@3<90
6.00, 2.00	5.00, 0.00	@3<-30	@4<-30
6.00, 1.00	6.00, 1.00	@2<0	@3<30
8.00, 1.00	7.00, 0.00	@4<90	@1<0
8.00, 4.00	8.00, 1.00	0,4	@3.24<230
5.00, 4.00	9.00, 0.00	@4<-90	@4<180
4.00, 5.00	10.00, 1.00		
1.00, 5.00	10.00, 3.00		
1.00, 1.00	9.00, 4.00		
	8.00, 3.00		
	7.00, 4.00		
	6.00, 3.00		
	5.00, 4.00		
	4.00, 3.00		

19. Using the information provided on the drawing, determine the coordinates you would use (absolute, relative, or polar) and the order in which you would enter them to create the following figures.

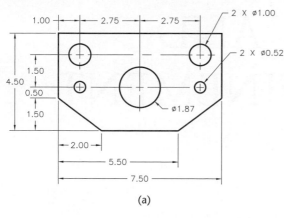

(a)

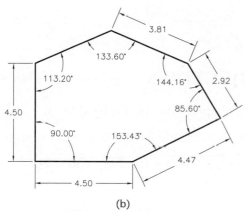

(b)

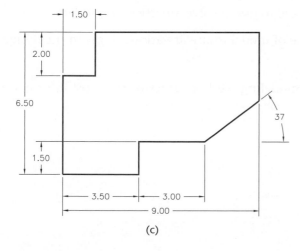

(c)

5

MODELING AND REFINEMENT

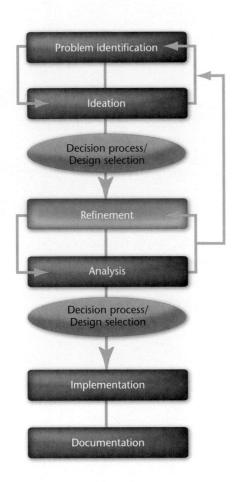

Problem identification

Ideation

Decision process/
Design selection

Refinement

Analysis

Decision process/
Design selection

Implementation

Documentation

——————— OBJECTIVES ———————

When you have completed this chapter you should be able to:

1. Describe modeling methods available to represent your design.

2. List a set of qualities models have that can help you select among them.

3. Describe how models can be used in refining designs.

4. Select which model to use to solve a particular problem.

5. Select which type of model to use at various stages in the design process.

6. Describe which modeling method contains the most information about a design.

The ENV hydrogen fuel cell motorbike was designed from the fuel cell outward. During its development many refinements were made both technologically and visually. Models in this phase must represent the design accurately enough to be used for testing as well as to convey the details of the design for implementation. *(Courtesy of Seymourpowell.)*

REPRESENTING YOUR CHANGING DESIGNS

During the ideation phase of the design process, graphic tools are selected for their ability to capture a design idea and to help team members visualize solutions to the design problem. Freehand sketches and conceptual representations are the best way to generate and explore a wide range of options. Remember, you cannot model what you have not planned and thought about.

When you move into the refinement stage, the criteria for the best means of describing your design change. During this middle stage of the design refinement, you need to include the detail necessary to define it for manufacture, as well as to test it against the design specifications. At the same time, you need to preserve your ability easily to modify the design as it evolves.

5.1 REFINEMENT AND MODELING

From the solutions generated during ideation, only a few are selected for further consideration. The design review process determines whether money will be committed for further development of a design. When a design concept is selected, it moves to the next stage of the process: refinement.

The process of refining and analyzing the design is iterative. As the refined design is tested the test results suggest modifications to the design, and once the design is changed, it needs further analysis.

Creating a CAD model is an important step in the refinement process. A huge amount of design information can be stored in CAD models that are easily modified. The CAD model can be used to test and evaluate the design. By making changes easy—and in some cases, automating the update process—CAD modeling can eliminate some barriers to a thorough review and refinement of the design.

CAD models also encourage refinement by making it easier and less costly to get feedback about a design while it is in development. Accurate part descriptions allow you to test how well parts will fit together. Customer acceptance and styling issues can be assessed early in the design process using realistically shaded models. Analysis programs can import 3D CAD files or run as part of the CAD software, saving much time. Some analysis packages suggest optimizations to incorporate directly in the CAD model. Animation and kinematic analysis uses CAD data to determine whether the design will function properly before any parts are actually built.

Many different methods are used to create CAD models, each with its strengths and weaknesses for capturing design information. Each item shown in Figures 5.1 and 5.2 may be modeled differently, possibly requiring different software and techniques.

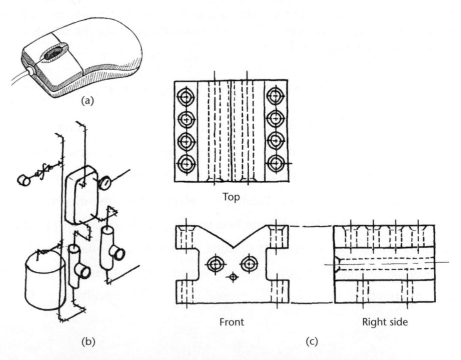

(a)

(b)

(c)

Top

Front   Right side

5.1   The same modeling software may be used to model (a) the ergonomic mouse, (b) the piping system, and (c) the V-block, but different modeling methods may be best suited to each.

5.2 These complex mathematical surfaces are realistically rendered using 3D CAD. *(Courtesy of Professor Richard Palais, University of California, Irvine, and Luc Benard.)*

In this chapter, you will learn how different modeling methods represent and store design information, and the advantages and disadvantages of each. This information will aid you in choosing the most effective method for a particular design problem.

What Is a Model?

In general, a **model** is a representation of a system, device, or theory that allows you to predict its behavior. A dictionary definition for *model* is as follows:

1. A small object, usually built to scale, that represents in detail another, often larger object.
2. (a) A preliminary work or construction that serves as a plan from which a final product is to be made: *a clay model ready for casting.* (b) Such a work or construction used in testing or perfecting a final product: *a test model of a solar-powered vehicle.*
3. A schematic description of a system, theory, or phenomenon that accounts for its known or inferred properties and may be used for further study of its characteristics: *a model of generative grammar; a model of an atom; an economic model.*

This definition defines three key qualities that are shared to varying degrees by models used in engineering design:

- They represent in detail another object: the product or system to be produced.
- They are a plan from which a final product will be made.
- They are used for studying and testing the design to provide an approximation of how it will behave.

Kinds of Models

Under the broad umbrella term *model* are several types. **Descriptive models** represent a system or device in either words or pictures. Descriptive models sometimes use representations that are simplified or analogous to something that is more easily understood. The key function of a descriptive model is to *describe,* that is, to provide enough detail to convey an image of the final product.

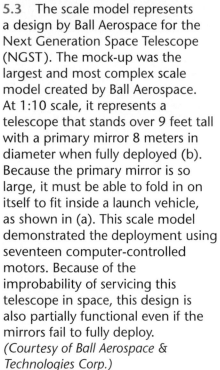

(a)

(b)

5.3 The scale model represents a design by Ball Aerospace for the Next Generation Space Telescope (NGST). The mock-up was the largest and most complex scale model created by Ball Aerospace. At 1:10 scale, it represents a telescope that stands over 9 feet tall with a primary mirror 8 meters in diameter when fully deployed (b). Because the primary mirror is so large, it must be able to fold in on itself to fit inside a launch vehicle, as shown in (a). This scale model demonstrated the deployment using seventeen computer-controlled motors. Because of the improbability of servicing this telescope in space, this design is also partially functional even if the mirrors fail to fully deploy. *(Courtesy of Ball Aerospace & Technologies Corp.)*

A set of written specifications for a design is a descriptive model. If all the specifications are followed, the system will perform correctly. Sketching is also a type of descriptive model for your design ideas on paper. 2D and 3D CAD drawings are also descriptive models. A physical model or prototype is another type of descriptive model, although sometimes physical models are made to a smaller scale (called a *scale model*).

Figure 5.3 shows a scale model of a design for the Next Generation Space Telescope (NGST) developed by NASA. A model helps you visualize the design. It has the added benefit of being easy for nonengineers to understand while decisions are being made about the next step in the project.

An **analytical model** captures the behavior of the system or device in a mathematical expression or schematic that can be used to predict future behavior. The electrical circuit model shown in Figure 5.4 is an example of an analytical model. The circuit and its properties are represented in the model by equations and relations that simulate the way the component would behave in a real circuit. The designer can change values to make predictions about what will happen in a real circuit that is wired the same way as the model. This makes it easier, less expensive, and faster to test the design.

Part of creating an effective analytical model is determining which aspects of the system's behavior to model. In the circuit model, some information, such as interference between some components, is left out because it is too complex to represent effectively. A **finite element analysis** (FEA) model, such as that used to generate the stress plot shown in Figure 5.5, simplifies the CAD model in a similar way. The FEA model breaks the model into smaller elements; reducing a complicated system to a series of smaller systems allows the stresses more easily to be solved. Understanding and using analytical models effectively requires knowing how the model differs from the actual system so the results can be interpreted correctly.

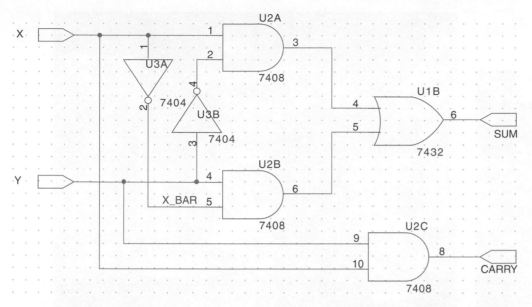

5.4 A simple circuit design created in ORCAD allows simulation of its function. *(Courtesy of Cadence.)*

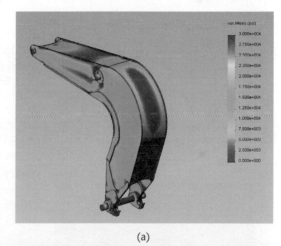

(a)

In the design process, use different types of models where they are appropriate. During the ideation phase of the design, sketching is probably the best technique to use. As you begin to refine the design, you will use both descriptive and analytical models to represent the design more accurately and to provide insight into its behavior.

A 3D CAD model combines qualities of descriptive and analytical models. Because a 3D CAD model accurately depicts the geometry of a device, it can fully describe its shape, size, and appearance as a physical or scale model would. Additional information about the final product, such as the materials from which it will be made, can also be added to the model

(b)

5.5 Finite element analysis (a) is used to calculate the stresses on the gooseneck shown in (b) in designing this heavy-duty equipment. *(Courtesy of TowHaul Corp.)*

(a)

(b)

5.6 This rendered view of the preliminary CAD model of the SIRTF assembly, shown in (a), is difficult to distinguish from the photograph of its 1:10 scale model, shown in (b) next to a 1:10 model of the Hubble Space Telescope. *(Courtesy of Ball Aerospace and Technologies Corporation.)*

description stored in the computer database. Figure 5.6 shows a CAD model of NASA's Space Infrared Telescope Facility (SIRTF).

A 3D CAD model can also be used for analytical modeling. The 3D model of the device can be used to study its characteristics (see Figure 5.7). Sophisticated software allows you to analyze, animate, and predict the behavior of the design under various physical conditions. The 3D CAD model is a detailed representation of the final object, suited for testing and study.

Model Qualities

Each modeling method has strengths and weaknesses. Models have general qualities that make them more or less useful for certain purposes and phases in the design process. Because design is an iterative process, being able to change the model easily can be important. Also, there can be a significant investment in equipment and effort in creating CAD models, so it is important for them to be useful for a variety of purposes. What are the qualities that can make models more or less useful?

Good models are

- visual (presenting information graphically);
- understandable (contain detail in a format suited to the audience);
- flexible (allow quick and easy updates and changes);
- cost-effective (provide benefits worth the cost of creating the model);
- measurable (able to extract size, shape and other information);
- accurate (provide information with useful precision); and
- robust (contain information depicting the necessary aspects of the system and reflecting the design intent).

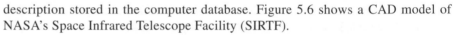

5.7 These welding robots are programmed with code developed with 3D model data. The motion of the robots is simulated with a 3D model of the part to be welded and 3D models of the robots. Motion paths are then exported and used to program the actual robots. *(Courtesy of FANUC Robotics America Corporation.)*

5.2 2D MODELS

Paper Drawings

2D sketches and multiview drawings are representations of the design. All the information defining the object may be shown in a paper drawing (see Figure 5.8), although it may require many orthographic views.

Multiview drawing techniques were designed to improve the robustness, measurability, and accuracy of paper drawings. These same techniques require some skill to be understood, however, and make multiview drawings harder to understand than a 3D model.

Equipment costs for paper drawings are minimal, but they take as long or longer to create than CAD drawings. Changes involve considerable erasing and redrawing, making them difficult to modify as the design changes. Because paper drawings are difficult to change, the labor costs associated with them usually outweigh the equipment savings.

Paper drawing accuracy is about plus or minus one fortieth the drawing scale. For example, a paper drawing of a map drawn at a scale of 1 inch = 400 feet has good accuracy if you can measure from it to plus or minus 10 feet. This makes paper drawings not particularly measurable or accurate—which is why proper dimensioning technique was developed. Proper dimensioning overcomes this limitation and provides measurements to the desired accuracy on the drawing. It is not good practice to make measurements from paper drawings.

Paper drawings can be very effective for quickly communicating the design of small parts for manufacturing. A properly dimensioned sketch can quickly convey all the information needed to make a part that is needed only one time (see Figure 5.9). Also, manufacturing facilities occasionally are not able to read electronic files and require paper drawings.

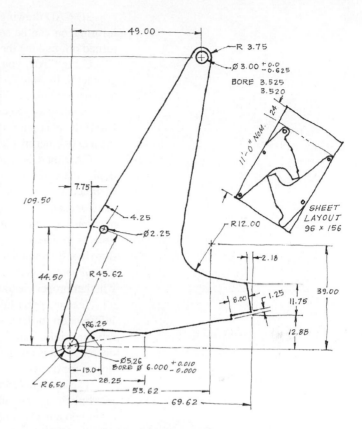

5.8 This fully dimensioned paper drawing contains all the information needed to manufacture this part, which will be flame cut from a 96 × 156-inch sheet of steel. *(Courtesy of Smith Equipment, USA.)*

2D CAD Models

2D CAD models share the visual characteristics of paper drawings but are more accurate and much easier to change. CAD systems have a large variety of editing tools that allow quick editing and reuse of drawing geometry. Standard symbols are easy to add and change.

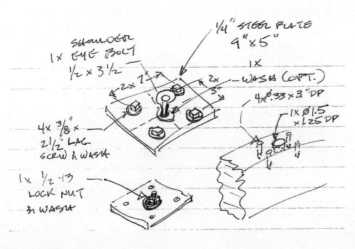

5.9 Computation Sketch Detail (*Courtesy of Jeffrey J. Zerr.*)

2D CAD drawings can quickly be printed to different scales. Different types of information can be separated onto different *layers* that can either be displayed or turned off, making the model more flexible than a paper drawing.

Using CAD, you can accurately define the locations of lines, arcs, and other geometry. In AutoCAD, for example, you can store these locations to at least fourteen decimal places. You can query the database, and the information will be returned as accurately as you originally created it. Of course, the adage Garbage In, Garbage Out (GIGO) applies. If your endpoints do not connect, or you locate distances by eye instead of entering them precisely, you will not have accurate results.

CAD models represent the object full size, unlike paper drawings, which often depict the object to a smaller or larger scale. Because features are represented using their actual sizes, you can make measurements and calculations from 2D CAD models. For example, from a 2D CAD layout showing equipment locations, you can make accurate measurements to determine clearances. Also, you can "snap" to locations on objects to determine sizes and distances that may not be dimensioned on a paper drawing. If there must be a clearance of 10 feet from the center of a tank to the location of another piece of equipment, you can measure from the CAD file to determine whether proper clearance is provided in the design.

Text and other information can be saved in a database and linked to the drawing information for later retrieval. The 2D database is limited, however, in its ability to represent information that depends on a 3D definition (such as the volume inside the object), so it is not generally useful for determining mass and other physical properties of the object.

The strengths and weaknesses of the multiview projection used to define an object fully in a paper drawing are shared by 2D CAD models. To create them and read them you must be able to interpret the 2D views to "see" the 3D object. Your ability to make measurements also depends on a good understanding of orthographic projection. To measure from a 2D CAD model, you must show distances and angles true length in a view. Orthographic projection and descriptive geometry are needed to create these views. *Descriptive geometry* is the study of producing views of an object that show true lengths, shape, angles, and other information about an engineering design. Mastering these subjects enables you to use auxiliary views created with a 2D CAD system along with the traditional methods of descriptive geometry to solve many engineering problems.

The flexibility and accuracy offered by the 2D CAD database make 2D CAD systems a cost-effective tool in a wide variety of businesses. For many civil engineering projects, the difficulty of capturing all the 3D information needed to model irregular surfaces, such as terrain, may not be worth the benefits of working in 3D. For

---

TIP

Lines and shapes in a 2D CAD drawing may appear to connect on the screen, but if the drawing elements are not drawn using methods that connect exact geometry, they may not actually connect. This can cause a number of different problems. For example, when trying to crosshatch inside a boundary, even a small gap may cause the hatching to "leak" out, when the software cannot determine a closed boundary.

Look at the drawing of the geneva cam shown in (a). It appears fine here as well as on the computer screen. But when the corner is enlarged, it is clear that the line that should be tangent to the arc does not actually connect (see (b)). At times, because of the algorithm used to generate circles and arcs, line segments used to represent them may not appear to connect when zoomed. But when the CAD drawing is regenerated from the data in the file, the line will—if tangent—clearly touch the arc, as shown in (c).

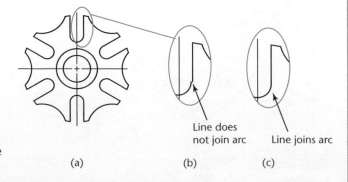

Line does
not join arc   Line joins arc

(a)   (b)   (c)

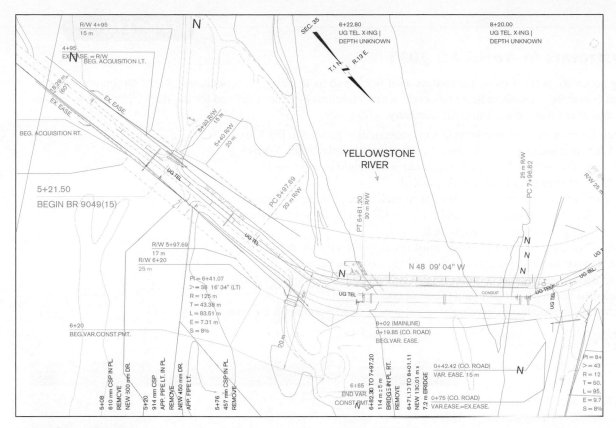

5.10 Large-scale projects such as this highway plan are often modeled in 2D CAD. Most 2D CAD systems associate dimension values with the entities they describe, so dimensions can be updated if the drawing changes. *(Courtesy of Montana Department of Transportation, MSU Design Section, Bozeman, MT.)*

projects such as highway design, civil mapping, electrical distribution, and building systems, 2D CAD models may provide enough information and be created more quickly from the information at hand (see Figure 5.10). Areas and perimeters can be calculated accurately, and drawings can be revised quickly. 2D CAD models are also more accurate than paper drawings.

2D Constraint-Based Modeling

Constraint-based modeling originally started as a method for creating intelligent 3D models. Constraint-based 2D models provide a mechanism for defining a 2D shape based on its geometry. Relationships like concentricity and tangency can be added between entities in the drawing. For example, once a concentric constraint is added between two circles, they will remain concentric, and you will be alerted if you attempt to make a change that will violate this geometric constraint. Dimensions in the drawing constrain the sizes of features.

Relationships defined between parts of the 2D model are maintained by the software when you make changes to the drawing. Geometric constraints can be a valuable tool, but to benefit from them, you must apply them with a good understanding of basic drawing geometry and the behavior desired when changes are made to the shape.

2D Constraints in AutoCAD 2011

2D drawing geometry defined using constraints must behave so as to satisfy the conditions placed on the drawing elements. Small constraint symbols indicate on-screen what conditions have been defined for the shape. The following constraints are available for defining AutoCAD geometry:

- *Horizontal:* Lines or pairs of points on objects must remain parallel to the X-axis.
- *Vertical:* Lines or pairs of points on objects must remain parallel to the Y-axis.
- *Perpendicular:* Two selected lines must maintain a 90° angle to one another.
- *Parallel:* Two selected lines must remain parallel.
- *Tangent:* Two curves must maintain tangency to each other or to their extensions.
- *Smooth:* A spline must be contiguous and maintain G2 continuity with another entity.
- *Coincident:* Two points must stay connected.
- *Concentric:* Two arcs, circles, or ellipses must maintain the same center point.
- *Collinear:* Two or more line segments must remain along the same line.
- *Symmetric:* Two selected objects must remain symmetrical about a specified line.
- *Equal:* Selected entities must maintain the same size.
- *Fix:* Points, line endpoints, or curve points must stay in fixed position on the coordinate system.

The following size constraints are also available:

- *Linear:* The distance between two points along the X- or Y-axis must be maintained.
- *Aligned:* A distance between two points must be maintained.
- *Radius:* The radius for a curve must be maintained.
- *Diameter:* The diameter of a circle must be maintained.
- *Angular:* The angle between two lines must be maintained.

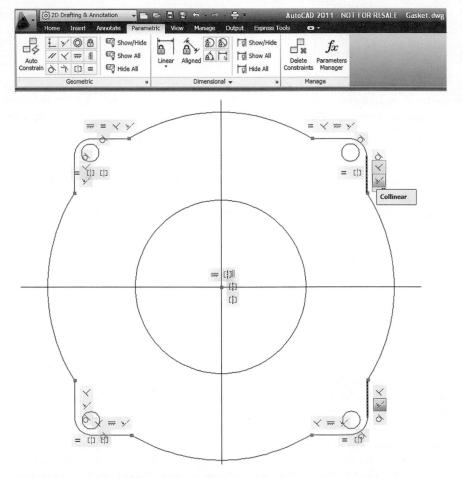

(Autodesk screen shots reprinted with the permission of Autodesk, Inc.)

5.3 3D MODELS

2D models must be interpreted to visualize a 3D object. To convey the design to individuals unfamiliar with orthographic projection—or to evaluate properties of the design that are undefined in 2D representations—3D models are used.

Physical Models

Physical models provide an easy visual reference. Physical models are called ***prototypes*** when they are made full size or used to validate a nearly final design for production. People can interact with them and get a feel for how the design will look and how it will function. Many problems with designs are discovered and corrected when a physical prototype is made.

The robustness and cost effectiveness of physical models are often linked. A simple model made of clay or cardboard may be enough for some purposes (see Figure 5.11), but determining the fit of many parts in a large-scale assembly, or producing a model realistic enough for market testing, may be cost prohibitive until very late in the design process. A full-size prototype of the BMW 850I cost more than $1 million to produce. The amount of information and detail built into the model adds to its cost.

5.11 This physical clay model of the Romulus Predator was used to evaluate the aerodynamics of its lines in a wind tunnel at the University of Michigan. *(Courtesy of M&L Auto Specialists, Inc.)*

The accuracy of a physical prototype also drives up its cost. Many objects are designed to be mass produced so that the cost of individual parts is reduced. When each part in the model must be produced one at a time, the cost of the prototype can be many times greater than the manufacturing cost of the final product. The more the model must match the final product in terms of materials used and final appearance, the greater the cost. Rapid prototyping systems offer quick and relatively inexpensive means of generating a physical model of smaller parts without the cost of machining or forming parts one at a time (see Figure 5.12).

Physical models are a very good visual representation of the design, but if they are not made from the materials that will be selected for the design, their weight and other characteristics will not match the final product. Sometimes, owing to the size of the project, the physical model must be made to a smaller scale than the final design. The accuracy of the final part also may not be possible with the materials and processes used for the prototype.

Probably the least attractive feature of a physical prototype is its lack of flexibility. Once a physical prototype has been created, changing it is expensive, difficult, and time-consuming. Consequently, full-size physical models are not usually used until fairly late in the design process—when major design changes are less likely. This can limit the usefulness of the model even though it provides important feedback about items that are not working well. When the information comes late in the design process, it can be too expensive to return to

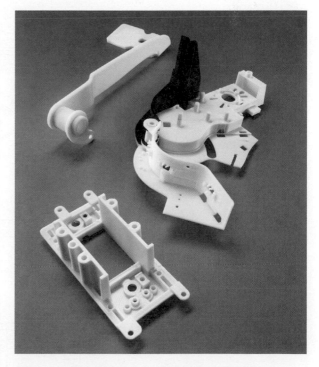

5.12 Prototype parts created using a fused deposition modeling system are useful for design verification. *(Courtesy of Stratsys, Inc.)*

a much earlier stage of the design to pursue a different approach. Only critical problems may be fixed. Solutions to problems found late in the process are generally constrained to cause the least amount of redesign while fixing the problem. When a prototype is created late in the process, there may not be time or resources to create a new prototype for the new solution. New problems introduced by the change may not be seen until actual parts are produced.

Despite the trade-offs, even very costly physical models have been a cost-effective way for companies to avoid much more costly errors in manufacturing.

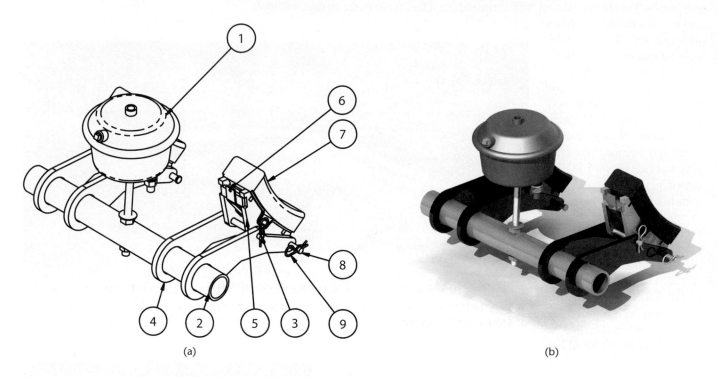

(a)

(b)

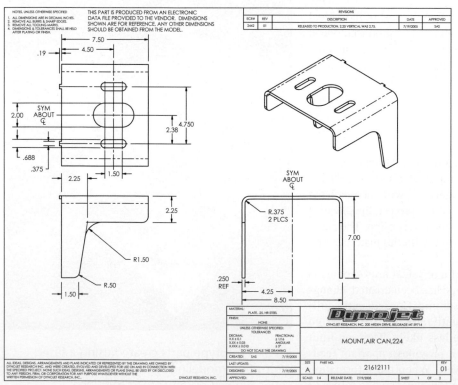

(c)

5.13   The 3D model, shown as outline (a), can also be rendered to produce the realistic view shown in (b) and to generate accurate 2D views for multiview drawings (c). *(Courtesy of Dynojet Research, Inc.)*

3D CAD Models

A 3D CAD model offers all the benefits of a 2D model and a physical model. As a visual representation, 3D CAD models can generate standard 2D multiview drawings as well as realistically shaded and rendered views. Because a 3D CAD model accurately depicts the geometry of the device, it may eliminate the need for a prototype—or make it easy to create one from the data stored in the model. Options for viewing the model make it understandable to a wide range of individuals who might be involved with the design's refinement (see Figure 5.13).

Virtual Reality

Virtual reality (VR) refers to interacting with a 3D CAD model as if it were real—the model simulates the way the user would interact with a real device or system. Using a virtual reality display, users are immersed in the model so that they can move around (and sometimes through) it and see it from different points of view. The headset display shown in Figure 5.14 uses two displays set about 3 inches apart (the typical distance between a person's eyes). Each display shows a view of the object as it would be seen from the eye looking at it, which creates a stereoscopic view similar to the one sent to the brain by our two eyes. Some headsets let you control the viewpoint by moving your eyes or head, so that as you look around, a new view is created to correspond with the new direction.

Some 3D interfaces, such as a 3D mouse or controller, let the user interact with the items in the 3D model in a way that enhances the illusion of reality. Some gloves (similar to those shown in Figure 5.15) even use model data to provide physical feedback to the user when an object is encountered. If a virtual object is squeezed, feedback to the glove provides the feeling of the resistance a solid object would give when squeezed. Systems may even interpret how much force would crush the object and provide this sensation to the user.

The term *virtual prototype* describes 3D CAD systems that represent the object realistically enough for users, designers, and manufacturers to get the same type of information they get from creating a physical prototype.

As more sophisticated software becomes available to analyze, animate, and predict the behavior of the design under various physical conditions, the 3D CAD model becomes a better representation of the final object—and better suited for testing and study.

5.14 Cybermind's Visette45 has both closed and see-through versions offering an effective screen size of 80 inches at 2 meters. *(Courtesy of Cybermind NL.)*

5.15 These 5DT Ultra series of data gloves are controlled via a belt-worn wireless kit designed to transmit data for two gloves simultaneously. *(Courtesy of 5DT/Fifth Dimension Technologies.)*

3D CAD models offer a high degree of accuracy and measurability and a high degree of flexibility. Each new generation of CAD software automates more common tasks, allowing the designer to focus more on the design and less on the mechanics of changing the model. This flexibility allows the CAD model to become a preliminary work that is refined and tested until it is ready to be used as a plan for the final product.

The ability of 3D CAD models to be used throughout the process, serve in lieu of a physical model, and be reused and modified indefinitely makes them very cost effective, especially in designing mechanical assemblies. However, not all 3D CAD models are the same. Familiarity with different types of 3D modeling systems lets you select the method and software most suitable for your design project.

5.4 TYPES OF 3D MODELS

Each of the major 3D modeling methods—wireframe modeling, surface modeling, solid modeling, and constraint-based modeling—has its advantages and disadvantages. Many CAD systems incorporate all these modeling methods into one single software package. Other software may provide only one or two of these methods.

Wireframe Modeling

Wireframe modeling represents the edges and contours of an object using lines, circles, and arcs oriented in 3D space. 3D wireframe drawings can be easy to create, are handled by many low-cost software packages, and provide a good tool for modeling simple 3D shapes. The method gets its name from the appearance of the model, which resembles a sculpture made of wires, as shown in Figure 5.16.

You create a wireframe model in much the same way that you create a 2D CAD drawing. Each edge where surfaces on the object intersect is drawn in 3D space

5.16 A wireframe model represents edges and contours of an object using lines, circles, and arcs oriented in 3D space. 3D wireframe is a suitable method for modeling this snow-melting system.

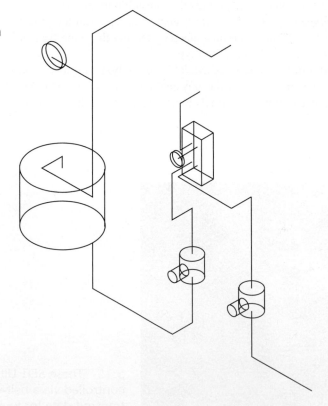

using simple geometric tools, such as a 3D line or arc. The X-, Y-, and Z-coordinates for the endpoints are stored in the database along with the type of entity.

3D wireframes can quickly display process control, piping, sheet metal parts, mechanical linkages, and layout drawings showing critical dimensions that use relatively simple geometric shapes. Interferences and clearance distances are difficult to visualize in 2D piping drawings. It can be critical that they are not overlooked in the design. For example, to quickly check clearances for the design of an electrical substation, the engineer may create a wireframe model of the high-voltage conductors to ensure that conductors are not too close to other equipment that could produce an electrical arc, causing a short.

A 3D wireframe can be useful for piping layout around other 3D equipment, which can be difficult to visualize in 2D. The centerline of the pipe is easy to represent in a 3D wireframe, as shown in Figure 5.16. This prevents problems with clearances that are acceptable in two dimensions, but not in a third. For a system with moving parts or where complex shapes need to fit, a different 3D modeling method may better allow you to assess interferences.

Wireframe models do not include surfaces that can be shaded, so they are not realistic looking. Because you can see through the model, some shapes cannot be represented unambiguously, and it may be difficult to see from a single view which areas are holes and which are surfaces. The model in Figure 5.17 shows a wireframe model. Without rotating or checking coordinate locations, you may visualize the part as showing the hole from the top or the bottom (Figure 5.18).

Before you can measure oblique edges and surfaces in a 2D drawing, you must first create an auxiliary view showing that feature true size. The true size is already contained in the 3D wireframe and can be measured directly. For example, the length of edge A in Figure 5.19 can be determined from this 2D AutoCAD drawing only in the auxiliary view. The 3D wireframe model of the same part can show the length of the edge directly.

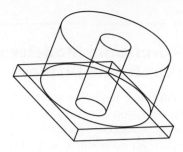

5.17 Are you looking at the model from below or from the top?

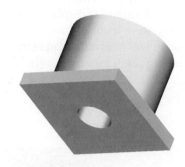

5.18 Adding faces to a wireframe model removes the ambiguity about the features.

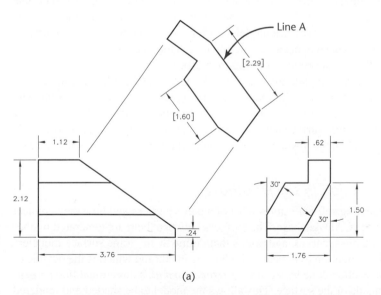

(a)

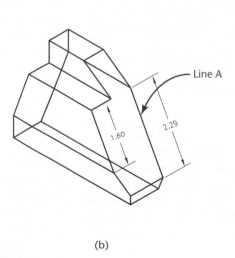

(b)

5.19 (a) The 2D CAD drawing displays the true length of a line in an auxiliary view. (b) The length of the line in the 3D wireframe model can be determined directly from the model.

Wireframe Modeler versus Wireframe Display

Wireframe modeling was the first 3D modeling method. Its depiction of edges and contours grew out of 2D modeling practices. The term *wireframe* is also used to describe an economical way of displaying a model on the computer screen in which just the edges and contours are shown. Representing objects on the screen using wireframe display is simpler and faster than showing shaded views. The computing power needed to generate and display a model was a constraint in the design of CAD modeling systems. Wireframe representation was developed when multimedia displays common on home computers today were available only on powerful mainframes. As computers evolved, computationally intensive modeling techniques became feasible. Display methods such as shaded views, shaded with edges, hidden line–removed views, and others can be generated from the data. There is still a trade-off between complexity and speed; more complex and detailed information means more processor time.

CAD modelers address this trade-off by representing the model data in two ways: first, as it is stored in the database (the coordinate locations for each entity in the model, for example) and second, as it will be displayed on the screen. The economy of wireframe display makes it a good choice for working with computer models—whatever the underlying modeling method. Surface, solid, and parametric modelers all offer a wireframe display and a range of tools for controlling that display. Each of them, however, creates a database with more information than can be found in that created by a wireframe modeler, which stores only vertex and edge information.

5.20   The industrial designer for the Timex TurnAndPull alarm was concerned that the inner rings would look too deep or too busy. A rendered surface model allowed team members to see how it would look when manufactured. *(Courtesy of Dana Rockel, Timex Corporation.)*

Surface Models

CAD surface models define a shape by storing its surface information. A surface model is similar to an empty box. The outer surfaces of the part are defined (although unlike an actual box, these surfaces do not have any material thickness). Because the surfaces are defined, they can be shaded to provide a realistic appearance.

Surface modelers are traditionally used by industrial designers and stylists who create the outside envelope for a device. These designers need modeling tools that convey a realistic image of the final product. Software tools developed for their use focused first on the challenge of modeling the exterior of the product, not the mechanical workings within. Many surface modelers offer specialized tools for lighting and rendering a model, so designers can create photorealistic images that cannot be distinguished from the actual product (see Figure 5.20).

Concurrent engineering and the use of the 3D CAD database as the product description contributed to the development of modeling software that satisfies the aesthetic needs of industrial designers as well as the functional information needed by engineers. Understanding surface modeling techniques will help you assess the kinds of features that are available in your modeler—and whether a dedicated surface modeler is required.

Surface Information in the Database

Most surface modeling software stores a list of a part's vertices and how they connect to form edges in the CAD database. The surfaces between them are generated mathematically from the lines, curves, and points that define them. Some surface modelers store additional information to indicate which is the inside and which is the outside of the surface. This is often done by storing a **surface normal**, a directional line perpendicular to the outside of the surface. This allows the model to be shaded and rendered more easily.

Storing the definitions of the surfaces in the CAD digital database is termed **boundary representation** (BREP), meaning that the information contained in the database represents the external boundaries of the surfaces making up the 3D model. Remember that these surfaces have no thickness.

The following are basic methods used to create surface models:

- extrusion and revolution
- meshes
- spline approximations

Extruded and Revolved Surfaces

You can define a surface by a 2D shape or profile and the path along which it is revolved or extruded.

Figure 5.21 shows a surface model created by revolution, and the profile and axis of revolution used to generate it. Regular geometric entities can be revolved or extruded to create surface primitives such as cones, cylinders, and planes. Surface primitives are sometimes built into surface modeling software to be used as building blocks for complex shapes.

Meshes

Mesh surfaces are defined by the 3D location of each vertex stored in the CAD database. Each group of vertices is used to define a flat plane surface. Figure 5.22 shows a mesh surface and a list of some of its vertices (in c).

Some mesh surfaces are called **triangulated irregular networks** (TINs), because they connect sets of three vertices with triangular faces to serve as the surface model. A mesh surface can be useful for modeling uneven surfaces, such as terrain, where a completely smooth surface is not necessary. A sufficiently large matrix will result in correspondingly smaller triangles that can better approximate a smooth surface. Refining the mesh will produce a smoother surface, but the size of the CAD file will increase and result in slower software performance. Some modeling packages allow you to define a smoothed representation of the surface instead of the somewhat bumpy appearance of a strictly mesh surface.

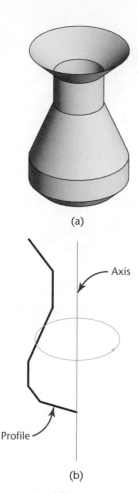

(a)

(b)

5.21 The surface model in (a) was created by revolving the profile about the axis shown in (b).

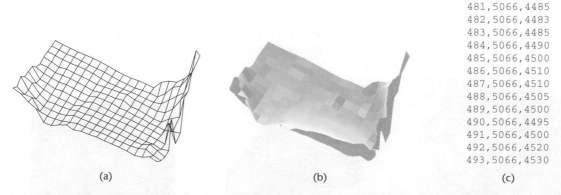

```
481,5066,4485
482,5066,4483
483,5066,4485
484,5066,4490
485,5066,4500
486,5066,4510
487,5066,4510
488,5066,4505
489,5066,4500
490,5066,4495
491,5066,4500
492,5066,4520
493,5066,4530
```

(a)   (b)   (c)

5.22 A mesh surface is composed of a series of planar surfaces defined by a matrix of vertices, as shown in (c). The wireframe view of the mesh in (a) appears more like a surface in the rendered view shown in (b).

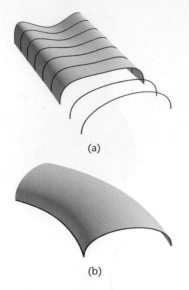

5.23 A lofted surface (a) blends a series of curves that do not intersect into a smooth surface. A swept surface (b) sweeps a curve along a curved path and blends the influence of both into a smooth surface model.

NURBS-Based Surfaces

The mathematics of nonuniform rational B-spline (NURBS) curves underlies the method used to create surfaces in most surface modeling systems. A NURBS surface is defined by a set of vertices in 3D space that are used to define a smooth surface mathematically.

Rational curves and surfaces have the advantage that they can be used to generate not only free-form curves but also analytical forms such as arcs, lines, cylinders, and planes. This is an advantage for surface modelers that use NURBS techniques, as the database does not need to accommodate different techniques for surfaces created using a surface primitive, mesh, extrusion, or revolution.

Spline curves can be used as input for revolved and extruded surfaces that can be lofted or swept. **Lofting** is the term used to describe a surface fit to a series of curves that do not cross each other. Sweeping creates a surface by sweeping a curve or cross section along one or more "paths." In both cases, the surface blends from the shape of one curve to the next (see Figure 5.23).

NURBS surfaces can also be created by meshing curves that run perpendicular to each other, as illustrated in Figure 5.24.

Reverse Engineering

Most reverse engineering software traces a physical object to generate mesh data, or data points for a surface model. Points on the object are captured as vertices, then translated into a digital surface representation. Reverse engineering can be an easy way to capture the surface definition of an existing model. Designers might use traditional surface sculpting methods and then digitize the physical model to create a

Visible Embryo Heart Model

By modeling embryonic hearts at different stages, researchers were able to use the models to examine the way different tissue layers expand and move as the heart develops. To study blood flow in developing hearts that are only 0.8 millimeter across, Kent Thornburg and Jeffrey Pentecost created models from which they generated a 5-centimeter-wide stereolithography physical model.

(a) Using cross sections of embryos from the Carnegie Collection of Human Embryos, the team made digital photomicrographs of each slide. Using the computer, they traced the outlines of heart tissues at each level.

(b) Lofting combined the cross sections into a surface model of the heart, shown in wireframe.

(c) The model made by stereolithography.

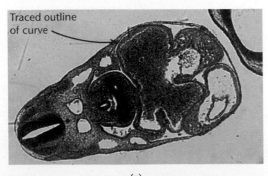

Traced outline of curve

(a)

(b)

(c)

(Images courtesy of Dr. Jeffrey O. Pentecost, Director, Visible Embryo Heart Project, and Dr. Kent Thornburg, Director, Congenital Heart Research Center, at Oregon Health Sciences University, with the cooperation of Alias Wavefront and Silicon Graphics, Inc.)

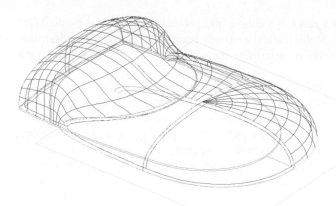
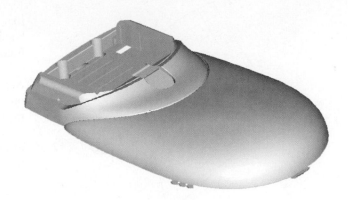

5.24 When the spline curves shown here are used to generate a NURBS surface, the functions defining each curve are blended. *(Courtesy of Robert Mesaros.)*

digital database. An existing part might be reverse engineered so that it can be added to the CAD database (Figure 5.25 shows data for the surface model in Figure 5.26 being created using a reverse engineering process.)

Complex Surfaces/Combining Surfaces

To create a surface model, you do not create the entire surface at once—just patches to combine into a continuous model. Just as curves can be made of individual segments that are smoothed into a continuous curve, surfaces can be made of entities referred to as *patches*. Like a spline curve, a patch can be *interpolated*, or approximated.

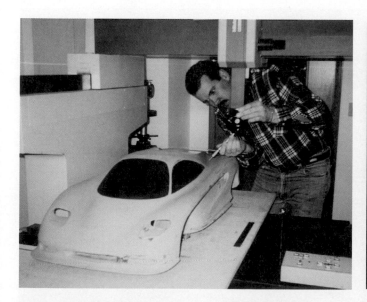

5.25 A coordinate measuring system was used to digitize the clay model of the Romulus Predator. Seventeen hundred data points were captured by digitizing half of the quarter-scale clay model of the Predator. Digitizing the data and mirroring it in the model ensured side-to-side symmetry. *(Courtesy of M&L Auto Specialists, Inc.)*

5.26 The lines on the surface model of the Predator prototype indicate individual surface patches that make up the model. Areas of the car with drastic curvature changes are divided into smaller patches, which keeps the mathematical order of the patches lower and reduces the overall complexity of the surface. *(Courtesy of M&L Auto Specialists, Inc.)*

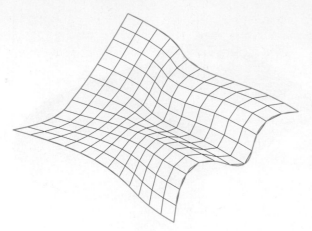

5.27 The lines between the boundary curves in this Coon's patch surface represent the shape of the surface and illustrate the influence of the boundary curves on the interpolated surface between them.

A ***Coon's patch*** is a simple interpolated surface that is bounded by four curves. Mathematical methods interpolate the points on the four boundary curves to determine the vertices of the resulting patch (see Figure 5.27).

Surface patches are joined by blending the edges of the patches. Areas created by blending, or the surfaces created by fillets, corners, or offsets, are called ***derived surfaces***. They are defined by mathematical methods that combine the edges of the patches to create a smooth joint.

Sometimes, complex surface patches are created by trimming. For example, a circular patch may start with a rectangular patch, then be trimmed to a circle, and finally be blended with other surface patches.

Some surface modeling systems include the use of Boolean operations, whereas others do not. Systems that do not include Boolean operations or good tools for trimming surfaces can be difficult to use to create a feature such as a round hole through a curved surface because the exact shape of the surface, including the hole, must be defined by specifying its edges.

Editing Surfaces

Once a surface has been defined, editing depends on the method used by the surface modeler to create and store the surface. Usually, the information describing the locations of vertices is editable, and each vertex can be changed. If the surface modeling software retains the original 2D profile that was extruded or revolved to produce the surface, it can be used to facilitate changes to these surfaces. If not, it can be difficult to efficiently edit the surfaces.

For NURBS surfaces, each control point can be edited to produce local changes in the surface model. (Editing Bezier surfaces, by the same token, produces global changes to the surface.) Grabbing and relocating vertices is a highly intuitive way to edit a surface model that contributes to its usefulness in refinement. The term ***tweaking*** is used to describe editing a model by adjusting control points individually to see the result. For example, the NURBS surfaces used to define the shape of the Predator's exterior made it easy to use the surface model to explore design options in several areas (Figure 5.28).

5.28 The roof scoop on the Predator's clay model was not visible enough, so the designer adjusted the control points in the surface model in real time until the scoop looked just right on-screen. *(Courtesy of M&L Auto Specialists, Inc.)*

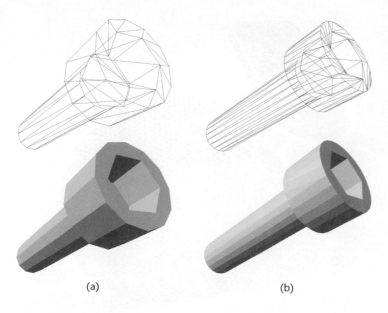

5.29 The mesh in (a) has fewer vertices and larger facets than the mesh used in (b). The finer mesh in (b) more accurately represents the cylindrical shape of the bolt.

(a)   (b)

Surface Model Accuracy

Surface model information can be stored with more or less accuracy. Compare the meshes shown in Figure 5.29. The mesh at left contains fewer vertices than the mesh shown at the right. Although the surface shown at the right is more accurate, the additional information makes its file size larger. This requires more computer processing to store and interact with the model. Depending on your purpose, the smaller number of vertices may be satisfactory.

Surface models used for computer-aided manufacturing (CAM) require a high degree of accuracy to produce smooth surfaces. The trade-off between speed and accuracy may be resolved by setting the surface display to a less accurate faceted representation and storing a smoothed or more highly defined surface definition in the database. It is important to distinguish between the screen display of the model and the surface definition stored in the database.

Splines used to create surfaces can also vary in accuracy. Splines can have more or fewer control points that you can use to shape the surface. To model surfaces smoothly, fewer control points may allow more fluid curves. A mesh surface used to calculate area may not be as accurate as one created using a smoothing algorithm such as NURBS or Bezier.

Tessellation lines are used to indicate surfaces in a wireframe view, and they may or may not reflect the accuracy of the surface (see Figure 5.30). It is important to distinguish when tessellation lines represent the accuracy of the *view* of the model, and when they represent the accuracy of the model stored in the database. If the software accurately represents the actual geometry of the model in the model database, the tessellations simply represent the model on the display in the on-screen view.

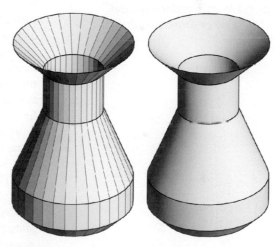

5.30 The faceted representation on the left uses planar surfaces to approximate smooth curves.

Using Surface Models

The primary strength of a surface model is improved appearance of surfaces and the ability to convey these complex definitions for computer-aided manufacturing. Customers purchase products not only on their function but also on their styling. A realistically shaded model can be used with potential customers to determine their reactions to its appearance. Lighting and different materials can be applied in most

5.31 This docking station illustrates the complex curves possible in a surface model. *(Courtesy of Mark Biasotti.)*

surface modeling software to create very realistic looking results (Figure 5.31). Many consumer products often start with the surface model, then the interior parts are engineered to fit the shape of the styled exterior. When surface models can be used in place of a physical prototype—or in place of the product itself for promotional purposes—the savings add to their cost effectiveness.

Because the relative locations of surfaces from a particular direction of sight can be calculated, surface modeling systems can automatically remove back edges (or represent them as hidden lines). Surface definitions remove the ambiguity inherent in some wireframe models and allow you to see holes and front surfaces by hiding the nonvisible parts of the model.

The complex surfaces defined by a surface model can be exported to numerically controlled machines for part manufacture, making it possible to manufacture irregular shapes that would be difficult to document consistently in 2D views. Both surface and solid models can be used to check for fit and interference before the product is manufactured. Often, surface models can be converted to solid models and vice versa.

Because surface models define surfaces, they can often report the surface area of a part. This information can be useful in calculating heat transfer rates, for example, and can save time, particularly when the surface is complex and doing calculations by hand would be time-consuming. The accuracy of the calculations may depend on the method used by the software to store the surface data.

Complex surfaces can be difficult to model. The cost-effectiveness of surface modeling depends on the difficulty of the surface, the accuracy required, and the purpose for which the model will be used. Many modeling software systems offer a combination of wireframe, surface, and solid modeling capabilities, making it possible to weigh the benefits against the difficulties and the time required to create each type of model.

Solid Models

Solid models go beyond surface models to store information about the volume contained inside the object. They store the vertex and edge information of the 3D wireframe modeler, the surface definitions of the surface modeler, plus volume information. Many solid modelers also store the operations used to create features, which makes it possible to edit them quickly (see Figures 5.32 and 5.33). Solid models closely approximate physical models, making them especially useful for defining, testing, and refining the designs they represent.

--- TIP ---

Understanding basic modeling methods will help you assess the modeling software of tomorrow. Some new modelers, referred to as "direct" or "explicit" modelers, such as CoCreate and SpaceClaim, offer a broad command set that adopts features from traditional surface *and* solid modelers.

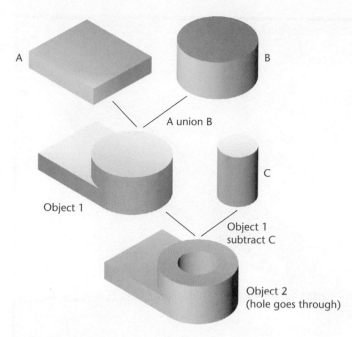

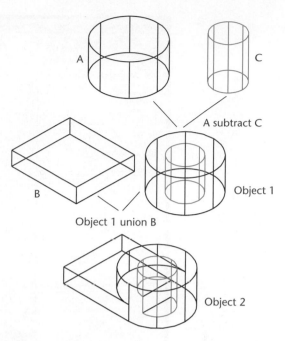

5.32 The order of operations matters when you use Boolean operations to create a solid model—and when you edit it (see Figure 5.33).

5.33 Joining A and B after the hole is created by subtracting C changes the design—the hole no longer goes through the entire part.

Using 3D Solid Models

Many of the examples so far illustrated how accurate representation of a 3D object provides valuable engineering information. Solid models are highly visual and easy to understand and measure. They are also highly accurate if modeled accurately. Solid models are often able to replace physical models, as well as control CAM equipment, which adds to their cost-effectiveness as a modeling method.

In addition, 3D solid models are particularly suited for use with analysis packages. They contain all the information about an object's volume, and volume is important to many engineering calculations. Mass properties, centroid, moments of inertia, and weight can be calculated from the solid model as often as needed during the refinement process. The system behavior can be simulated from the wealth of information available in the model.

Solid models provide information needed for other analyses. Finite element analysis (FEA) methods break up a complex object into smaller shapes to make stress, strain, and heat transfer easier to calculate. Because the solid model defines the entire object, FEA software can use this information to generate an FEA mesh automatically for a part. This allows inclusion of this type of analysis earlier in the design process, incorporation of changes, and testing of the new model with FEA analysis again. Some FEA and solid modeling packages are integrated to provide this analysis within the modeling software. Others even allow direct model optimization based on the FEA results, creating a new version of the model that the designer can then revise further.

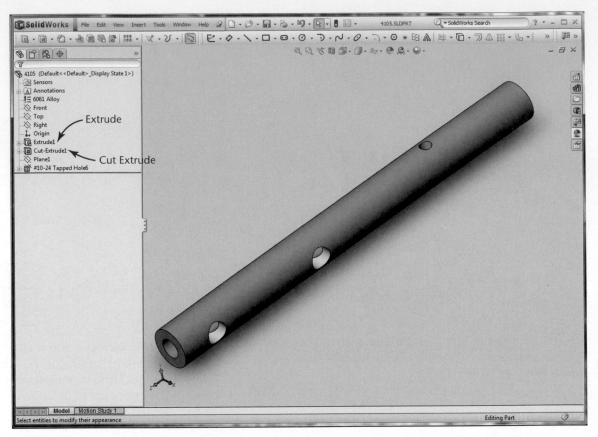

5.34 The **Extrude** and **Cut Extrude** (a subtracted extrusion) operations used to create this model are stored in the **Model Tree.** These operations can be edited later as needed.

Constraint-Based Models/Intelligent Models

Constraint-based or parametric modeling is a special modeling method that captures information about the relationships among geometric features. Constraint-based models store the steps and relationships used to create the model, so they can be revisited and used to edit the model (see Figure 5.34). When changes are made, the model is updated to preserve these relationships. Assembly models use constraints to define how parts relate to one another and how they should fit together (see Figure 5.35). Drawings created from constraint-based 3D models are linked to the model itself, so that changes to the 3D model cause its 2D drawing to be updated. Also, in many systems changes to the drawing will update the model.

To make the best use of constraint-based modeling, capturing the design intent and an object's fit with other parts in the overall design are key. Often easier to update than other modeling methods, simple constraint-based models can be designed early on. As additional information becomes available and the design evolves, updating the model with new detail and dimensions or changing geometric relationships is relatively easy. This requires an initial investment of time, but the advantages of better planning may be one of the largest rewards of constraint-based modeling.

Constraint-based modeling methods can be applied to 2D, surface, and solid models, and constraint-based modeling software may offer 2D, solid, and surface tools in the same package. You will learn more about constraint-based modeling in the next chapter.

(a)

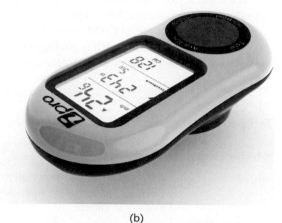

(b)

5.35 The parts that make up the iBike Pro cycling computer were modeled individually as solids, then combined into a single assembly model to represent the finished product. The exploded view in (a) shows the individual parts clearly; the assembled view in (b) shows how the parts will fit together. *(Courtesy of iBike and Salient Technologies, Inc.)*

Modeling Kernels

Although every CAD package operates slightly differently, some of them share the same *modeling kernel*. ACIS and Parasolid are two kernels used in many popular programs. The differences between modeling kernels are a result of their development history. The ACIS kernel was developed first for regular solid geometry that was well suited to CAM applications. The Parasolid kernel offered more options for the blends and curves needed for consumer product modeling. Today, the differences between the kernels are disappearing rapidly as the technology matures. Not every modeling package uses a third-party kernel such as ACIS or Parasolid. Pro/ENGINEER and CATIA are two that use their company's own, proprietary modeling engine.

The differences among the kernels themselves are less important than understanding which one your modeler uses and how that affects translation of data from one package to another. CAD software that uses the same modeling kernel can often maintain all the information or intelligence built into the model when it is translated to different software using the same kernel. CAD files translated between software using different modeling kernels may contain less information after translation. Paying attention to the formats available for storing and exporting model data helps you choose a modeling package that will minimize data loss. If you are considering changing software and want to convert your existing files, make sure to test some real examples to determine how your current models will import into the new system.

5.5 CHOOSING THE RIGHT MODELING METHOD

As you design, you will want to choose a modeling method that makes sense for your product, the stage in the design process, and the cost. Table 5.1 recaps the modeling methods according to the model qualities identified earlier in this chapter.

Key considerations are the time required to model the part and the purposes for which it will be used. A simpler method that provides all the needed information is more cost-effective than one that takes longer to model. However, time invested in a complete digital model of the product can pay for itself if it is used to generate visuals that help shape the design, reduce manufacturing difficulties, foster concurrent engineering—or promote the product later.

The accuracy of the modeling method is another key consideration. What types of analysis must be or could be completed with a model? What level of accuracy will be required to interface with CAM programs?

As you have seen, 3D modeling packages that are strong in surface modeling are needed to model the geometry of smoothly contoured features such as the ergonomic

Table 5.1 Characteristics of Modeling Methods.

Characteristic	Paper Drawings	2D Wireframe	Physical Model
Visual	2D views require interpretation and may be less understandable to a nonengineer.	2D views require interpretation and may be less understandable to a nonengineer.	Similar to actual object. Can be seen from all angles. May be smaller scale, unrealistic material.
Understandable	May require experience interpreting multiview drawing technique to be understood.	May require experience interpreting multiview drawing technique and descriptive geometry to be understood.	Can interact in a way similar to real object to aid in understanding. May be different scale, material than real object.
Flexible	Changes require erasure; difficult to accommodate large changes without redrawing.	Editing capabilities and layers make models more flexible than paper drawings.	Can be modified to a degree, but substantial changes may require a new model.
Cost-effective	Low equipment costs, but man-hours may be greater than with other methods. Cost effective for visualization and documenting simple parts. May be necessary if electronic formats cannot be read.	Useful for projects where the information gained from a 3D model is outweighed by the cost of capturing it, as in civil mapping and electrical circuit and distribution design.	Varies with the information contained in the model (where more information is usually more expensive) and the cost of manufacturing errors that could be prevented by the model.
Measurable	Not suited to be measured from.	Because objects are drawn full size, measurements may be taken from the model; knowledge of descriptive geometry required to derive information not shown in standard views.	Depends on how closely the model matches the materials and processes to be used to create the actual product.
Accurate	Requires proper dimensions to reflect design accurately.	Offers a high degree of accuracy when orthographic drawing techniques are understood.	Depends on how closely the model matches the materials and processes to be used to create the actual product.
Robust	May require multiple views to fully define the shape of an object.	More robust than paper drawings and can include links to large amounts of data; lacks information about volume found in 3D models.	Can include as much information as the actual product.
Application	Effective for sketches and quickly conveying the design of simple parts for manufacturing.	Useful for largely 2D information such as maps, layout drawings, and electrical circuits.	Good for testing fit of parts, interaction with other products and people, and analyzing aspects that cannot be simulated.

mouse shown earlier in the chapter. Piping systems and other structures of simple geometric shapes can sometimes be modeled very effectively using 3D wireframe modeling. Many inexpensive CAD packages support this type of modeling. However, a solid modeling package is required to create a model that can be used to check parts for fit and interference; calculate the weight of a final assembly from the many individual parts; generate a rendered view of the assembly that can be used in manufacturing, marketing, service and repair; and drive the computer-aided machinery to manufacture parts. The same assembly model created using constraint-based modeling methods would have the added advantage of being updated automatically when key dimensions change. It would also make it easy to create a family of similar parts.

Whether you choose the modeling package or learn to use the tools provided by your employer, you should be aware of the strengths and limitations of the software so you can use it most effectively.

Characteristic	3D Wireframe	Surface	3D Solid
Visual	Can be viewed from any direction and used to create standard 2D views; lack of surfaces may make views ambiguous.	Usually includes lighting and background options that can be used to create photo-realistic images of the model.	Shaded and rendered views present a realistic view of the object; offers automated 2D view generation; with most equipment only 2D view is available on monitor; part must be rotated to picture clearly.
Understandable	Requires viewer to mentally add surfaces to see the 3D shape.	Shaded views are easily understood by the viewer.	Shaded views are easily understood by the viewer.
Flexible	Editable, but changing individual objects more time-consuming than in other methods.	Each vertex is editable, but ease of changes depends on method used to create and store the surface.	CSG or hybrid models offer the most flexibility; can edit at the level of the solid object instead of individual geometric entity.
Cost-effective	Requires less computing power, and software is generally lower priced; good for modeling geometrically simple shapes.	Most cost-effective for modeling irregular surfaces that must be conveyed to manufacture with a high degree of accuracy; photo-realistic displays can offset modeling costs by adding value in marketing.	Completeness of the solid model cost-effective when model can be used for multiple purposes, such as testing, presentation, and CAM.
Measurable	Full-size entities are measurable without auxiliary views; can only measure between wireframe entities, however.	Can be used to calculate surface area.	Can be used to measure not just size but also weight, mass, and other physical properties.
Accurate	Offers a high degree of accuracy for the entities represented.	Depends on method used to store model data; smoothed surfaces generally more accurate than mesh surfaces.	Depends on method used to store model data; high degree of accuracy possible.
Robust	Model stores only vertices and edges; lacks surface or volume information; may be ambiguous as to voids or holes.	Fully defines the surfaces of a part; lacks volume information.	Includes volume as well as surfaces and vertices; can eliminate need for physical model.
Application	Good for applications where relationships in 3D space must be modeled, simple geometric shapes are sufficient for the objects to be modeled, and realistic views are not required.	Good for designs that include irregular or free-flowing surfaces and for those where the appearance of the product is a critical design criterion.	Good for applications where fit between parts and other engineering properties need to be evaluated before manufacture.

CHOOSING THE RIGHT METHOD: SURFACE MODELING

The Stryker SmartPump tourniquet system continuously monitors and controls tourniquet pressure during surgeries (Figure 5.36). When Kent Swenseid, Design Director at Strategix Vision, designed the enclosure for the product, he used surface modeling techniques to obtain the smooth curves envisioned for the case and built-in handle.

Before these outside surfaces were modeled, the internal components and user interface were worked out to determine the shape and volume required. Concept sketches were done (including rough models in 3D) to show various options for the enclosure. Once a concept was chosen and the internal configuration was known, surface modeling started in earnest.

Swenseid used a modeling approach that works well for complex molded assemblies: a "base part" controlling the overall shape and dimensions of the assembly is modeled first, then the component parts are derived from the base part. This way, if something in the design has to change, only the base part has to be modified to alter the exterior. The derived parts are updated accordingly. This approach also allows various parts to be modeled in parallel by a team rather than a single designer.

The case concept was a clamshell of two large molded parts with a molded-in handle. The first step was to identify key features, such as where and how the two molded parts would come together, and where mounting positions for things like

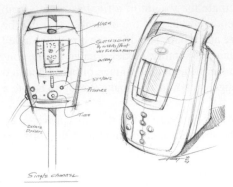

5.36 A Concept Sketch for the "Smart Tourniquet"

the display would be. These were represented as planes in the model. Shapes and contours important to the shape were also created early in the modeling and are visible in the feature tree for the part (see Figure 5.37). These changed as modeling progressed, but because the model was built around them, it was simple to update.

The front face was modeled as a loft—a surface that morphs from one curve to another. The curvature was controlled by either guide curves or tangency constraints at the ends (Figure 5.38). Early surfaces are almost always modeled larger than the face that will end up on the part to simplify manipulation of the curvature—and to avoid unwanted edge effects.

The next surfaces added do not show up in the final model directly. These "helper surfaces" can be used to control the angle where two surfaces meet (Figure 5.39). To ensure that the surfaces of the mold will let go of the part when it is ejected, the surface has to have what's known as "draft"—the

5.38 The front surface lofts between guide curves.

surface has to sit at an angle to the direction the mold opens. Creating a simple helper surface at the parting plane provides a draft constraint to which to tie a more complex surface: if the surface is tangent to a surface with enough draft, it will have enough draft too. Helper surfaces are trimmed off, deleted, or hidden when no longer needed.

Additional faces were built using swept, extruded, and revolved surfaces to define half of the symmetrical shape of the case. Using the symmetry does more than save work; it also guarantees that the model will actually *be* symmetrical. For a surface model, it also ensures that there will be no visible artifact where the two parts join. This avoids distracting highlights and waviness in the finished part.

In Figure 5.40, you can see that the part surfaces were modeled to extend beyond their intersections with other surfaces, thus allowing the software to calculate the intersections mathematically to create the edges between surfaces. It also means that simpler profiles can be used to loft or sweep the surfaces. The mathematics of surface modeling favors using the fewest possible constraints, because more constraints require higher-order polynomials to describe the surface. If the spline defining the intersection were used to generate the surface, tiny perturbations in the driving geometry could result in ripples or other undesired effects in the surface.

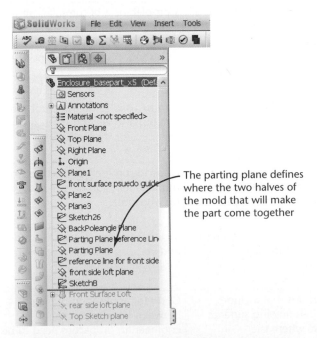

The parting plane defines where the two halves of the mold that will make the part come together

5.37 The feature tree for the base part consists of planes and reference lines that other features are derived from and constrained to.

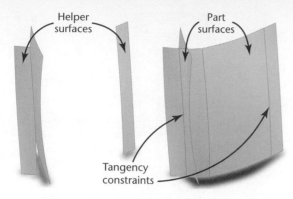

5.39 Helper surfaces help position the surfaces and are trimmed off when no longer needed.

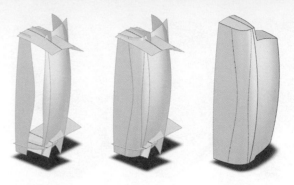

5.41 Trimming and Lofting a Surface to Create the Corner

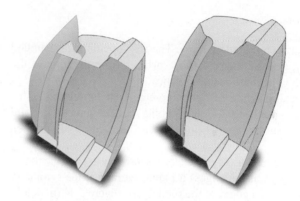

5.42 Adding the Clearance Surface for the Mounting Pole

5.40 Half the case is modeled to take advantage of the symmetry of the part.

Trimming forms the boundary between surfaces. This operation can be used to create hard edges, but in this case the shape required a rounded edge. This requirement could be met with a fillet, but the designer wanted to avoid the sudden change of curvature where the fillet meets the two surfaces it joins. This is a common desire in product modeling, especially with glossy surfaces that reveal the change of curvature as an unwanted highlight instead of a smooth 3D curve. The designer also wanted to "bulge" the rounded edge toward the top and bottom, which would have required a variable-radius fillet—at the time, an unreliable feature.

To create the rounded edge, the front and side surfaces were trimmed to helper surfaces that described the boundary (Figure 5.41 (left)) and the gap was filled with another loft, which allowed the result to be curvature-continuous. Note that the top and bottom of the two surfaces in Figure 5.41 (middle) still extend beyond where they will eventually end, again to minimize the number of constraints on the fill surface.

The order in which operations are done affects modeling efficiency. Sometimes it makes sense to leave trimming and joining surfaces to the very end, and sometimes it makes sense to join them earlier to simplify selecting them for other operations.

For example, the device mounts by fitting around a vertical pole behind it, so the case needed an indentation to clear the pole. Because the pole clearance surface crossed several others, modeling this feature was delayed until the rest of the device had taken shape (see Figure 5.42).

When it came time to add the handle opening, ergonomic concerns and balance issues made it difficult to use the main parting plane of the base part. The handle does not symmetrically straddle the main parting plane, so where the two halves

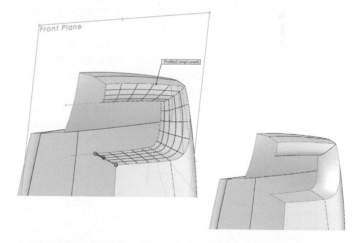

5.43 Modeling the Surface for the Front Half of the Handle Opening

of the clamshell would come together would be very close to the front edge of the handle. Because this would make it uncomfortable to hold the unit, the boundary of the part inside the handle was shifted rearward toward the middle of the handle. The handle opening was given its own parting plane.

The front opening for the handle area was then modeled in one feature (Figure 5.43). The face at the front of the handle was deleted to open up the edges, and those edges were captured as a compound curved profile. The inner opening was sketched on the handle parting line plane, and a surface was

(continued)

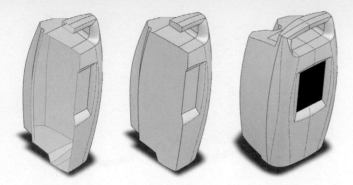

5.44 Closing the Model and Mirroring the Solid Part

5.45 Nonsymmetrical Detail Added to Back

lofted between the two curves. The rear opening was handled similarly, as were the transition surfaces from the front of the case to the opening for the screen.

Once the surface half is complete, it can be "closed." This turns the surface model into a solid model. Actually, a solid model is a surface model, just an "airtight" one with no open edges.

With a symmetrical model, the easiest way to close the model is to knit (join at the edges) a planar surface at the symmetry plane to the open edges of the model (see Figure 5.44 (middle)). Some modelers then automatically assume the model is a solid, but some have to be told to check for edge gaps and missing faces before changing the status of the body.

Once the model is solidified, it can be mirrored non-symmetrical features can be added, and the surface can be used in later design stages (see Figure 5.45). In some cases, it can make sense to mirror the surfaces themselves, but that can increase the likelihood of small mathematical errors owing to the more complex nature of the boundary.

Building a model from a base part allows detail work to be started even before modeling of the base part is complete.

5.46 Adding Features to the Shelled Model

As long as the surface can be closed and a solid formed, that solid can be shelled and split up into components. As additional features are added they are derived from the base part geometry so they will be updated if the base part changes. Figure 5.46 shows this kind of detailing in the rear half of the assembly.

(All figures courtesy of Strategix Vision team members Kent Svendseid, Marty Albini, and Aki Hirota.)

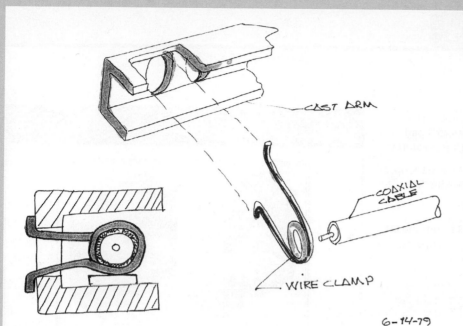

5.47 Solid modeling was Albert Brown's choice for refining the design of the parts shown in this sketch. The coaxial cable was formed by revolution; the cast arm was extruded and the holes formed with a Boolean operation. The wire clamp was created by sweeping its cross-sectional shape along a curved path. The resulting solid model allowed Brown to check the fit of the wire clamp. (*Courtesy of Albert W. Brown, Jr., Affymax Research Institute.*)

5.48 Concept sketches for this surgical device reveal several options for the enclosure, which was modeled using surface techniques. (*Courtesy of Strategix Vision.*)

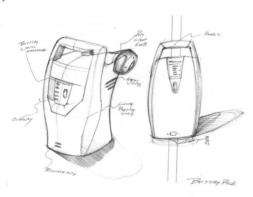

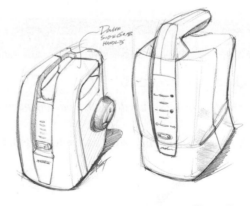

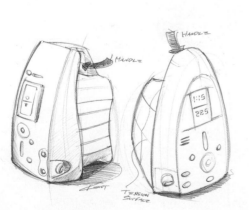

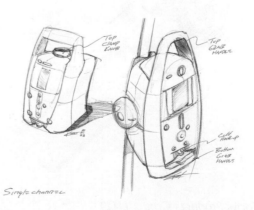

HANDS ON TUTORIALS

SolidWorks Tutorials

Parts

In this tutorial, you will create a new part. You will learn about some of the tools for adding features to a part and will explore the effect of editing some of those features.

- Launch SolidWorks from the icon on your desktop.
- Click to expand the topics listed below the "?" help icon.
- Click: **SolidWorks Tutorials.**
- Click: **All SolidWorks Tutorials (Set 1)** from the list.
- Click: **Lesson 1 - Parts** from the list that appears, and complete the tutorial on your own.

Lesson 1 - Parts - Overview

Begin with the first section or skip to a later section to bypass tasks you already know how to do.

(Image courtesy of DS SolidWorks Corp.)

REVERSE ENGINEERING PROJECT

The Can Opener Project

In this segment of the project, you will begin modeling parts of the can opener and considering the relationships that your design must preserve.

1. *Measure the pin, bushing, and rivet.* Measure the pin, bushing, and rivet labeled in the photos.

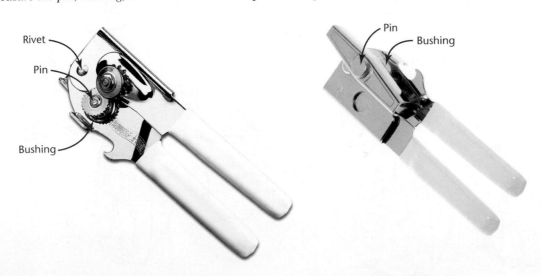

(Courtesy of Focus Products Group, LLC.)

2. *Model the pin, bushing, and rivet.* Create models for these parts.

What other features must mate to fit the pin?

Which other features must mate to fit the bushing?

Which features must fit to the rivet?

Pin

Bushing

Rivet

Pin

Bushing

Rivet

KEY WORDS

Analytical Model

Boundary Representation

Constraint-Based or Parametric Modeling

Coon's Patch

Derived Surfaces

Descriptive Geometry

Descriptive Models

Finite Element Analysis

Interpolated

Layers

Lofting

Model

Modeling Kernel

Patches

Prototypes

Surface Normal Vector

Tessellation Lines

Triangulated Irregular Networks

Virtual Prototype

Virtual Reality

Wireframe Modeling

SKILLS SUMMARY

Now that you have completed this chapter you should be able to list the qualities that models have that make them useful in the design process. You should be able to describe four CAD modeling methods that you might use to refine your design ideas. Not all modeling methods capture the same amount of information about the parts or device. You should be able to select the best modeling method to use for a particular design problem. The exercises allow you to test your understanding of the modeling process, ask you to apply a particular CAD modeling process to a design problem, provide practice with your modeling package, and ask you to decide which modeler you would choose to design a particular object.

EXERCISES

1. You are a member of a team designing an automatic door-opening device so wheelchair users can enter and exit buildings on campus. (Reference: Chapter 2, Exercise 5.) Which of the modeling methods available to your design team are appropriate in each of the following situations:
   a. Preliminary design review with your immediate engineering (technical) supervisor
   b. Discussions of installation details with the campus architectural staff
   c. Meeting with a group of nontechnical advocates for accessibility by persons with disabilities
   d. Meeting with machine shop technicians who will fabricate your prototype
2. Which modeling method contains the most information about a design? Identify a situation in which this would not be the best model to use.
3. Recent developments in Global Positioning System (GPS) technology have resulted in small, relatively inexpensive, handheld GPS receivers (Figure 5.49).

   A need exists for a mount to attach the system to mountain bikes in such a way that the display is visible while the rider is mounted. It is not desirable to modify the GPS to provide attachment. Consideration should be given to protecting the electronics and display from damage during rugged use while providing a clear sky view for the antenna. What types of CAD models could be used to develop this mounting attachment?

   At what stage of the process would each type of model be appropriate?
4. For the following situations, choose a modeling method and explain why you would use it.
   a. Designing a prosthetic limb
   b. Routing a piping system for a building
   c. Designing a one-of-a-kind machined part
   d. Verifying clearances for an electrical substation located near an airfield
   e. Mapping an oil reservoir
5. Create a 2D CAD layout of the antenna mount shown. Use the approximate dimensions shown in the orthographic drawing to define the shape of the mount. List some of the design constraints that you face as you consider alternative designs. (*Figure reprinted with the permission of Garmin Corporation.*)

5.49 A handheld GPS receiver. (*Courtesy of Garmin International.*)

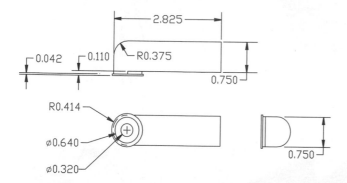

6. Using a CAD system, create a solid model of the antenna shown in Exercise 5.
7. Not all machines have moving parts. A wedge is an example of a simple machine. Use each of the modeling methods discussed in the chapter to model a wedge. Use the same size wedge for each model.

Exercises 8–11 Use your modeling package to create solid, surface, or 3D wireframe models of the items in Chapter 4, Exercises 4 a–d.

Exercises 12–39 Use your modeling package to create solid, surface, or 3D wireframe models of the objects depicted in Chapter 3, Exercises 7–34. Use the units indicated to model the objects full size.

Exercises 40–45 Use a solid or surface modeling package to revolve the shapes in Chapter 4, Exercises 7–12 about the axes shown.

Exercises 46–49 Use a solid or surface modeling package to extrude the shapes in Chapter 4, Exercises 13–16.

50. For each part shown, list the modeling method you would choose to create it if you were creating the original product definition for mass-producing these items. Give the reasons for your choices.

a. *(Shutterstock.)*

b. *(Shutterstock.)*

c. *(Sergei Devyatkin/Shutterstock.)*

d. *(Shutterstock.)*

e. *(Shutterstock.)*

f. *(Photos.com.)*

g. *(Shutterstock.)*

h. *(Shutterstock.)*

i. *(Shutterstock.)*

j. *(Photos.com.)*

k. *(Shutterstock.)*

l. *(Shutterstock.)*

m. *(Photos.com.)*

n. *(Photos.com.)*

o. *(Photos.com.)*

p. *(Shutterstock.)*

q. *(Photos.com.)*

r. *(Photos.com.)*

s. *(© stocksnapp/ Shutterstock.)*

t. *(Shutterstock.)*

u. *(Shutterstock.)*

6

CONSTRAINT-BASED MODELING AND DESIGN

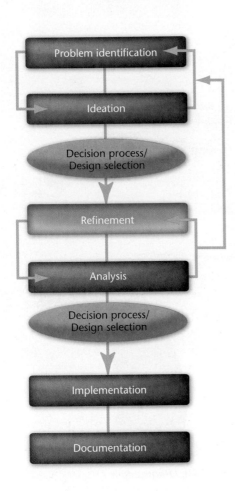

Problem identification

Ideation

Decision process/
Design selection

Refinement

Analysis

Decision process/
Design selection

Implementation

Documentation

OBJECTIVES

When you have completed this chapter you should be able to:

1. Describe how a constraint-based modeler differs from a traditional solid modeler.

2. Define design intent and how design intent is reflected in a constraint-based model.

3. Define the basic functions of a constraint-based modeler.

4. Identify the features and the role of a base feature.

5. Identify constraint relationships and driving dimensions.

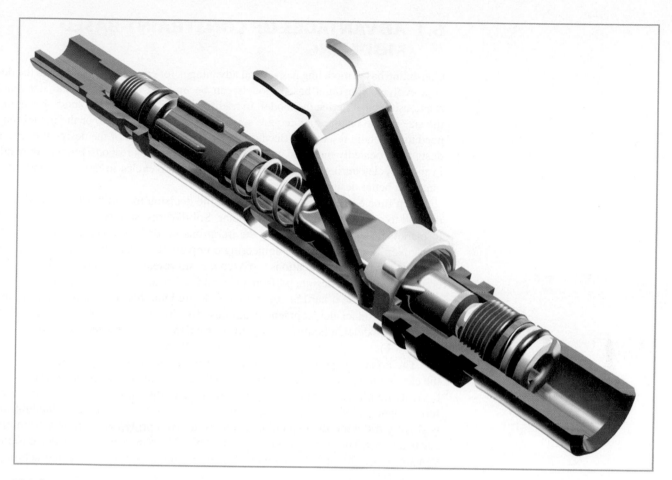

This low-pressure regulator valve assembly has been rendered in perspective and the top shell cut away to reveal the inner mechanism. *(Courtesy of Leo Greene, www.e-Cognition.net.)*

INTELLIGENT MODELS

In general, 3D models provide more information than 2D models, and 3D solid models provide more information than 3D wireframe or surface models. Constraint-based (or parametric) models add "intelligence" to the design database.

Essentially, a constraint-based model is a set of rules for how to build parts and how to join them into assemblies. Features of the constraint-based model are built by defining the geometric relationships and dimensions they must obey. When one element changes, so do any other elements that are defined in terms of that element. Through the relationships built into the model, the designer captures the design intent for the part, so that as changes are made, its model is updated in a logical way consistent with the design solution.

6.1 ADVANTAGES OF CONSTRAINT-BASED MODELING

Constraint-based modeling has several advantages for engineering design. As the design evolves, constraint-based models can be updated by changing the sizes and relationships that define the model. In traditional solid modeling methods, changing the size of one feature may require several others to be changed, or the model may need to be totally re-created. Because of the iterative nature of the design process, a design is repeatedly modified as it is refined. The model's responsiveness can result in more cycles during the refinement stage (or more cycles in less time) and, ultimately, a better design.

Nep'tune Sea Technology, Ltd. made the decision to switch from a 2D design method to a constraint-based 3D modeler, SolidWorks, so it could get its products to market faster while preserving the iterations needed for complex, accurate designs. Nep'tune is a Finnish engineering company that specializes in the design of vehicles for subsea applications—passenger and research submarines. To meet increasingly rigorous design, performance, and safety standards, the company found 2D methods made it hard to complete the desired number of design iterations and still meet product development deadlines. Its first project using the constraint-based 3D modeler, a floating bridge for ferries, took only 18 weeks to produce (see Figure 6.1).

The company enjoyed two other benefits of constraint-based modeling. First, the ease with which constraint-based models can be updated also makes it possible to create families of designs. Nep'tune's floating bridge allows cars to pass from the ferry to land and vice versa during loading and unloading. Because each landing site is slightly different, the main dimensions of a given bridge vary slightly from one site to another. The bridge manufacturer that hired Nep'tune had several designs previously created for specific landing sites. To eliminate confusion in manufacturing,

6.1 Nep'tune's parametric model of the floating bridge for ferries was modified to create a family of designs for the variations in ferry landings. *(Courtesy of Nep'tune Sea Technology, Ltd.)*

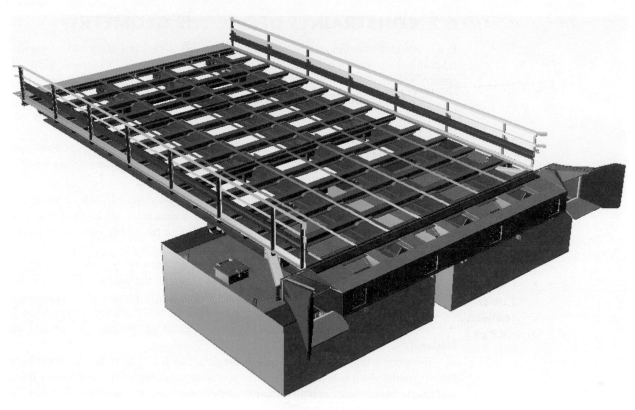

6.2 Nep'tune's bridge model, created in SolidWorks, involved more than 1700 parts. *(Courtesy of Nep'tune Sea Technology, Ltd.)*

the customer wanted a design that could be updated for new bridges. Nep'tune proposed to design one bridge, then produce the others by changing the length and width factors. The constraint-based model of the bridge made it easy to change the pertinent dimensions and have the software update the related parts to a new size. The investment in the bridge model was used to refine the original bridge design, then used again as a starting point for variations of differing sizes in the product family (see Figure 6.2).

Nep'tune also enjoyed the software's ability to analyze mass properties. The weight and volume data for a floating bridge need to be evaluated during design so that the resulting bridge floats at the correct level. Calculating the weight of a floating bridge design would have taken 1 to 2 weeks using Nep'tune's previous methods, but the built-in capabilities of SolidWorks made it possible to monitor this data throughout the design process. The ease with which the model could be modified made it possible to analyze the model, make changes, and analyze the model again. The software also made it possible for analysis to occur earlier in the design process, allowing more time to optimize the design.

Constraint-based modeling also improves designs by focusing the modeler on the design intent for the product. If a constraint-based model is to be updated successfully when changes are made, the rules for building the model must capture the intended design solution for the system or device. The extra attention and planning required to capture design intent in the model makes the designer carefully consider the function and purpose of the item being designed, which in turn results in better designs.

In this chapter, you will learn techniques for constraint-based modeling to reflect and maintain your design intent and allow the model to be updated with the expected results.

6.2 CONSTRAINTS DEFINE THE GEOMETRY

In a constraint-based model, an object's features are defined by sizes and geometric relationships stored in the model and used to generate the part. Two basic kinds of constraints are used to drive the model geometry.

- *Size constraints* are the dimensions that define the model. The choice of dimensions and their placement is an important aspect of capturing the design intent.
- *Geometric constraints* define and maintain the geometric properties of an object, such as tangency, verticality, and so on. These geometric constraints are equally important in capturing design intent.

In constraint-based modeling, a *parameter* is a named quantity whose value can change; it is similar to a variable. Like a variable, a parameter can be used to define other parameters. If the length of a part is always twice the width, you can define its size as "2 * Width," where *Width* is the name of the dimension.

Unlike a variable, a parameter is never abstract: it always has a value assigned to it, so that the model can be represented. For example, the parameter *Width* is defined by a value, such as 3.5. The value for the length parameter is calculated based on the *Width*, so in this case *Length* is 7. If the width value is changed to 5, the length is automatically updated to the new value of "Width * 2," or 10 (see Figure 6.3).

Even something as simple as this rectangular block may have design intent built into its model. As it is being updated, do you want the center of the part to stay fixed and the part to elongate in both directions, or do you want it to lengthen all in one direction? If so which end should stay put?

Figure 6.4 shows a part that has been modeled in Pro/ENGINEER, a 3D constraint-based design package. The parameters driving the geometry of the model are indicated on the drawing. When the parameter value for the length of the part was changed, the part was updated automatically, as shown in part (b) of the figure. Notice that the threaded length did not increase. If you wanted the thread length to increase as the bolt was lengthened, you would need to define the constraints and parameters differently.

Equation	Evaluates To
"Width"=3.5	3.5
"Length"="Width"*2	7

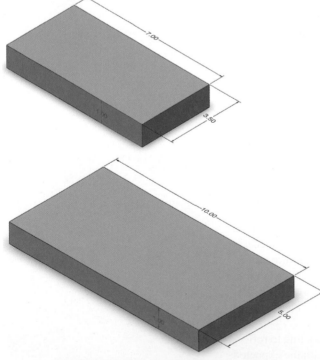

6.3   Once the parameter changes, the shape is updated.

Feature-Based Modeling

Constraint-based modeling is also called *feature-based modeling* because its models are combinations of features. Creating a constraint-based model is similar to creating any other solid model. However, in constraint-based modeling the resulting part is derived from the dimensions and constraints that define the geometry of the features. Each time a change is made, the part is re-created from these definitions. Individual features and their relationships to one another make up the constraint-based model.

A *feature* is the basic unit of a constraint-based solid model. Each feature has properties that define it. When you create a feature, you specify the geometric constraints that apply to it, then specify the size parameters. The modeler stores these properties and uses them to generate the feature. If an element of the feature, or a related part of the model, changes, the modeling software regenerates the feature in accordance with the defining properties assigned to it. For example, an edge that is defined to be tangent to an arc will move to preserve the tangency constraint if the size of the arc is changed.

Some features, such as holes and fillets, may use predefined features with additional properties that are maintained as

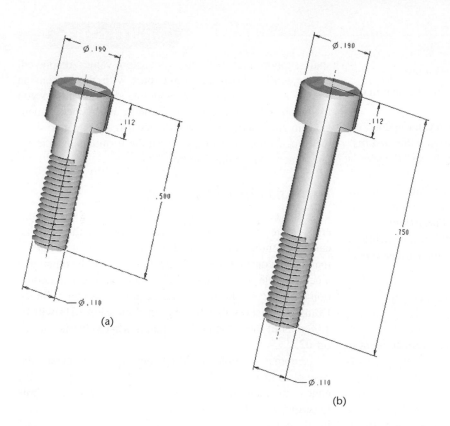

(a)

(b)

6.4 When the length dimension increased to .750, the part was updated to reflect this new size constraint. Notice that the threaded length remained the same and the bolt head remained attached after the update. The modeler maintained the size and geometric relationships that were defined.

the model is updated. For example, a "through hole" (one that goes completely through the part feature) will be extended if the part thickness increases.

When you look at the fixed-height shaft support shown in Figure 6.5(a), can you identify the features that make up its shape? Figure 6.5(b) shows the major cylinder, hole, plate, end plate, slot, and rounds that form the model of the shaft support.

Being able to update models easily to reflect changes in the design is a key ability of constraint-based modelers. If you are not using a constraint-based solid modeling program, you may have to re-create the feature just to change a size. This can result in considerable time and effort, especially for interrelated features.

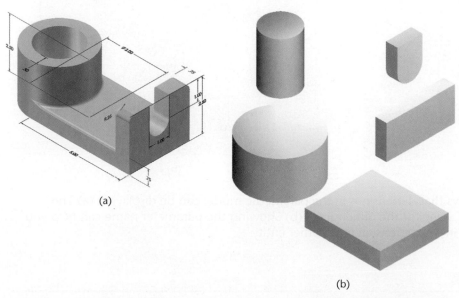

(a)

(b)

6.5 The features that make up the support in (a) are shown individually in (b).

Feature Dimensions

Dimensions in a constraint-based model can behave in different ways. A dimension can be:

- A size constraint: changing the dimension value updates the model feature. These are often called *driving dimensions*.
- A parameter used in an equation that "drives" the size of a model feature.
- A reference dimension, which gets its value from the model geometry, but is not a size constraint for the model. These are often called *driven dimensions.*
- A dimension that is purely a text note on a drawing or model view and whose value may not relate to the model geometry.
- A combination of the above.

Driving Dimensions

Driving dimensions control the size of a feature element in the model. Each driving dimension has two components: its name and its numerical value. Naming the dimensions allows them to be used in equations or relations that define other parts of the model geometry. The numerical value may be entered as a number in the definition of the dimension, or it may be derived from an equation. Figure 6.6(a) shows the

shaft support part with its dimension values displayed; Figure 6.6(b) shows the same part with the dimension names. The constraint-based modeling software allows you to switch between display of numeric and named dimensions in your model. Some modelers also allow you to show the dimensions so that they reveal the formulas used to relate dimensions to one another.

Formulas in Dimensions

Like the geometric constraints in the model, the size parameters create relationships between features of the part that can be captured in formulas. For example, if you wanted to maintain a constant material thickness around the outside of a central hole with diameter d2, you could enter the equation d1 = d2 + 1.00 for the diameter of the outer cylinder. Doing so, ensures that the driven dimension (d1) will be 1.00 overall more than d2 no matter what value is entered for d2.

Another example is adding four holes to a rectangular base plate that should be centered 0.75 inch away from the edge of the plate. If the width of the plate changes, the hole locations should also. Defining the location of the holes using the equation *h_location = width / 2 − .75,* where *width* is the plate's width dimension and *h_location* is given from

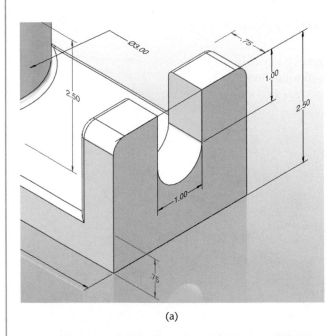

(a)

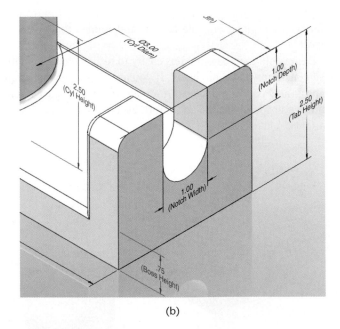

(b)

6.6 These figures illustrate two ways that dimensions in the parametric model can be displayed. (a) The numerical display shows the current value for the dimensions. (b) Showing the parameter name can help you locate dimension names to be used in the dimension for another entity.

the center of the part, will make the holes stay in the same relative positions. To change the distance from .75 from the edge to 1.00, you would just update the equation and regenerate the model.

Equations in constraint-based dimensions generally use operators similar to those used in a spreadsheet or other programming notation. Many modelers make it easy to import to and export dimension values from an outside application, such as a spreadsheet. Complex formulas can be used to calculate sizes in the spreadsheet or other program, and the resulting values can be imported back into the modeling package. Each modeling package has its own notation and syntax. Table 6.1 lists operators that can be used in equations in SolidWorks. Most modelers have a similar list.

It is important to keep track of the relationships you create. Most software allows you to name and document your constraint-based dimensions so you (and your colleagues) can interpret parameter names more readily. The software will generate a name for each dimension by default, but you should give key dimensions recognizable names so they are easier to interpret.

Parameters that are common to more than one part in an assembly are called **global parameters.** Naming global parameters that can be used throughout an assembly is a powerful means of ensuring that parts work together and that critical dimensions are updated in all parts of an assembly. With global parameters, it is even more important to choose names that are clear and to document their purpose in the notes field.

Driven and Cosmetic Dimensions

Driven or *reference* dimensions are associated with the model geometry but are not used to create the constraint-based model. Sometimes it is necessary to add a dimension to a drawing that was not needed to create the model. A driven dimension is a one-way link from the database; it cannot be used to change the model, but it will be updated if the model changes.

A **cosmetic dimension** has no link to the model; it is simply a text label. To dimension a feature at a size different from what is in the model, you must use a cosmetic dimension or otherwise break the link to the database value for the feature.

You should not often need driven dimensions. A well-created constraint-based model contains all the dimensional relationships needed.

Table 6.1 Operators.

Operator	Name	Notes
+	plus sign	addition
−	minus sign	subtraction
*	asterisk	multiplication
/	forward slash	division
^	caret	exponentiation
sin (a)	sine	a is the angle; returns the sine ratio
cos (a)	cosine	a is the angle; returns the cosine ratio
tan (a)	tangent	a is the angle; returns the tangent ratio
sec (a)	secant	a is the angle; returns the secant ratio
cosec (a)	cosecant	a is the angle; returns the cosecant ratio
cotan (a)	cotangent	a is the angle; returns the cotangent ratio
arcsin (a)	inverse sine	a is the sine ratio; returns the angle
arccos (a)	inverse cosine	a is the cosine ratio; returns the angle
atn (a)	inverse tangent	a is the tangent ratio; returns the angle
arcsec (a)	inverse secant	a is the secant ratio; returns the angle
arccosec (a)	inverse cosecant	a is the cosecant ratio; returns the angle
arccotan (a)	inverse cotangent	a is the cotangent ratio; returns the angle
abs (a)	absolute value	returns the absolute value of a
exp (n)	exponential	returns e raised to the power of n
log (a)	logarithmic	returns the natural log of a to the base e
sqr (a)	square root	returns the square root of a
int (a)	integer	returns a as an integer
sgn (a)	sign	returns the sign of a as −1 or 1; For example: sgn(−21) returns −1
pi	pi	ratio of the circumference to the diameter of a circle (3.14...)

6.7 The window at the left side of the screen called the "model tree" (enlarged here) lists the features in the order in which they were created. The icon next to the feature indicates the operation used to create the feature. Extrusions and extruded cuts were used to form the cylinder, baseplate, and hole. Then, the end feature was extruded and the slot created using an extruded cut. Finally, fillets were added to round the edges.

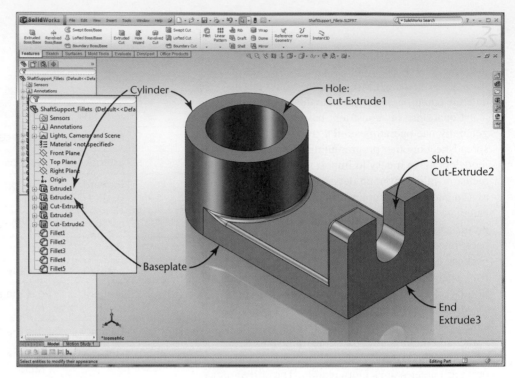

Using the shaft support as an example again, a relationship can be defined so the diameter of the hole is always 1 inch smaller than the diameter of the cylinder. Changing the size of the cylinder will cause the diameter of the hole to change to preserve the size relationship.

In addition to defining relationships between features, constraint-based modeling software also allows you to use constraints and parameters across parts in an assembly. Thus, when a part changes, any related parts in the assembly can also be updated. Because the software regenerates the features and parts from the relationships stored in the database (see Figure 6.7), planning constraint-based relationships that reflect the design intent of the part or product is the key to efficient and useful constraint-based models.

6.3 PLANNING PARTS FOR DESIGN FLEXIBILITY

Design intent refers to the key dimensions and relationships that must be met by the part. Where does the part need to fit into other parts? Do certain features need to align with other features? How will the part change as the design evolves? By thinking about these relationships, you can organize the features so that the structure of the part and its relationships are the foundation for later features. Four key aspects of planning for design intent are summarized in Figure 6.8. Of course, you cannot know everything about a design before you begin, and there will be changes that will require more work than others; but starting your model with the geometric relationships in the design in mind will allow you to benefit from the power of constraint-based modeling. To see how a constraint-based model can reflect design intent, you should understand some basics about constraint-based modeling software.

Constraint-based modeling software in many ways parallels the design process. To start creating a part, you often start with a 2D sketch of the key shape of a feature, such as that shown in Figure 6.9. The initial sketch may be rough in appearance because the software will apply constraints to define the relationships between the simple 2D geometric elements of your sketch. This process is similar to the way one of your engineering colleagues interprets your hand-drawn

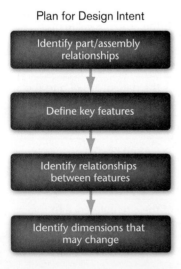

6.8 Four Key Considerations When Planning a Parametric Model

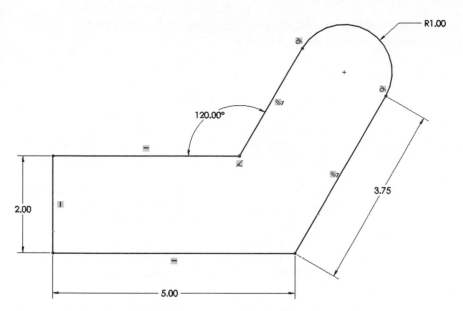

6.9 A constrained sketch is used as the basis for a model feature.

sketches. If lines appear to be perpendicular or nearly so, your colleague assumes that you mean them to be perpendicular, without your having to mark the angular dimension between the lines. In a similar way, constraint-based modelers apply constraints to your rough sketch, so that lines that appear nearly perpendicular, parallel, vertical, horizontal, concentric, or collinear will have a constraint relationship added to them so that they function that way in the model. This process is called "solving" the sketch in some modeling software, because the software is interpreting and assigning dimensions and constraints to generate a 2D "profile" of the feature. Before you generate the actual feature, you can review and change the dimensions and geometric constraints so they are what you intended. This step is key to defining your design intent. The sketch constraints in Figure 6.9 have already been added.

To generate a 3D feature, you then select a command to extrude, revolve, sweep, or blend the profile geometry. Figure 6.10 shows the profile in Figure 6.9 as it would

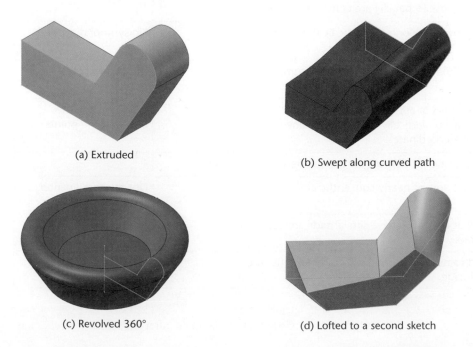

(a) Extruded

(b) Swept along curved path

(c) Revolved 360°

(d) Lofted to a second sketch

6.10 The same sketch but a different operation was used to create each of these features. The sketch in (a) was extruded, in (b) swept along a curved path, in (c) revolved 360° about an axis, and in (d) lofted between the sketch and a second sketch with angled lines and a straight top.

appear after being (a) extruded, (b) swept, (c) revolved, and (d) lofted to create different types of features from the 2D sketch.

Sketch Constraints

Constraint-based modeling software starts by automatically interpreting your sketch, to define the constraints that it will hold constant. To do so, it evaluates your sketch against a set of rules stored in the software. A line, for example, that is drawn within 3° or 4° of being horizontal will be constrained to remain horizontal. Your sketch will be altered so the line you drew is, in fact, horizontal. The software will apply the constraints it needs to solve the sketch.

Constraint-based modeling programs allow you to review the constraints that are applied to the sketch geometry so you can change them to those needed to reflect your design intent. Table 6.2 describes some possible sketch constraints and shows symbols often used to indicate them on screen. (Different modelers use different symbols and means of displaying the constraints—see Figure 6.11 and Table 6.3). These constraints may create relationships to other lines in the sketch or to geometry in an existing feature.

Table 6.2 Sketch Constraints.

Constraint	Rule	Symbol
Horizontal	Lines that are close to horizontal are constrained to be horizontal (usually you can control the number of degrees for this assumption, typically set to 3° or 4°).	H
Vertical	Lines that are close to vertical are constrained to be vertical (usually within 3° or 4°).	V
Equal length	Line segments that appear equal length are constrained to be equal.	L_1, L_2, etc.
Perpendicular	Lines that appear close to 90° apart are constrained to be perpendicular.	⊥
Parallel	Lines that appear close to parallel are constrained to be parallel.	$//_1$
Collinear	Lines that nearly overlap along the same line are assumed to be collinear.	No symbol displayed
Connected	Endpoints of lines that lie close together are assumed to connect.	Dot located at intersection
Equal coordinates	Endpoints and centers of arcs or circles that appear to be aligned horizontally or vertically will be constrained to have equal X-or equal Y-coordinates.	Small thick dashes between the points
Tangent entities	Entities that appear nearly tangent are constrained to be tangent.	T
Concentric	Arcs or circles that appear nearly concentric are concentric.	Dot at shared center point
Equal radii	Arcs or circles that appear to have similar radii are constrained to have equal radii.	R
Points coincide	Points sketched to nearly coincide with another entity are assumed to coincide.	No symbol displayed

Constraining a Sketch

Like a hand-drawn sketch, the sketch for a constraint-based model captures the basic geometry of the feature as it would appear in a 2D view.

First, sketch the basic shapes as you would see them in a 2D view. Many modelers will automatically constrain the sketch as you draw unless you turn this setting off in the software.

Second, apply geometric constraints to define the geometry of the sketch. If it is important to your design intent that lines remain parallel, add that constraint. If arcs must remain tangent to lines, apply that constraint. Here, lines A and B have been defined to be parallel; note the parallel constraint symbol.

Finally, add dimensional constraints. The length of line B was sketched so that the software interpreted the dimensional constraint to be 3.34. The designer changed this dimension to 3.75 (the desired length), and the length of the line was updated to the new length (c).

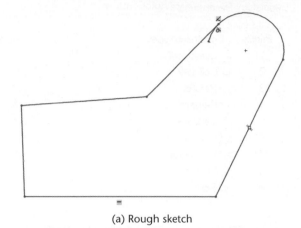

(a) Rough sketch

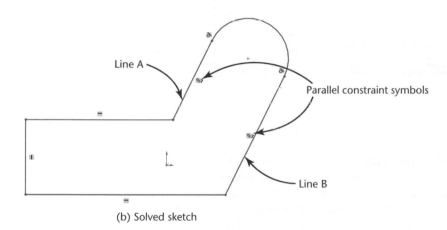

(b) Solved sketch

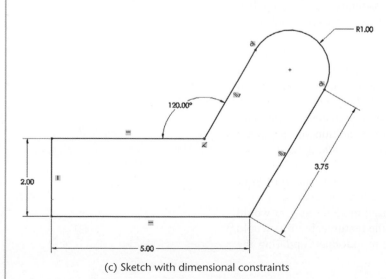

(c) Sketch with dimensional constraints

TIP

It is often useful to start drawing the feature near the final size required. Otherwise if the software is automatically constraining your sketch, a line segment that is proportionately much shorter may become hard to see or even considered effectively zero length and deleted by the software.

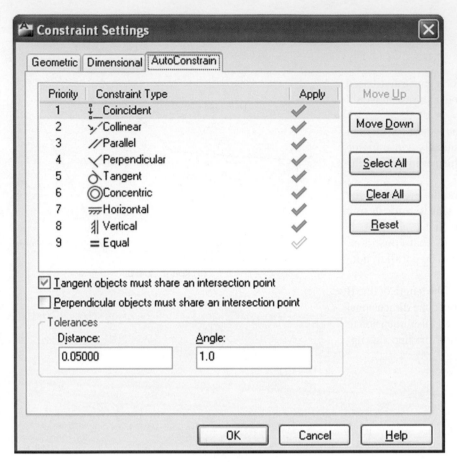

6.11 The **Constraint Settings** dialog box in AutoCAD lets you change the priority in which the constraints are applied. You can also set the tolerance for sketched endpoints to be considered coincident, and lines to be assumed horizontal or vertical. Understanding which assumptions are being applied to your sketch can help you define the exact geometry you require. *(Autodesk screen shots reprinted with the permission of Autodesk, Inc.)*

Not all constraint-based modeling programs apply all these constraints, nor do they apply them in the same fashion. Study the constraints available in the software you use and determine their effect on the sketch. You also may be able to set the threshold values for various constraints. For example, instead of having lines at 1° be automatically constrained as horizontal lines, you may wish to increase this value to 3° (Figure 6.11). If you are having difficulty getting the constraints you desire applied to the sketch, you may want to turn off automatic constraint application and select each specific constraint to apply to your sketch.

Most constraint-based modeling programs let you remove or override a constraint that is applied, or specify that the geometry is exact as drawn. Sometimes, you may have to force the sketch closer to the desired relationship. For example, if two lines are too far off in the sketch to be constrained as perpendicular by the software, you could add an angular dimension of 90° between them to force them to that relationship. With the dimensions driving the geometry to the condition you want, the software may apply the perpendicular constraint. Once the constraint is applied, you often can delete the dimension you added.

It is usually beneficial to let geometric constraints determine much of your sketch geometry, then dimension the sizes and relationships that cannot be defined using constraints. You could use a constraint-based modeler as you would a solid modeler, adding dimension values to define all the features, but this would not build in the intelligence that allows the software to automate updating the model.

Table 6.3 Selected Constraint Relations in SolidWorks.

Symbol	Relation	Entities to Select	Result
	Horizontal or Vertical	One or more lines or two or more points	The lines become horizontal or vertical (as defined by the current sketch space). Points are aligned horizontally or vertically.
	Collinear	Two or more lines	The items lie on the same infinite line.
	Coradial	Two or more arcs	Items share the same center point and radius.
	Perpendicular	Two lines	The two items are perpendicular to each other.
	Parallel	Two or more lines. A line and a plane (or a planar face) in a 3D sketch	The items are parallel to each other. The line is parallel to the selected plane.
	ParallelYZ or ParallelZX	A line and a plane (or a planar face) in a 3D sketch	The line is parallel to the YZ (or ZX) plane with respect to the selected plane.
	Tangent	An arc, ellipse, or spline, and a line or arc	The two items remain tangent.
	Concentric	Two or more arcs, or a point and an arc	The arcs share the same center point.
	Midpoint	Two lines or a point and a line	The point remains at the midpoint of the line.
	Intersection	Two lines and one point	The point remains at the intersection of the lines.
	Coincident	A point and a line, arc, or ellipse	The point lies on the line, arc, or ellipse.
	Equal	Two or more lines or two or more arcs	The line lengths or radii remain equal.
	Symmetric	A centerline and two points, lines, arcs, or ellipses	The items remain equidistant from the centerline, on a line perpendicular to the centerline.

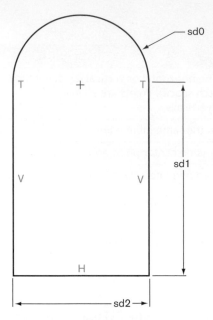

6.12   This sketch is overconstrained.

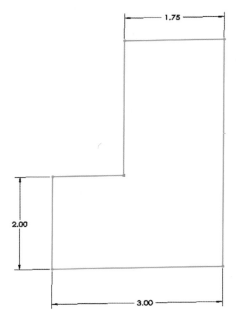

6.13   This sketch is underconstrained.

Overconstrained Sketches

Sketches used in constraint-based modeling may not be overconstrained. An overconstrained or overdimensioned sketch is one that has too many things controlling its geometry.

The sketch shown in Figure 6.12 is overconstrained. Because the top arc is constrained to be tangent to both vertical lines, and its radius is dimensioned (sd0), the width of the sketch is defined. The horizontal dimension shown at the bottom (sd2) defines this width again. If both the arc radius dimension and the overall width dimension were allowed in the model and one of them changed, it would be difficult or impossible for the part to be updated correctly. If the overall width changed and the tangent constraint or arc radius did not, the sketch geometry would be impossible.

Underconstrained Sketches

An underconstrained or underdimensioned sketch is one that has too few dimensions or constraints and is therefore not fully defined.

The sketch shown in Figure 6.13 is underconstrained. It lacks a height dimension controlling its vertical size. Without this dimension the sketch is not fully defined. Some software will allow you to create a 3D feature from an underconstrained sketch as long as the sketch follows basic rules—for example, the sketch does not cross itself, and the sketch encloses an area. If the sketch is underconstrained, the software assumes values for unspecified dimensions based on the sizes drawn. This can be helpful when you are trying out ideas, but it is best to plan to create a model that reflects your design intent.

Dimensions can easily be changed by typing a new value, so go ahead and guess the size and change it later if needed; but keep in mind that choosing which dimensions to use to define the sketch is a critical aspect of making sketches and features that will be updated as expected when a dimension changes.

Applying Constraints

You should select dimensions that define the proper relationships between the sketched geometry and other existing features in your model. The surfaces between which you choose to create the dimensions are a large factor in creating a model that will be updated in the desired manner. You should not let the rules of your software package determine the dimensions, as it is unlikely that they will reflect your design intent.

When you are dimensioning a sketch used to create a feature, place the dimensions where you can view them clearly. Later, if you create a drawing from your model, you will be able to clean up the appearance of the dimensions if needed. You can move dimensions, change their placement, and arrange them so they follow the standard practices. Good placement practices such as placing dimensions outside the object outline and keeping the dimensions a reasonable distance from one another will make the dimensions in your model easier to read.

3D models are accepted as final design documentation. When you are going to use the model as the design database and not provide drawings, it is even more important to ensure that dimensions and tolerances shown in the model are clear. You will learn more about practices for documenting inside the 3D model in Chapter 9.

If you do a good job selecting the dimensions that control the drawing geometry, you will have little cleanup work to do later when you are creating drawing views.

Setting the Base Point

Some constraint-based modeling software assumes a **base point** in the sketch for the first feature you create. The base point is fixed on the coordinate system; the other sketch geometry is located on the coordinate system based on its relationship to this base point. The resulting feature will stay fixed on the coordinate system at this base point. When dimension values change, they will change relative to that point. There is only one base point per model.

The base point is indicated in Figure 6.14(a) by the × in the lower right corner of the sketch. When the width dimension changes, the sketch is updated to the new size while leaving the base point fixed on the coordinates. The rest of the sketch geometry is reoriented, as shown in part (b).

If your software uses a base point, it should become the base point for the entire model. When the model is updated, all dimensions will be updated from this point. A good rule of thumb is to identify a fixed point on your model that you would use as a starting point for measuring the part for inspection purposes. Use this point as the base point, or locate your dimensions from it as if it were the base point.

Locating most of your dimensions from a common edge or major surface can be effective for two reasons. First, it helps you create dimensions that will be useful for inspecting the part after manufacture; the dimension values in your model will measure from the same point that an inspector will use to measure the finished part. Second, it helps you anticipate how the model will be updated. When a value is changed, the features on the model will be updated (move) relative to the edge referenced in their dimensions. When you locate dimensions relative to other features you may

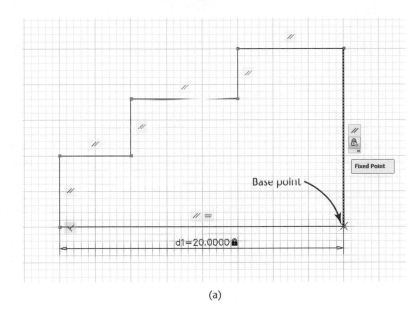

(a)

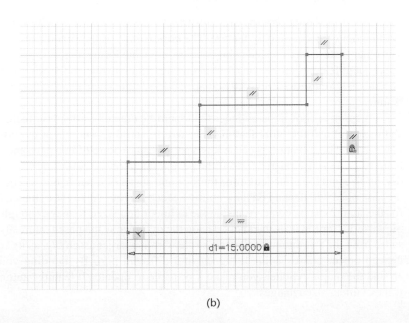

(b)

6.14   The base point fixes the sketch on the coordinate system in this underconstrained sketch. The base point serves as a fixed point when the model is updated. In AutoCAD, adding a Fixed Point constraint lets you identify the base point; otherwise, it is determined by software rules.

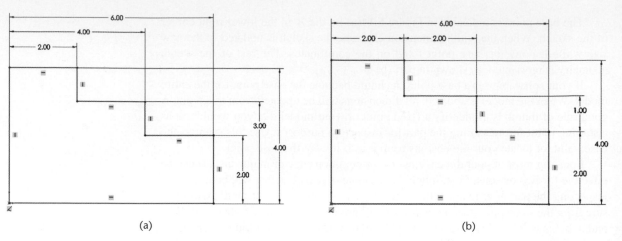

(a)   (b)

6.15   These two sketches of the same part have very different relations built into the dimensions for the sketch. The sketch in (a) relates all dimensions to the left and bottom edges of the part. The sketch in (b) has a chain of relationships built into the dimensions.

have difficulty predicting how the model will be updated because you have to add the effect of changes to the intermediate features.

More important than any rule of thumb governing dimension selection or placement is to consider thoroughly the parts you are designing and to make the drawing geometry, constraints, and dimensions reflect the design intent for the part. In some cases, dimensioning a part from another feature is required to reflect the design intent. Figure 6.15 shows two sketches that have the same geometry but are dimensioned differently. When the 2.00 height dimension is changed to 2.50, how will each of these drawings be updated?

In the first case, shown in (a), the height of the 2.00 "step" will change to 2.50, narrowing the gap between it and the 3.00 step. In (b), the height of the 2.00 step will change to 2.50, and the height of the next step will be updated to 3.50 units so it remains 1.00 unit above it. In this case, the narrow gap will occur between the 3.50 and the 4.00 step. Either is correct to the extent that it reflects the intended design.

6.4 THE BASE FEATURE

To create useful relationships in a constraint-based model, it is helpful to start out your model with a good base feature. The **base feature** is the first feature you create. The other features are created using relationships that locate them to the base feature. Subsequent features are updated based on their relationships to the base feature or to one another. You should generally create a fairly large, significant feature as the base feature.

Figure 6.16 shows the cylinder that could be created as the base feature for the shaft support and the other features that will relate to it. The major cylinder would make a good base feature because it is a significant feature of this part, and the other features make sense related to it. If the design intent of the part centers on the cylinder, its size and location determine the size and shape of other features. The width of the base plate will depend on the stability needed for the size of the cylinder. The hole for the shaft must be centered inside the cylinder and sized to leave enough remaining material for the wall thickness. Using the cylinder as the base feature also makes sense from the standpoint of the part's role within a larger assembly, as its size could depend on the shaft's being designed in a different part. By building your model around the major cylinder, you make it possible to update all features automatically if the cylinder size changes.

Adding Features to the Model

When you add features to the model, the way you constrain the sketch determines how it relates to the other features. Just as the base point serves to locate the base feature, at least one of the constraints applied to subsequent sketches must tell the software how the new feature relates to an existing feature.

To create the second feature, you can use a dimension to locate it relative to an existing edge or use a geometric constraint to align one of the sketch entities to the existing geometry. For example, you could dimension the center point of a hole to be a certain distance from the end of an existing feature or constrain the hole to be concentric to a cylinder or arc on an existing feature.

The way you dimension, constrain, or align your second feature determines how it will be updated with respect to the first feature. Use the things you know about the design to add features. Where are the fixed points on the model, and how does the feature you are adding relate to them? What are the conditions that the new feature needs to satisfy at all times? If existing features were to change, how would the feature you are adding need to change? In some cases, a simple dimension from an existing edge may be all that is needed to add the feature. In other cases, you will want to build in geometric or size relationships needed in the design.

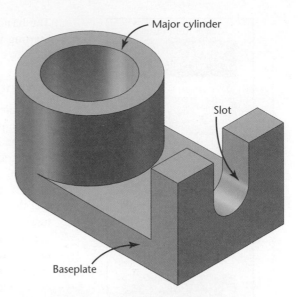

6.16 The major cylinder for the shaft support makes a good base feature because the cylindrical shaft to be supported is an important aspect of this design.

Use an Existing Edge

Pro/ENGINEER (as well as many other modelers) has a sketch command that allows you to specify the edge on an existing feature to project as a line in your sketch. This command creates the line in the sketch and constrains the sketched line to be collinear with the existing geometry in a single step. Even better, if the edge on the feature moves, the sketched entity moves with it to stay in alignment.

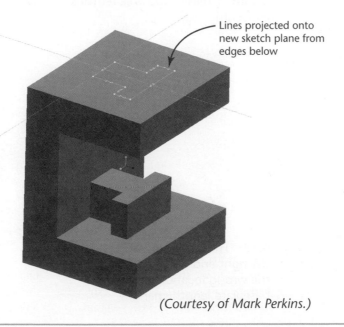

(Courtesy of Mark Perkins.)

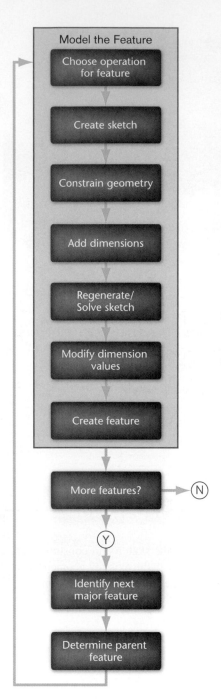

6.17   Flowchart of Constraint-Based Modeling

The basic process of constraint-based part modeling is illustrated in Figure 6.17. Starting with the base feature, you build a model by adding features and relating them to existing features. For each feature, the same basic process applies.

Parent-Child Relationships

Constraint-based modeling software stores information about each feature and its relationship to its parent feature. The term *parent-child relationship* is used to describe how one feature is derived from another feature. This relationship is defined by the geometric constraints applied to the sketch as well as its dimensioned size constraints. Planning when creating a parent-child relationship is important.

The base feature is the parent feature for the next feature created from it. Just as real children depend on their parents, child features are dependent on their parent features. If a parent feature moves, the child feature moves in relation to it. For example, when the hole feature was added to the shaft support part, its location was defined as concentric with the major cylinder. When the cylinder's location changes, the hole feature remains concentric with the cylinder (Figure 6.18). If the hole is located relative to the baseplate or to some other feature and the cylinder is moved, the hole will not move with it (Figure 6.19).

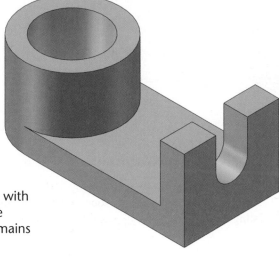

6.18   The hole is concentric with the cylinder. As the baseplate length increases, the hole remains concentric.

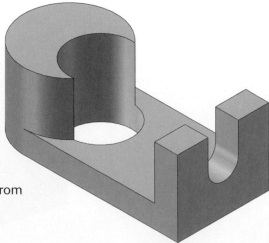

6.19   The hole dimensioned from the right end of the part is in the wrong location when the baseplate length changes.

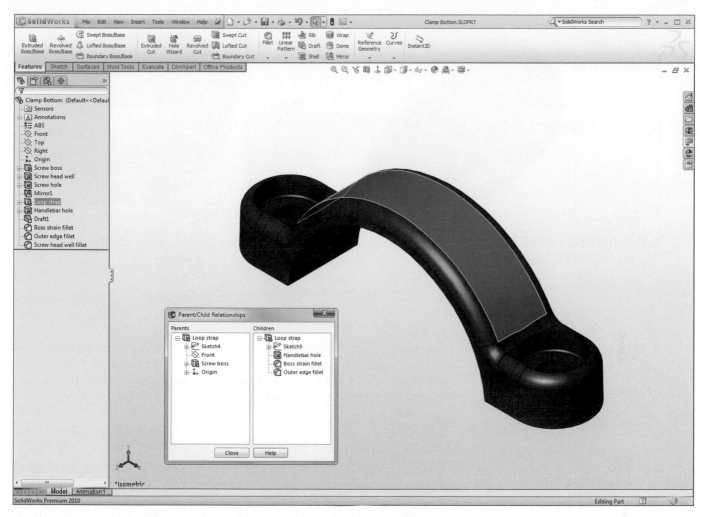

6.20 A tree diagram showing the dependencies between features can help you plan your parametric model. Notice how the list of parent and child features from this SolidWorks model shows the dependencies between features *(Courtesy of Salient Technologies, Inc. (www.salient-tech.com))*

If you delete a feature, you will also delete its child features unless you redefine them so that they are related to a different parent feature. This is another reason for selecting a good base feature; you do not want the first feature you create to be something that might be deleted later in the design. Not all features will have child features, but all features other than the base feature must be related to some parent feature.

One way to visualize the parent-child relationships as you plan your model is to build a tree diagram. The relationships in the Clamp Bottom model in Figure 6.20 are illustrated in the **Parent/Child Relationships** window. The feature named Loop strap has Handlebar hole, Boss strain fillet, and Outer edge fillet as its children, along with the sketch that formed its cross-sectional shape. These features may or may not have children (or "grandchildren") of their own. The parent features for Loop strap are the Sketch4, the Front datum plane, Screw boss feature, and the Origin of the coordinate system.

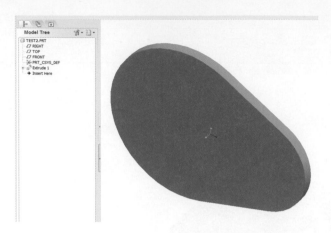

6.21 The base feature is the first feature created. *(Courtesy of Mark Perkins.)*

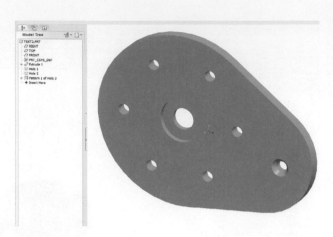

6.22 Subsequent features are children of previous features. *(Courtesy of Mark Perkins.)*

Some constraint-based modelers, such as Pro/ENGINEER, allow you to "play back" the model to see how it is being generated by the underlying parameters. The sequences shown in Figure 6.21 and Figure 6.22 illustrate how the features used to create the parts are ordered in relationships that reflect the design intent for the parts.

Datum Planes and Surfaces

In engineering drawing, a *datum* is a theoretically exact point, axis, or plane. A datum provides a reference surface, axis, or point used for inspecting the manufactured part. Theoretically exact planes used in modeling are called *datum planes*. They are planes used as a reference for model geometry. You can also create datum axes and datum points in 3D models. Features can be constrained and dimensioned relative to these planes, axes, and points, even though they are not solid objects and add no mass to the model. Some modeling packages provide a set of three mutually perpendicular datum planes, located so that they pass through the origin of the coordinate system, as a default. You typically may also create additional datum planes at other orientations and locations.

Often, it is very useful to use a set of datum planes (three mutually perpendicular planes) as the base feature in your drawing. In fact, most software provides them as the default starting features. Figure 6.23 shows the shaft support part with a set of datum planes used as the base feature.

One of the benefits of starting with a set of datum planes is that they can provide a set of normal surfaces that you can use to orient sketch geometry. This is particularly helpful when there are no planar surfaces on the part that you can use to build subsequent features easily. With a cylinder as the base feature for the shaft support, for example, there is no flat surface to which the end of the base plate can be constrained to be parallel. The datum plane Right Plane in Figure 6.23 provides a reference to which the plate is parallel.

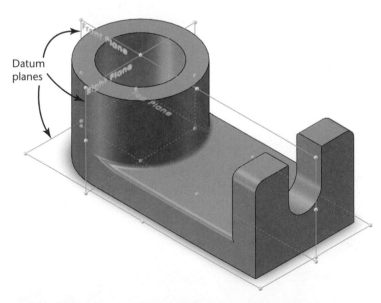

Datum planes

6.23 The three mutually perpendicular planes in this drawing serve as the base feature for the model.

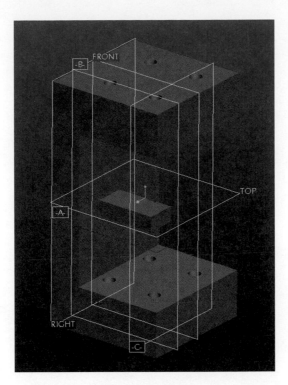

6.24 Surface C on this fixture is a datum surface or reference location that will be used to inspect the location of a part feature. *(Courtesy of Mark Perkins.)*

Whether or not you start with a set of datum planes as a base feature, consider the surfaces that will serve as datum surfaces on your part. Parts are inspected to see whether they were manufactured acceptably based on a similar set of three mutually perpendicular, theoretically exact planes. Measurements of the finished part are taken from surfaces coincident with one of the datum planes. As you define your features consider the dimensions that will determine whether the part will fit. How will the part be measured after manufacture? If there is a surface from which measurements will be taken, it can be advantageous to identify it as a datum surface in your part (see Figure 6.24).

Like the base point in a sketch, a datum plane serves as a reference from which the other features build. When you define a set of datum planes as the starting point for your model, you will be well on your way when deciding which datum surface to identify for use in inspecting the finished part.

Datum surfaces can also be helpful when defining tolerances for features. Defining your model using the dimensions that will be used to inspect the finished part will make it easier to determine the impact of variations from the stated dimension that occur during manufacture. (You will learn more about tolerances in Chapter 10.)

Datum planes also provide a common reference point for parts in an assembly. When parts share a set of datum planes, it is easier to align the part with other parts when you bring them together in an assembly model. If you do not use a set of datum planes as a base feature, you can locate new features from existing surfaces on the base feature.

6.25 This constrained sketch acts as the skeleton for the bike assembly. It is made only of circles, lines, and points, which have no mass. *(Courtesy of Santa Cruz Bicycles.)*

6.26 Dimensions are added to the constrained sketch defining the basic size relationships. These dimension values can easily be changed to try out different relationships. *(Courtesy of Santa Cruz Bicycles.)*

6.27 Parts can be assembled to, and even get their sizes directly from, the skeleton. This way parts can be viewed as assembled, even though all parts are not yet modeled. *(Courtesy of Santa Cruz Bicycles.)*

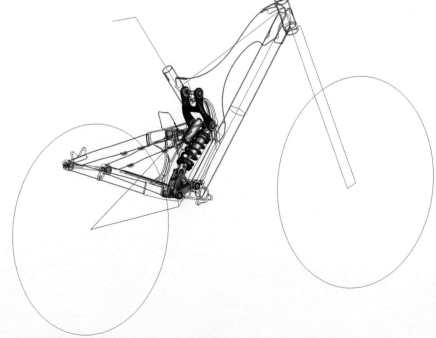

Creating datum planes, lines, and points as base features can be used as a strategy called "skeleton modeling" (see Figures 6.25–6.27). Because the datums have no mass, they do not affect mass property calculations. They are easy to model and often do not change. For example, if you create a datum axis that is the centerline of the part, this is unlikely to change, whereas the shape of the part might. Even when some parts are not fully modeled or are represented only as a centerline, they can still be assembled using the datums as references.

Datum planes and datum surfaces can be added and identified at any time, but considering them as you plan your model will help you use them more effectively in capturing design intent in your model.

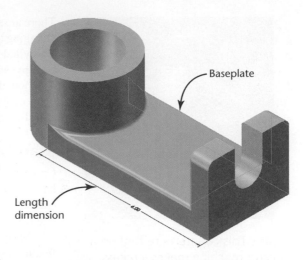

6.5 EDITING THE MODEL

One of the biggest advantages of constraint-based modeling is that you can edit the features after they are created. If you have several parts open at a time in your software, you may have to identify which is the active part, then select the feature you want to edit. Once you select the feature you wish to edit, the dimensions for the feature are displayed in a way that lets you change their values. Alternatively, you can also return to the original sketch that you used to create the feature and modify the sketch. Figure 6.28 shows the shaft support drawing with a change being made to the overall length of the part.

6.28 The length of the shaft support is changed by editing the dimension value and updating the model.

Depending on how the model was created, you may have more or less success in updating the model. For example, the baseplate that connects to the cylinder (indicated in the figure) can be created many different ways. Each way may have advantages or disadvantages for how the part will be updated.

Figure 6.29 shows three ways the baseplate could be generated:

a. Sketch the shape of the bottom surface on a sketch plane aligned with the bottom of the cylinder and extrude it upward.
b. Sketch the cross section on the midplane and extrude to both sides.
c. Sketch the rectangular end view on a plane offset parallel to the center datum plane and extrude the plate to meet the cylinder feature.

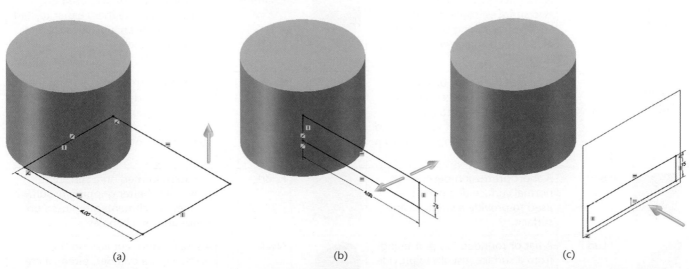

(a)   (b)   (c)

6.29 (a) Extruding upward; (b) extruding to both sides of a center plane; (c) extruding to the cylinder from an offset plane. All these methods could be used to generate the plate feature for the shaft support, but they update differently.

6.30 This counterbore tool on a milling machine makes a recess around the top of a hole. The center of the tool fits inside the hole to keep the counterbore concentric.

There are several other ways, too. Each of these methods works about the same when changing the length of the part. They vary if you decide to make the plate wider than the cylinder. Often, when a feature is extruded to an existing feature as in (c), the new feature cannot be updated so that it is wider than the surface to which it is extruded. The boundary, or end, of the feature is limited by the next surface. Where there is no portion of the surface at which the feature may end, it will continue to infinity and therefore be undefined.

Standard Features

Certain useful features are part of many engineering designs. Common features often have specific manufacturing processes developed to form them efficiently. For example, a **counterbore** is often used to create a recess for a bolt head or other fastener. Specific tools like the one shown in Figure 6.30 make it easy to form this feature.

Many constraint-based modeling programs use the same terminology for these features as that used in manufacturing and design. Review the standard feature types in Table 6.4.

Working with Built-in Features

Many standard features, such as counterbored and countersunk holes, fillets, lugs, slots, and others may be available in the constraint-based modeling program as **built-in** or **placed features**. These features can be placed with a dedicated command on part surfaces. Some features are solids, and others are negative solids that are subtracted from the model (for example, a cylindrical hole). The features have properties built into them that the modeling software will preserve, just as it preserves the geometric and size relationships you set.

Many constraint-based modelers provide a number of options for quickly placing holes, often through a "hole wizard" or other dialog box. Properties that can often be set for holes are listed in Table 6.5.

Table 6.4 Typical Placed Features.

	Fillet	A rounded interior blend between two surfaces; some uses are to strengthen joining surfaces or to allow a part to be removed from a mold		Chamfer	A small angled surface, used on cylinders to make them easier to start into a hole, or on a plate to make it easier to handle
	Round	A rounded exterior blend between two surfaces; used to make edges easier to handle, improve strength of castings, and make parts easier to remove from a mold		Keway/ Keyseat	A shaped depression cut along the axis of a cylinder or hub (shown) to receive a key, used to attach hubs, gears, and other parts to a cylinder so they will not turn on it
	Boss	A short protrusion beyond the normal surface of a part, often used to provide a strong bearing surface		Knurl	A pattern formed on a surface to provide for better gripping or surface area for attachment; often used on knobs and tool handles
	Lug	A flat or rounded tab protruding from a surface, usually to provide a method for attachment		Neck	A small groove cut around the diameter of a cylinder, often where it changes diameter
	Flange	A flattened collar or rim around a cylindrical part to allow for attachment		Bushing	A hollow cylinder that is often used as a protective sleeve or guide, or as a bearing

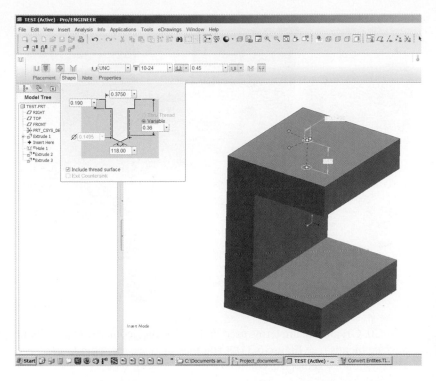

6.31 Pro/ENGINEER's Hole Wizard has many placement options as well as types of hole features available. *(Courtesy of Mark Perkins.)*

Table 6.5 Typical Hole Properties.

	Through	The hole goes all the way through the part or feature.
	Blind	Material is removed to a specified depth to form the hole.
	Countersink	A conical shape is also removed to allow for a countersunk tapered screw head.
	Counterbored	A cylindrical recess is formed, usually to receive a bolt head or nut.
	Spotface	A shallow recess like a counterbore is formed, used to provide a good bearing surface for a fastener.

A Hole command may also allow you to choose the specialized constraints for locating holes, such as concentric, placed by edge, and on point, as illustrated in Figure 6.31.

A *fillet* is a common built-in feature that rounds the edge formed by two surfaces. Typically, a fillet can be created by selecting the edges to be filleted (several can be selected at once) and specifying the radius of the fillet. Rounds are another type of standard feature created in the same way, usually using the same Fillet command. When the surfaces move or are lengthened, the fillet or round adjusts automatically. Figure 6.32 shows a typical fillet and a typical round with a uniform radius.

Sophisticated constraint-based modelers allow you to create uniform radius fillets and rounds, chained fillets and rounds that blend around a corner of a surface, constant-chord fillets and rounds, and variable-radius fillets and rounds (shown in Figure 6.33) using different methods. A constant-chord fillet is used in situations similar to (b), where a constant-radius fillet would result in thicker material on the uphill and downhill side of the blend. The constant-chord distance creates a fillet with a uniform amount of added material. More sophisticated blends between surfaces are also available in some software.

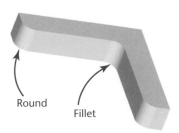

6.32 Fillet and Rounds

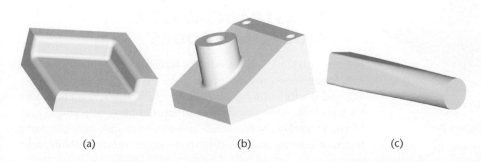

(a)   (b)   (c)

6.33 (a) Chained Fillets and Rounds; (b) Constant-Chord Fillet; (c) Variable-Radius Round

6.34 (a) The iPhone at right was modeled with tangent surfaces: C1 continuity. (b) The smoother blends of the model at the right use C2 continuity. *(Courtesy of Mark Biasotti.)*

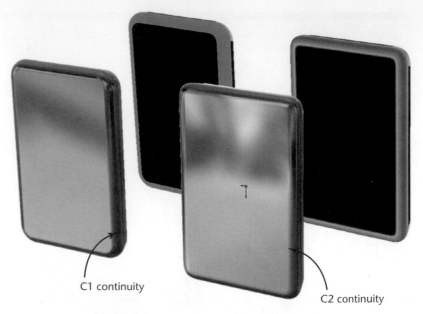

C1 continuity

C2 continuity

C0, C1, and C2 are terms used to describe the geometric continuity of surface blends (see Figure 6.34). (You may also see the letter G used as a designation for geometric continuity.)

*   *C0:*   Coincident surfaces meet but are not tangent. These are essentially intersecting surfaces, not ones that have a fillet.
*   *C1:*   Tangency surfaces meet and are tangent. This is the typical tangency produced by Fillet commands, and are suitable for most rounded edges.
*   *C2:*   Curvature continuous surfaces meet and are tangent, and where surfaces blend, they share a common center of curvature. Surface blends with this continuity do not have any abrupt change in radius. These blends are important for styling of surfaces such as automobile hoods and exterior plastic parts.
*   *C3:*   The next level above C2, except that the rate of change for the curvature is also continuous.

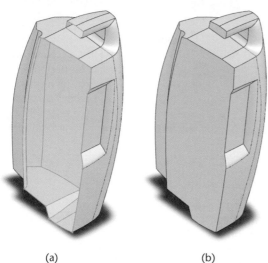

(a)   (b)

6.35 The open-surface model in (a) is patched and then "solidified" into a solid model (b). *(Courtesy of Strategix Vision.)*

Complex Shapes

Complex shapes can often be made using special surface modeling commands that create surfaces that cannot easily be made with regular extrusion, revolution, sweeping, and blending. These surfaces can then be added to the model (see Figure 6.35). Once a surface has been added to the model and the edges completely matched (rather like being watertight), the object then can be "solidified" into a solid, which allows mass property calculations.

Because the complex surface can slow down model regeneration, an approximation of the surface may be used as a placeholder until later in the refinement process.

6.6 CONSTRAINT-BASED MODELING MODES

Most constraint-based modelers have three main modes that link together to provide a full range of functionality in creating and documenting your designs: ***part mode,*** which allows you to create individually manufactured parts of a single material; ***assembly mode,*** which allows you to link multiple parts together into assemblies (Figure 6.36) or ***subassemblies;*** and

6.36 An Assembly Model of a Dynamometer *(Courtesy of Dynojet Research, Inc.)*

drawing mode, which allows you to create orthographic, section, and pictorial drawings to document the design.

Assemblies

Assembly capabilities in constraint-based modelers make concurrent engineering feasible. Companies can use the single design database as the hub of a collaborative environment for a cross-functional design team. Planning the critical relationships that capture the design intent of the assembly—just as would be done for each individual part—allows the constraint-based assembly to be used to coordinate efforts of several different individuals. The entire assembly and each part are stored in the design database for members of the design and manufacturing team to view, measure, or change (if authorized) while the parts are being designed.

The assembly provides a framework for the model. In the assembly, global parameters can be used to build intelligent assemblies. For example, if several parts assemble onto a shaft, the shaft diameter may be made a global parameter. The hole sizes on the parts that fit onto the shaft may be created by adding or subtracting a value from the global parameter for the shaft diameter. If the shaft diameter is changed, all the parts that fit with it may automatically be updated. Using global parameters can be an effective way to coordinate the design effort for a team. You will learn more about using assemblies to refine your design in Chapter 7.

Drawings from the Model

An advantage of solid modeling software is that drawings can be generated from the solid model (Figure 6.37). Creating drawings using constraint-based modeling software adds a significant dimension to this advantage: *associativity*.

Bidirectional Associativity

When a set of drawing views have been created from the model (Figure 6.38) and a change is made to the model, the drawing views will automatically be updated

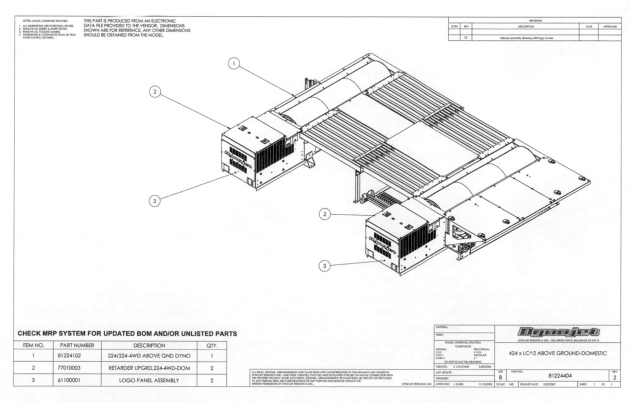

6.37 A Drawing of the Dynamometer Assembly *(Courtesy of Dynojet Research, Inc.)*

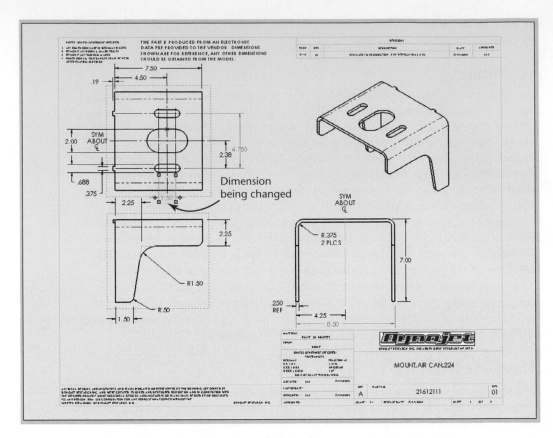

6.38 The 1.50 dimension for the slot highlighted in green in the top view is changed in the drawing to the new size, 2.00. *(Courtesy of Dynojet Research, Inc.)*

Before

After

6.39 The model for the mounting bracket is shown before (slot at 1.50) and after the dimension on the drawing was changed to 2.00. *(Courtesy of Dynojet Research, Inc.)*

because they are associated with the actual model. Dimensions in most constraint-based modeling software control the sizes of the model features. The process of dimensioning a drawing can be automated because the software reports the dimension value for the entities in the CAD database.

If the size of a feature changes in the model, the value for the dimension in the drawing is automatically updated to show the correct values. When this is a one-way process—when changing the model causes the dimension value to be updated on the drawing—the dimensions are said to be ***associative.*** When this is a one-way process, changing the dimension value on the drawing has no effect on the model geometry. In fact, it may break the link between the dimension value and the model geometry.

Most constraint-based modelers are ***bidirectionally associative:*** changing a dimension in the drawing changes the model geometry, and vice versa (see Figures 6.38 and 6.39). Constraint-based modelers store and use the dimensions and constraints to generate the model. When dimensions are shown in a drawing, the values are the dimensions used to create the model. Whether these dimensions are changed in the part or on the drawing, the new values are stored in the database, and the model and all associated drawings are updated to reflect the new dimension values. Bidirectional associativity allows drawing changes to update the model, and model changes to update the drawings.

CONSTRAINT-BASED MODELING: CAPTURING DESIGN INTENT

When William Townsend was a Ph.D. student at MIT, he designed a four-axis robot arm that simulates the motion of the human arm. This human-scale dexterous arm works with the subsequently developed Barrett wrist and robot hand produced and manufactured by Townsend's engineering firm, Barrett Technology. To prepare the design for larger-scale production, the original 2D drawings were converted to constraint-based solid models, which enhanced Barrett Technology's ability to quickly make design changes and evaluate the results. By redesigning the robot arm, the company was able to make it the lightest one on the market and save 50% over previous manufacturing costs.

To benefit from constraint-based modeling, it was important to design the parts with the design goals in mind. "You need to keep your design intent in mind. Even though the software is a powerful tool, if you tie up a feature or overconstrain it, you are limiting the amount of flexibility you have in changing it easily later," said Brandon Larocque, project manager at Barrett Technology.

The top plate of the BarrettHand is a good example of how Larocque approaches constraint-based design (see Figure 6.40). The part has a total of fourteen features and took about 30 minutes to model in SolidWorks. When a new part is created, SolidWorks provides a default set of three orthogonal planes that intersect at the origin. The user can add more planes, but SolidWorks allows any surface to be used as a sketch plane. Larocque estimates that 80% to 90% of the parts he models do not need additional sketch planes.

The top plate of the hand began as half of a symmetrical sketch that was mirrored across the centerline (see Figure 6.41). The constraint-based relations in the mirrored half of the part were set up so they would be the same as those in the original half. To create the sketch, three radii were dimensioned to form the curved end of the plate: one in each corner and a larger one to form the curved end of the plate. The center of the largest radius is constrained to always lie on a horizontal construction line inserted into the base sketch. The other end-radii have tangency relationships where they meet the large radius and the straight lines that define the other edges of the plate. These tangency constraints were automatically created by SolidWorks when the radii were sketched in place. The three radii were constrained to be mutually tangent so the tool path generated by the model would be a single smooth curve through the three radii. The only other dimensions needed to define the part were those for the overall length of the part and the rectangular shape at the top of the part. All distances (dimensional constraints) were defined from the center of the part (which also coincided with the origin in SolidWorks). The sketch was then extruded to become the base feature for the part.

When Larocque thinks about selecting a base feature, he chooses one that will support additional features and reflect the way the part may be modified and eventually built. "I think

6.40 This rendered view of the BarrettHand clearly shows the top plate in relation to the rest of the hand.

modeling in 3D is similar to the way a machinist shapes a part. The modeling operations give you some intuition about the steps required to make the part and how many different setups a part will require. If you can make all of your features from the same surface of a part, chances are the machinist will be able to make them from a single setup as well. If you can form a part with a single revolution, you can probably make the part on a lathe in a single setup. Each time you add a sketching plane that is not normal or parallel to the default planes, you can assume the part may require an additional setup in order to machine it. Reducing a part's design from four setups to three can reduce the machining costs by 25%." In the case of the top plate, all its features except the bevel and the counterbores were created from the same side of the part.

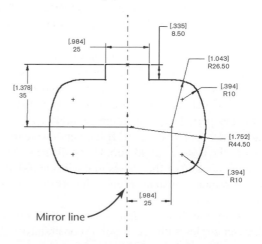

6.41 The sketch for the base feature is symmetrical about the centerline. Dimensions are in SI units with inch units in brackets.

(continued)

Because the plate is symmetrical, its center point also served as the design center of the part. The plate is designed to mate with two bearings set a specified distance apart. Dimensioning the part about a point at the intersection of the mirror line of the plate and a line through the center points of the two bearing bores means that any change in the distance between the bearings requires few changes to update the part.

The next step was to create the bevel around the outside the base feature. Using the mirror plane (across which the first sketch was mirrored) as the sketching plane, Larocque sketched a triangular shape with a 30° angle (see Figure 6.42). This sketch was the profile for the bevel. A contour was sketched on the back surface of the plate to serve as the path for the triangular profile. After selecting both sketches, he used a sweeping operation to create the feature. SolidWorks prompted him to identify the profile (the triangle), the path (the contour sketch), and the sweep operation (a cut). The profile was dragged normal to the sweep path to cut the bevel on the plate. A machinist cutting this feature into actual metal would similarly use a 30° tapered mill to sweep out the same path defined in the model.

The next feature hollowed out the part. Sketching on the surface of the part that lies on the original sketch plane, Larocque created half of the shape of the area to be removed, dimensioned it from the centerline, and mirrored it to create the complete outline. The sketch was then extruded to the appropriate depth to be cut from the base feature (see Figure 6.43).

The fourth feature was a semicircular cutout to provide clearance for a motor pinion (see Figure 6.44). This circle was sketched on the original sketching plane, its diameter was dimensioned, and its center point was constrained to the center of the part (which corresponds to the distance from the centers of the bearings). Using a constraint to position the feature on the part's centerline guaranteed that the feature would stay centered even if the overall size of the part changed. To form the feature, Larocque specified a blind extrusion to a specified depth.

With the part hollowed out, the new interior surface was used as the sketch plane for the bosses that would be the bearing bores (see Figure 6.45). The bosses could have been formed on the initial sketching plane in the same operation that hollowed out the part, but their height would have been constrained by the depth of the part. If the depth were to change, the height of the bosses would, too. Using the new surface ensured that the relation of the bosses to the inside surface of the plate would remain constant, even if the depth of the surface changed.

To create the bosses, Larocque created symmetrical 20-millimeter circles with center points constrained to the horizontal centerline of the part. The circles were extruded, and two new concentric circles were sketched on the tops of the

6.42 The triangular shape of the bevel is a sketch on the mirror plane that was swept along the contour.

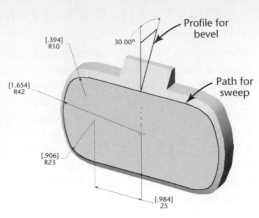

6.43 The sketch for hollowing out the plate included two distinct outlines that were extruded as a single feature.

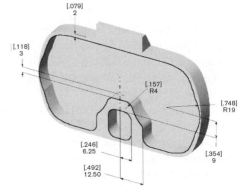

6.44 The sketch for the motor pinion cutout was constrained to the center of the part.

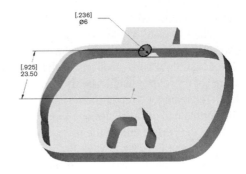

6.45 The sketch for the bosses was drawn on the new surface created by the hollowing operation.

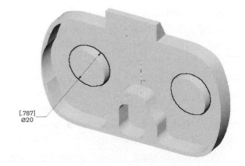

(continued)

new features (see Figure 6.46). From the tops of the bosses, Larocque extruded cuts back toward the bottom. Because the boss is to serve as a bearing bore, the depth of the hole in the middle must correspond to the width of the bearing, not the height of the boss. Extruding the hole from the top of the boss means that the hole will remain at the same depth even if the boss becomes taller or shorter. The same process was repeated to make the two smaller bosses.

The next feature was a chamfer on the inside edges of the bearing bores. SolidWorks allowed Larocque to multiselect all the bosses at once, then insert a ***chamfer*** (see Figure 6.47). A chamfer is a built-in feature with its own dialog box for setting its angle and depth. The single operation created the four chamfers, all to the same specifications. They constitute a single feature that cannot be decoupled into individual chamfers. Independent control of each chamfer would require creating four separate features.

Another built-in feature, a fillet, was added next. Larocque selected the two corners where the rectangular protrusion intersected the body of the part and added both fillets with the same operation.

The H-shaped pocket at the top of the plate was sketched next as a symmetrical feature. This feature is an alignment pocket that should move with the bearing bores. Its location and radii were dimensioned from the center of the part (to coincide with the centers of the two 20-mm bosses), and its length constrained to the centerline of the part (see Figure 6.48). "Most of the features after the base feature are dimensioned from the bearing bores because they are the defining characteristic of the part. I set up a relationship between the bearing bores and the alignment pocket because the alignment pocket determines where the bearings will fall."

The final features were the six holes and the counterbores to accommodate the fasteners to go into them. The first hole was given a diameter dimension, then the other holes were constrained to be the same size as the first (see Figure 6.49). All six were sketched on the sketch plane used for the bevel contour and extruded with the "through all" setting so they will always go all the way through the part, even if the part thickness changes. From the same side of the part, the counterbores were added to the holes as the final features. Larocque specified counterbores at the appropriate depth that are concentric with the hole (see Figure 6.50).

When Larocque picked the center point of the existing hole as the location for the counterbore, SolidWorks interpreted his action as a concentric constraint. This feature can be toggled off to make it easier to place an object near another without snapping to a constrained relationship. SolidWorks makes it easy to review and modify any existing constraints. A feature can be selected, and Display or Delete Relations will display all the relations for the part so they can be edited. When two features are selected, Add Relations displays a list of all the possible relations those two geometric entities could have. For two circles, for example, the only possible constraints would be concentric or coradial.

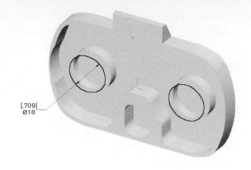

6.46 The bearing bores were sketched on top of the bosses.

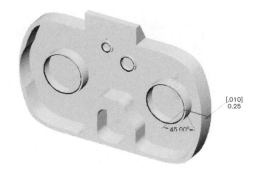

6.47 All four chamfers were added at once.

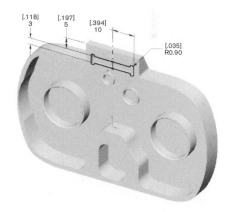

6.48 The alignment pocket sketch was dimensioned from the center of the part.

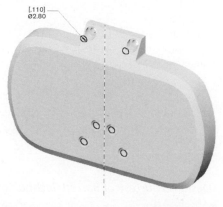

6.49 Only one of the holes displays a dimension here; the other five were constrained to have the same diameter as the first.

(continued)

When asked about the degree to which he uses constraint-based relations, Larocque explained, "We like to keep the model somewhat segmented so we have more control over the updates, but we make sure to add relations where they make sense. For example, constraining symmetrical parts to a centerline as we did with the top plate of the hand can save enormous amounts of time when the part can update automatically. This aspect of the model was instrumental in our ability to make the BarrettHand lighter and less costly to manufacture."

6.50 The counterbores were the last features added.

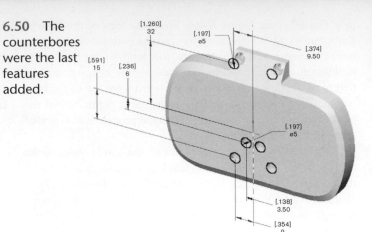

(Designs courtesy of Barrett Technology, Inc., Cambridge, MA.)

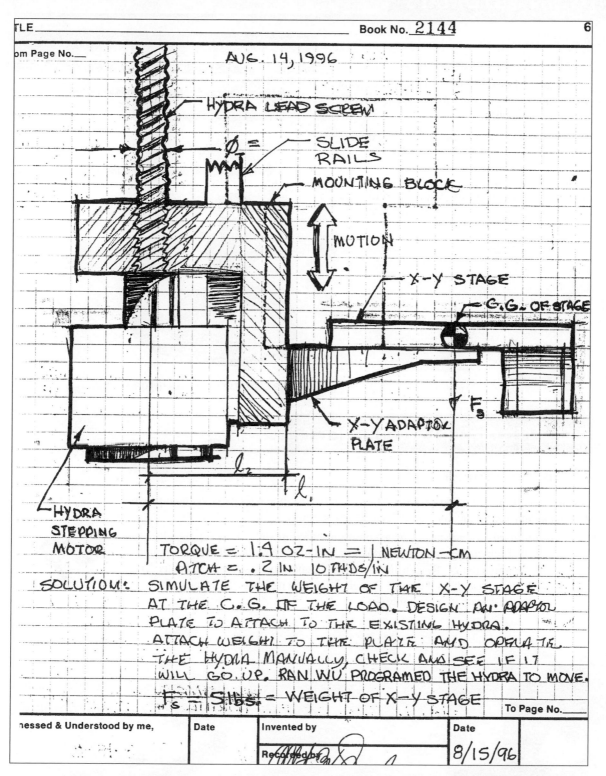

om Page No.___

AUG. 14, 1996

HYDRA LEAD SCREW

ø =

SLIDE RAILS

MOUNTING BLOCK

MOTION

X-Y STAGE

C.G. OF STAGE

X-Y ADAPTOR PLATE

F_s

l_2

l_1

HYDRA STEPPING MOTOR

TORQUE = 1.9 OZ-IN = 1 NEWTON-CM
PITCH = .2 IN 10 THDS/IN

SOLUTION: SIMULATE THE WEIGHT OF THE X-Y STAGE AT THE C.G. OF THE LOAD. DESIGN AN ADAPTOR PLATE TO ATTACH TO THE EXISTING HYDRA. ATTACH WEIGHT TO THE PLATE AND OPERATE THE HYDRA MANUALLY, CHECK AND SEE IF IT WILL GO UP. RAN WU PROGRAMED THE HYDRA TO MOVE.

$F_s = 5$ lbs. = WEIGHT OF X-Y STAGE

To Page No.____

nessed & Understood by me,	Date	Invented by		Date	
		Recorded by		8/15/96	

6.51   This sketch from Albert Brown's design notebook shows the relationship between the centerline of the lead screw and the center of gravity of the stage. The critical dimensions shown in the sketch need to be captured in the parametric model to reflect the design intent for this device. (*Courtesy of Albert W. Brown, Jr., Affymax Research Institute.*)

DESIGNER'S NOTEBOOK

HANDS ON TUTORIALS

SolidWorks Tutorials

Using Revolve, Sweep, and Extrude

In this tutorial, you will create a candlestick using some of the common tools for feature creation: revolving, sweeping, and extruding 2D profiles.

- Launch SolidWorks from the icon on your desktop.
- Click to expand the topics listed below the "?" help icon.
- Click: **SolidWorks Tutorials.**
- Click: **All SolidWorks Tutorials (Set 2)** from the list.
- Click: **Revolves and Sweeps** from the list that appears, and complete the tutorial on your own.

Revolve and Sweep Features

In this lesson, you create the candlestick shown below. This lesson demonstrates:

- Creating a revolve feature

- Creating a sweep feature

- Creating an extruded cut feature with a draft angle

(Image courtesy of DS SolidWorks Corp.)

REVERSE ENGINEERING PROJECT

The Can Opener Project

In this segment of the project, you will continue modeling the can opener's parts, with emphasis on the dimensional and geometric constraints that are appropriate.

1. *Make constraint-based part models.* Create constraint-based models of the top handle, bottom handle, cutter, knob, and gears. Consider part fit, and make your model so that the major dimensions can flexibly change. Review the dimension names you assigned in Chapter 3 for the lower handle, and revise them if needed. Assign dimension names to these parts wherever it makes sense for the can opener assembly.

(Courtesy of Focus Products Group, LLC.)

Knob

Lower Handle

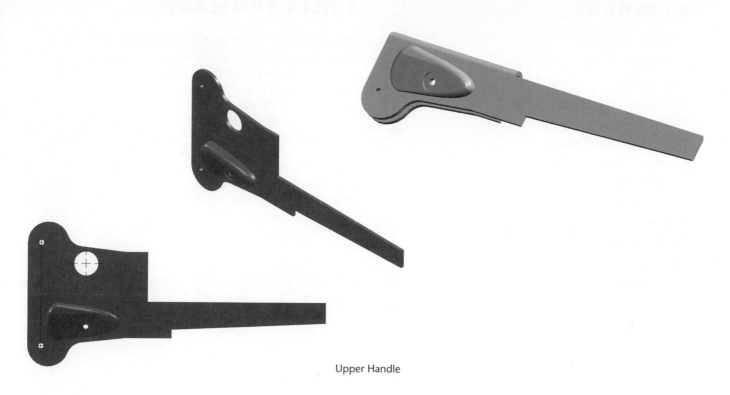

Upper Handle

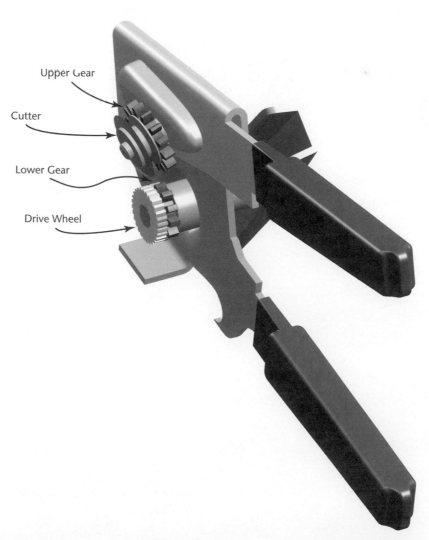

Upper Gear

Cutter

Lower Gear

Drive Wheel

KEY WORDS

Assembly Mode

Associative

Base Feature

Base Point

Bidirectionally Associative

Built-in or Placed Features

Chamfer

Cosmetic Dimension

Counterbore

Datum

Datum Planes

Design Intent

Drawing Mode

Driven Dimensions

Driving Dimensions

Feature

Feature-Based Modeling

Fillet

Geometric Constraints

Global Parameters

Parameter

Parent-Child Relationship

Part Mode

Size Constraints

Subassemblies

SKILLS SUMMARY

Having finished this chapter, you should be able to list advantages of constraint-based modeling over solid modeling. You should be able to correctly use the term *parent-child relationships* as applied to part features. Associativity between drawing views and the 3D model is an important characteristic of sophisticated constraint-based modeling programs. Selecting the base feature of the model is important, as it is the first parent feature. The exercises ask you to demonstrate your understanding of parent-child relationships and base features by capturing design intent in a number of part models. They also provide additional practice with your constraint-based modeling package.

EXERCISES

1. Constraint-based, or feature-based, modeling uses defined relationships between model elements to control various features. What two basic types of parameters are used to control features?

2. Why are the linear, angular, or other physical dimensions of a constraint-based model feature given a name? Give an example where this practice might be useful.

3. An acronym sometimes used to describe the process of preparing for the creation of a constraint-based model is TAP, representing the steps Think, Analyze, and Plan. Consider what it means to think, analyze, and plan a constraint-based model. Why are these steps especially important prior to beginning the constraint-based model?

4. What is meant by *parent-child relationship* in constraint-based modeling?

5. Define *bidirectional associativity* with respect to a constraint-based modeling program's drawing mode. How does this program feature enhance ease of use of the program?

6. You are considering modifying a model by changing dimensions in a drawing view. What problems might you encounter when using a constraint-based modeler *without* bidirectional associativity? What problems might you anticipate when using a constraint-based modeler *with* bidirectional associativity in this task?

7. The toy building blocks in the photos are assembled by inserting the posts on top of one block into the space around the central circular extrusion on the bottom of another block The parts are made of molded plastic. Consider the design intent for these blocks. What dimensions are critical so these blocks will fit together in different arrangements? Create a model of a six-post block (2 × 3). Update your model to contain eight posts (2 × 4), as in the block at right

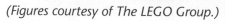

(Figures courtesy of The LEGO Group.)

8. Model each of the parts shown. Before starting, carefully consider which feature to use as the base feature and how you will create the model so it will be updated as specified.

a. Design intent: The length and width of the slot must be able to change to accommodate fit with a different part. The four holes must remain equally spaced around the wider/longer slot.

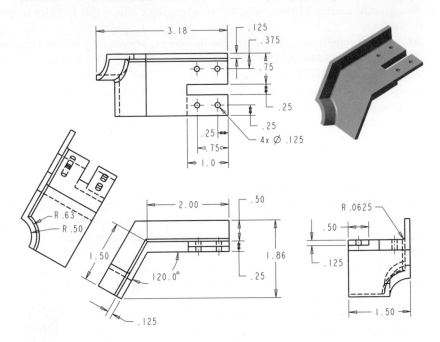

b. Model the hand weight shown. Determine the weight if this part is cast from ASTM class 20 gray iron. Update the model to create a set of three hand weights of different weights, but maintain the diameter and length of the crossbar grip.

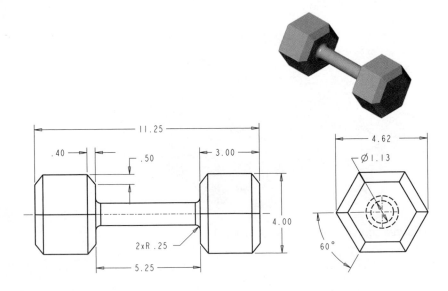

c. Model the part so that as the overall height, width, depth of the part are changed the other features are updated proportionally to them.

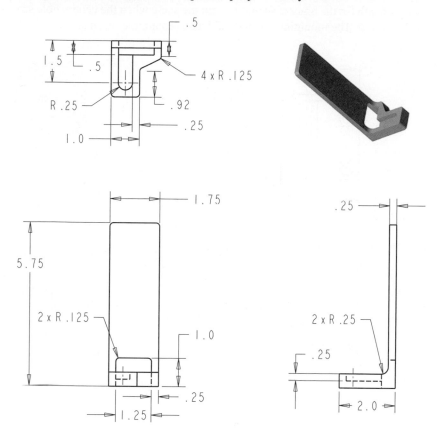

d. Create a model for the decorative corner cap shown.

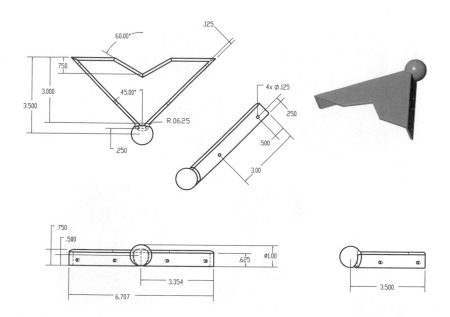

e. Model the rail bracket shown so that the diameter for the central hole can change to accommodate different diameters of railing. The height, width, and depth for the bracket should remain the same when the central hole size is updated. The mounting holes should also remain the original size.

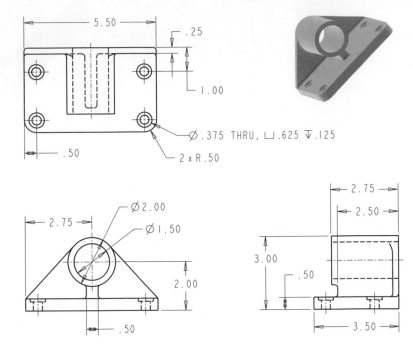

ALL ROUNDS = .125

f. Model the plastic speaker housing shown.

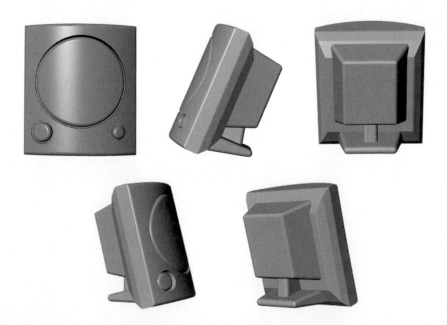

9. Create a constraint-based model of the four-spoke hand wheel shown such that it can be resized to match the dimensions in the table.

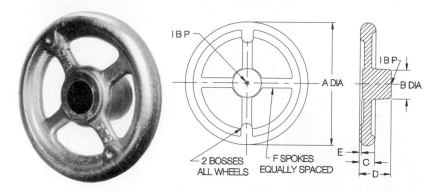

STRAIGHT SPOKES

PART NO. CAST IRON	A DIA	B DIA	C	D	E	F
CL-4-HWSF	4	1-1/4	5/8	1-11/16	1/8	
CL-5-HWSF	5	1-1/2	3/4			
CL-6-HWSF	6	1-5/8		1-13/16		4
CL-8-HWSF	8	1-7/8		2-1/16	5/32	
CL-10-HWSF	10	2-1/4	1	2-1/2		
CL-12-HWSF	12	3		3-1/4	1/8	8
CL-14-HWSF	14			3-3/4	3/32	

10. Create a constraint-based model of the swing washer shown such that it can be resized to match the dimensions in the table. Capture size relationships between features in the constraint-based dimensions wherever possible.

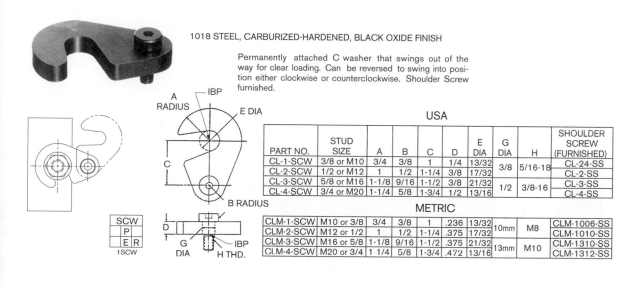

1018 STEEL, CARBURIZED-HARDENED, BLACK OXIDE FINISH

Permanently attached C washer that swings out of the way for clear loading. Can be reversed to swing into position either clockwise or counterclockwise. Shoulder Screw furnished.

USA

PART NO.	STUD SIZE	A	B	C	D	E DIA	G DIA	H	SHOULDER SCREW (FURNISHED)
CL-1-SCW	3/8 or M10	3/4	3/8	1	1/4	13/32	3/8	5/16-18	CL-24-SS
CL-2-SCW	1/2 or M12	1	1/2	1-1/4	3/8	17/32			CL-2-SS
CL-3-SCW	5/8 or M16	1-1/8	9/16	1-1/2	3/8	21/32	1/2	3/8-16	CL-3-SS
CL-4-SCW	3/4 or M20	1-1/4	5/8	1-3/4	1/2	13/16			CL-4-SS

METRIC

PART NO.	STUD SIZE	A	B	C	D	E DIA	G DIA	H	SHOULDER SCREW (FURNISHED)
CLM-1-SCW	M10 or 3/8	3/4	3/8	1	.236	13/32	10mm	M8	CLM-1006-SS
CLM-2-SCW	M12 or 1/2	1	1/2	1-1/4	.375	17/32			CLM-1010-SS
CLM-3-SCW	M16 or 5/8	1-1/8	9/16	1-1/2	.375	21/32	13mm	M10	CLM-1310-SS
CLM-4-SCW	M20 or 3/4	1 1/4	5/8	1-3/4	.472	13/16			CLM-1312-SS

11. Create a constraint-based model of the electronics bracket shown.

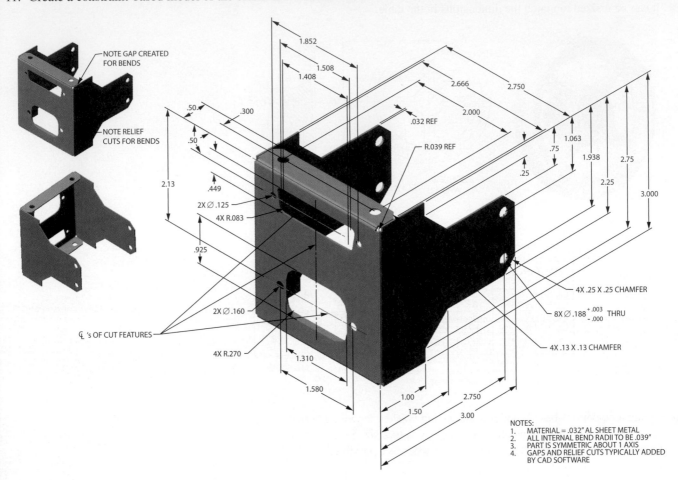

NOTE GAP CREATED FOR BENDS

NOTE RELIEF CUTS FOR BENDS

℄ 's OF CUT FEATURES

2X ⌀ .125
4X R.083
.925
2X ⌀ .160
4X R.270
1.310
1.580
.449
2.13
.50
.50
.300

1.852
1.508
1.408
2.666
2.750
2.000
.032 REF
R.039 REF
1.063
.75
.25
1.938
2.25
2.75
3.000

4X .25 X .25 CHAMFER
8X ⌀ .188 $^{+.003}_{-.000}$ THRU
4X .13 X .13 CHAMFER

1.00
1.50
2.750
3.00

NOTES:
1.   MATERIAL = .032" AL SHEET METAL
2.   ALL INTERNAL BEND RADII TO BE .039"
3.   PART IS SYMMETRIC ABOUT 1 AXIS
4.   GAPS AND RELIEF CUTS TYPICALLY ADDED
      BY CAD SOFTWARE

12. Create a constraint-based model of the gyroscope base shown.

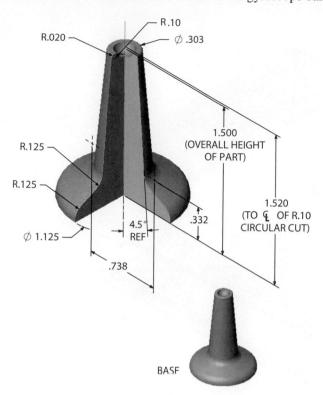

R.10
R.020
⌀ .303
1.500
(OVERALL HEIGHT
OF PART)
1.520
(TO ℄ OF R.10
CIRCULAR CUT)
R.125
R.125
⌀ 1.125
4.5°
REF
.332
.738

BASE

13. Create a constraint-based model of the pry bar shown.

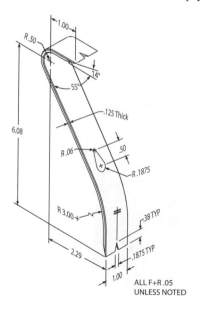

1.00
R.50
14°
55°
6.08
.125 Thick
R.06
.50
R.1875
R 3.00
.38 TYP
2.29
.1875 TYP
1.00
ALL F+R .05
UNLESS NOTED

14. Create a constraint-based model of this ice cube tray.

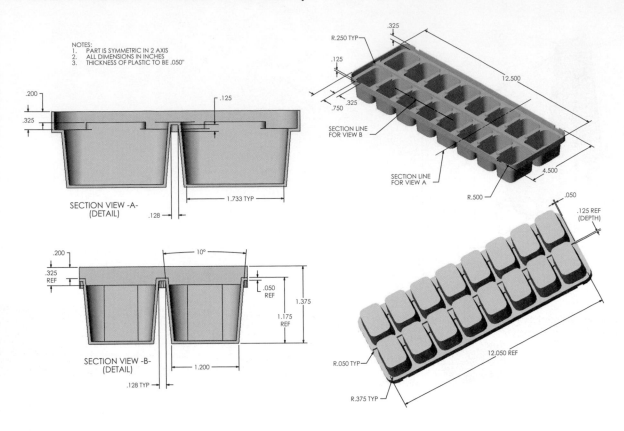

NOTES:
1. PART IS SYMMETRIC IN 2 AXIS
2. ALL DIMENSIONS IN INCHES
3. THICKNESS OF PLASTIC TO BE .050"

SECTION VIEW -A-
(DETAIL)

SECTION VIEW -B-
(DETAIL)

15. Create a constraint-based model of this simple knob.

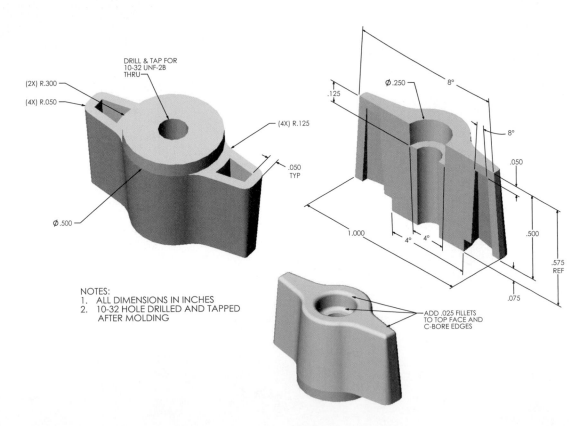

NOTES:
1. ALL DIMENSIONS IN INCHES
2. 10-32 HOLE DRILLED AND TAPPED
   AFTER MOLDING

16. Create a constraint-based model of the first part in each set, save the part, then edit the part to create the other configurations in each set. Save all part files.

a.

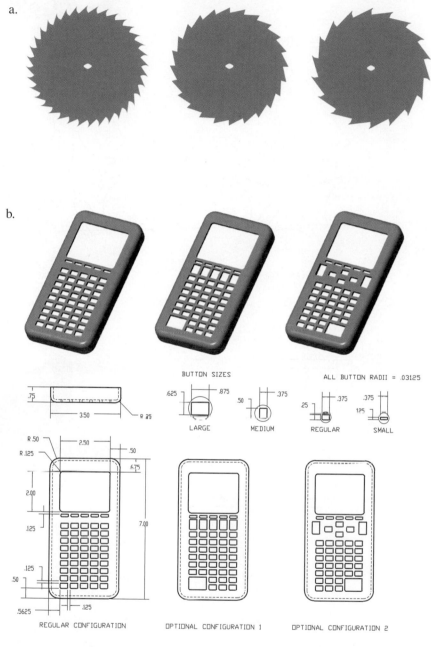

b.

BUTTON SIZES

ALL BUTTON RADII = .03125

.625   .875
        LARGE

.50   .375
       MEDIUM

.25   .375
      REGULAR

.375
.125
    SMALL

.75   3.50   R .25

R .50   2.50   .50
R .125   .675
2.00
.125   7.00

.125
.50

.5625   .125

REGULAR CONFIGURATION   OPTIONAL CONFIGURATION 1   OPTIONAL CONFIGURATION 2

NOTE: CALCULATOR BODY THICKNESS = .125

c.

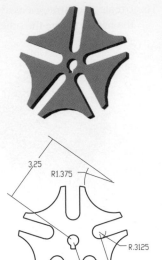

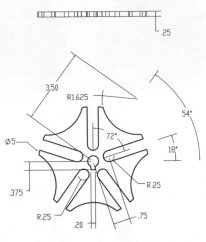

.25

3.50

R1.625

54°

72°

18°

Ø5

R.25

.375

R.25

.20

.75

GENEVA GEAR CONFIGURATION # 1

3.25

R1.375

R.3125

1.25

GENEVA GEAR CONFIGURATION # 2

d. BASE FEATURE W/ REGULAR RIB

4.00

.75

2.25

ALL RADII = .125

.25

4x Ø.375

R.50

4.00

5.00

.750

2.00

2.50

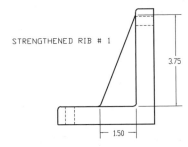

STRENGTHENED RIB # 1

3.75

1.50

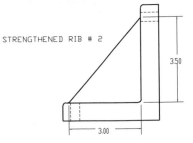

STRENGTHENED RIB # 2

3.50

3.00

NOTE: RIB THICKNESS REMAINS .25

17. Create a constraint-based model of the wall hanger.

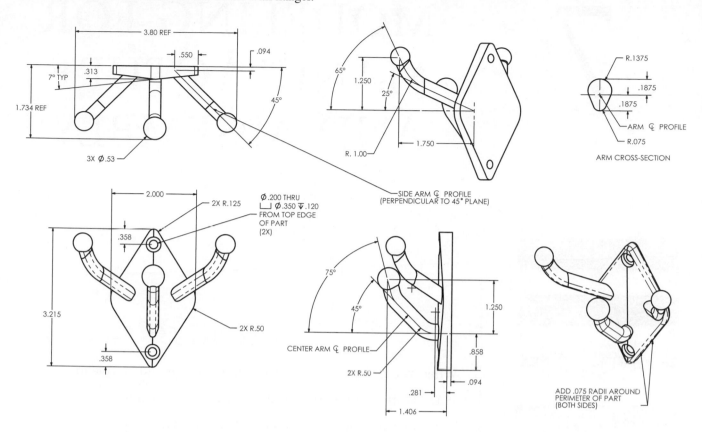

3.80 REF
.550
.094
7° TYP
.313
1.734 REF
45°
3X Ø.53
65°
1.250
25°
R. 1.00
1.750
SIDE ARM ℄ PROFILE
(PERPENDICULAR TO 45° PLANE)

R.1375
.1875
.1875
ARM ℄ PROFILE
R.075
ARM CROSS-SECTION

2.000
2X R.125
Ø.200 THRU
⌴ Ø.350 ▽.120
FROM TOP EDGE
OF PART
(2X)
.358
3.215
2X R.50
.358

75°
45°
CENTER ARM ℄ PROFILE
2X R.50
1.250
.858
.094
.281
1.406

ADD .075 RADII AROUND
PERIMETER OF PART
(BOTH SIDES)

7 MODELING FOR MANUFACTURE AND ASSEMBLY

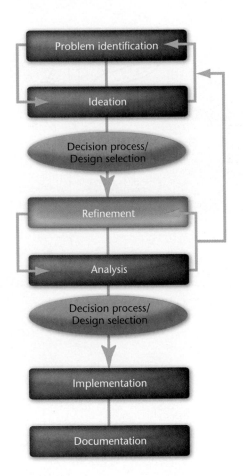

OBJECTIVES

When you have completed this chapter you should be able to:

1. Distinguish between static and dynamic assemblies.

2. Explain how parts can be added to an assembly.

3. Describe the role of intelligent assemblies in top-down design.

4. Identify how standard and static parts are used in assembly models.

5. Describe issues in modeling fasteners and springs.

6. Use an assembly model to test for fit and interference.

7. Describe manufacturing considerations to be reflected in a model.

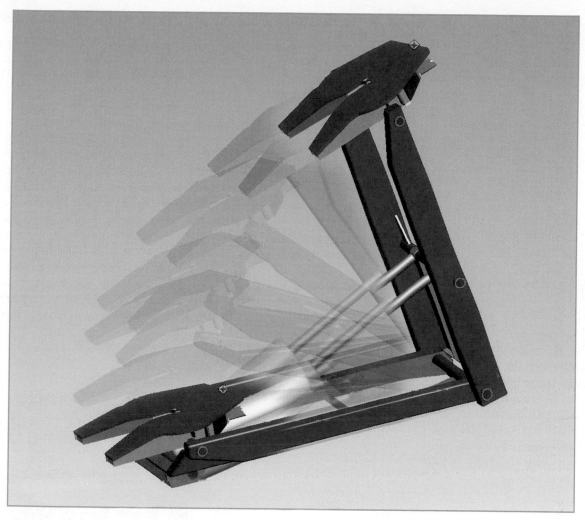

This 3D assembly model can be used to simulate the mechanism's range of motion. *(Courtesy of Leo Greene, www.e-Cognition.net.)*

PUTTING IT TOGETHER

Modeling accurate parts that reflect your design intent is the foundation of a good digital model database. In addition to modeling parts, you can use the design database to see how parts fit and function in the final product. Individual parts can be checked at a range of size variations to see if they interfere with others in the assembly. The assembly can be defined to behave as a mechanism that simulates the function of the design.

Variations in manufacturing can be anticipated and evaluated in the model to see whether they will affect the function of the design—and whether parts will be rejected because they cannot be assembled. Finally, considering the design intent of the entire assembly up front can make it easier to work with others concurrently and to model complex designs effectively.

7.1 COMBINING PARTS IN AN ASSEMBLY MODEL

Individual part files make up a large part of the digital database for a design. Parts within the database may be organized by levels. At the highest level, all the parts are shown in the final assembly. (Figure 7.1 shows the many individual parts in a coffee brewer assembly.) Within a large assembly, subgroups may be broken out into subassemblies. *Subassemblies* are groups of parts that fit together to create one functional unit, often one that can be preassembled in some fashion to fit into the assembly as a unit. Breaking a design into levels can make it easier to divide tasks among different groups working on the same project. It can help in presenting, managing, and understanding large, complex assemblies. Some subassemblies may even be reused in other designs as a unit.

You can combine individual parts into assemblies using your 3D CAD software in different ways. One method is to insert copies of your solid parts into a single assembly model. Assemblies that are created by copying 3D solid parts like these are *static assemblies*; that is, they do not change to reflect alterations in the individual parts. Because there is no link established between the part files and the assembly, if the individual parts are changed, they must be recopied or reinserted into the assembly model to update it. The two are completely separate, and changes made to either one will not affect the other. In contrast, in a *dynamic assembly*, parts are imported into the assembly through a linking process that allows the software to update the assembly as individual part files are modified.

7.1   The exploded view of the Zuma coffee brewer in (a) shows the individual parts that make up the assembly, shown assembled in (b). Each part is stored in its own part file and linked through external references to the dynamic assembly model. *(Courtesy of VKI Technologies, Inc.)*

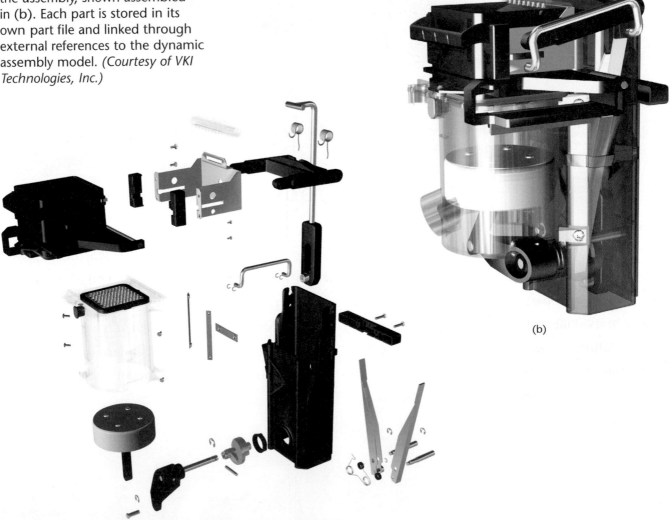

(b)

(a)

External References

External references (as used in AutoCAD) are one way to create a dynamic assembly. An *external reference* (called an XREF in AutoCAD) creates a link to the part file that imports a viewable representation of the part into the assembly file. If the part file changes, the assembly automatically shows changes that have been made to the part. The data displayed in the assembly are stored in the externally referenced files and imported through the link each time you open the assembly. If the part files are not available, you cannot view them in the assembly.

Figure 7.2 illustrates the difference between assemblies created by copying parts from one file into another and those created using external references. When part file A is copied and placed into the static assembly file, changes to the original part file do not change the assembly file. When part file A is externally referenced by the assembly file, as in Figure 7.2(b), changing part file A also changes the assembly file. An advantage of external references is that they do not copy the contents of the part file to the assembly model; this helps keep the size of the assembly file from getting overly large.

The links between assembly and part files are important when you are managing your work. If a part is changed and it is linked to two different dynamic assemblies, it will change in both assemblies. If that is your intention, it can be a great time-saver. If you are modifying the part to change it in only one of the assemblies, you will get unwanted results the next time you open the second assembly; and if you delete a part that is referenced by other parts, the referenced information will not be available.

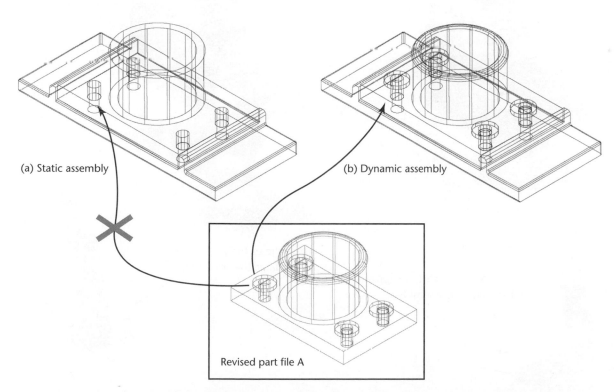

(a) Static assembly   (b) Dynamic assembly

Revised part file A

7.2   (a) Static Assembly; (b) Dynamic Assembly. Changes made to part file A are reflected in the dynamic assembly shown in (b) because they are linked to the assembly model file. To update the static assembly in (a), the revised part would have to be copied into the assembly file again.

Constraint-Based Assemblies

Constraint-based assemblies are similar to assemblies that are created using externally referenced parts. When you create a constraint-based assembly, the part file is linked, not copied, to the assembly model. Like assemblies with external references, constraint-based assembly files rely on subassembly and part files to be available for you to view them or make changes to the assembly. If you lose the file for a part, it will not show up in the assembly.

The bidirectional associativity built into many constraint-based modelers adds an additional dimension to constraint-based assemblies. This type of associativity lets you alter a model by making changes to a drawing view of the model. In the same way, constraint-based modeling software allows you to make changes to parts while viewing the assembly model. These changes update and modify the individual part files. This two-way updating capability is illustrated in Figure 7.3.

With constraint-based modeling software, you use assembly constraints to create the relationships between parts. The first part you add to the assembly is the parent part. As you add parts, you specify the constraints between the parent part and the added child part that reflect the design intent for the assembly. Mating parts have features that should fit together in an assembly. For example, if you want two holes to line up, you can use an assembly constraint to align them. If a part changes, it will still be oriented in the assembly so that the holes align.

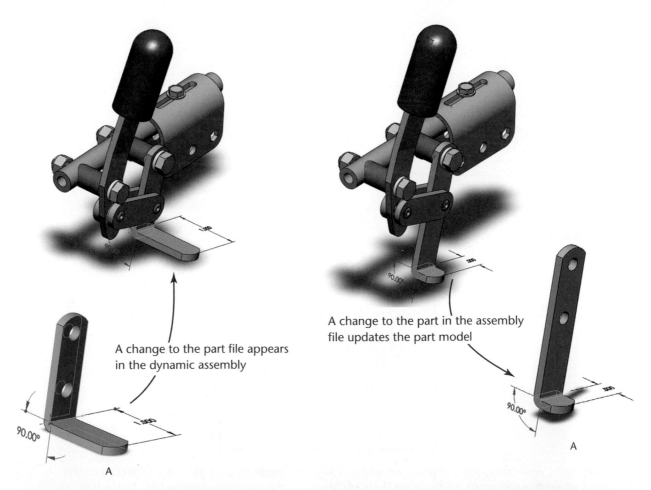

A change to the part file appears in the dynamic assembly

A change to the part in the assembly file updates the part model

7.3   The links between part and assembly files are analogous to the bidirectional associativity between model and drawing. A change made to part A in the assembly updates the part file, and vice versa.

Choosing the Parent Part

Constraint-based assemblies also differ in the parent-child relationships that are built into the model. The first part that is added to the assembly acts as the parent part for other parts you add to the assembly, just as the base feature is the parent for any other features in the part model. Remember that these parent-child relationships in a constraint-based model are critical. If the parent is moved, the children move along with it. This can be a great advantage: if one of the parts in the assembly moves or changes, the child parts move to update their positions relative to it.

If you delete a part that is the parent of some other part in the assembly, then the location of the child part is undefined. As the modeling software tries to generate the assembly from the parts and relationships stored in the database, undefined locations require you to redefine the placement of the child parts and may, in the worst case, cause the file to fail to open or the software to crash.

One strategy is to insert a standard set of datum planes as the first part in your assembly so that all other parts can be defined as children of this blank part that never changes (see Figure 7.4). Using a set of datum planes as the parent part in the assembly also helps fix your assembly on the coordinate system. Assemblies that are not fixed on the coordinate system may move in undesired ways when you are later trying to animate assembly motions.

When choosing the parent part to start out your assembly:

- Consider adding a blank set of datum planes as the first part in the assembly, particularly if your modeler does not already start with assembly datums. These can often be handy even for small tasks like reorienting subassembly parts. Additional datums are rarely ever detrimental. Manage the display of the datums to hide them when they are not needed.
- Start with major parts to which other parts connect.
- Start with key parts that are unlikely to be eliminated from the design or to be drastically changed.
- Mate the first part inserted to datum planes to fix it on the coordinate system, so that it does not move.
- Assemble functional units into subassemblies first before adding them to the main assembly.

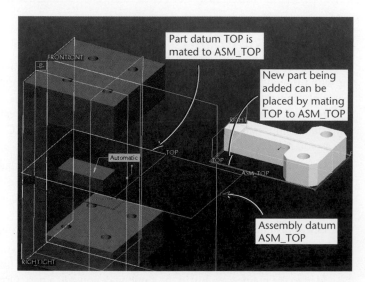

7.4 Assembly datums and establishing logical parent-child relations are key to assemblies that are updated easily. *(Courtesy of Mark Perkins.)*

Dynamic Assemblies: The Zuma Brewer

Dynamic assemblies provide an advantage to design teams sharing files over a network. The assembly model coordinates individual part design with the overall product effort. Most systems let you dynamically update your view of a part in the assembly even while someone else is actively changing it. This keeps the assembly current and ensures that the part you are designing fits with the other parts in the assembly.

When VKI Technologies switched to constraint-based modeling, many benefits were directly related to the assembly capabilities. VKI Technologies' single-cup coffee brewers require many moving parts and sophisticated electronics all contained in a compact, easy-to-use coffee maker. Using constraint-based modeling helped them fit the many components into smaller machines while reducing design time almost 50%.

The Zuma brewer assembly pulled together 62 different parts, shown in Figure 7.5. The assembly allowed the designer to be sure the components fit compactly yet did not interfere with one another. Moving parts were rotated through their range of motion in the assembly model to analyze and eliminate conflicts with other parts.

A big advantage of an assembly model is its role as a virtual prototype. For the Zuma brewer, the ability to visualize parts in 3D and evaluate them for fit with other parts almost eliminated the need to create physical models of the parts. When the brewer design was complete, the model was used to generate molds for the 15 plastic parts in the brewer. When the first real molded parts were available, they fit together well enough to make a functional brewer.

In the past, with many parts and a limited amount of time, it was not uncommon for fit problems to be discovered after manufacture, not before.

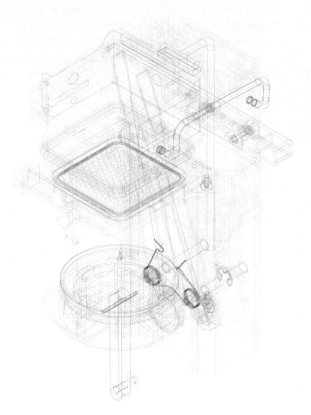

7.5 The 62 individual parts in the assembly model of the Zuma brewer shown in a wireframe view. *(Courtesy of VKI Technologies, Inc.)*

Assembly Constraints

Different software packages offer a similar set of constraint options. Become familiar with those available to you in your software. Table 7.1 lists some of the common assembly constraints and their definitions.

Just as feature relationships are important to the way you create a model, assembly relationships can make your assembly model work for you. Consider the following as you add parts to the assembly:

- Use constraints that orient the new part using relationships that will persist in the assembly.
- Think about the mechanism and the types of connections that parts should have to be a working device.
- Leave fasteners until last and insert them in manageable groups.

You can create a subassembly in much the same way that you create an assembly: by making an assembly of the subassembly components. This subassembly can be added to the main assembly in the same way you add a part. Organizing the model so that it comes together as it will on the assembly line aids in visualizing assembly difficulties. If a group of components are likely to be changed or replaced, linking all the subparts to a main component can make it easy to substitute an alternative design for that group of parts.

Table 7.1 Common Assembly Constraints.

Name	Definition	Illustration
Mate	Mates two planar surfaces together	
Mate Offset	Mates two surfaces together so they have an offset between them	
Insert (Concentric)	Inserts a "male" revolved surface into a "female" revolved surface, aligning the axes	
Align (Coincident)	Aligns two surfaces, datum points, vertices, or curve ends to be coplanar; also aligns revolved surfaces or axes to be coaxial	
Align Offset	Aligns two planar surfaces with an offset distance between them	
Parallel (Orient)	Aligns two surfaces, edges, or axes to be parallel (equal distance apart over their entire length)	
Perpendicular	Aligns two surfaces, edges, or axes to be perpendicular (at 90° to one another)	
Tangent	Aligns plane and curved surfaces, or two curved surfaces or edges to be tangent to one another	
Offset	An option for an assembly constraint that allows you to specify the relationship with distance between the entities	
Flip	An option for an assembly constraint that allows you to choose the opposite orientation for a plane, axis, or other entity	

Planning is essential to creating assemblies efficiently and getting the most out of them.

Managing Assembly Files

The links between files that make dynamic assemblies possible also make it important to manage your files to prevent deleting needed files. Use directory structures and naming conventions to keep your files organized on disk. The extra time it takes to name and store your file in the appropriate directory can save hours of searching for the file at a later date.

File management is especially important when you are working in a design team using networked computers. You must take responsibility to manage your files so that everyone can easily find them. Without good communication and an organized system for storing files, it is not unusual for one engineer to spend considerable time modifying a part only to have the latest modification not be used in the assembly because it was stored in the wrong directory.

It can happen that two different people modify the same part, undoing each other's changes. If you work as part of a team, make it clear who has the responsibility for changing different parts. One person should have "ownership" and the ability to make changes to the part file. If a change in another part requires a change to a part for which you are responsible, you should be notified of the changes so you can update your part accordingly. Some systems allow other team members to work with the file in read-only mode. This way, everyone can view the part, and perhaps even indicate changes, but changes are not made until the owner of the part approves them. This prevents different designers from concurrently changing the part.

Your software may manage how files are linked or externally referenced and how others access the files (see Figure 7.6). As you create assembly models, be aware of these options and how your work group uses them. Some companies invest in Product Data Management (PDM) systems that organize the design database and control the work flow process.

7.6 The SolidWorks **Pack and Go** feature lets you gather the assembly, parts, drawings and other model information to copy or move a project, without losing the linked files. If you did not use this feature, you would need to remember to copy or move each of the individual files to open the assembly file. *(Image courtesy of DS SolidWorks Corp.)*

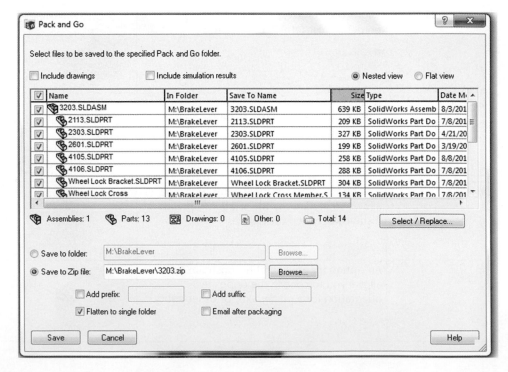

7.2 ASSEMBLIES AND DESIGN

So far, we have talked about assemblies as a means of pulling the parts of your design together. To work efficiently and concurrently, it is important that each designer be clear on the design intent so that problems of fit are minimized at later stages. Constraint-based modeling software also allows you to start with an assembly framework that can be used to define the design intent or parameters for individual parts and to help coordinate the work of different team members.

With this method, parts are designed so they link to a skeleton framework that defines major relationships in the assembly using lines, arcs, curves, and points. When you create the framework for each part up front, all parts do not have to be finished before they can be assembled. Parts can be assembled onto the skeleton at any stage of completion. Allowing the assembly to evolve as the parts are designed and refined allows each designer to see the parts the others are creating—or at least the critical relationships between parts—by looking at the assembly.

Top-down design starts the design by examining the function of the entire system, breaking that down into subassemblies or component groups based on their major functions, and, finally, defining each part that must be manufactured and assembled to create the design.

Bottom-up design starts at the part level, sizing individual components and building the design up from them. This is typically the approach used when components are standardized parts.

Middle-out design combines these two methods: some major standardized parts are used in the assembly to begin the design. The new components are designed to fit with these parts and function in the overall assembly.

An example of middle-out design is the Romulus Predator prototype, which was planned to combine racing capability with everyday street driving (Figure 7.7). This vehicle was never intended for mass production; the design and production costs would be recouped from a small manufacturing run. The design team used commercially available components such as brakes, steering column, steering wheel, and engine components to keep development costs down. These off-the-shelf parts were modeled in 3D and added to the assembly model. The custom parts were designed to work with these stock parts.

7.7 The Predator's use of off-the-shelf components is an example of middle-out design. This view of the Predator was taken from the surface model created in ICEM Surf. *(Courtesy of M&L Auto Specialists, Inc.)*

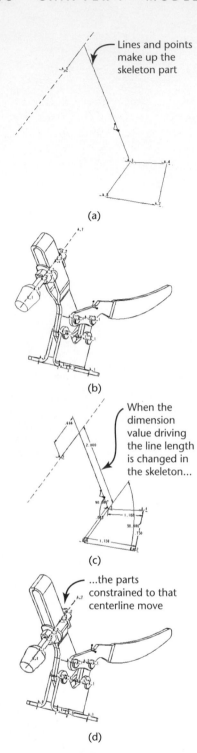

Lines and points make up the skeleton part

(a)

(b)

When the dimension value driving the line length is changed in the skeleton...

(c)

...the parts constrained to that centerline move

(d)

7.9 (a) Skeleton model for the clamp assembly; (b) parts assembled onto the skeleton; (c) changing dimensions of the skeleton controls the positions in the assembly; (d) resulting change in the assembly. *(Courtesy of Mark Perkins.)*

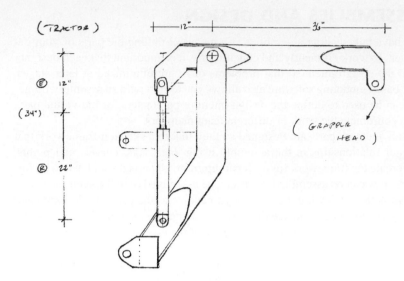

7.8 Layout Drawing. *(Courtesy of Implemax Equipment Co., Inc.)*

Layout Drawings

Traditionally, top-down design uses *layout drawings* to show the relationships between major functional items in the design. Layout drawings are especially useful when the product or system being designed has to fit with existing equipment. By defining the critical distances for fit with the existing equipment, the layout drawing documents the size constraints for the new equipment.

A typical layout drawing is shown in Figure 7.8. Note that the drawing is not complete in every detail but shows the major centerlines, sizes, and relationships between parts in the assembly.

Assembling to a Skeleton

A wireframe *skeleton* in a 3D model serves the same purpose as a layout drawing. It is a framework on which the individual components can be located. Figure 7.9 shows a skeleton model used to control the location and position of parts in the assembly model of a clamp.

To create a skeleton model, first define the critical dimensions in your assembly. What dimensions are fixed—because of a physical space requirement, sizing requirement, or some other relationship? What dimensions are likely to change later in the design process? How will each part relate to the others in the assembly? Begin your skeleton as a part model made up of a constraint-based framework of 3D planes, lines, curves, and points that identify the basic relationships between parts in the assembly.

Figure 7.10 (a) shows a skeleton model created for a laser printer. Laser printers, like most "mechatronic" devices, include a large number of mechanical and electronic components organized into subassemblies. The printer division creates skeleton models of each subassembly, which can then be combined into a top-level assembly. The dimensions between the planes, lines, points, and other entities making up the skeleton are parametric; they can be changed as needed as the design evolves to preserve the interfaces between parts. Changing the dimensions of the skeleton also allows the clearances and interferences between parts to be checked when the device is in different positions.

The skeleton drawing defines the interfaces where parts must come together. For the laser printer, which has many moving parts, the skeleton can also define the limits within which moving parts must operate. By establishing this framework initially, each designer can upload a part to the assembly at any time to see how it will operate

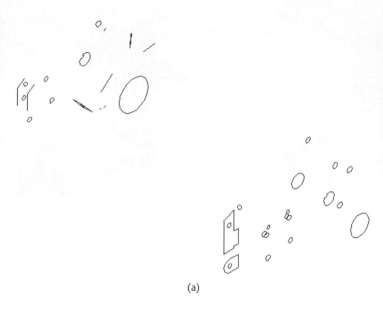

(a)

7.10 (a) The skeleton model for this laser printer subassembly establishes a 3D framework for relationships between parts or groups of parts. (b) As parts are designed, their center points or edges are matched to the datum planes and axes of the skeleton. (c) In the completed subassembly all the parts are matched to their locations. (d) The completed printer assembly is shown with the covers removed. (e) The finished printer assembly. *(Courtesy of the Color Printing and Imaging Division of Tektronix, Inc.)*

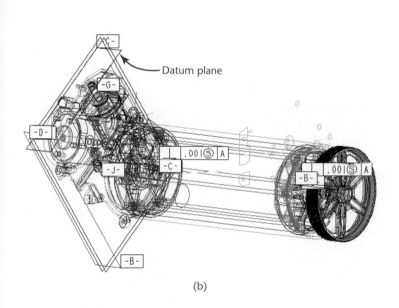

Datum plane

(b)

(d)

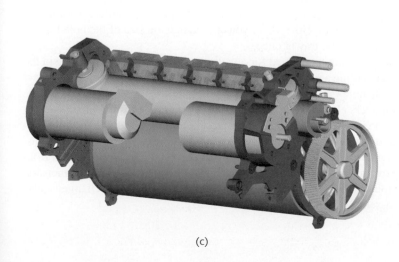

(c)

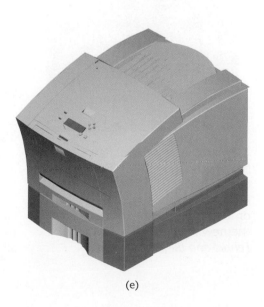

(e)

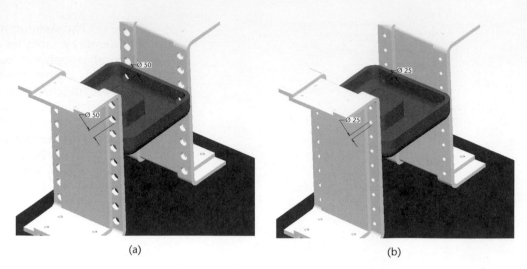

(a)   (b)

7.11 (a) The holes in the brackets and tray are defined by the global parameter Hole_Size that has the value .50 to fit a .50-diameter round head machine screw. (b) The holes in both parts are changed to .25 by updating the global Hole_Size parameter value.

7.12 IRI International has parametric models for different derrick styles. The basic rig structure is quickly reshaped to begin work on the design of custom parts for the new rig. *(Courtesy of Wayne Fehres and IRI International.)*

within the constraints established by the framework. Because the skeleton is made up of lines, planes, points, and other entities that do not have volume, using a skeleton will not change the mass property analysis for the assembly.

Global Parameters

As you define the skeleton framework, you may also define *global parameters* for the assembly. Parts created directly from the skeleton inherit the parameters used to define the skeleton, but global parameters may be needed to capture other aspects of the design intent.

A global parameter is one that is the same across multiple parts. You can use global parameters to control the size of a feature so that mating features can be re-sized as a group.

Consider the fit between the mating parts shown in Figure 7.11. The holes on the mounting brackets and tray must align and must be roughly the same size so that a machine screw will fit through both parts. By creating a global parameter for the size of the hole, called in this case Hole_Size, you can change the size of the hole feature on both the brackets and the tray at once.

Global parameters can be defined for any dimension to be shared by parts in the assembly. You may define a global parameter from the skeleton itself to drive critical distances on part features.

Global parameters also make it possible to build relationships into the assembly as you would for an individual part. The drilling rig in Figure 7.12 contains more than five thousand parts but was designed to be resized to fit different drilling situations. Changes can be made at the assembly level (the top level) to change the width from any corner, the overall height, and the height of any bay. Equations in the dimension parameters for individual parts reference global parameters that when changed cause all parts in the assembly to be updated to the new size.

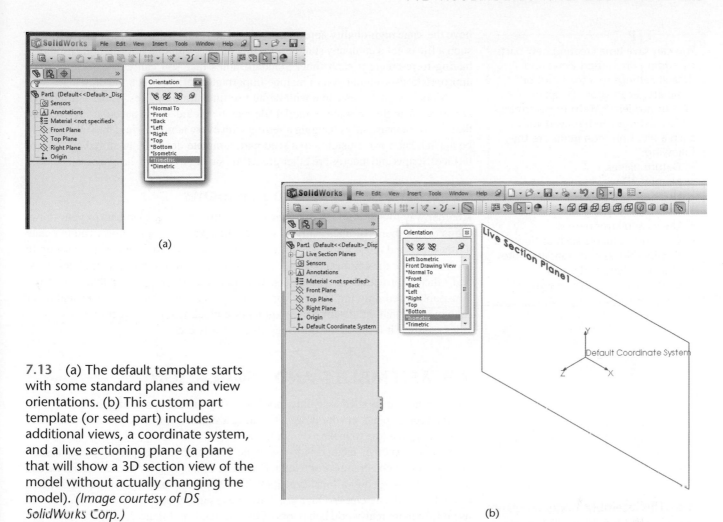

7.13 (a) The default template starts with some standard planes and view orientations. (b) This custom part template (or seed part) includes additional views, a coordinate system, and a live sectioning plane (a plane that will show a 3D section view of the model without actually changing the model). *(Image courtesy of DS SolidWorks Corp.)*

(a)

(b)

Seed Parts

Seed parts, also called *templates* or *prototype drawings*, are another technique for assembling parts effectively and for starting new drawings quickly and systematically. Seed part files contain elements you want every model to contain. A seed part might have a set of datum planes defined according to the company standard, named view orientations matched to the datum planes, unit settings, coordinate systems, layer names, and other items. Starting your new parts from a seed part saves you the time it would take to create these settings and elements in a new file.

Seed parts can help you assemble parts by providing a standard orientation. For example, when you assemble two parts, you may want a common datum surface to face the same direction on each part. Using a seed part, where standard datum planes have already been set up and named to make them easy to identify, can help you quickly insert your part in the correct orientation with respect to the rest of the assembly. Using standard view names as a part of your seed drawing helps you quickly produce views from your models (Figure 7.13).

Other members of the design team save time in creating a drawing from your model when you use consistent view-naming conventions.

Seed parts with consistent layer names can eliminate confusion in viewing assemblies. By starting parts with the same basic set of layers, you can keep the assembly organized and make it easy for other users to identify on which layer a feature would typically be shown.

If your company has a standard title block, tolerance block, notes, and set of required views, seed parts or prototype drawings can help produce drawings that all

— **TIP** —

You can save time creating new parts by adding often-used views and default settings to a seed part or template. Read about template files in your modeler's **Help** to determine which settings can be saved with such a file. Common items are the following:

- Datum planes
- Named 3D views
- Unit settings
- User coordinate systems
- Layers and layer names
- Drawing elements such as title and revision blocks and standard notes
- Drawing styles and views
- Customized workspace settings

have the same high-quality appearance even when created by different users. Even if such a file is not a company standard, setting one up with this information eliminates having to re-create it each time. Standard notes that you can edit or delete if not appropriate also remind you to include important information on the drawing.

Modeling packages come with default settings for new files. Some settings that are stored with the drawing or model file may be included in a seed part. If you find that you are consistently changing a setting with every new drawing, the setting might be a candidate for inclusion into a seed part, template drawing, or stored custom interface. Scripts and macros are other great time savers when you have a repeated task.

Constraint-Based Drawing Elements

Another way to add borders, title blocks, and other drawing elements to your drawing is to link them parametrically as you would a part file. This allows you to create a title block and reuse it for all your drawings. This title block can even be set up to prompt you for the engineer's name, approval date, material information, tolerances, and other information as parameters. It is important to remember that if you use constraint-based drawing features (or external references) for title blocks and other drawing information, deleting the original title block from the directory where it resides removes it from all the drawings where it was used.

7.3 ASSEMBLIES AND SIMULATION

An accurate and detailed assembly model is a step toward simulating the interactions between components in your design. Creating a skeleton model to drive the assembly geometry is one way to make a dynamic assembly that allows you to study component interactions. Another method is to assemble your parts using special "mechanism"-style assembly constraints. For example, instead of aligning the centers of a hole and a shaft, you can specify a "pin"-type joint. These joints restrict the degrees of freedom of the part in the assembly. Once you have the assembly defined in this way, you can use it to explore real-world behaviors of the mechanism. Figure 7.14 shows an example mechanism created in Pro/ENGINEER.

7.14 This assembly model simulates the real-world forces on a Maserati MC12 race car suspension. Notice the yellow symbols showing the forces that are applied to the model. *(Courtesy of PTC.)*

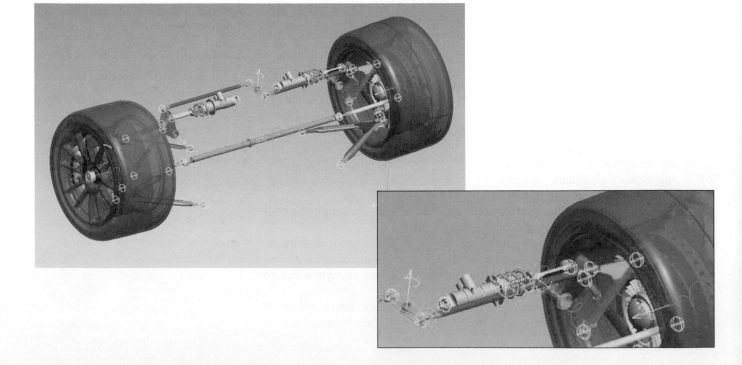

Depending on the software you have, you may be able to simulate part contacts; measure interference between parts; simulate gravity, springs, dampers, belts, gears, friction, and ergonomics; as well as perform kinematic analysis directly in your modeling environment.

7.4 PARTS FOR ASSEMBLIES

To fully enjoy the benefits of a digital assembly model, your design database should include all the parts designed for the assembly. Most constraint-based modeling software allows you to add and import non-constraint-based parts to the constraint-based assembly so they can be represented during the design refinement process (see Figure 7.15).

Standard Parts

Using standard company parts (or purchased parts that are readily available) can be cost-effective for two reasons:

- First, they do not have to be designed. Many suppliers provide models of their parts in a variety of CAD formats that you can insert into your assembly model. Using standard company parts that were designed for another project saves design effort.
- Second, volume manufacture may be less expensive than manufacturing the quantity needed for a particular project. Parts manufactured in quantity have a lower per-piece price. Some companies maintain a library of parts on the network that can easily be used. Sharing the part among more products helps increase the volume used. A parts database is useful even when manufacturing quantities are not high, as the company saves design time for every part reused.

When a company part is not available, a part that meets your need may be available from a supplier. The Web is a good place to check for stock parts. Suppliers often provide solid or constraint-based models of their parts for free. Some engineers spend up to 20% of their time redrawing standard parts to show how they will fit with the newly designed parts in the assembly. Finding a model available from the supplier lets you spend your time designing the new components needed to get the job done.

Many sites provide their parts in standard 2D and 3D formats (such as .dxf, IGES, or STEP,) and in the native formats for different modelers (such as AutoCAD, SolidWorks, and ProENGINEER). ThomasNet.com is a good place to start when looking for standard parts. Their interface makes it easy to locate and view parts from many different manufacturers. Figure 7.16 shows a plastic knob available in 2D and 3D CAD formats from the vendor Davies Molding.

Common parts are also available in libraries that ship with your software or in third-party part libraries. These parts often have the advantage of being modeled using constraints so their sizes can easily be changed.

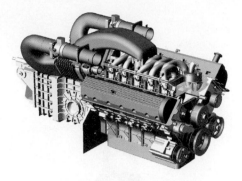

7.15 All the components of this engine assembly were modeled as solids—whether they were new or off-the-shelf—so that the Predator assembly model would be complete and could serve as a virtual prototype for the car. *(Courtesy of M&L Auto Specialists, Inc.)*

7.16 CAD models of this plastic knob are available in a variety of formats. The CAD Drawings selection from ThomasNet.com makes it easy to find vendor part models like this one. *(Courtesy of Davies Molding, LLC.)*

7.17 The helical shape of thread is similar to the striping on a barber pole. *(Shutterstock.)*

7.18 This screw-type jack uses thread to transmit power to raise the car. *(Shutterstock.)*

Adding Static Parts to a Constraint-based Database

Some part models from outside vendors or ones imported from other modeling systems may not contain the constraint-based information you need for editing them. These parts are sometimes called *static parts*, because they were not created with parametric dimensions that allow them to be updated. They contain just the solid or surface information from the software in which they were created. Ordinarily, you will make little or no modification to stock parts, so the inability to edit the files is no disadvantage. You can call them into your assembly drawing and *add* new constraint-based features to these parts, but you cannot delete or resize the existing features easily.

7.5 THREAD

Thread is the term for the helical shape, similar to the striping on a barber pole, that you commonly see on screws, bolts, and other devices (see Figure 7.17). Thread has three typical uses:

- To hold parts together, as do the threads on fasteners such as screws or bolts
- To provide adjustment between parts, for example, the thread on the adjusting screw on a simple carburetor
- To transmit power, as the driving screw in a screw-type jack does (see Figure 7.18)

Figure 7.19 shows a threaded shaft with major features identified. Familiarize yourself with thread terminology:

- *Major diameter:* the largest diameter of a threaded hole or shaft
- *Minor diameter:* the smallest diameter of a threaded hole or shaft
- *Thread angle:* the angle between the sides of the thread
- *Thread axis:* the centerline of the cylinder where it appears rectangular
- *Crest:* the top surface of a thread where the two sides join
- *Root:* the bottom surface of a thread where the two sides join
- *Side:* the surface of the thread connecting the crest and root
- *Pitch:* the distance from a point on a thread to the corresponding point on the next thread
- *Lead:* the distance the thread advances when turned one complete turn (single thread advances one pitch distance in one complete turn)
- *Thread depth:* the distance from the crest to the root of the thread measured perpendicular to the thread axis

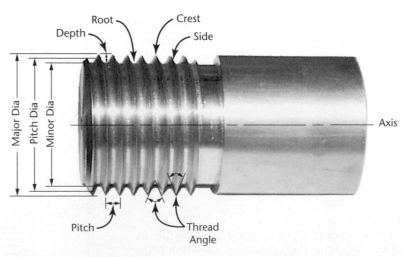

7.19 Thread Features

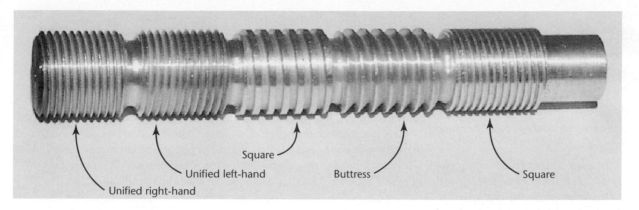

Square

Unified left-hand   Buttress   Square

Unified right-hand

Thread Form

The shape of the thread is called the ***thread form***. Most thread is based on the sharp-V thread form, which has 60° angles for its roots and crests. This thread form is still used for some purposes, such as brass pipe fittings.

Other thread forms are common in different applications. Some of these forms are illustrated in Figure 7.20.

Unified thread is based on a sharp-V thread form, but the root is rounded and the crest may be rounded or flattened. This is the form that was agreed on by the United States, Great Britain, and Canada during World War II. The Allies found it hard to find replacement parts because they did not have an agreed-upon standard, so they adopted the unified thread form in 1948 to promote interchangeability of parts.

Metric thread is similar to the unified thread, but it has a shallower depth of thread. This is the agreed-upon standard for international fasteners today.

Square, ***acme***, ***standard worm***, and ***buttress*** thread are all thread forms that are used to transmit power.

Knuckle thread is rolled from sheet metal or cast from aluminum or steel. It is used on bottle tops and lightbulbs and for various other purposes.

Double and Triple Thread

Double and triple thread are used for adjusting screws, to provide for quick opening and closing of valves, and for quick assembly of parts that do not need to withstand large forces. Typically, thread is single; single thread has one single helix formed on the shaft or inside a hole. Double thread has two helixes that parallel each other along the shaft or inside the hole; triple thread has three helixes paralleling each other. Double thread advances two pitch distances in one complete turn; triple thread advances three.

Figure 7.21 shows examples of double and triple thread. Notice that you can determine whether it is double or triple thread by looking at the number of thread starts at the end of the bolt (Figure 7.22). Double and triple thread are noted by the

7.20 This shaft shows examples of right- and left-hand unified thread, square, and buttress thread forms. Note how the profile of the thread shows its form most clearly.

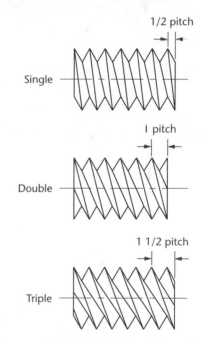

7.21 Like the colored striping on the barber pole, double and triple thread have two or three helixes running parallel around the shaft.

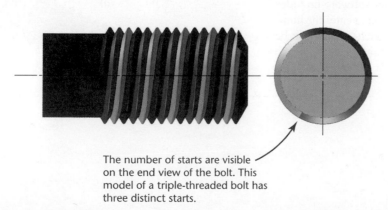

The number of starts are visible on the end view of the bolt. This model of a triple-threaded bolt has three distinct starts.

7.22 Count the number of thread starts on the end view of the bolt to determine if a bolt is single, double, or triple thread.

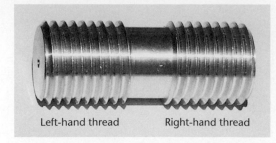

Left-hand thread   Right-hand thread

7.23 The right end of this shaft has right-hand thread, which advances when turned to the right. The left-hand end has left-hand thread, which advances when turned to the left.

7.24 This bottom bracket is used to attach the cranks on a bicycle to the frame. One end has right-hand thread and the other side left-hand thread so the pedal motion won't unscrew the bottom-bracket bearing cups.

wording 2 STARTS, or 3 STARTS in the note specifying the thread. If not designated, thread is understood to be single thread.

Right- and Left-Hand Thread

Most thread advances when turned toward the right or in a clockwise direction. Left-hand thread advances when turned in a counterclockwise direction—toward the left. It is labeled LH on the drawing. Thread that is not designated as left-handed with the note LH is always considered to be right-hand thread. A shaft with right- and left-hand thread at opposite ends will advance in both directions when the shaft turns—for example, as a turnbuckle does. Figures 7.23 and 7.24 show examples of right- and left-hand thread.

Fasteners

Fasteners are a common threaded part sometimes added to assembly models and drawings. It can be desirable to show the fasteners in the assembly to make it easy to track the fasteners specified. It is easier to make detailed assembly drawings showing fastener locations and sizes and to help keep the bill of materials structured when they are a part of the model. Standard fasteners are usually purchased.

Modeling Thread

Thread is not typically modeled, because this complex helical surface adds greatly to the complexity and file size of the model without providing much useful manufacturing information. Many 3D modeling packages allow you to model thread accurately, but it is better to include this type of information as a note, or if you require this level of detail in the model, wait to add it after you have worked out the major details of the design.

Sometimes, the major and minor diameters of the thread are represented so they can be used in checking clearances and interferences. If not, the nominal size (the general size used to identify the fastener) of the threaded hole or shaft is modeled. A detailed list of fasteners and their proportions is included in Appendixes 18–33.

SPOTLIGHT

Sweep a Thread Form

To model thread:

a. Draw the cross-sectional shape of the thread.
b. Draw the helical path.
c. Sweep the cross section along the helical path to form the shape of the thread.

Some modeling software allows you to define the helical path curve using an equation. Other software provides a helical sweep command that allows you to enter the number of revolutions and diameter parameters, which can be changed later. Surface modeling techniques can also be used to model thread, but it can be difficult to edit later.

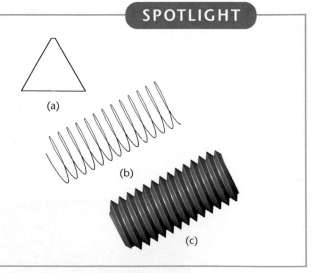

STEP by STEP

THREE WAYS TO SKETCH THREAD

Detailed Representation

Detailed representation is the most realistic looking, although it is a symbolic way of representing thread and does not show the helical shape accurately. Detailed representation is rarely ever sketched because it is time-consuming to draw the large number of lines making up its shape. Detailed representation does not reproduce well when the thread diameter on the sheet of paper is smaller than 1 inch or 25 millimeters.

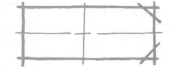

1 Lightly block in the threaded shaft. Draw a light vertical line to define the threaded portion. Add lines angled at 45° to represent the chamfer.

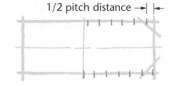

2 Make light ticks to mark the pitch distances.

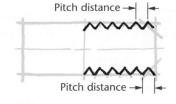

Pitch distance

Pitch distance

3 Sketch 60° V's representing the thread form.

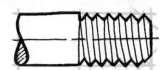

4 Connect crest to crest and root to root and darken in the final lines.

Simplified Representation

Simplified representation uses hidden lines to represent the depth of the thread. Because the thread depth is different for different pitches or thread series, and can sometimes be quite small, this hidden line is often shown at 1/16 inch or 2 millimeters so it can easily be identified on a sketched or printed drawing. Notice the lines that represent the chamfer and the end of the threaded length. The chamfer makes it easier to get the bolt started in the hole.

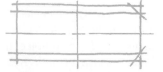

1 Lightly block in the threaded shaft. Make a light vertical line to define the threaded portion. Use a light line to mark thread depth.

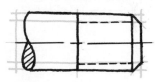

2 Sketch short dashes for the thread depth, then darken in the final lines.

Schematic Thread Representation

Schematic thread representation uses shorter thick parallel lines to represent the roots of the thread and longer thin lines to represent the crests. The ends of screws and bolts are shown chamfered, as they are in detailed and simplified representation. Distances from crest to crest or root to root are not always shown accurately; instead they are often shown about 1/8 inch or 3 millimeters apart to give the appearance of thread.

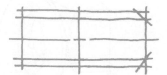

1 Lightly block in the threaded shaft. Make a light vertical line to define the threaded portion. Use a light line to mark thread depth.

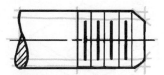

2 Use short thick lines to represent the roots and long thin lines to represent the crests.

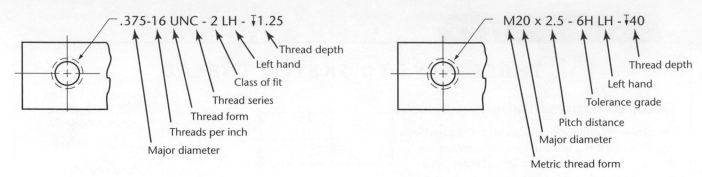

7.25 Thread Notations

Thread Note Abbreviations	
UN	Unified form
N	American national form
M	Metric
C	Coarse
F	Fine
LH	Left hand
RH	Right hand
A	External thread
B	Internal thread

Thread Notes

Thread notes give all the information needed to purchase or manufacture the thread fastener or part indicated (see Figure 7.25). Instead of dimensioning thread with many separate notes, using a standard notation provides this information in the simplest fashion. The thread note is usually placed where a threaded hole appears round or where a threaded shaft appears rectangular.

Bolt Heads

Bolt heads are proportioned to the diameter of the threaded shaft. Figure 7.27 shows the proportions for a typical hex head bolt and nut. When sketching bolt heads, represent the shape so that it looks in proportion, but do not worry about getting the dimensions exact. Often, in a quick sketch, as shown in Figure 7.26, all bolt heads are shown as hex type heads even if another type may be used.

The shape of the bolt head can be called out in the note. For example, the note

.375 − 16 UNC × 1.875 HEX HEAD

indicates a .375-diameter threaded shaft (16 threads per inch, Unified coarse form) on a hex head bolt that is 1.875 inches long. For most bolts the length does not include the head portion of the bolt. Countersunk machine screws do include the head in the specified length. Refer to the appendixes for the proportions and dimensions for typical bolt heads and other useful information such as thread tables for standard series, bolt proportions, and typical proportions for other fasteners.

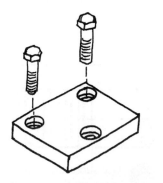

7.26 These bolts are sketched as hex heads, but another bolt may be called out in a note.

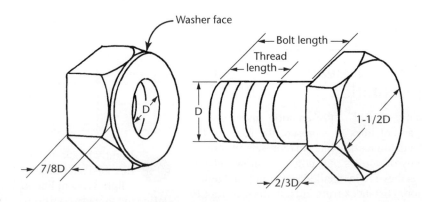

7.27 Bolt and Nut Proportions

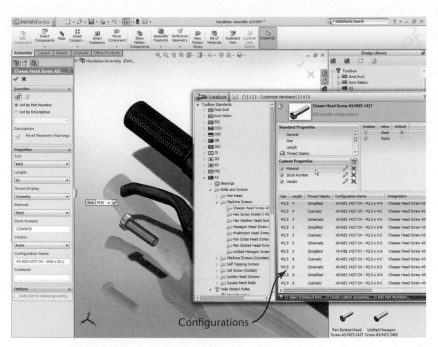

7.28 SolidWorks provides drag-and-drop fasteners as a part of the Toolbox parts library. Notice the list of configurations for the fastener shown. These allow one model to serve different sets of parameter values stored. Because the basic shape of the fastener is the same, it can quickly be resized to different sizes. After a configuration is chosen, it can be further customized, as shown in this screen. *(Image courtesy of DS SolidWorks Corp.)*

Fastener Libraries

Many modeling packages provide part libraries for fasteners, bearings, and other similar content that is a frequent part of many designs (see Figure 7.28). If your software does not, you may want to build a library of typical fasteners used by your company so that you can quickly place them into assembly models. Many suppliers provide 3D models of their parts to make it easy to show them in your assemblies (Figure 7.29).

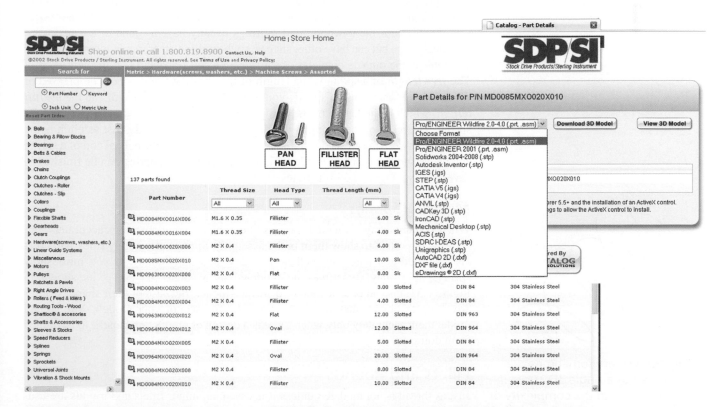

7.29 CAD models of standard parts are available for download from companies like Stock Drive Products/Sterling Instrument. *(Courtesy of Stock Drive Products/Sterling Instrument, sdp-si.com.)*

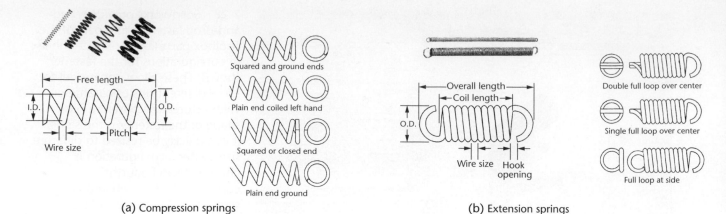

(a) Compression springs   (b) Extension springs

7.30   Compression springs (a) and extension springs (b) differ in how they are specified and how their ends may be finished. Basic terminology and common end types are illustrated here. *(Courtesy of Reid Tool Supply Company.)*

Springs

A spring designed for the purpose of being compressed is called a ***compression spring***. ***Extension springs*** are designed to be extended. Both are usually formed in the shape of a helical coil that can be compressed or extended and still return to its original shape.

The properties that determine the behavior of a spring are its material, thickness, major diameter, length, number and direction of the helical coils, and type of ends. To select a spring for a particular purpose, or to specify a standard part in your drawings, you need to specify these properties.

Springs may be made of materials such as hard-drawn steel, spring steel, stainless steel, or brass. The thickness of the spring may also be referred to as the ***wire size***; it is the width of the unwound coil. Springs usually have a round or square cross-sectional shape but can have other shapes, such as elliptical or rectangular. The ***major diameter*** is the outside diameter of the spring's coil (labeled O.D. in Figure 7.30). The ***free length*** of a spring is the length of the spring when it is not compressed or extended. For an extension spring, you would specify the coil length and the ***overall length***, which is the length of the coil plus the hooks at each end used to attach the spring. The ***pitch*** of the spring is the distance between the coils, just as pitch is the distance between the crests on a threaded item. Like thread, the helical coil of a spring can be wound right- or left-handed. Finally, the ends of the spring can be finished in several ways. Figure 7.30 illustrates several of these options.

Modeling Springs

When you are using 3D modeling, it is uncommon to model springs accurately, unless it is important to show them in the assembly. Springs are usually stock parts that are specified and purchased rather than designed individually. If you must show springs, they may sometimes be modeled as a single 3D curve, as shown in Figure 7.31, instead of as a solid or a surface model. Because of their shape, springs are complex objects that greatly increase the file size of your models. You should model them accurately only when there is a design or production benefit to be gained from doing so.

7.31   Springs are sometimes represented as a single 3D curve in a 3D model when the complexity of the shape outweighs the benefits of modeling it.

Fillets and Rounds

Just as thread is not modeled unless it is essential, minor fillets and rounds are usually best left until late in the design process. These features make the object more complex and increase the file size, usually without adding much necessary

7.32 The inside pocket was formed by cutting away material with a rotating endmill. The rounds in the inside corners can be smaller or larger, depending on the cutting tool used, but they cannot be square using this process.

information to the design. One way to define fillets or rounds is to specify a radius and select two surfaces. Each time a surface changes, the fillet or round is recalculated for the new location. If the angle of the surface changes (such as when draft angle is added to a part), another element is added to the calculation needed to regenerate the feature. For this reason, fillets and rounds may not be updated properly when the design changes and the model is regenerated.

To document your design and convey it accurately to manufacturing, you may want to show fillets and rounds in the final version of the parts that are released to manufacturing, but you should generally wait until just before then to add them.

In some cases, it may not be necessary to add these features to your model. Often, a note on the printed drawing that accompanies the CAD file stating "round all sharp edges to R3" (or some other value for the radius of the round) may be perfectly acceptable. The smaller, sometimes cleaner, file that results when you do not show the minor fillets and rounds may even work better if you plan to use the file for generating numerically controlled (NC) machine code to be used to manufacture the part.

In other cases, part manufacture may dictate fillets and rounds on the part, even if they did not exist in the model. Certain manufacturing processes, such as milling, use rotating cutting tools that cannot make a square corner. These processes create small rounds on the part that correspond to the size and type of the cutter used. Figure 7.32 shows a rounded inside corner formed by a milling machine cutter. The radius of the round is equal to half the diameter of the cutter. Figure 7.33 shows some milling machine cutters.

Whether to include small fillets and rounds in the model is an issue you should discuss with the person responsible for manufacturing the parts. If you are unsure, you may want to include the features.

Square

Ball nose

7.33 An endmill is a cutting tool for removing metal and forming an edge. The endmill can have a square or ball end, to form a square or rounded corner, at the base of an inside pocket or trough. *(Courtesy of Reid Tool Supply Company.)*

7.6 USING YOUR MODEL TO DETERMINE FITS

When parts are accurately modeled in 3D and organized into an assembly, you can use the assembly to perform fit and interference checking on the models. Doing a thorough study of how the parts will fit once assembled can save time and money. After tooling has already been produced to manufacture parts, changes to the design become very expensive.

Because the parts in the assembly can be shaded and viewed from any direction on the computer monitor, you can visually inspect the fits and clearances. You can also make measurements and list dimension values to compare parts with one another. Most solid and constraint-based modeling software also provides a command for checking the interference between two parts.

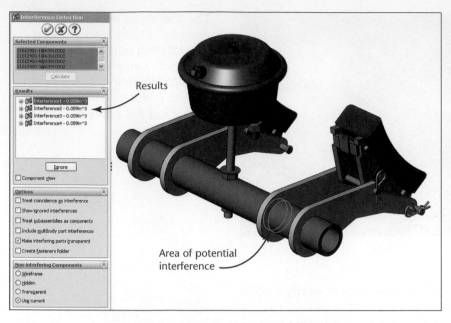

7.34   The shaft is checked against the support arms to see whether these parts interfere. *(Courtesy of Dynojet Research, Inc.)*

Interference Checking

Interference is the amount of overlap between one part and another. When you use a command to check the interference between two parts in an assembly, the solid modeling software will report that the solids do not interfere or will indicate the amount of overlap—sometimes by creating a new solid to represent the overlap (see Figure 7.34).

Because constraint-based modeling software makes it easy to assemble parts early in the design process, interference checking tools can also facilitate more effective concurrent engineering. For example, Boeing's designers are responsible for checking their work in the digital preassembly model of the aircraft to identify places where the system or component they are designing will interfere with work in progress in other areas of the design. Boeing's proprietary visualization tools and assembly database work together to pinpoint problems with the assembly during refinement. Each designer then takes steps to resolve the interference with the appropriate designers for the systems involved. Interference checking with an assembly model created in SolidWorks, Pro/ENGINEER, or another solid modeling package can serve the same role.

Tolerances

The manufactured parts for your design must fit together when assembled, but no part can be manufactured exactly. Some manufacturing processes are very accurate, but there is always some slight variation between the actual manufactured parts and the dimensioned drawings or 3D models. *Tolerance* is a statement of the total amount a dimension on the manufactured part may vary from the specified value in the drawing or the model.

There are a number of ways to state tolerance values for dimensions; you will learn more about them in Chapter 10. Figure 7.35 shows a part drawing dimensioned with limit tolerances for the size of the holes. A *limit tolerance* specifies the upper and lower allowable value for the dimension when measured on the actual part.

7.35   The limit tolerance shown here indicates that the size of the hole in the manufactured shaft support may be as small as 4.99 and as large as 5.01 and still be an acceptable part.

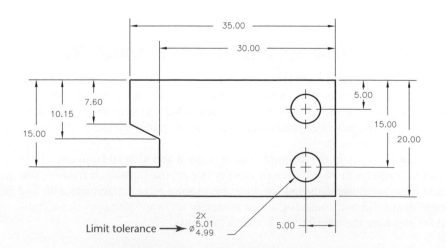

Manufacturing processes have certain ranges of accuracy. Requiring a high tolerance—one with very little allowable deviation—can limit the choice of manufacturing process to those that can be expected to meet that level of accuracy. Specifying very small tolerance values can increase the cost of the finished piece. In general, tolerances should allow as much variation as possible without affecting the functionality of the design.

Material selection also plays a role in the accuracy of the finished parts. Parts can be manufactured more accurately from some materials than others. In addition to determining the tolerance range for a part based on its function in the assembly, you should talk to personnel in manufacturing or the vendors who will make the parts. You should fully understand the issues in manufacturing the part to the tolerance you have specified and the effect it will have on the piece price for the finished parts.

Pro/ENGINEER and some other solid modeling software allow you to specify a tolerance range for a dimension and then apply that to the feature at either the upper or lower end of the dimension range. By doing so, you can run a check on how the parts will fit at the minimum and maximum clearance values. The options available for doing so are illustrated in Figure 7.36.

Even if your software does not provide a special function for checking fits at the lower and upper limits of the tolerance range, you can systematically change the dimensions for mating features to represent the minimum clearance, then inspect the model to see how the parts fit. You can change the values as necessary to represent the maximum clearances and inspect the fits. This ability to inspect clearances between mating parts is one of the benefits of using solid modeling to create your parts.

Fit Between Mating Parts

Tolerances for individual features are determined by their fit requirements with mating parts in the assembly. As you might expect, the maximum clearance between two parts (or loosest fit) occurs when the external member is at its largest size and the

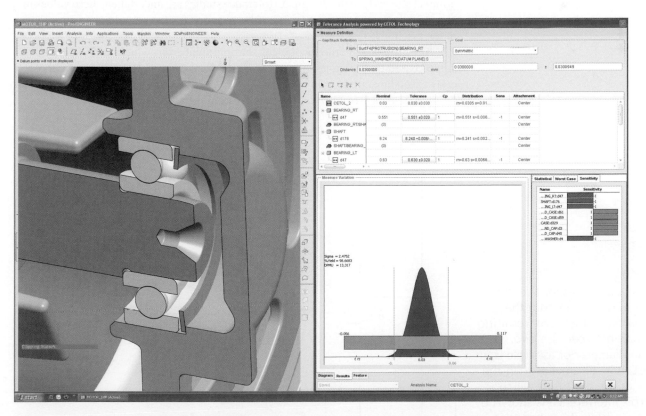

7.36 Pro/ENGINEER provides a sophisticated interface for performing fit studies for a range of part tolerances. *(Courtesy of PTC.)*

internal member is at its smallest size. The minimum clearance (or tightest fit) happens when the external member is at its smallest size and the internal member is at its largest size. *Allowance* is the term for the minimum clearance.

Four categories of fit between mating parts are clearance fit, interference fit, transition fit, and line fit.

1. A *clearance fit* requires a space or clearance between the internal and external member.
2. An *interference fit* requires that the internal member always be larger than the external member. The parts must be forced together or the outer part heated to enlarge it enough to fit the internal part before it shrinks back to size. The tolerances for this kind of fit, often called a *force fit*, are illustrated in Figure 7.37.
3. A *transition fit* indicates that either a clearance or interference fit is acceptable. That is, some parts will fit with a clearance and others will have to be force fit.
4. A *line fit* indicates that the ranges for the dimensions ensure that either a clearance fit or no more than surface contact will result between the parts.

Nominal Size and Basic Size

Two terms you should understand in connection with tolerance and fit are *nominal size* and *basic size*. These terms are often confusing because in general use they mean nearly the same thing. However, these are two different terms when used correctly in engineering graphics. ANSI states that *nominal size* is "the designation used for purposes of general identification." The nominal size is often expressed in common fractions, such as 1/2 inch. The nominal size is similar to the way lumber is typically sized. A two-by-four board is the general designation (nominal size), however, the actual size of a two-by-four is approximately 1-1/2 × 3-1/2 inches. The general designation of a hole as 1/2 inch in diameter is useful early in the design before accurate tolerance studies have been undertaken, but it does not always mean that the final specified size will be exactly 0.5000 inch. Typically, features are modeled at their nominal size early in the design. As tolerance studies are undertaken, these values may change to reflect the more accurate sizes required for the parts to function properly in the assembly.

The *basic size* is the theoretical starting point from which the tolerance is assigned based on design requirements. *Basic dimension* is the term for the theoretically exact dimensions used in *geometric dimensioning and tolerancing* (GD&T) to define a boundary from which tolerance variations are established.

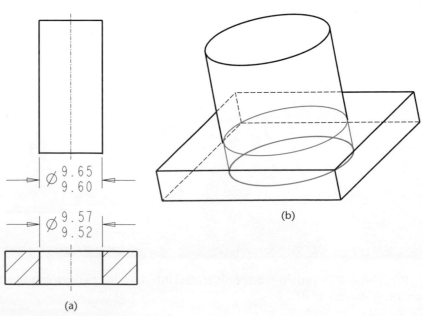

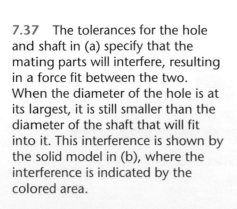

7.37 The tolerances for the hole and shaft in (a) specify that the mating parts will interfere, resulting in a force fit between the two. When the diameter of the hole is at its largest, it is still smaller than the diameter of the shaft that will fit into it. This interference is shown by the solid model in (b), where the interference is indicated by the colored area.

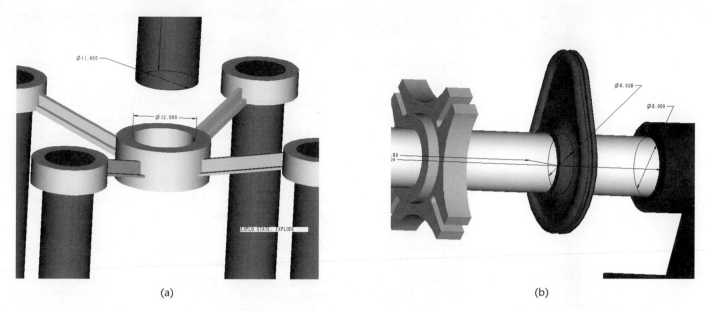

(a)

(b)

7.38 (a) The basic hole system is the most commonly used method for determining tolerances because holes are often formed with standard sized tools. To use it, start with the minimum size of the hole and determine the tolerances for the shafts using fit tables. (b) When a shaft is available in standard sizes and is difficult to machine, or in an assembly where many parts must fit on a single shaft, the basic shaft system may be preferred for determining its tolerances.

Thus, if you were manufacturing two-by-fours, the basic size for a theoretically exact two-by-four would be 1.50 by 3.50 inches. However, no two-by-fours would meet this exact specification, so a tolerance range would be specified around that perfect size to describe allowable manufacturing variation.

Two systems for calculating a tolerance based on the desired type of fit and the basic size of the feature are the **basic hole system** and the **basic shaft system**. The basic hole system is the most common because standard machine tools, such as drills and reamers, are often used to create the hole to the specified size. Then, the shaft is machined down to the specified tolerance so that it fits inside the hole, providing the desired type of fit.

In the basic hole system, the lower limit or minimum hole size is used as the basic size for the hole feature. Designing around the hole size as the basic size allows you to use standard fit tables to look up the tolerance values you can use to produce a desired type of fit. (Refer to Appendixes 6–17 for a set of standard fit tables.)

The basic shaft system uses the maximum size of the shaft as the basic size to determine the tolerance. This is a less common method, but it is sometimes used when many features are to fit on a shaft made of a material that is hard to machine without changing its desired properties. In the basic shaft system, the holes in mating parts are toleranced to fit with the shaft size, which is used as the basic size. See Figure 7.38 for examples of designs that might use each method.

Tolerances per Manufacturing Process

As the engineer, you are responsible for fully defining exactly what you want manufactured and for prescribing a scheme by which the final manufactured parts can be inspected to determine whether they are acceptable. The term *as per manufacturing process* means that the typical tolerance for the specified manufacturing process is acceptable for the part. This designation may be used when specific fits do not matter, such as a part exterior with no mating part, or when you have modeled the part accurately and NC manufacturing processes will be used to make the part, with an acceptable allowance.

If the fit between mating parts is important, however, you should specify the allowable tolerance range for the dimension, by either providing an unambiguous set of drawing views or using the CAD digital design database. You should also perform a fit study to determine whether the assembly will function at the size ranges you have

7.39 Digital Calipers. *(Courtesy of L. S. Starrett Company.)*

No-go plug

Go plug

7.40 Go/No-Go Gage. *(Courtesy of Tom Jungst.)*

specified. If no tolerances are specified, you give up the ability to later reject parts that do not fit in your assembly.

Measurement and Inspection

During and after manufacture, parts are inspected to determine if they meet the specified design requirements. Measurement and inspection can be performed a number of different ways, depending on the requirements for the accuracy and fit of the parts. Parts can be measured with a machinist's scale, calipers, micrometer, or other graduated measuring devices (ones that have markings that indicate the measured size) (see Figure 7.39).

If it is not necessary to measure the size of the features but only to determine whether parts will fit, gaging can be used. One of the most common types of gage is a *go/no-go plug gage*. This type of fixed gage is typically used for checking hole sizes. One end of the gage is machined to the size of the minimum cylinder that the hole should accept; the other end is machined to be just larger than the maximum size that should fit an acceptable hole. A part is acceptable if the small end of the plug gage fits into the hole, and the slightly oversized cylinder at the other end does not fit into the hole. Figure 7.40 shows an example of a go–no go gage.

To inspect parts accurately, measurement and inspection devices are designed with tolerances that are one tenth of the tolerance being measured on the part. Their operation depends on measurements taken consistently from one part to the next. To establish a starting point for accurate measurement, datum surfaces are defined and specified. Because the manufactured part may be rough or have only curved or angled surfaces, a datum surface provides a plane from which measurements can begin. A part may rest on an accurately finished surface (often, a granite slab, as it does not rust and is dimensionally stable). The actual surface of the part may be very rough, in which case it may contact the inspection surface at only three points. These three points are enough to define a plane that can be used as the datum surface. This surface, determined by three points and from which the measurements will be made, is referred to as the *primary datum plane*. The secondary datum plane can be established from two contact points on the next surface perpendicular to the primary surface. A tertiary datum can be established from one additional point located perpendicular to the other two datum planes. This combination creates a framework of three mutually perpendicular planes that can be used to accurately locate measurements on the manufactured part.

Datum planes can be established in the model from which to locate principal dimensions (Figure 7.41). Features on the part that will be used for measurement may be located from these planes to facilitate using the model to evaluate the effect of manufacturing deviations on fit and function. Remember that the solid model is geometrically perfect, but a manufactured part will only approximate the geometric shape. You must specify the degree of accuracy that is required for proper function of the part. Even a "flat" surface on the part may not contact the datum surface at every

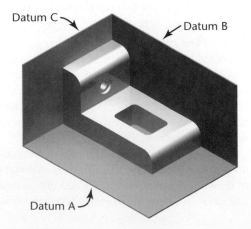

Datum C

Datum B

Datum A

7.41 Datum planes in the model can represent datum planes used for inspection.

(a)

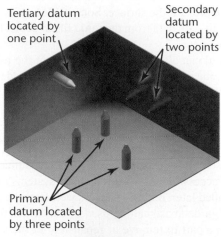

Tertiary datum
located by
one point

Secondary
datum
located by
two points

Primary
datum located
by three points

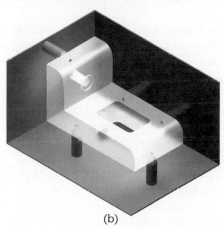

(b)

(c)

7.42 (a) Targets on the drawing specify the locating points where the fixture will support the part. (b) The primary datum is located by three points. The secondary datum surface is perpendicular to the primary datum (two points and the primary plane define this plane). The tertiary datum is located perpendicular to other datum surfaces (one point and the two existing datum planes define this plane). (c) An inspection fixture is designed to hold the manufactured part to be inspected so that it can be measured relative to the datums in the design. *(Courtesy of Classic Moulds & Dies.)*

point on its surface. For this reason, and especially for rough or irregular parts, datum targets, illustrated in Figure 7.42(a), may be identified on the drawing to locate the control points on the part that will be used for inspection. These points, called *principal locating points,* allow the inspection team to determine the theoretical planes from which measurements are to be taken.

In addition to the location of features with respect to datum surfaces, various other geometric properties of the part may need to be inspected. It may be important that the overall flatness of a surface not vary more than a certain amount (e.g., .002 inch) from that of a perfectly flat surface, or perhaps two features must be within some tolerance range of perfectly concentric for the part to work. The tolerances for these geometric properties can be specified on the drawing using special symbols for geometric dimensioning and tolerancing, about which you will learn more in Chapter 10. Being able to visualize important relationships of how parts fit together in a 3D assembly can aid you in determining key geometric features and the effect that manufacturing variations will have on the fit between parts.

Modern inspection equipment can produce very accurate measurements of the finished part. Inspection can be contact or noncontact inspection. The use of gages, probes, and other measurement devices that touch part surfaces is termed *contact inspection*. *Noncontact inspection* includes various optical methods of inspection,

7.43 Coordinate Measuring Machine. *(Used by permission of Brown & Sharpe.)*

laser triangulation, ultrasound, and other techniques. Coordinate measuring machines (CMMs) can provide accurate measurement and inspection using both contact and noncontact methods. Figure 7.43 shows a coordinate measuring machine. 3D point locations on the inspected part as measured by a CMM can be compared to determine geometric properties of the part. For example, if three points are used to establish a theoretical plane on the surface, other points can be compared with those to determine the variation in flatness for the surface. Some CMM software can compare the measured part with data from the CAD file to depict the variation from the perfect geometry.

Temperature and material can also affect the accuracy of the part. If it is not stated otherwise, measurements are intended to be made at 68° Fahrenheit (20° Celsius).

Accessibility Checking

The assembly model can also be used to ensure that the design can be assembled. Whether the assembly will be done by robots or humans, parts and fasteners must be accessible. Many devices and systems also need to be repaired or upgraded later, requiring access well after the assembly phase. The 3D design database can be used to check whether people will be able to reach a part to remove or repair it after the system is assembled.

Ergonomic analysis software can help determine accessibility for assembly and repair. *Ergonomics* studies the ability of humans to use a system. Many ergonomic analysis packages have 3D models of human beings (anthropomorphic data) that you can position inside the designed system to see how people will fit. Sophisticated ergonomic analysis systems allow you to choose various ranges of human sizes. This can help you see how your design will work when used by an average-size woman, a tall man, or a child. Some systems even have demographic populations built in so you can evaluate how well a design will work for the average French woman or the average American man.

7.44 Planning access to an airplane for service and repair is an important part of its design. This digital mannequin was imported into the CATIA software used by Boeing to evaluate the accessibility of its parts. *(Copyright © Boeing.)*

Even if robots will assemble the components, people will almost always perform repairs. Evaluating repair issues can be even more difficult than assembly issues, because the assembly can proceed in such a way that interior parts are fitted together before their exteriors obscure access and visibility. You can use ergonomic analysis or human models to check whether parts can be reached to repair them, as shown in Figure 7.44.

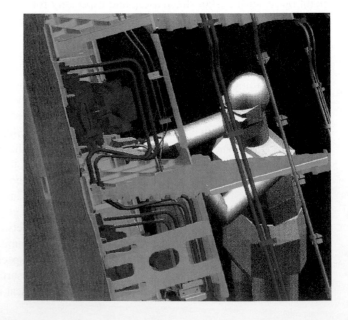

Usability is another ergonomic consideration: operator visibility, the complexity of controls, and the ability of the operator to correctly map the control to its function may be critical to the operation of the device you are designing.

Without additional software tools, you can still use the assembly model to consider how the parts will be assembled and whether they are accessible. By assembling the model in the order in which the parts will be assembled in the factory, you can visualize the location of the components as they are being built. Measurements from the model can be used to determine whether a tool can be applied as needed, and whether a distance is greater than the reach of the average person.

7.7 DESIGN FOR MANUFACTURING

To do a good job designing any part, you should understand the manufacturing processes that are involved in producing your design. Even though you may not need to understand how to produce the parts yourself, you should understand the basic capabilities and limitations of various production methods, as well as the range of tolerances available from that process. You should also understand material selection and the role it plays in manufacturing processes and tolerances. The more you know about manufacturing processes, the better you will be able to design low-cost parts that function as you need them to.

In this section you will learn some basic information about the impact that three manufacturing processes may have on the design of individual parts. Keep in mind that this is not intended to be a thorough presentation of everything you need to know about a particular process or all processes. You can learn more about manufacturing in engineering courses and from on the job experience.

Design for manufacturability (DFM) is a key phrase that implies the importance of designing parts that can be manufactured inexpensively. If companies are to compete in today's world market, they must design and manufacture parts that are functional and inexpensive. Sometimes seemingly small decisions and variations in the design can add to the cost of the part. Even a small amount per part can be important when thousands or millions of the part will be produced. During the process of designing a part, communicate with the people responsible for having it manufactured and work together to produce the best design possible.

The processes used to manufacture parts may change over the production lifetime of a part. As a device becomes more and more accepted by consumers and production volumes increase owing to demand, it may become cost-effective to produce the device in new ways. Molded plastics, cast parts, machined parts, and sheet metal parts are common low-cost production methods. Design and modeling for these types of parts will alert you to issues you should consider and investigate as you model parts.

Modeling Injection-Molded Plastic Parts

Many plastic parts are formed by injecting plastic into a mold. Injection-molded plastics are popular for low-cost parts for several reasons: the material is attractive, non-corrosive, colorful, lightweight, and can provide any number of surface textures and graphics. Molded plastic parts allow for good design control overall, high-volume production, and relatively low cost per part.

Most molds have two halves. The side of the mold into which the plastic is injected is called the *active side* of the mold. After the plastic is injected and allowed to cool, the two halves of the mold separate, and the part (or multiple parts) is removed.

7.45 Injection molds have two halves that separate to release the plastic part. This image is taken from a cartoon education series on plastics manufacturing sponsored by GE Plastics. Visit their website at www.ge.com/plastics to learn more about plastics manufacturing. *(Courtesy of GE Plastics.)*

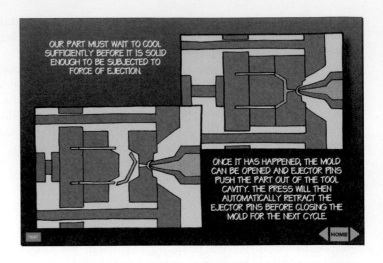

Parting line

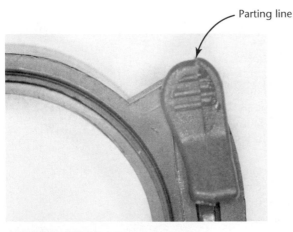

Ejector pin mark

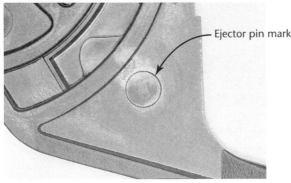

7.46 Injection-molded parts may have a visible line where the mold halves joined, or marks from ejector pins.

Figure 7.45 shows an illustration of an injection molding machine with the two mold halves separated.

Injection-molded plastic parts have characteristics you must consider when designing for them. Molded parts in general must include *draft*, or taper, that allows the part to be removed from the mold. Plastic shrinks as it cools, and the shrinkage plus the draft allows the parts to be removed from the mold into a collection bin. If the shape of the part requires it, ejection pins may be necessary to force the part out of the mold. Usually, the mold designer will determine the necessary cooling lines, ejection pins, and other features, but you should be aware of the general process. Ejector pins may leave a slight mark on the finished part. If these marks would negatively affect the cosmetics of your design, you may need to change the part shape so that it is removed from the mold easily and does not require ejector pins. You can identify important cosmetic surfaces on the part when you send drawings or files to the manufacturer so the mold maker can use this information when designing the mold.

The *parting line* is where the two mold halves come together. If you look at common molded plastic parts, such as the one shown in Figure 7.46, you may see a fine line in the plastic where the two halves of the mold came together. You may also see marks where ejector pins pushed the part out of the mold.

When you model parts for injection molding, consider how the part will remove from the mold. Although there are some molding methods (such as slides) that allow you to make interior holes, the part cannot have a shape that will not allow it to be removed from the mold. Look at the part shape in Figure 7.47; the lip of the bowl makes it impossible to remove from the mold.

7.47 Can you see why this part could not easily be removed from a mold?

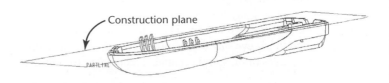

Construction plane

7.48 The construction plane in this part identifies the location of the parting line for the mold. The draft angle is added to the part on both sides of the parting line. *(Courtesy of Rob Mesaros.)*

As you design molded parts, plan for the parting line and design it so that both sides will be removed from the mold. You may want to add a construction plane to your CAD model, then project its edge onto the part as shown in Figure 7.48 to represent the parting line. Add draft to the surfaces of the part on each side of the parting line.

The amount of draft is decided by the size of the part, the finish of the mold, and the type of material and its shrink rate, among other factors. The Society of the Plastics Industry publishes guidelines to help you determine the exact amount of draft to add to the part. Usually, the mold designer will determine the shrinkage for the part and size the mold cavity so that the final parts will be to the size specified. Part shrinkage is not always uniform along the X-, Y-, and Z-directions. Many CAD packages can be used to size the part for shrinkage as well as add the draft.

The following are four guidelines for designing injection molded plastic:

1. *Try to maintain a constant wall thickness.* See Figure 7.49(a). The thickness of the plastic affects the rate at which it cools. Uniform wall thickness helps prevent sink marks in the plastic and keeps the part flat and uniform once molded. If wall thicknesses must change, make the change gradually instead of abruptly.

2. *Round all inside and outside corners* so parts will easily be removed from the mold. Rounded corners also improve uniform mold filling and help relieve stress concentrations in the mold. See Figure 7.49(b).

3. *Use the maximum allowable draft angle, and never less than 1° per side.* See Figure 7.49(c). Draft allows the parts to be removed from the mold. When parts are designed without enough draft, a vacuum can be created between the part and the side of the mold during removal. In extreme cases, this can damage the mold. When molded parts will have a surface texture, the draft angle must be increased to allow the part to be removed without damaging the texture.

4. *Projections should be not more than 70% of the normal wall thickness.* See Figure 7.49(d). For best results, projections should be no more than two and a half to four times the wall thickness in length.

7.49 (a) Maintain a constant wall thickness on injection-molded parts.

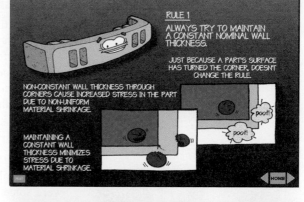

(b) Round all inside and outside corners to improve release from the mold and relieve mold stress concentrations.

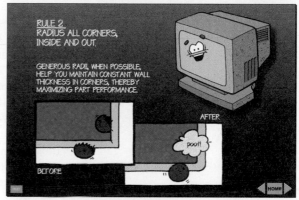

(c) Use the maximum draft angle.

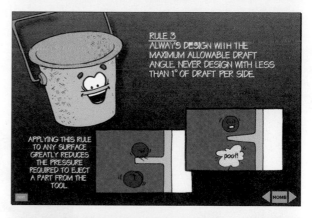

(d) Size projections so they are no more than 70% of the wall thickness and no longer than four times the wall thickness. *(Courtesy of GE Plastics.)*

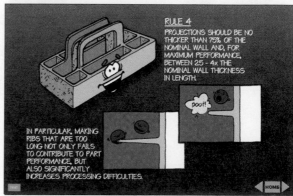

7.50 Milled Part (top); Part Turned on a Lathe (bottom) *(Courtesy of PENCOM.)*

Cast Parts

Casting is the process of forming a part by pouring molten metal into a hollow mold. Many of the design issues for cast parts are similar to those for plastic parts. For example, surfaces must have draft and rounded corners to make it possible to remove the part from the mold. The Aluminum Association publishes the *Standard for Aluminum Sand and Permanent Mold Castings,* which is a useful guide to design requirements for these types of parts.

Modeling Machined Parts

Machining processes remove material to produce the part shape desired. A rotating cutting tool shaves away material to form the shape of the part. These processes were some of the first to be computer controlled and are a common way to form metal parts.

Many machined parts start from common stock shapes such as round, square, plate, hex, block, and bar stock steel that come in standard sizes. When it is possible to use a standard size in your design, it may lower the cost to produce the part.

Some features, such as a perfectly square interior corner, are difficult to produce with this process. If you are planning your part for NC machining, avoid features that are difficult to machine or require multiple machine setups, unless they are needed. (You can make square interior corners using the electrodischarge machining process.) Fillets and rounds that result from machining methods provide additional strength and smooth the corners so they are not sharp when the part is handled. Figure 7.50 shows a part that has been milled (top) and a part that has been turned on a lathe (bottom).

Modeling Sheet Metal Parts

Sheet metal parts, such as the one shown in Figure 7.51, are laid out as a flat pattern that is then bent into shape.

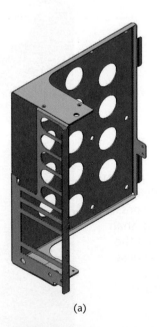

(a)

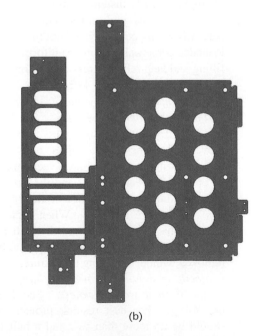

(b)

7.51 The 3D part in (a) will be formed by cutting the flat pattern (b) from a piece of sheet metal and bending it into shape. *(Courtesy of J. E. Soares.)*

7.52 Brake Press *(Getty Images—IStock Exclusive RF.)*

When modeling sheet metal parts, you may start by modeling them in 3D to visualize how they fit with mating parts. Eventually, the flat pattern for the part will have to be laid out for the part to be manufactured. Each surface on the part must be shown true size in the flat pattern. Additional material for hems or overlaps to be welded must be added and the bend allowance factored in. The bend allowance is determined by how much the metal compresses and stretches when it is bent. Its value depends on the thickness and material of the metal.

Other sheet metal parts may be difficult to manufacture because of the way they must be bent. The brake press in Figure 7.52 is commonly used to form sheet metal parts by bending the metal along a straight line. Without special tooling, some parts may be impossible to bend without deforming other sections of the part.

Another consideration in sheet metal part design is arranging the flat pattern so that the parts nest together to waste the least amount of material. Software packages that can aid in the development of the flat pattern from a 3D wireframe, solid, or surface model (see Figure 7.53) oftentimes will help with pattern nesting to reduce waste material.

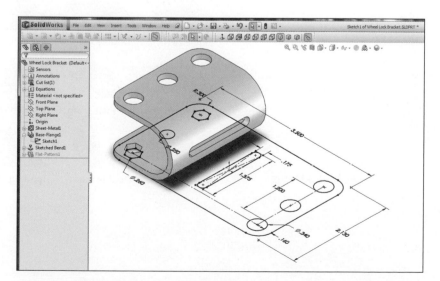

7.53 The flat pattern sketch is bent to create the 3D sheet metal feature. *(Courtesy of Salient Technologies, Inc. (www.salient-tech.com.))*

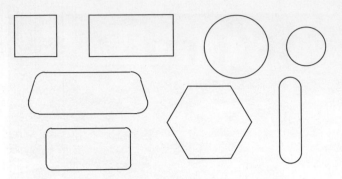

7.54 These 2D outlines are a library of shapes that correspond to standard punches.

Holes in sheet metal parts are often created by a punch, such as the automated punch press. Standard punches can quickly create openings in the sheet metal part. Figure 7.54 shows a library of shapes corresponding to standard punches.

To help in designing sheet metal parts, you may want to keep a library of 2D outlines of standard punch shapes. You can then use these standard shapes to create 3D model geometry quickly by importing the 2D section into your file and extruding and subtracting the feature to create the opening.

Some parts are possible to design but not possible to develop into a flat pattern. Figure 7.55 shows an example. This part cannot be created from a single flat sheet, as it would overlap itself.

Sheet metal parts are joined together by welding, soldering, or using hems. Figure 7.56 shows standard sheet metal hems.

As you have seen, the design database can serve as a proving ground for the digital assembly. A comprehensive database of the design can be used for testing, analysis, and documentation. Analysis will be explored in the next chapter.

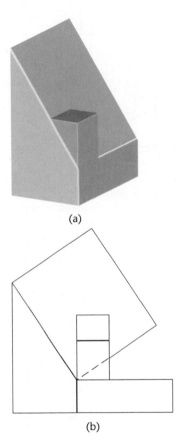

(a)

(b)

7.55 (a) This 3D shape cannot be made of a single piece of sheet metal. (b) The flat pattern overlaps itself.

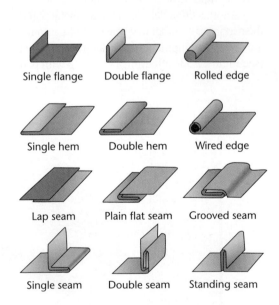

Single flange   Double flange   Rolled edge

Single hem   Double hem   Wired edge

Lap seam   Plain flat seam   Grooved seam

Single seam   Double seam   Standing seam

7.56 Sheet Metal Hems

MODELING SHEET METAL PARTS: THINK FLAT

When Stan McLean designed the Zuma coffee brewer manufactured by VKI Technologies, he used Autodesk's Mechanical Desktop (a precursor to Inventor) to model the brewer's 62 parts into a compact design that brews one cup of coffee at a time (see Figure 7.57(a)). The Zuma brewer, like all the VKI vending machines, uses a number of sheet metal parts—brackets and three- or four-sided boxes—to hold the various components together. McLean's approach to modeling sheet metals parts reflects their role in the brewer and is well suited to the design challenge of optimizing the fit of the various components.

"A vending machine is essentially a sheet metal box. Once I have the rough dimension of the machine, I start placing components into 3D model space within the outline of the machine. Our vending machines are built around a series of standard parts that we purchase, so we've created a model library of motors, pumps, valves, and fasteners. I place these into the assembly, then start moving them around to see how they will fit."

To do this, McLean usually has two files open at the same time: the assembly model (as shown in Figure 7.57) and the individual part model he is working on. By using external references to link individual parts into the assembly, McLean can update the assembly each time he changes an individual part. "I'm always loading the parts and checking the fit, then going to the part file and making modifications, then reloading the part to see if it is fitting the way I want it to."

With the individual components in place, McLean can begin modeling the sheet metal parts needed to hold them. "A sheet metal part is a series of flanges or a box with various bends at various angles. I could create a rectangle, dimension it, extrude it to the gauge of the sheet metal I'm using, then repeat the process in another plane to create the flanges. I could then fillet the corners to add the bend radius. [The modeler] simplifies the process by automating many of the steps. For example, when I add a flange, it automatically adds the radius at the bend and calculates clearance at the corner based on the parameters I've entered for that gauge sheet metal."

The design of the brewer's sheet metal parts is largely dependent on the components they secure or support. For that reason, McLean generally models them one face at a time. He can add screw holes, clearance holes, or key slot holes as needed to match the component that will attach to that face of the sheet metal part. He also inserts clinch nuts from the library of standard parts where fasteners will go into the sheet metal. "By designing them one face at a time, I can adjust the faces as component locations change until everything is where it is supposed to be."

The solid modeling software offers another means of creating sheet metal parts via its **Clad a solid** command. To create a sheet metal part, McLean could create a solid box with the dimensions of the part, add holes and slots as needed, and then issue the **Clad** command. He would be prompted to select the faces of the solid box to be retained as sheet metal faces. Any holes in the box would be retained on the faces of the sheet metal part. Corner reliefs, overlapping corners, and corner fillets could then be added to the sheet metal enclosure created by the **Clad** command, and the box would be changed into a solid

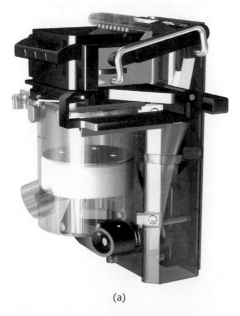

(a)   (b)   (c)

7.57 These three views of the Zuma brewer rendered some parts to be transparent so the internal structure would be more visible. The three-quarter rear view in (b) shows the sheet metal bracket in Figure 7.58. It holds the plastic part at the top through which water is added to the brewer.

(continued)

sheet metal part. Although the command is quick, McLean finds it less useful for designing parts such as his, in which hole locations and flange dimensions must evolve with the overall design of the assembly.

McLean enters parameters for the bend allowance, also called the bend *setback*, and the gauge of the metal in use. Although standards exist for these values, VKI Technologies does all its own sheet metal manufacturing and has developed accurate values for the dies and materials used there. For 20-gauge satin coat porous sheet metal to be bent with a 1/32-inch nose die, the machine shop uses a bend allowance of 0.065. For 22-gauge metal, which is slightly thinner, the bend allowance would be 0.054. For 18-gauge, it would be 0.085.

Once the parts have been designed, McLean creates multiview, fully dimensioned 2D drawings of the 3D sheet metal part and a bill of materials for any clinch nuts it includes. He then uses the **Unfold** command to create the flat pattern for the part (see Figure 7.58). The software prompts him for a reference face for the part, then uses the parameters for the bend allowance to create the flat. For an L-shaped folded part with a 2-inch and 3-inch flange (their outside dimensions) to be made of 20-gauge metal, the overall length of the flat would be $2 + 3 - 0.065$, or 4.935 inches.

Because the sheet metal parts will be cut with a numerically controlled machine, McLean saves the flat pattern as a separate drawing file that can be exported to the software VKI uses to generate the NC code for the part. The result is a complete documentation drawing of the folded part and an electronic file of the flat shape that can be used to cut the shape from the sheet metal.

Although the software speeds up the modeling of sheet metal parts, it does not prevent the creation of impossible parts. "Because you're creating this bent-up enclosure, you have to think about the flat all the time," says McLean. "You can create a solid or sheet metal enclosure with flanges that when unfolded, will overlap each other. You also have to know how the metal is going to be bent. If you have a part with several bends,

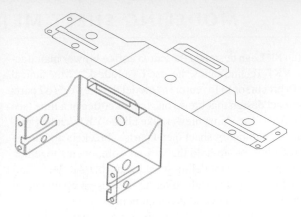

7.58 This sheet metal bracket appears here in its 3D folded form and as the flat pattern that will be used to cut the sheet metal.

and you do them in the wrong order, you can get to a point where you can't complete the bend because the part will either hit the machine or hit the die."

McLean's manufacturing background helps him "think flat" and allows him to optimize his designs in ways that others cannot. "Sometimes I look at a flat where the gap between two flanges is about 0.35, and I know that the punch operator will want to use a 0.2 punch. I might go back to the model and make some modifications so that the gap in the flat pattern will be exactly 0.2 so the punch operator won't have to punch up and down both sides to make the part."

Short of manufacturing experience, however, McLean's advice for sheet metal parts is to "keep it simple." "Instead of welding three parts together, is it better to make a single, more complicated part? This is a trade-off with no simple answer. The question is, Where are you actually saving? If you get to the point where a part has 10 to 15 bends in it, you're probably better off going to two parts and welding them together."

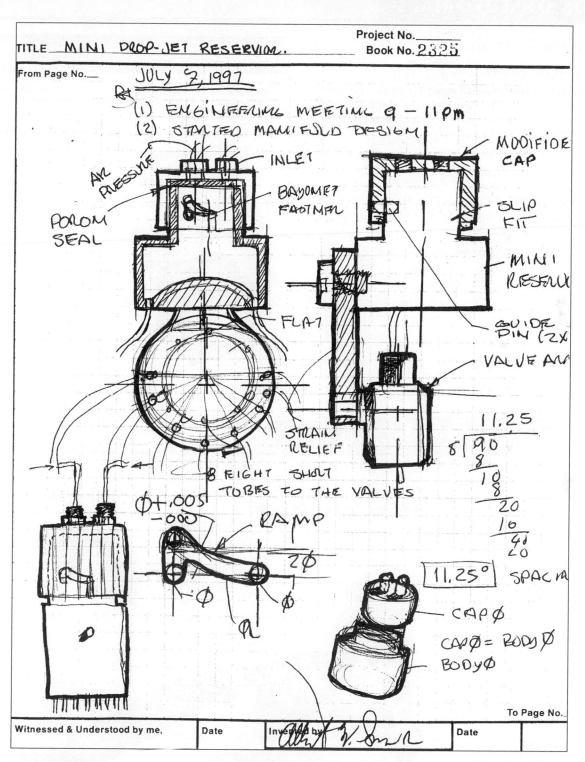

7.59 Albert Brown's sketches for this Mini Drop-Jet Reservoir define some of the important relationships for features in the design. Notice the equations in the lower right defining the cap diameter equal to the body diameter (as shown in the revised sketch at the bottom left). When the parts are modeled parametrically this relationship can be used to build intelligence into the model. Then, if the body diameter changes, the cap diameter can be updated to the correct size automatically. (Courtesy of Albert W. Brown, Jr., Affymax Research Institute.)

HANDS ON TUTORIALS

SolidWorks Tutorials

Creating an Assembly

In this tutorial, you will create an assembly mode and learn shortcuts for moving and rotating the parts in the assembly. You will create a display state allowing you to manage the appearance of your assembly.

- Launch SolidWorks from the icon on your desktop.
- Click to expand the topics listed below the "**?**" help icon.
- Click: **SolidWorks Tutorials**.
- Click: **All SolidWorks Tutorials (Set 1)** from the list.
- Click: **Lesson 2 - Assemblies** from the list that appears, and complete the tutorial on your own.

Lesson 2 - Assemblies

An assembly is a combination of two or more parts, also called components, within one SolidWorks document. You position and orient components using mates that form relations between components.

In this lesson, you build a simple assembly based on the part you created in Lesson 1.

This lesson discusses the following:

- Adding parts to an assembly

- Moving and rotating components in an assembly

- Creating display states in an assembly

(Image courtesy of DS SolidWorks Corp.)

REVERSE ENGINEERING PROJECT

The Can Opener Project

1. *Assemble it.* Using the parts you created in earlier chapters, create an assembly of the can opener.

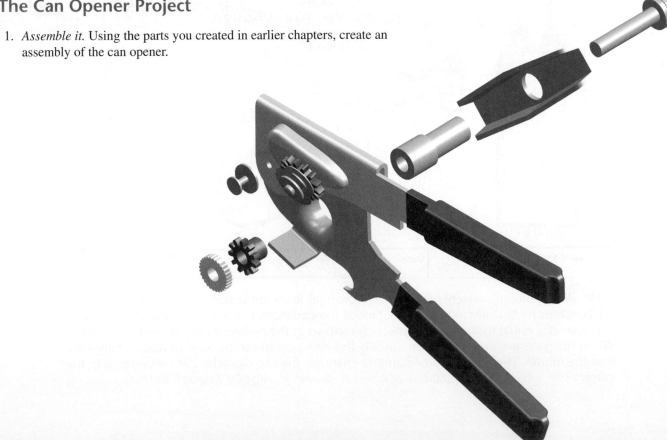

2. *Make it move.* Use the capabilities of your software to animate the motion of the can opener. What is the minimum distance between the cutting wheel and the lower gear?

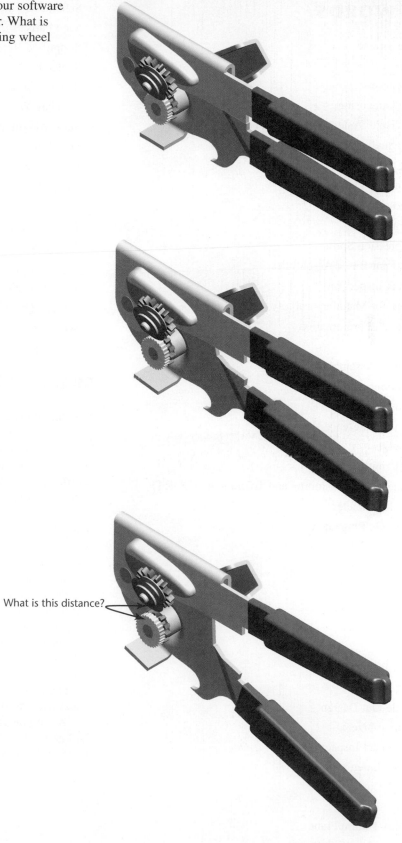

What is this distance?

KEY WORDS

Acme Thread

Active Side

Allowance

Basic Dimension

Basic Hole System

Basic Shaft System

Basic Size

Bottom-Up Design

Buttress Thread

Clearance Fit

Compression Spring

Constraint-Based Assemblies

Contact Inspection

Design for Manufacturability (DFM)

Detailed Representation

Draft

Dynamic Assemblies

Ergonomics

Extension Spring

External Reference

Force Fit

Free Length

Geometric Dimensioning and Tolerancing (GD&T)

Global Parameters

Go/No-Go Plug Gage

Interference

Interference Fit

Knuckle Thread

Layout Drawings

Limit Tolerance

Line Fit

Major Diameter

Metric Thread

Middle-Out Design

Nominal Size

Noncontact Inspection

Overall Length

Parting Line

Pitch

Primary Datum Plane

Principal Locating Points

Prototype Drawings

Schematic Thread Representation

Seed Parts

Simplified Representation

Skeleton

Square Thread

Standard Worm Thread

Static Assemblies

Static Parts

Subassemblies

Templates

Thread

Thread Form

Tolerance

Top-Down Design

Transition Fit

Unified Thread

Wire Size

SKILLS SUMMARY

Now that you have completed this chapter you should be able to distinguish between static and dynamic assembly models. Assembly constraints are used in constraint-based modeling to create "intelligent" assemblies. You should be familiar with several types of assembly constraints you may encounter in your software package.

Often, it is desirable to show stock parts in assemblies. Sometimes these parts can be acquired from a vendor. Thread, standard fasteners, and springs are among the stock parts that you may not want to model, as they can easily be indicated by a standard note.

Fit between mating parts in an assembly is an important design consideration. You should be able to list four general categories of fit between mating parts and be familiar with ways that a CAD database can be used to check fit.

You should be able to list three manufacturing processes that are commonly used to manufacture parts and some of the considerations that may arise when modeling a part that will be manufactured using one of these methods.

When you create an assembly model, you may add standard parts, sheet metal parts, or use it to simulate the range of motion it is capable of.

EXERCISES

1. Parent-child relationships are important for assemblies built using a constraint-based modeler. What problems might occur, and how could you prevent drawing management problems when deleting or changing component parts in an assembly?

2. Define the differences between a *static assembly* and a *dynamic assembly*. Which method would offer advantages to design teams who access a common database of files through a network?

3. Many of the recent generation of computer modeling programs have capabilities for solid modeling and constraint-based modeling. Research the capabilities of the CAD program, solid modeler, or constraint-based modeler at your school or workplace. What nomenclature, commands, and procedures are used by the program to define capabilities in the following:
   a. external references
   b. global parameters
   c. interference checking
   d. dynamic assemblies

4. A constraint-based modeler is used to create a gear cluster assembly by inserting modeled bicycle gear cogs and spacers onto a splined hub. Axial and radial alignment of the elements is critical to maintaining an accurate model. Define a sensible base point to use when adding individual cogs (gears) to the assembly.

5. Drawing templates, prototype drawings, and seed parts are one means to help assemble parts effectively. What are three advantages of using these file types when beginning a new model?

6. As directed by your instructor, create a dimensioned sketch or skeleton model that defines the assembly relations for the following devices:
   a. Vise grip

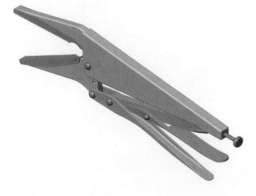

a. Vise grip *(continued)*

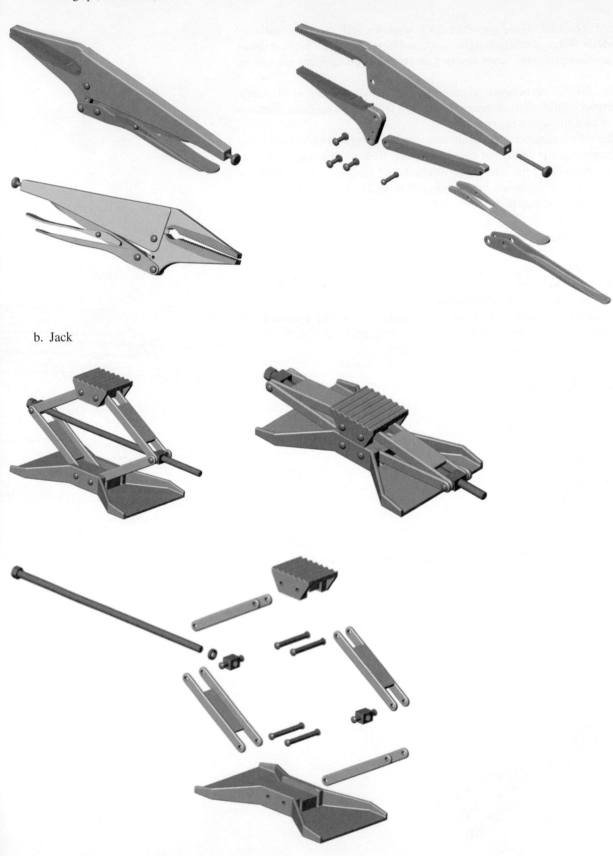

b. Jack

b. Jack parts *(continued)*

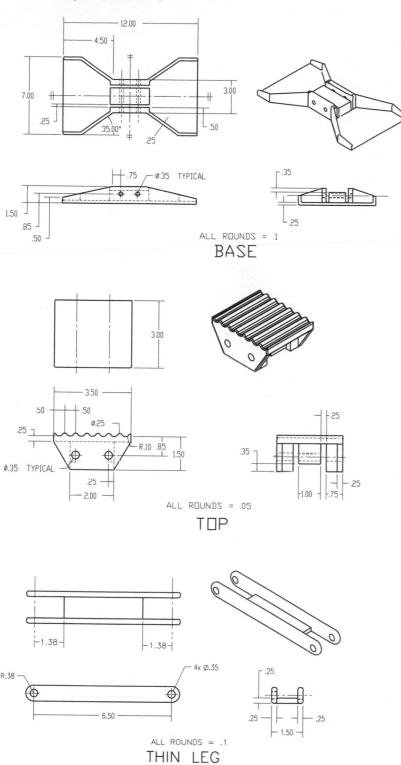

BASE

TOP

THIN LEG

b. Jack parts *(continued)*

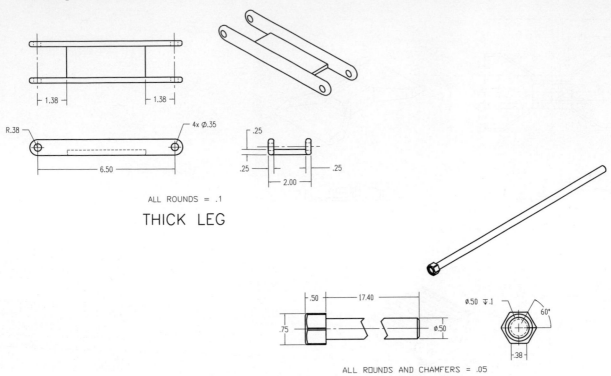

ALL ROUNDS = .1

THICK LEG

ALL ROUNDS AND CHAMFERS = .05

THROUGH BOLT

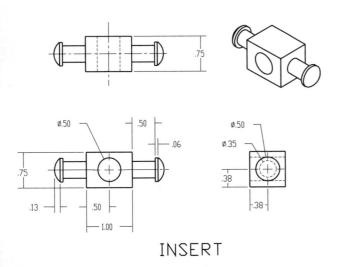

INSERT

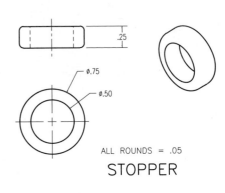

ALL ROUNDS = .05

STOPPER

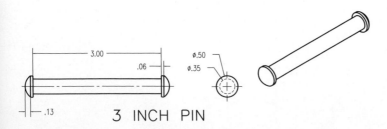

3 INCH PIN

c. Remote control

d. Compass

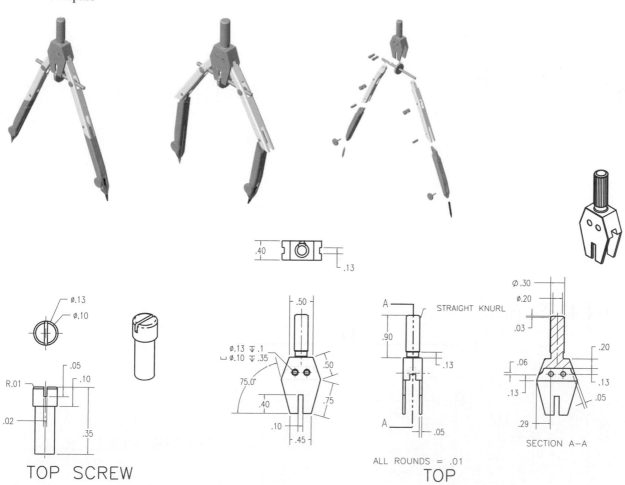

.40 .13

Ø.13
Ø.10

R.01
.05
.10
.02
.35

TOP SCREW

.50
Ø.13 ⍊ .1
⌐ Ø.10 ⍊ .35
.50
75.0°
.40
.75
.10
.45

A
STRAIGHT KNURL
.90
.13
A
.05

ALL ROUNDS = .01
TOP

Ø.30
Ø.20
.03
.06
.20
.13
.13
.05
.29

SECTION A–A

d. Compass parts *(continued)*

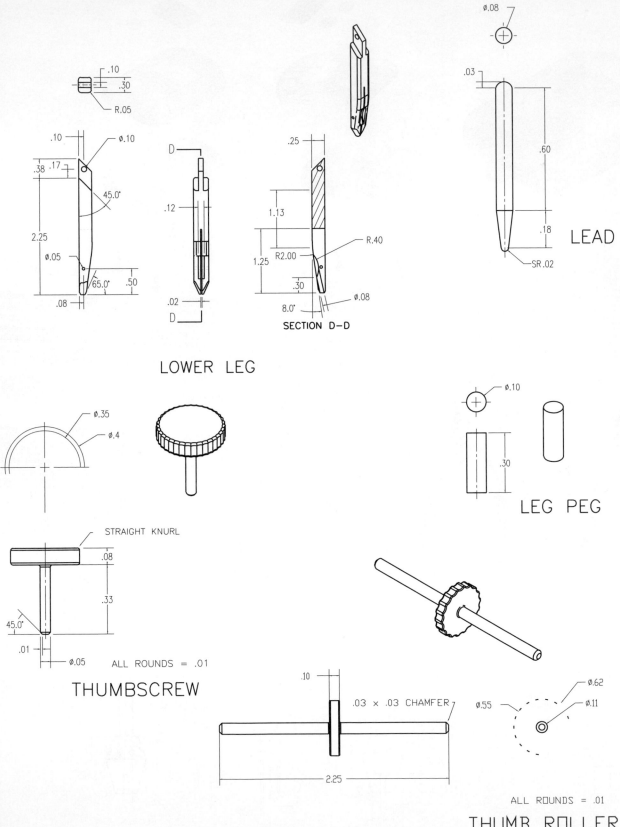

LOWER LEG

SECTION D–D

LEAD

LEG PEG

THUMBSCREW

STRAIGHT KNURL

ALL ROUNDS = .01

.03 × .03 CHAMFER

ALL ROUNDS = .01

THUMB ROLLER

d. Compass parts *(continued)*

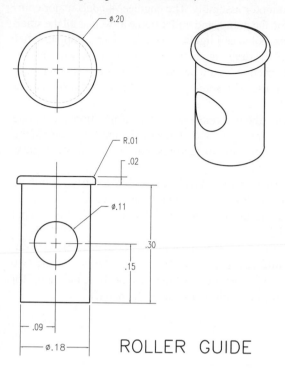

ROLLER GUIDE

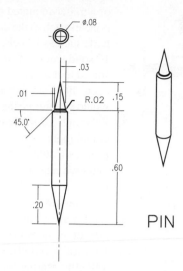

PIN

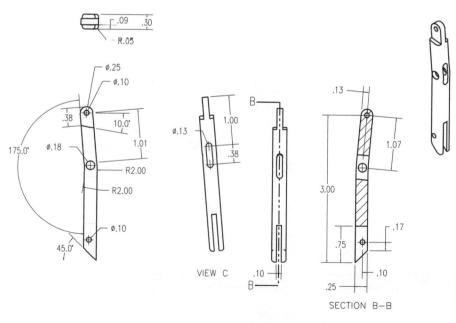

VIEW C

SECTION B–B

UPPER LEG

(Courtesy of REED.)

7. An entry-level engineer is put on the job of creating a detailed assembly drawing of a trailer-mounted pump and motor assembly. The engineer models major components such as the diesel motor, the pump, and trailer frame as well as all the small parts such as wheels, tires, nuts, bolts, and lights. When the job is complete (after 10 solid weeks of work) and all the drawing elements are correct, the work is presented to the supervisor. What reaction would be expected from engineering management? Was this an efficient use of resources? Do you have suggestions that would streamline the process?

8. Although 3D modeling can describe part and assembly configuration in accurate detail, many parts or features are rarely included in a model until very late in the process if included at all. List four parts or features that fall into this category, and give reasons for omitting detail.

9. Component models of detail parts or assemblies can be obtained from suppliers and manufacturers. These models save the designer time and reduce the potential for error but can be difficult to use in a constraint-based model. Why? What advantage does the supplier gain from this service?

10. Constraint-based assemblies often have global parameters defined to assist in developing mating features of different parts. Identify which features of the two (mating) parts described in the following drawings would be candidates for shared parameters. (Note the different plotted scale of the parts!)

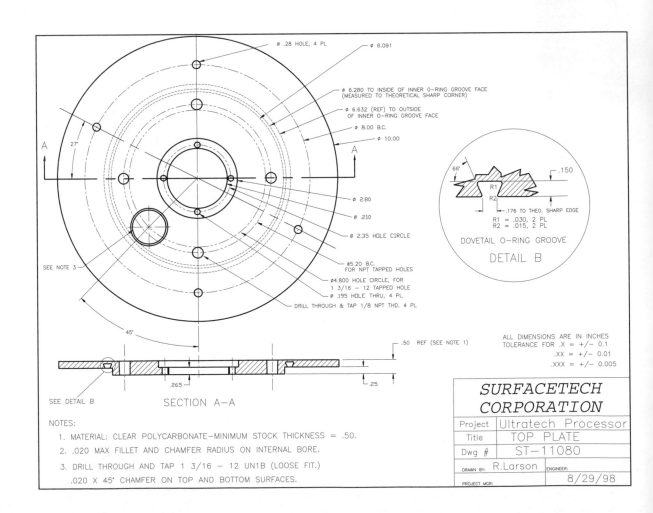

NOTES:

1. MATERIAL: CLEAR POLYCARBONATE—MINIMUM STOCK THICKNESS = .50.

2. .020 MAX FILLET AND CHAMFER RADIUS ON INTERNAL BORE.

3. DRILL THROUGH AND TAP 1 3/16 - 12 UN1B (LOOSE FIT.) .020 X 45° CHAMFER ON TOP AND BOTTOM SURFACES.

SECTION A–A

ALL DIMENSIONS ARE IN INCHES
TOLERANCE FOR .X = +/- 0.1
.XX = +/- 0.01
.XXX = +/- 0.005

DOVETAIL O-RING GROOVE

DETAIL B

SURFACETECH CORPORATION

Project	Ultratech Processor
Title	TOP PLATE
Dwg #	ST–11080
DRAWN BY:	R.Larson

ENGINEER:

PROJECT MGR: 8/29/98

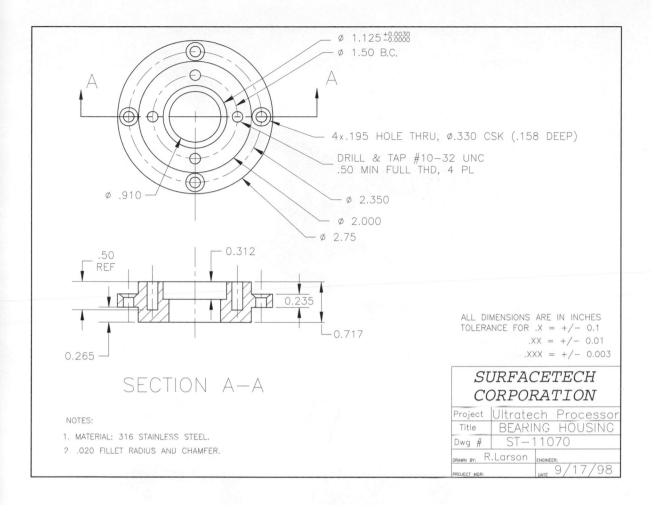

Ø 1.125 $^{+0.0030}_{-0.0000}$
Ø 1.50 B.C.

A   A

4x.195 HOLE THRU, Ø.330 CSK (.158 DEEP)

DRILL & TAP #10−32 UNC
.50 MIN FULL THD, 4 PL

Ø .910

Ø 2.350

Ø 2.000

Ø 2.75

.50
REF   0.312

0.235

0.717

0.265

SECTION A−A

ALL DIMENSIONS ARE IN INCHES
TOLERANCE FOR .X = +/− 0.1
.XX = +/− 0.01
.XXX = +/− 0.003

NOTES:

1. MATERIAL: 316 STAINLESS STEEL.
2. .020 FILLET RADIUS AND CHAMFER.

SURFACETECH CORPORATION	
Project	Ultratech Processor
Title	BEARING HOUSING
Dwg #	ST−11070
DRAWN BY: R.Larson	ENGINEER:
PROJECT MGR:	DATE 9/17/98

11. A designer must anticipate and understand the manufacturing processes that will be used to fabricate the created components. Describe differences in the model for a part designed for polyethylene (plastic) injection molding versus one for machined aluminum. Can you anticipate unique model requirements for a part created using hand lay-up resin-transfer molded composite (a process similar to that used for fiberglass)?

12. Purchase a clamp similar to the one shown from your local hardware store.

   a. Make measurements and model each part.

   b. Create an assembly model from the parts.

   c. Create a skeleton and assemble the part models to the skeleton. Change the angle in the skeleton and investigate how the model changes.

   d. Build the assembly as a mechanism so that you can simulate the motion of the mechanism by dragging the parts.

   e. Investigate your software's capabilities for simulating the motion of a mechanism.

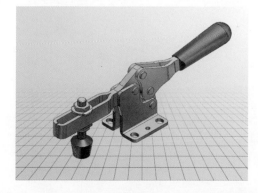

(Courtesy of Reid Supply via ReidSupply.com.)

13. Create part and assembly models for the clamp shown.

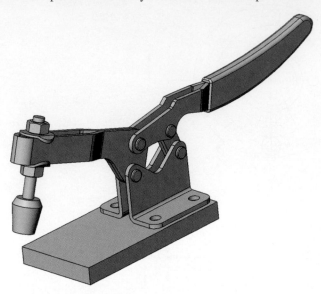

14. Use the sheet metal features of your software to model the drill bit stand. Create different configurations of hole sizes, to accommodate different drill bits.

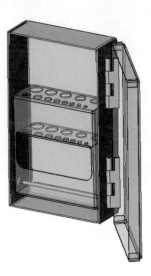

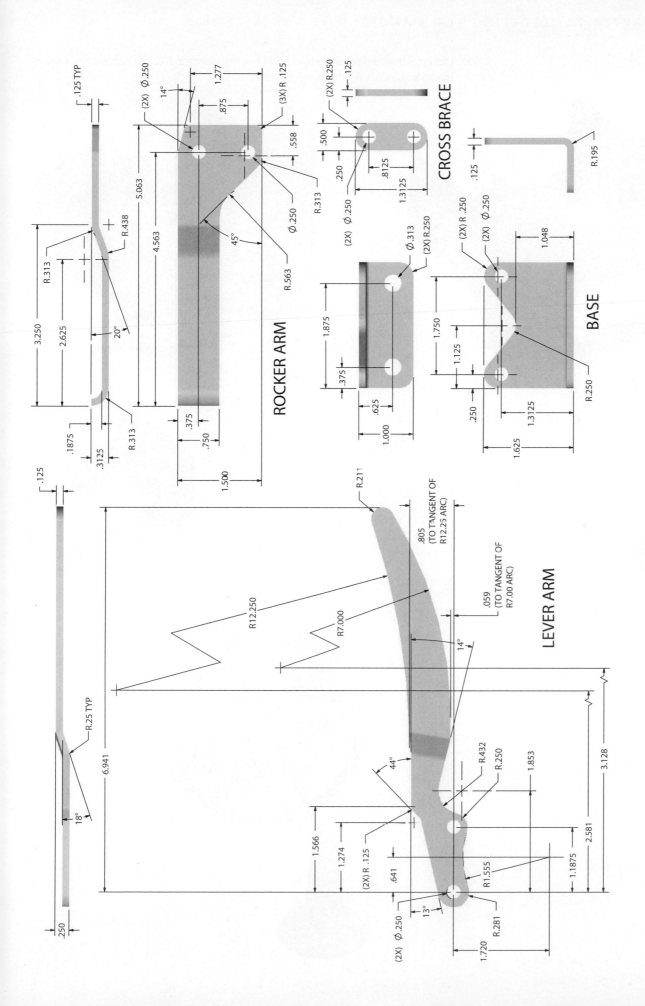

ROCKER ARM

CROSS BRACE

BASE

LEVER ARM

319

15. Design a housing for the power and D-sub connector, like the one shown. Download part models for the standard parts.

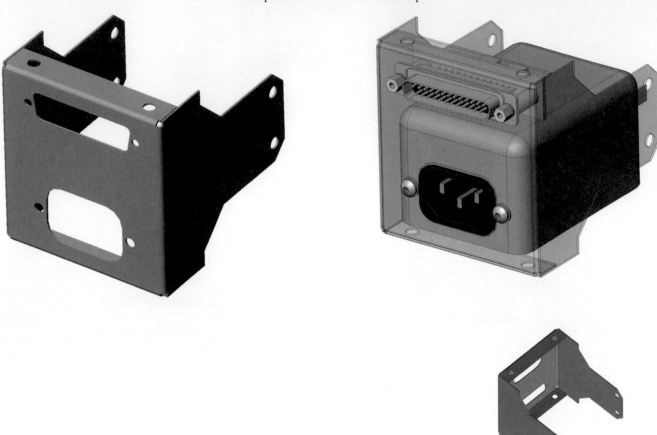

16. Model and assemble the parts of the gyroscope.

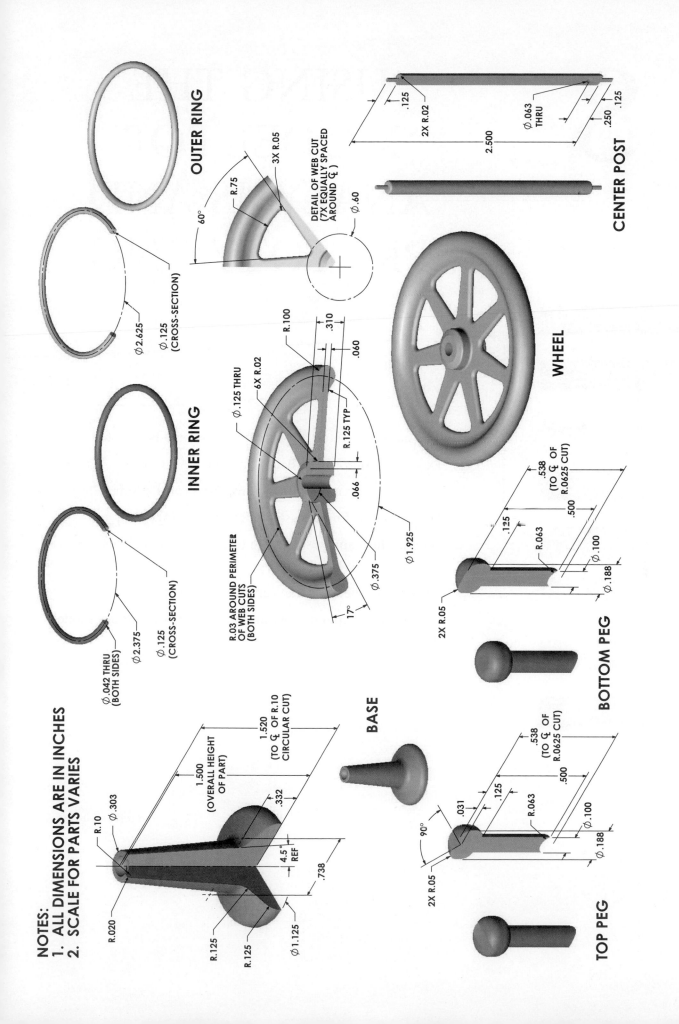

NOTES:
1. ALL DIMENSIONS ARE IN INCHES
2. SCALE FOR PARTS VARIES

OUTER RING

INNER RING

CENTER POST

WHEEL

BASE

BOTTOM PEG

TOP PEG

321

8

USING THE MODEL FOR ANALYSIS AND PROTOTYPING

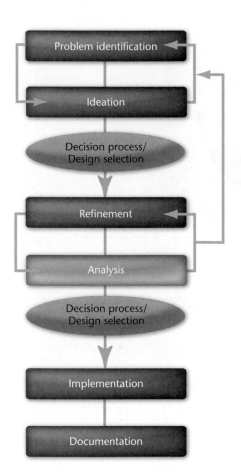

OBJECTIVES

When you have completed this chapter you should be able to:

1. Extract mass properties data from your CAD models.

2. Evaluate the accuracy of mass properties calculations.

3. Define the file formats used for exporting CAD data.

4. Describe how analysis data can be used to update parametric models.

5. List analysis methods that can use the CAD database.

6. Describe how rapid prototyping systems create physical models from CAD data.

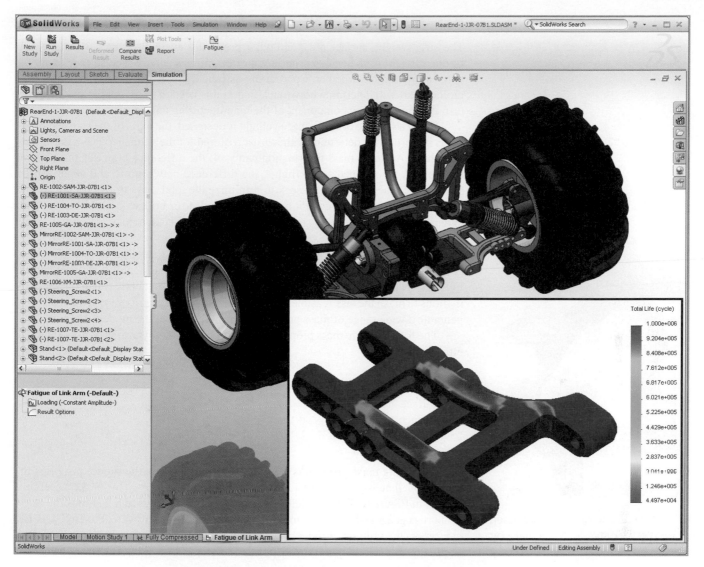

Fatigue analysis uses peak loads to estimate the lifetime of critical components. "Hot" colors in the link arm indicate sections that will fatigue soonest. *(Image courtesy of DS SolidWorks Corp.)*

TESTING, TESTING, TESTING

The refinement phase forms an iterative loop with the analysis phase of design. Test results feed back into model changes, and the revised model is used for further testing. Your 3D solid model database defines the geometry of the design and provides information for engineering analysis, such as determining mass, volume, surface area, and moments of inertia.

You can use tools such as a spreadsheet, equation solver, motion simulator, and finite element package with model data to check for stress concentrations; determine deflections, shear forces, bending moments, heat transfer properties, and natural frequencies; perform failure analysis and vibration analysis; and make many other calculations.

You can use rapid prototyping equipment to create physical parts as well as "virtual prototypes" that are computer simulations of model behavior. Both kinds of prototypes allow customers and others to interact with and evaluate the design.

8.1 DETERMINING MASS PROPERTIES

The size, weight, surface area, and other properties available from a 3D model are frequently part of the design criteria your design must satisfy. The surface area of a part can determine whether a part cools as quickly as it should. The volume of a molded part will determine how much material is needed to fill the mold—which in turn determines a key component of the manufacturing cost.

When a satellite or other space vehicle is designed, the mass of the finished product is important for determining the thrust required for the vehicle to leave Earth's atmosphere. Accurate mass information can spell the difference between success and failure and, for manned space flights, the difference between life and death. At the same time, spacecraft designs seek to minimize the total mass of the system, so the built-in factors of safety cannot be as high as those for bridges built of heavy steel. When the margin for error is small, the importance of accurate mass properties information is even higher.

Calculating the mass of a model as complex as a satellite by hand is time-consuming and might add years to the design process. Traditionally, engineers wrote FORTRAN programs to perform such calculations and frequently relied on approximations to make the task feasible. Today, most 3D CAD systems allow you to directly generate a wide range of mass property information about the model.

The following are mass property calculations commonly available in CAD modeling software:

- volume
- centroid or center of gravity
- surface area
- moments of inertia
- mass
- radii of gyration

Acquiring mass properties information directly from the model provides information that would be difficult to calculate accurately. The brake assembly shown in Figure 8.1 was modeled in Pro/ENGINEER. Figure 8.2 shows the mass properties generated from the model. Like other modeling software, Pro/ENGINEER can report the volume of the part—regardless of its shape—with accuracies of up to plus or minus one hundred-millionth of a percent ($\pm$ 0.00000001).

8.1 The volume of this brake assembly would be difficult to calculate by hand.

Factors of Safety

A *factor of safety* expresses the ratio of two quantities, such as *maximum safe speed/expected operating speed* or *failure load for a part/expected highest load in use*, that will be used as a design guideline. For example, if the failure load for a part is 250 pounds per square inch, and the expected highest load in use is 125 pounds per square inch, the safety factor for the part would be 2. Factors of safety for commercial aircraft are within the range of 1.2 to 1.5. Factors of safety for elevators may be as high as 14. For some types of equipment, such as that used in public buildings, codes specify safety requirements that must be met. You should be familiar with any codes or standards that apply to your design area.

Understanding Mass Property Calculations

Before you make use of the mass properties generated by your modeling software, it is important to understand how the values are derived. Modeling methods, algorithms, system variables, and units used by your CAD software make a difference in the accuracy of mass property calculations. For example, models that store the faceted representation of the part may produce less accurate calculations than models that store the accurate geometry.

Different modeling software uses different algorithms to calculate the volume of a solid. Some of them have error ranges of plus or minus 20%. One algorithm used to calculate the volume of a solid simply encloses the solid in a bounding box, then breaks it up into rectangular prisms parallel to the bounding box. (A bounding box is the smallest box inside which the solid will fit.) The lengths of the rectangular prisms are determined by the surface boundaries of the solid. The volumes of the prisms are then summed to find the volume of the solid. The accuracy of this method depends on the number of subdivisions into which the object is broken. The direction from which the prisms are created also affects the ability of the prisms to approximate the shape. Typically, both these settings are variables with ranges that you can set prior to calculating the mass properties.

Other systems use an iterative process to approximate the integral that describes the volume. These systems often let you specify a range of accuracy for the calculations. The more accurate you require the calculations to be, the more time they will take.

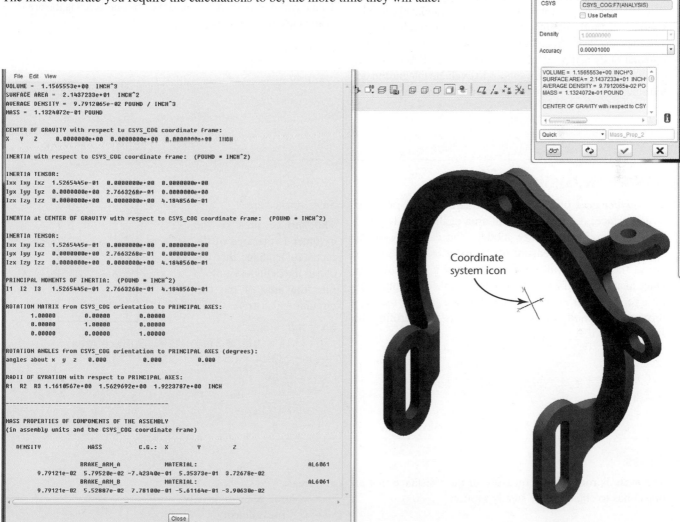

8.2 Highly Accurate Mass Properties Data for the Brake Assembly Generated Using Pro/ENGINEER

Mass Property Calculations

Mass Properties of a Right Cylinder

You will learn more about using mass property calculations, such as moments of inertia, in engineering statics and dynamics, machine design, and design of structures courses. The equations and illustrations in this example illustrate these properties as they would apply to a 1-inch diameter *right cylinder* (a cylinder in which the base is perpendicular to the height), shown in Figure 8.3.

For this example cylinder:
π is approximately 3.14
$r = .5$ inch, the base's radius
$d = 1$ inch, the base's diameter
$l = 2$ inches, the height of the cylinder.

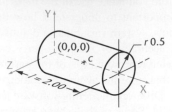

8.3 Right Cylinder

Volume (*v*)

The *volume* of a solid is basically the amount of space it takes up; volume is measured in cubic units. For many basic geometric solids, like a cylinder, volume can be calculated using simple formulas. Complex shapes can be calculated either by breaking them down into simpler geometric solids—whose volumes are easier to calculate—and adding those volumes together or by using calculus-based methods. CAD software usually uses approximate methods to evaluate integrals to determine the volume and mass properties of any shape.

$v = $ volume

The volume of a 1-inch-diameter cylinder, in which the 2-inch length is perpendicular to the circular base, is the area of the base times the length.

$$v = \pi \times r^2 \times l$$
$$= 3.14 \times (.5 \text{ in.})^2 \times 2 \text{ in.}$$
$$= 1.57 \text{ in.}^3$$

Surface Area (*Sa*)

The *surface area* of a solid is the square measure of its exterior surface. It is also the total area of the boundary surfaces of a solid or surface model. You can think of the surface area as the measure of the exterior surface of the object flattened out into a plane (as shown in Figure 8.4). Once again, it is easy to calculate for simple cases, but complex for irregular shapes. CAD software usually uses approximate methods to evaluate integrals to determine the surface area of any shape.

$Sa = $ surface area

The surface area of the cylinder is the area of the rectangular shape that encloses the cylinder ($\pi d \times l$) plus the area of the two circular ends (πr^2).

$$Sa = \pi dl + 2(\pi r^2)$$
$$= (3.14 \times 1 \text{ in.} \times 2 \text{ in.})$$
$$+ 2 [3.14 \times (.5 \text{ in.})^2]$$
$$= 7.85 \text{ in.}^2$$

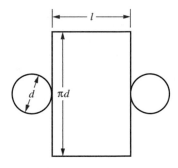

8.4 Surface Area

Mass

The *mass* is roughly the measure of the resistance that an object has to changing its steady motion.

$m = $ mass
$= $ volume $\times$ mass density
$= v \times$ mass density

Mass Property Calculations

Mass Density

Mass density for a given material is usually found by looking up the value in an engineering materials table. This is typically information you provide to calculate the mass for a given object. It is very important to make sure the units for your model (millimeters or inches, for example) match the units for the mass density that you provide (milligrams per cubic millimeter, for example). If they do not, you must convert the volume value reported from the software, or you will be off by a large factor in your calculations.

The mass of this cylinder, if made from copper, is its volume times the mass density of the material, which is 0.295.

$m = 1.57 \text{ in.}^3 \times 0.295 \text{ lb/in.}^3$

$\quad = 0.46315 \text{ lb/in.}^3$

Centroid or Center of Gravity

A *centroid* is a point defining the geometric center of an object. The centroid and center of gravity coincide when the part is made of a uniform material in a parallel gravity field. Most CAD software reports the centroid relative to the coordinate system that is selected. Make sure that you understand how the coordinate system is oriented when interpreting this value. Approximations of double integrals are used by the CAD software to calculate centroids.

C = centroid

The centroid for the cylinder is stated in terms of the three axes of the 3D coordinate system. Given the location of the origin at the center of one end of the cylinder, the centroid for this part is at $(1,0,0)$ inches.

$C \quad = C_x, C_y, C_z$

$C_x \quad = l/2 = 2/2 = 1$

$C_y \quad = 0$

$C_z \quad = 0$

$C \quad = 1,0,0$

Moments of Inertia

A *moment of inertia* is the measure of the resistance that an object has to changing its steady motion about an axis. A moment of inertia depends on the mass of the object and how that mass is distributed about the axis of interest. A moment of inertia is the second moment of mass of the object relative to the axis. Products of inertia are similar to moments of inertia, but they describe the mass distribution relative to two axes of interest. Make sure you understand how the coordinate system is oriented when interpreting this value.

I = moment of inertia

I = second moment of mass about an axis

For this cylinder,

$I_x = \int (y^2 + z^2)dm$

$I_y = \int (x^2 + z^2)dm$

$I_z = \int (x^2 + y^2)dm$

Radii of Gyration

The *radius of gyration* is the distance from the axis of interest where all the mass can be concentrated and still produce the same moment of inertia. Make sure you understand how the coordinate system is oriented when interpreting this value.

k = radius of gyration about an axis

For this cylinder,

$k_x = \sqrt{\dfrac{I_x}{m}}$

$k_y = k_2 = \sqrt{\dfrac{I_y}{m}}$

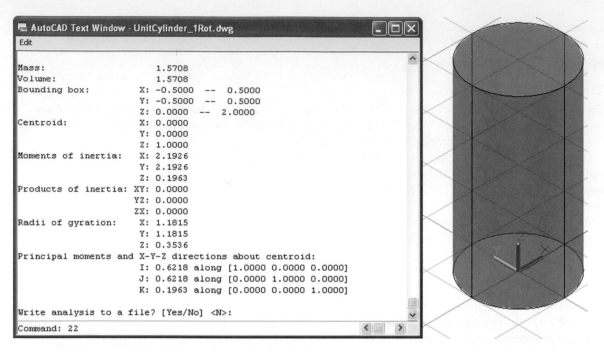

```
AutoCAD Text Window - UnitCylinder_1Rot.dwg

Edit

Mass:                   1.5708
Volume:                 1.5708
Bounding box:      X: -0.5000  --  0.5000
                   Y: -0.5000  --  0.5000
                   Z:  0.0000  --  2.0000
Centroid:          X:  0.0000
                   Y:  0.0000
                   Z:  1.0000
Moments of inertia:    X:  2.1926
                   Y:  2.1926
                   Z:  0.1963
Products of inertia: XY: 0.0000
                   YZ: 0.0000
                   ZX: 0.0000
Radii of gyration:  X:  1.1815
                   Y:  1.1815
                   Z:  0.3536
Principal moments and X-Y-Z directions about centroid:
                   I: 0.6218 along [1.0000 0.0000 0.0000]
                   J: 0.6218 along [0.0000 1.0000 0.0000]
                   K: 0.1963 along [0.0000 0.0000 1.0000]

Write analysis to a file? [Yes/No] <N>:

Command: 22
```

8.5 Mass Properties for the Cylinder from AutoCAD 2011
(Autodesk screen shots reprinted with the permission of Autodesk, Inc.)

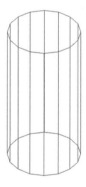

8.6 The faceted models stored by AutoCAD Release 12, as shown here, produce a less accurate volume calculation than the more accurate form of the model stored by AutoCAD 2011.

Verifying Accuracy

A good way to verify the results you are getting for mass properties is to model a shape in which the geometry is known (such as the 1-inch-diameter cylinder used in the preceding example), then compare the properties calculated by hand with those given by the CAD system. Use your CAD software to model the cylinder shown in Figure 8.5: it has a diameter of 1 inch and a height of 2 inches. Calculate the volume by hand using the formula.

$$v = \pi \times r^2 \times l$$

$$= 3.14 \times (.5 \text{ in.})^2 \times 2 \text{ in.}$$

$$= 1.57 \text{ in.}^3$$

Then, list the same information from your solid modeling software. Is it the same? A modeling package that stores an accurate representation, such as AutoCAD 2011 (see Figure 8.5), will yield different results from one such as AutoCAD Release 12, which used faceted models and an approximate method to calculate volumes (Figure 8.6).

Even after checking the results with a simple object, you should routinely estimate the values for your models to make sure that the computed values from your system are reasonable. If there is a significant difference between the software report and your hand calculation, take time to determine what is producing the variation. Do not just believe the values reported by the software and base important decisions on those results. (There is a tendency to think the software values are right because they are printed neatly on the screen. Make sure you understand the values and their range of accuracy before you use them.)

Units and Assumptions

Some information reported under mass properties, such as the centroid, moments, and radii of gyration, is related to the coordinate system in which the model is stored or to a coordinate system you have selected. You must make sure that you understand the model orientation with respect to the coordinate system, or you may misinterpret the numbers reported.

The type of units specified or assumed is especially important in calculating mass properties. Using mismatched types of units is a common reason mass property values may not match your expected values. The common unit systems in the United States are the *inch-pound-second* (ips) and *foot-pound-second* (fps) systems. Elsewhere, the *Système International* (SI) is used, with meters, kilograms, and seconds (mks) as the basic units,

Pounds Mass versus Pounds Force

Don't confuse **pounds mass** with **pounds force** when using U.S. units. An object on Earth that has a mass of 1 pound also has a weight of 1 pound. If you send that object to the moon, where the gravitational force is about one sixth what it is on Earth, the object will still have 1 pound mass; however, its weight in pounds will be about one sixth of what it weighed on Earth. Mass equals weight divided by gravitational acceleration ($m = W/gc$).

although the **centimeter-gram-second** (cgs) system is also sometimes used, and machine parts are often modeled using millimeters. You must make sure that you understand the units that are used in the calculation of the mass properties. If you misinterpret the units, or use units different from those used in the model when you enter the material properties, the value reported by the software for the mass may be wrong by a huge factor.

Materials

Many modeling packages allow you to assign material properties to the part, which will be used in calculating mass properties. This information is often stored in a separate file, called a **material file**. Once you have defined a material file, you can store and reuse it, saving you the effort of having to look up and redefine the material information each time it is needed. The calculated mass of your part depends on the density for the material that is entered in this file and the units for that density. The **Material** dialog box from SolidWorks in Figure 8.7 shows several standard materials defined for you in the software, and the kinds of properties you can set.

It is important when entering this information to match the units to the other units you are using. For example, if the mass density for the material is expected to be entered in slugs per cubic inch, and you enter the value using slugs per cubic *feet*, the resulting calculations for the mass will be off by a significant factor—1728. When using material files from an existing library, you must take particular care to notice the units in which the information was entered and make sure that it matches the units you are using for your model.

Some CAD software uses a density of 1.00 for all materials and reports the mass based on this value. To determine the actual mass of a part, you must first determine the density for the material, then calculate the true value for the mass. When a CAD system uses a density of 1.00 to calculate the mass of your model, you can calculate the mass for a specific material by knowing its specific gravity (which can be looked up in a materials table). You can use the specific gravity to calculate the actual mass of a part.

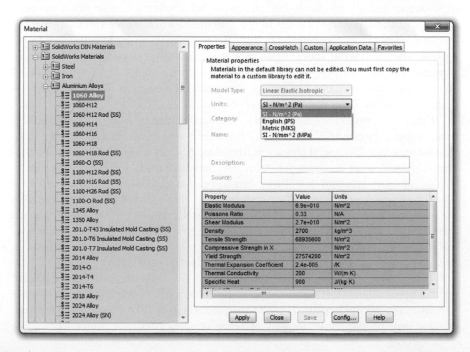

8.7 The material properties of 1060 aluminum alloy are shown in the dialog box. SolidWorks and other modelers provide a range of preexisting materials that you can assign to parts. You also can set the properties of a material and save them in a custom material file that you can apply to your parts.

Specific Gravity

Specific gravity is the value that can be used to relate the density of different materials by comparing them with the density of water. The mass density of distilled water at a temperature of 20° C is 1 gram per cubic centimeter, and its specific gravity is 1.00. Notice that specific gravity has no units. Gold has a specific gravity of 19. This essentially means that a volume of gold is 19 times as heavy as the same volume of water. The specific gravities for some other common materials are as follows:

Aluminum alloys	2.8
Stainless steel	7.8
Brass	8.6
Titanium	4.4

TIP
Preparing for Mass Property Calculations

- Check the accuracy of your modeler by calculating values for a simple part by hand.
- Routinely estimate the values you expect and compare them with the modeler's results.
- Be sure the units used for a calculation are consistent with your model units.
- Note the coordinate system orientation used in certain calculations.
- Set materials properties for each part, taking care that units used fit your needs.
- Confirm the density used for materials in your modeler: does it use a default value?

If you do not check the calculations by hand and inadvertently use the density of 1.00 (as though your model were made of water), your incorrect mass value may cause a significant error.

8.2 EXPORTING DATA FROM THE DATABASE

In some cases, mass properties information provided by the solid model is enough to determine whether a key criterion is met; in other cases, the information from the model needs to be analyzed further or combined with additional information. The surface area of a model, for example, can be used with an estimate of paint per square foot to gauge the amount and cost of paint required.

Most solid and parametric modeling software allows you to export a wide range of data from the CAD database so it can be used in other applications. *Exporting* means saving the desired data in a file format that can be read by other software applications. Mass properties data are commonly exported to text files that can be read by a spreadsheet, for example. Parameter data and the geometry of the model itself can also be exported to be used in other applications. Conversely, parameter information can be imported *into* parametric modeling software from spreadsheets or other analysis tools.

To use your CAD database with other applications, you should be aware of how data are transferred from one application to the other and which options are available to you in the package you are using.

File Formats

Each software application has its own *native file format*. A native format is designed to store any and all information created with that application in an efficient form. Each application has its own codes stored with the data that are interpreted by the application when the file is opened. For this reason, file formats are identified by an extension. Microsoft Word, for example, creates .doc (or .docx) files. AutoCAD's native format is .dwg. If you try to open a .doc file in a different application—and do not get an error message—you may see strange characters on the screen. These are elements in the file used by Word that the other application does not know how to interpret.

Other file formats are standardized so they can be read by many different applications. A text-only or *ASCII* file (which often uses a .txt extension) uses nothing more than the standard 256 ASCII characters (the older ASCII standard had only 128 characters). ASCII is an acronym for American Standard Code for Information Interchange, an early standard that allows data to be shared reliably among all different kinds of computers. Each letter of the alphabet plus other common characters are coded the same way on any machine that reads and writes text files. If you have ever saved a formatted word processing document as a text file, you know that it does not save special formatting (such as bold and italic) or any graphics in the file. It does, however, create a copy of the words in your document that can be read on any computer.

Today, there are many options for exporting data from and importing data into another application. Most software applications have filters that allow them to write or read nonnative file formats. These nonnative formats may be standardized formats, or they may be the native formats of other popular packages. AutoCAD's .dwg

TIP
Text-based tools and formats enabled the growth of the Internet and the World Wide Web. The HTML language used on the Web is a text-only format designed to allow formatted text and graphics to be displayed from a text-based file.

format, for example, can be read by many applications that make it easy for their users to work with these files.

Sometime, changing to a different format causes a loss of data—either in the new format or when you reconvert the data back into a form readable by your CAD software. You can never increase the amount of information in a file by translating it; you can only keep it the same or decrease it. You should explore the options for writing a file in the format used by the target application—as well as the ability of the target application to read your CAD files—before you export the data.

Common Formats for Export

Export formats for CAD data depend on what you are exporting. If you are exporting mass property data, attribute data, or dimension parameters, the export options offer a choice of formats, most of which are text-based. Some CAD packages export data to common spreadsheet formats, such as Microsoft Excel's .xls format. Almost all CAD software offers text export as an option.

When you export a table of data as text, you will commonly choose among *comma-delimited*, *space-delimited*, or *tab-delimited* text (see Figure 8.8). Tables of data are generally exported so that each record (or row of data in a table) is separated

8.8 (a) Spreadsheet Showing the Parameter Data for the Brake Part Exported from Pro/ENGINEER. The parameter data text file is shown exported in (b) comma-delimited text format; (c) tab-delimited text format; (d) space-delimited text format.

	A	B	C	D	E	F	G	H	I	J	K	L
	A1		fx	Pro/E Family Table								
1	Pro/E Family Table											
2	BRAKE_ARM_A											
3												
4	INST NAME	COMMON NAME	d12	d5	d7	d1	d19	d23	d24	d4	d2	
5	!GENERIC	brake_arm_a.prt_INST	9	82	25	16.75	60	20	36	50	14.5	
6	A_1001	brake_arm_a.prt_INST	10	90	25	16.75	68.73	20	36	60	14.5	
7	A_1002	brake_arm_a.prt_INST	13	110	30	20	85	28	44	70	20	
8	A_1003	brake_arm_a.prt_INST	11	95	30	17	80	25	39.5	65 *		
9	A_1004	brake_arm_a.prt_INST	9	82 *	*		60 *	*		60 *		
10												
11												
12												

Sheet1   Sheet2   Sheet3

(a)

```
Pro/E Family Table,,,,,,,,,,
BRAKE_ARM_A,,,,,,,,,,

,,,,,,,,,,
INST NAME,COMMON NAME,d12,d5,d7,d1,d19,d23,d24,d4,d2
!GENERIC,brake_arm_a.prt_INST,9,82,25,16.75,60,20,36,50,14.5
A_1001,brake_arm_a.prt_INST,10,90,25,16.75,68.73,20,36,60,14.5
A_1002,brake_arm_a.prt_INST,13,110,30,20,85,28,44,70,20
A_1003,brake_arm_a.prt_INST,11,95,30,17,80,25,39.5,65,*
A_1004,brake_arm_a.prt_INST,9,82,*,*,60,*,*,60,*
```

(b)

```
Pro/E Family Table
BRAKE_ARM_A

INST NAME        COMMON NAME     d12    d5     d7     d1      d19    d23    d24    d4     d2
!GENERIC                brake_arm_a.prt_INST  9      82     25     16.75   60     20     36     50     14.5
A_1001   brake_arm_a.prt_INST   10     90     25     16.75   68.73  20     36     60     14.5
A_1002   brake_arm_a.prt_INST   13     110    30     20      85     28     44     70     20
A_1003   brake_arm_a.prt_INST   11     95     30     17      80     25     39.5   65     *
A_1004   brake_arm_a.prt_INST   9      82     *      *       60     *      *      60     *
```

(c)

```
Pro/E Family Table
BRAKE_ARM_A

INST NAME    COMMON NAME     d12    d5     d7     d1     d19    d23    d24    d4     d2
!GENERIC   brake_arm_a.prt_INST   9     82     25     16.75  60     20     36     50     14.5
A_1001     brake_arm_a.prt_INST   10    90     25     16.75  68.73  20     36     60     14.5
A_1002     brake_arm_a.prt_INST   13    110    30     20     85     28     44     70     20
A_1003     brake_arm_a.prt_INST   11    95     30     17     80     25     39.5   65*
A_1004     brake_arm_a.prt_INST   9     82*    *      60*    *      60*
```

(d)

by a line break (signaling the software to go to the next line). Columns of data are separated (*delimited*) by a character—a space, comma, or tab. When the text file is read by the target application, you may have to specify which character was used to delineate the columns. If your data entries have many spaces in them, space-delimited would be a bad choice, as the target software would not be able to distinguish between the spaces in the entries and those between the columns. Tab-delimited is the most commonly used export format for tabular data. The comma-delimited format is also popular for importing to and exporting from CAD software.

If you plan to export your CAD model for analysis, to generate NC tool paths, or to do animation or kinematic analysis, you may need to use some kind of ***graphics exchange format***. These are standardized formats that capture graphics information in the same way that a text file captures text data. Each varies in the degree to which it captures the information in the model. The required accuracy for graphical data can also vary between graphics standards. This can be a problem because endpoints of lines that may have been considered connecting in one CAD package, where the accuracy indicates that endpoints connect if they are within 0.000001, may not connect when exported to a format that uses accuracies of 0.0000000001. The new format may not know how to interpret this difference in accuracy. The result may be that lines that previously connected no longer connect when translated.

The ***Initial Graphics Exchange Specification*** (IGES) is a graphics format capable of exporting wireframe, surface, or solid information. This format captures the 3D

```
                                                                       S0000001
1H,,1H;,23HMASTERCAM version 6.13a,10HSLTEST.GE3,9HMASTERCAM,1H1,16,8,
                                                                       G0000001
24,8,56,,1.,1,4HINCH,1,0.01,13H980608.150243,0.00005,100.,,,,8,0,;
                                                                       G0000002
        110       1       1       1       1             0   00000000D0000001
        110       0       3       1       0                         0D0000002
        110       2       1       1       1             0   00000000D0000003
        110       0       3       1       0                         0D0000004
        110       3       1       1       1             0   00000000D0000005
        110       0       3       1       0                         0D0000006
        110       4       1       1       1             0   00000000D0000007
        110       0       3       1       0                         0D0000008
        124       5       1       0       0             0   00010000D0000009
        124       0       3       1       0                         0D0000010
...
        100       0       3       2       0                         0D0000042
        110      34       1       1       1             0   00000000D0000043
        110       0       3       2       0                         0D0000044
        110      36       1       1       1             0   00000000D0000045
        110       0       3       1       0                         0D0000046
        110      37       1       1       1             0   00000000D0000047
        110       0       3       1       0                         0D0000048
        110      38       1       1       1             0   00000000D0000049
        110       0       3       2       0                         0D0000050
110,0.,-1.6974091509,-0.375,0.,-0.5,-0.375;                         1P0000001
110,0.75,0.,-0.375,0.5,0.,-0.375;                                   3P0000002
110,0.875,-1.8232830142,-0.375,0.875,-2.4167906074,-0.375;          5P0000003
110,0.75,-0.25,-0.375,0.5,-0.25,-0.375;                             7P0000004
124,1.,0.,0.,0.,0.,1.,0.,0.,0.,0.,1.,0.;                            9P0000005
100,-0.375,0.5,-0.5,0.5,0.,0.,-0.5;                                11P0000006
100,-0.375,0.75,-0.5,1.2457224307,-0.5652630961,0.75,0.;           13P0000007
110,1.2457224307,-0.5652630961,-0.375,1.0008780126,-2.425041092,   15P0000008
-0.375;                                                            15P0000009
..
S0000001G0000002D0000050P0000039                                       T00000
```

8.9 The IGES standard defines and codes entities in the CAD database using only comma- and space-delimited ASCII characters arranged in a standard format.

information and is commonly used to export the model for computer-aided manufacturing. It was developed to export engineering drawing data, with emphasis on the 2D and 3D wireframe model information, text, dimensions, and limited surface information typical of early CAD-drawn mechanical parts. The IGES standard is controlled by the National Computer Graphics Association (NCGA). The NCGA administers the National IGES User Group (NIUG), which provides access to information on IGES. The volunteer organization IGES/PDES (Product Data Exchange using STEP) helps maintain the evolving IGES standard and publishes a specification for IGES. Because it is evolving, there are various versions of IGES in existence, with various capabilities. Because IGES has capabilities for exporting a wide variety of information, different *flavors* of it exist. You can often select options for how you want the information translated. You can find references on the Web that list the types of entities and how the information for them is stored in the IGES standard. A segment from an IGES file is shown in Figure 8.9.

The *STandard for the Exchange of Product model data* (STEP) is another 3D format that was agreed on more recently by the International Organization for Standardization (ISO). STEP may eventually allow more information to be transferred between CAD platforms than IGES, but it is still relatively new.

Autodesk's *Drawing Exchange Format* (DXF) is popular for exporting 2D and 3D geometry between CAD platforms.

The *STereo Lithography* (STL) format is used to export 3D geometry to rapid prototyping systems. STL format translates the surface of the object into triangular facets. The size and number of these facets determine how accurately the STL file matches the original object (see Figure 8.10).

Native versus Neutral Formats

Many CAD systems now allow you to read the native format file from in another CAD system. This usually preserves the maximum amount of information in the translated file. When a native format is not available, only a more general format such as IGES may be available. These general formats are sometimes referred to as *neutral files*. Neutral files are useful both for exporting data and for importing files that have been created on a system other than your own.

Vector versus Raster Data

Graphics programs are generally of two main types: *raster* or *vector*. CAD systems use vector-type data. This means that the type of information stored in the database contains the endpoints of lines, centers, radii of arcs, and other information based on the geometry of the objects. The CAD software draws a particular entity on the screen

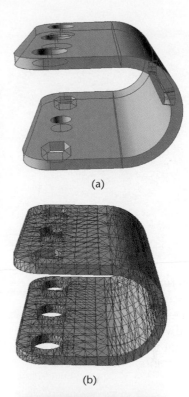

(a)

(b)

(c)

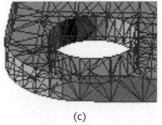

8.10 The bracket was modeled in SolidWorks, then imported into (a) Rhino 3D and saved to (b) STL format. The enlarged view in (c) shows the faceted appearance of the rounded end in this file format.

SPOTLIGHT

Standard for the Exchange of Product Model Data

The STEP standard was developed by ISO to provide a format for representing and exchanging computer-readable product data. The initiative provides a neutral mechanism for describing product data throughout the life cycle of a product—not just its design, but also its manufacture, use, maintenance, and disposal. This means that the standard not only provides a neutral file format for exchange of 3D model information but also addresses additional product information such as that handled by a variety of database systems or a product data management (PDM) system. U.S. industry is represented in the process by PDES, a voluntary activity coordinated by the National Institute of Standards and Technology (NIST) (www.nist.gov).

8.11 A rendered view of the model was exported to a raster-based file format, .jpg. This enlarged view of the raster image shows that it is made of pixels, not editable points, lines, and arcs.

based on this vector information. The definition of the entities allows them to be drawn, printed, or converted to any size or format. All the previously mentioned graphics formats are used to exchange vector-type information.

Raster graphics programs (such as Photoshop or PaintShop) store information about the discrete pixels, or dots, that make up a graphical image. Because this information is stored "dot by dot," it cannot contain useful engineering information about the individual entities in the image, such as the center of a circle. Some examples of these formats are .bmp, .pcx, .tif, .png, .jpg, and .gif. These formats contain the information about the color and intensity of the pixels that should be formed to display the image. Images produced by scanning a picture or document are raster-type images. You cannot directly scan a drawing and use it in a vector-based CAD program. First, its raster-type information must be converted to vector information using a conversion algorithm. This is often unreliable, because the connectivity of endpoints, line types, and other information can be difficult to translate reliably in every case.

Many vector-type CAD programs allow you to export and import raster-type images. For example, a shaded model may be exported to a raster format, as in Figure 8.11, for printing or inclusion in a text document, or a scanned image, such as a company logo, or picture of an actual part, may be imported and placed in a vector drawing. Raster data in the CAD drawing remain a set of pixels; the entities in the drawing cannot be edited individually.

Translating Data

Before selecting a particular CAD platform, you should investigate the export formats that are available and make sure that you can successfully transfer the information between the different software packages you may be using. This is not always an easy task. If you plan to use your CAD data with another package, export a test part first to be sure the conversion will suit your needs. Each time you transfer from one application, such as the CAD modeling package, to another, there is a possibility that some data will be lost. If you want to bring the data back into your CAD database from another application, this testing is even more important. Do not leave this testing to the last minute when you may be facing a critical deadline.

8.3 DOWNSTREAM APPLICATIONS

As the CAD model has evolved from a 2D drawing to a 3D database of design information, other software packages have evolved to facilitate engineering analysis of product and system design. These ***downstream*** applications may use data already created for the solid model. Increasingly, extensions for CAD packages offer analysis tools that operate within the 3D modeler so you don't have to define the model in another format. In this section, you will learn about analysis software that uses information from the CAD database.

Spreadsheets

You are probably already familiar with spreadsheets. Values, text, and equations are entered into cells arranged in rows and columns. Each cell can be defined to store text and numeric data or to store the results of an action on data stored in other cells. Most spreadsheets offer a wide range of built-in functions—from finding a simple sum to more complex mathematical functions. Advanced users can create sophisticated problem-solving tools using spreadsheets, but a spreadsheet is handy for any kind of calculation.

Spreadsheets are an easy way to do "what-if" analysis. Once relationships have been set up, it is easy to change values and test many different scenarios with little effort. This ability to evaluate, modify, and reevaluate is integral to the iterative process of engineering design. The rows and columns make it easy to check the calculations done by the spreadsheet and easy to see results.

Many constraint-based modelers allow you to link the driven dimensions of the model to spreadsheet cells. Thus, you can use the spreadsheet to evaluate different options and then update the model dimensions based on the results.

Spreadsheets are an extremely helpful engineering tool not only for calculations but also for projecting and keeping track of costs. Creating parts that can be manufactured cost effectively and on time is an important aspect of design.

Equation Solvers

Equation-solving software, such as MathCAD, TK Solver, Matlab, Maple, and Mathematica, is used to solve simultaneous equations and to visualize the results using built-in graphing functions. Equation solvers, like spreadsheets, allow you to change values easily and run different cases, but they also allow you to run a range of values to select an optimum solution from among several solutions. Their ability to optimize a solution makes them a valuable tool.

MathCAD's interface uses a standard notation, so you can type equations in the MathCAD editor the way you would ordinarily write them. MathCAD's interface is very simple to learn. You type in a way similar to how you write math expressions on paper. Most symbolic computation software lets you manipulate numbers, symbols, and math and logic expressions in a similar way.

You export data for use in an equation solver in the same way you would for a spreadsheet. In the example in Figure 8.12, information for the tank modeled in Pro/ENGINEER was exported and loaded into MathCAD, where the convective heat loss for the tank was calculated from the surface area.

TK Solver's interface is more similar to spreadsheet software. It uses a *variable sheet* to keep track of the variables, where you can list both the input variables and the outputs for variables. The *rule sheet* keeps track of the equations or rules. (The software is billed as a rule-based programming language, not just an equation solver.)

One of TK Solver's abilities is to *backsolve* for variables regardless of the position in an equation. This makes it easy to test different values. For example, if you entered the equation for Newton's second law, Force = mass × length / time², in most programs you would be able to solve this form only for force. In most computer programs, the equal sign is an assignment function and does not work in the way it does mathematically. In most software A = B means "assign the evaluated value of B to the variable or storage location called A." In TK Solver, you can define any of the variables listed in the equation, and the software solves for the remaining one. This makes it very easy to try different values or use a guess-and-check-approach to selecting initial values without constantly having to rewrite the equations.

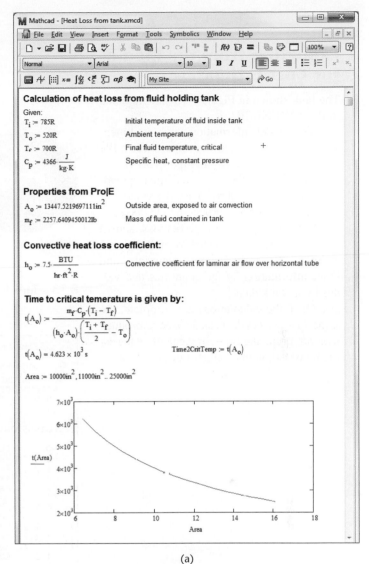

(a)

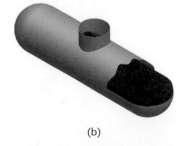

(b)

8.12 The assembly model of the tank in (b) gives the combined mass properties of the tank and the fluid, which are different materials. Using the combined mass properties allows you to calculate the convective heat loss more accurately (a).

Exporting from the Model for Cost Estimates

The tank shown in Figure 8.13 was modeled in Pro/ENGINEER.

The model information, such as dimensions and surface area, was exported to a file named tank3.pls (shown in Figure 8.14).

This file was then imported into a spreadsheet (tank inf file.xlsx) to use the values for estimating the cost of producing the tank.

A second spreadsheet, costs.xlsx, shown in Figure 8.15, was created to multiply cost estimates for weld lengths, materials, and other information by the quantities and values in the tank data.

Once the calculations are complete, the same spreadsheet data can be used to produce a report, graph the costs, or recalculate the estimate based on updated cost or tank data.

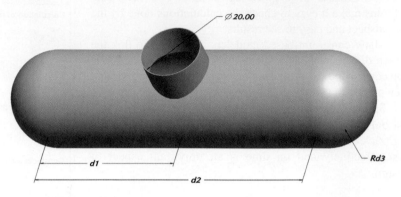

8.13 Tank Model

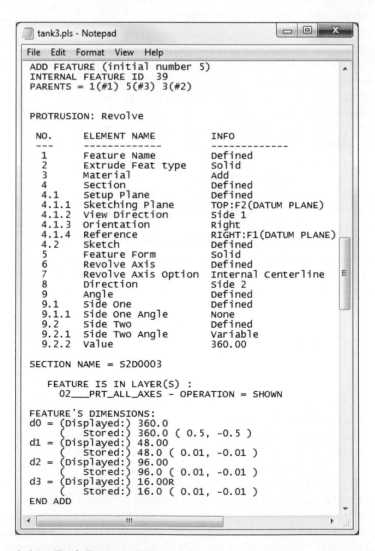

8.14 Tank Parameters

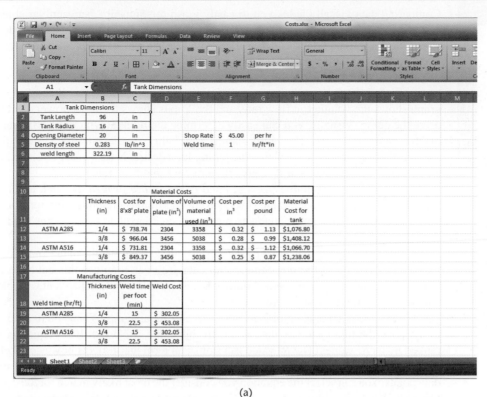

(a)

8.15 The parameters from the tank were used to calculate cost information in Microsoft Excel, as shown in (a). The formulas used in the spreadsheet are visible in the view of the same spreadsheet shown in (b). Note that the formulas in cells B2–B4 and B6 relate directly to data in the tank inf file.xlsx, the spreadsheet into which the tank data were imported. If updated tank parameters are exported and imported into a new version of the tank inf file, the calculations in costs.xlsx will be updated with this new information.

These formulas reference values from another spreadsheet

(b)

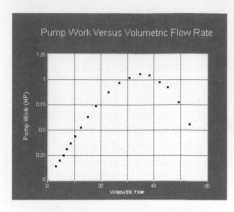

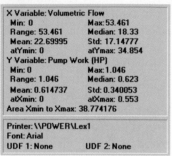

8.16   Empirical Data for Flow and Horsepower Plotted

Graphs

Comparing data visually helps you to easily grasp relationships between variables, compare results, and observe relationships between groups of data. Graphs provide a visual way to interpret data, which makes them easier for most people to understand than symbolic or numerical data.

Consider the example shown in Figure 8.16. The graph shows the relationship of empirical data (data actually measured) for a centrifugal pump. The X-axis of the graph shows the volumetric flow rate, and the Y-axis shows the horsepower used to drive the pump. Visually, you can easily determine the point at which the efficiency of the pump falls off. This information can help you design the system for efficiency—a combination of sizing the piping for a desirable flow rate and selecting a pump that will operate near its peak efficiency for that flow rate.

There are many graphing packages and **_curve-fitting programs_**, such as Easy Plot, TableCurve, Sigmaplot, and Maxsyma (in addition to graphing functions built into spreadsheets), that are available to aid you in creating charts and graphs used in engineering.

Curve-fitting programs let you generate an equation that fits the data from actual measurements such as the preceding pump data. You can then use this equation to plug in independent variable values and determine a value for the dependent variable, which you can use as design information.

Figure 8.17 is a plot of empirical data for a centrifugal pump. When you fit a curve to the data, the top equations that would fit this curve are listed. Each equation is ranked by its R^2 value. (The R^2 value is a statistical term that reflects how accurately the equation fits the data.) The higher the R^2, the better the equation captures all the data points. You can select the equation you consider the most likely match for the data. Figure 8.18 lists the top matching equations for the empirical centrifugal pump data.

For the centrifugal pump, once you have determined an equation giving the relationship between flow rate and horsepower, you can plug in a value for flow rate and determine the horsepower required to run the pump. This information allows you to determine the cost of operating your system at a particular flow rate, because the cost of power to run the pump is based on the horsepower. You can also make other cost-based decisions, such as whether to spend the money for a larger-size pipe to increase the flow rate, and the length of time needed to pay back the additional cost of the pipe through increased energy efficiency.

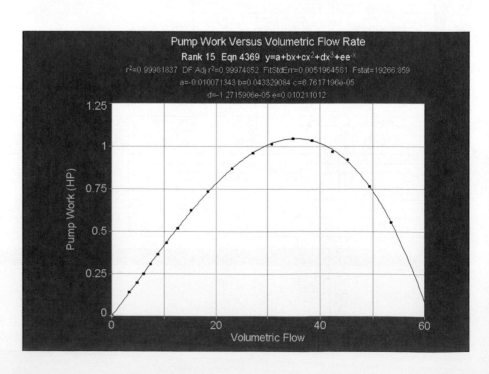

8.17   A curve is fit to the plotted empirical data for flow and horsepower.

2	0.9998586	0.9998042	0.0045852	2.475e+04	22	4342	$y=a+bx+cx^2+dx^{2.5}+ee^x$
3	0.9998576	0.9997864	0.0047754	1.825e+04	13	7004	$y=(a+cx+ex^2)/(1+bx+dx^2+fx^3)$
4	0.9998472	0.9997885	0.0047660	2.29e+04	21	4689	$y=a+bx+cx^3+de^x+ex^{0.5}$
5	0.9998349	0.9997715	0.0049537	2.12e+04	26	4698	$y=a+bx+cx^3+de^x+ee^{-x}$
6	0.9998251	0.9997578	0.0050999	2e+04	79	8054	$[Beta\_]$ $y=a((x-b+cm)/c)^{d-1}*(1-$
7	0.9998224	0.9997541	0.0051387	1.97e+04	22	4220	$y=a+bx+cx^{1.5}+dx^3+ee^x$
8	0.9998222	0.9997538	0.0051416	1.968e+04	20	4740	$y=a+bx+cx^3+dx^{0.5}+ee^{-x}$
9	0.9998215	0.9997323	0.0053453	1.457e+04	106	8059	$[ADC]$ $y=a+(b/4)(1+erf((x-c+d.$
10	0.9998209	0.9997520	0.0051598	1.954e+04	17	4585	$y=a+bx+cx^{2.5}+dx^3+ex^{0.5}$
11	0.9998207	0.9997518	0.0051628	1.952e+04	16	4224	$y=a+bx+cx^{1.5}+dx^3+ex^{0.5}$
12	0.9998207	0.9997517	0.0051637	1.951e+04	14	4360	$y=a+bx+cx^2+dx^3+ex^{0.5}$
13	0.9998194	0.9997500	0.0051810	1.938e+04	21	4233	$y=a+bx+cx^{1.5}+dx^3+ee^{-x}$
14	0.9998193	0.9997498	0.0051835	1.936e+04	15	4174	$y=a+bx+cx^{1.5}+dx^2+ex^3$
15	0.9998184	0.9997485	0.0051965	1.927e+04	19	4369	$y=a+bx+cx^2+dx^3+ee^{-x}$
16	0.9998178	0.9997478	0.0052043	1.921e+04	22	4594	$y=a+bx+cx^{2.5}+dx^3+ee^{-x}$
17	0.9998173	0.9997470	0.0052117	1.915e+04	18	4205	$y=a+bx+cx^{1.5}+dx^{2.5}+ex^3$
18	0.9998148	0.9997619	0.0050687	2.7e+04	11	2090	$y=a+bx+cx^3+dx^{0.5}$

8.18 The equations shown fit the data collected for volumetric flow rate and horsepower for a specific centrifugal pump. The equations are ranked by their R^2 values (in column 3), so the top ones listed fit the data most accurately.

Finite Element Analysis

Finite element analysis (FEA) is a method of breaking up a complex shape into discrete smaller parts (*finite elements*) for which properties such as stress (see Figure 8.19), strain, temperature distribution, fluid flow, and electric and magnetic fields can more easily be found. The collective results for these smaller elements are linked together to determine the overall solution for the complex shape. The combination of smaller elements needed to cover a 2D or 3D shape is called a *mesh*. Many finite element analysis programs are available that can use CAD model geometry as the basis for FEA mesh.

Some FEA software is linked to or runs inside the CAD interface. For example, Pro/Mechanica is available in Pro/ENGINEER; CosmosWorks, now named Solid-Works Simulation, is available within the SolidWorks interface. Other FEA software requires you to import or build your CAD model in the FEA software. Many FEA programs, such as ANSYS, will directly read CAD models from some of the more popular 3D software. Most FEA software will also import CAD geometry from neutral formats such as IGES, STEP, or DXF.

Meshing the Model

Selecting the proper element type when forming the mesh is essential to getting accurate results from the FEA program. Element types are shapes that will be used to make

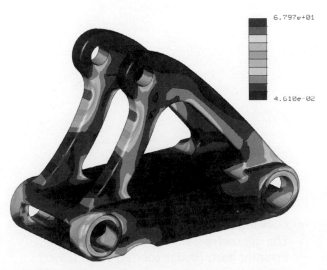

8.19 This FEA stress plot makes it easy to visualize the distribution of stresses for the conditions applied to the model. *(Courtesy of Santa Cruz Bicycles.)*

8.20   Different parts in this crankshaft assembly have different finite element types applied to them prior to analysis in the NX-Nastran advanced simulation package. *(© 2011 Siemens Product Lifecycle Management Software Inc. Reprinted with permission.)*

the mesh (see Figure 8.20). The choice of element type will depend on a number of factors, such as the material for the part, its geometric properties, and anticipated load.

A good understanding of material properties is necessary when using FEA methods. Material properties affect the appropriateness of the finite element type for your purposes. One property of a material is its *elasticity*. Only the simplest finite element types represent materials as entirely elastic. Most have capabilities that go far beyond simple elastic behavior. The element shape used in the mesh can determine how the material behavior is modeled. Some elements are used to represent elastic materials and small strain; some are elastic and include large strains; some are plastic; some are viscoelastic. Some elements even allow for coupled behavior. For example, temperature and stress can be coupled for thermoelastic problems in which temperature affects stress and vice versa. Similarly, electric current and stress couple for piezoelectric materials in which an electric current causes stress and vice versa. You must understand the range of material behavior likely to be encountered and select an appropriate element type to get good results from FEA analysis (see Figure 8.21).

Most modern finite element types have automatic mesh generation. The quality of your FEA results also depends on the quality (or refinement) of the mesh. For example, regions where stress gradients are steep require a finer mesh to give good results. For symmetrical parts, where results will be the same on both sides of the axis of symmetry, time can be saved by cutting the model in half (or quarter) and analyzing only a portion of it. This keeps the file size and computation time down, which is particularly important when generating fine meshes.

After the mesh has been created, you use the FEA software to enter the forces. Determining how to represent the forces on the object as they are applied to the mesh is another important consideration.

FEA software cannot tell when your assumptions, element type, or mesh density is not right for the problem you are analyzing. It will produce some type of result, but you must have the engineering background to interpret the validity of the result.

Total Deformation
Type: Total Deformation
Unit: in
Time: 1
11/15/2007 11:16 PM

1.015 Max
0.90219
0.78942
0.67664
0.56387
0.4511
0.33832
0.22555
0.11277
0 Min

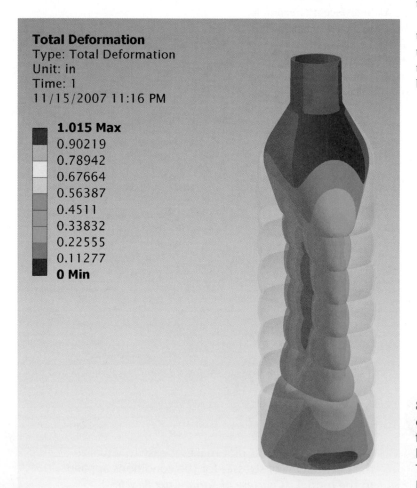

8.21   This color plot shows the buckling of a plastic bottle in which the sides collapse from a negative internal pressure. This linear buckling model was created in ANSYS. The geometry is from a sample Autodesk Inventor part. *(Image courtesy ANSYS, Inc.)*

STEP by STEP

ANALYZING THE BRAKE ASSEMBLY

The brake assembly in Figure 8.22 shows CAD data being used for FEA analysis. The brake arms are attached to the bike frame and to each other at the pivot point. Brake pads attach at the slots on each arm. In this example, Mechanica was used to analyze the brake arms to ensure that the arms can withstand the braking forces while still providing flexibility needed for smooth braking.

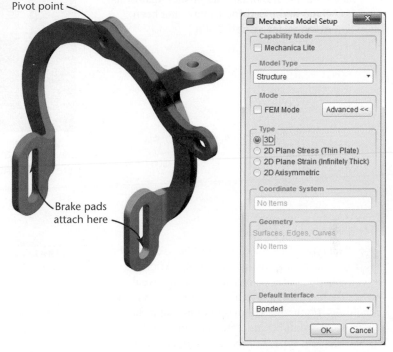

8.22 The Brake Assembly as Modeled in Pro/ENGINEER. Note the attachment points for the brake pads and the bike frame.

1 Boundary conditions are applied to the model to constrain the axis of the hole for the pivot. It is fixed in the X-, Y-, Z-directions to model how this surface is used to attach the brake arms to the bike frame. This still allows the rotational motion of the arm during the analysis. Mechanica lets you select an entire region (surface) to use, then constrains the nodes at each point in the selected region of the mesh automatically. Figure 8.23 shows where constraints are applied to the left brake arm model.

8.23 Constraints are applied to the part named BRAKE_ARM_A to represent the boundary conditions.

(continued)

2 Next, loads are added to the model. The arrows pointing away from the selected regions in Figure 8.24 represent the applied loads that result when the brake cable pulls upward on the end of the brake arm, and the brake pad pushes against the wheel. (All the brake pad forces have not been shown here for easier visualization.)

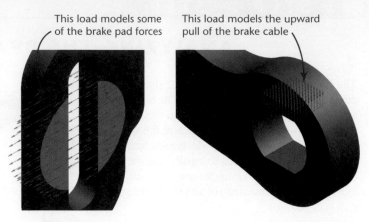

This load models some of the brake pad forces

This load models the upward pull of the brake cable

8.24 Loads are applied representing the actions of the brake cable and pads.

3 The type of element that will be used to form the mesh is set using the AutoGEM dialog box. Each element type has strengths and weaknesses for modeling particular features and material conditions. Figure 8.25 shows options that influence the mesh that will be generated automatically for the element type. In this case, the tetrahedron element type is selected for use with solids.

> ── **TIP** ──
> Pay close attention to units during any type of analysis. When values are input, make sure the units match the requested input as well as the units used in your model.

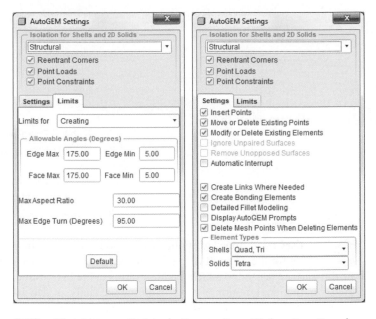

8.25 The Automatic Mesh Generation Dialog Box Panels. Allowable angles for the mesh are input into a dialog box.

4 Next, the material properties for the part are specified (if this has not already been done during modeling). In this case, AL 6061 has been chosen from a list of predefined materials (dialog box not shown.)

Mechanica's automatic mesh creation function fits a mesh to the part similar to that shown in Figure 8.26.

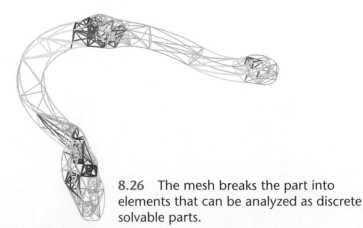

8.26 The mesh breaks the part into elements that can be analyzed as discrete solvable parts.

5 Now, the model is ready for setting up the analysis (Figure 8.27). Mechanica uses a method called *interpolating polynomials* (P-elements) to solve for the results. Instead of refining the model by increasing the mesh density, using P-elements allows the mesh to remain the same, and the order of the interpolating polynomial for the element is increased. The hotter colors shown in Figure 8.26 represent areas where higher polynomial orders are used.

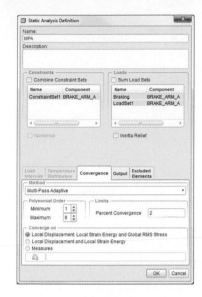

8.27 Setting Up the Analysis

6 In this case von Mises effective stress analysis was selected because it analyzes the combined results of stresses in all the different directions. This means that the analysis combines the various shear and normal stresses, then represents them as a single stress that you can relate to material properties. This is useful for determining maximum loading.

Next, the analysis is run and the results verified. Figure 8.28 shows the results of a quick check analysis. Hot colors (such as orange) are used to represent higher stresses. Cool colors (such as blue) represent lower stresses. Figure 8.29 shows a more refined analysis. The quick check might be used for base testing, but it is unlikely to yield useful analytical results.

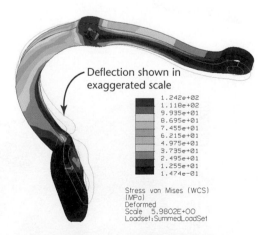

8.28 A Quick Check of the von Mises Plot for the Left Brake Arm. The key indicates values for the stress regions.

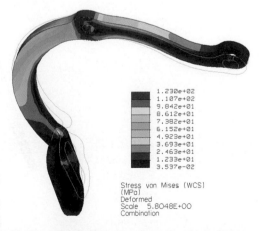

8.29 The Left Brake Arm Analysis Using Increased Model Density

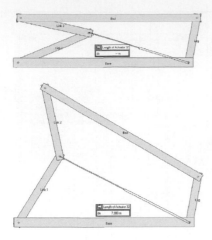

8.30 This four-bar linkage was modeled in a 2D version of Working Model.

Simulation Software

Simulation software allows you to simulate the function of a mechanism or mechanical system using rules that define how the parts will behave. Once you have modeled the mechanism's behaviors accurately you can set the simulation into motion to see how the model will behave under different conditions or in response to different inputs. Most simulations provide a visual representation of the action as well as numerical data about key behaviors. The ability to inspect the behavior of the device or system visually is a tool you can use to test designs early on—before prototyping—and to get a feel for how they will work. The data generated by the simulation can be used to assess how well the design meets criteria for the project. A simple 2D example is the four-bar linkage shown in Figure 8.30.

Many 3D modelers today offer a motion analysis feature that allows you to see and measure a mechanical device in action. Adams and Working Model 3D are two simulation software packages that can work with the CAD data. Other simulation software exists for specific purposes such as wind tunnel simulations and manufacturing simulations. The broad term *simulation* can be applied to any program that reasonably mimics the behavior of a system.

In mechanical design, simulations can be especially effective in the area of kinematics, or the study of motion. Figure 8.31 shows a virtual prototype of a wind turbine simulated in MSC Software's Adams Multiphysics software. Although parts may be rotated through a range of motion in the CAD assembly model, simulating complex interactions among parts in an assembly may better be created in a dedicated package such as Adams Multiphysics or Working Model.

In most cases, the CAD model (from software such as Autodesk Inventor, Pro/ENGINEER, SolidWorks, and Solid Edge) can be imported directly or after

8.31 Simulation allows manufacturers to design better performing and more reliable systems by predicting how wind turbines will behave throughout their lifecycle. *(Courtesy of MSC.Software Corporation.)*

being exported to a .dxf, STEP, or IGES file. Once the model is imported, part and assembly constraints are further constrained to model the kind of action allowed. Degrees of freedom define the directions in which a part can move. Materials and their properties are assigned to each part in the assembly (either in the CAD model or in the simulation software), as are motors and power sources. The appropriate forces are applied to the model, and other environmental settings (such as gravity) can be set to simulate action on the moon or in space.

Once the 3D model has been built, more complex analysis can be completed. The CAD model becomes a "virtual prototype" after joints and constraints are applied. In motorcycle design, for example, stability and safety of the design are a function of how the torque and stress induced by acceleration are handled by the vehicle. A designer can test the vehicle as if it were a real prototype and take accurate measurements of torque and tension. Variations on the design can be tested as the design is optimized.

In the example in Figure 8.32, robot actions are simulated in a 3D virtual world where process parameters critical to achieving system requirements are easy to visualize. For example, the transparent walls that define the working area of the robot are not visible in the real world, but they define the robot's work envelope in a way that is easy to understand. System users modify the virtual walls (and other parameters) and use ROBOGUIDE to optimize the performance of the system

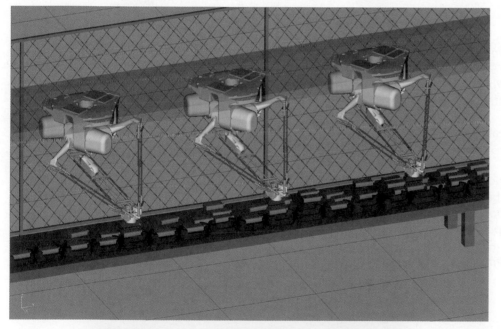

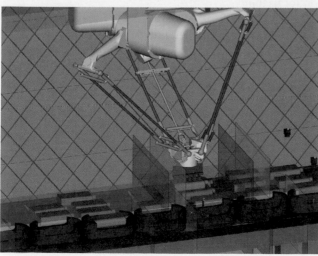

8.32 In these images, FANUC Robotics ROBOGUIDE software is being used to validate highspeed picking and placing of parts by the robots shown. In the simulation, parts flow randomly into the robot work cell at 300 parts per minute. Multiple robots share the material-handling pick process and then place the parts into an outfeed packaging container. In the bottom figure, you can see the virtual transparent walls that define the working area of the robot. *(Courtesy of FANUC Robotics America Corporation.)*

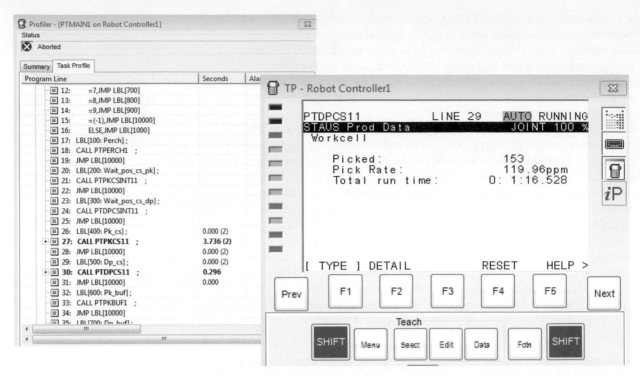

8.33 Using metrics provided by FANUC Robotics ROBOGUIDE, you can analyze and understand the result of process modifications. You can make changes to the simulated system and evaluate the result before implementing it in the real world. Changes to process parameters result in changes to characteristics such as motion cycle time and process throughput. The images above show detailed robot cycle time and process throughput information available in the software. *(Courtesy of FANUC Robotics America Corporation.)*

(Figure 8.33). The optimized parameters and motion paths can then be downloaded to the real robot system.

When a simulation has been created directly from a CAD model, any change to the CAD model can automatically update the simulation. This interaction makes it even easier to use the simulation to guide the iterative process of design.

Human Factors

Human factors analysis considers how people will interact with a design. As you are designing, consider the qualities of the users of the system or product. Human populations have general characteristics, such as average height and language, that may affect their ability to use or operate the device you are designing. Operator visibility (see Figure 8.34), the complexity of the controls, and the ability of the operator to correctly map a control to its function are all ergonomic considerations. In addition, you also need to consider the people who will assemble, service, and repair the product. You can use your CAD model to help determine the sizes for access openings, critical distances that affect users' reach, and other important information for usability.

A variety of software products work with the CAD data to test human factors. The CAD model can be imported into the software, or human models can be imported into the CAD environment to see how they fit with the model (shown in Figure 8.35).

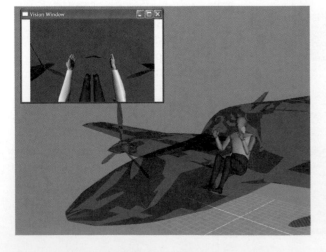

8.34 This view through the pilot's eyes illustrates the ability of the pilot to see the runway from the airplane cockpit. Operator visibility is an important factor in meeting safety requirements as well as functional goals. *(Courtesy of NexGen Ergonomics, Inc., www.nexgenergo.com.)*

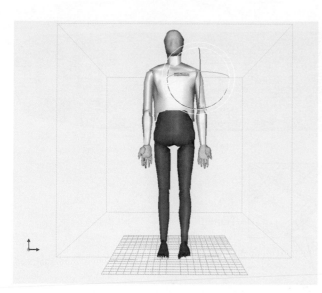
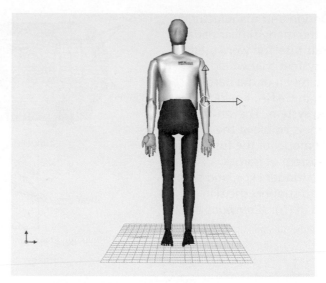

8.35 Human models with easy positioning controls can be added to your CAD models to test human interactions. *(Courtesy of NexGen Ergonomics, Inc., www.nexgenergo.com.)*

NexGen's HumanCAD software models typical humans from 11 different populations, including databases of human sizes such as 1988 Natick US Army and NASA-STD-3000. These populations provide a range of ethnic groups, size percentiles, body types, ages, cultures, and gender that you can then fit with your CAD model to see how different categories of people might interact with the design. The library includes typical predefined body and hand positions, linked for realistic human motion ranges, that you can import and export among your design software using common file formats. It even provides physically challenged mannequins that you can use to test accessibility. You can also use mannequin software to "see" the view from the mannequin's eyes as it moves along a path that you define, and to simulate lifting, pushing, and pulling by adding forces and torque in any direction on any body part (Figure 8.36).

Ergonomic analysis using similar models and databases is also an effective way to test work safety conditions for machine operators and others who will be near moving parts.

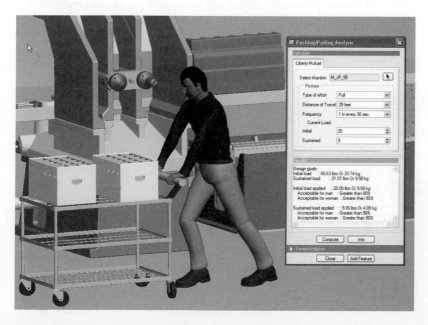

8.36 Pro/ENGINEER's Manikin Analysis Extension helps you analyze human movements such as pushing and pulling. *(Courtesy of PTC.)*

8.37 Design and manufacturing rules for the many similar parts in the aircraft fuselage were used to create a system that would automate much of the part modeling in CATIA, Boeing's 3D modeling system. The structural framework is visible in the assembly model of the fuselage in (a); the system of frames, stringers, stringer clips, and shear ties used to support the skin is illustrated in (b). *(Copyright © Boeing.)*

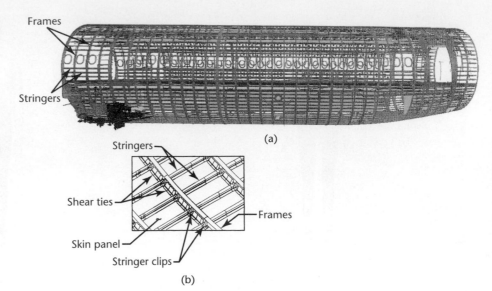

Integrated Modeling and Design Software

The iterative nature of design refinement is fueling software tools that will allow the design to be modeled and analyzed in one software environment. Many analysis packages already provide tools that make it easier to alternate between the analysis package and the CAD model.

Design rules that can be used for design optimization have also generated a closer link between parametric modelers and equation solvers. Using design rules to automate the optimization and/or creation of model geometry has been successfully used at Boeing in the design of certain aspects of the fuselage system for commercial aircraft. The central fuselage of the Boeing 777, for example, is made of 50 skin panels, each of which contains more than four hundred parts. These four hundred–plus parts are the framework to which the aircraft's "skin" is attached. Each frame, stringer, stringer clip, and shear tie, illustrated in Figure 8.37, is very similar to the next one, but not exactly the same. Each is modeled individually because each may vary in the holes added to accommodate wiring, fasteners, and so on. Each is also individually designed to weigh as little as possible. Chemical milling is used to remove all unnecessary material from each part to reduce the weight of the aircraft. Rule-based design has eliminated much of the work of modeling these 20,000 parts. The Boeing knowledge-based engineering system uses both design rules and manufacturing rules to ensure that each part is structurally sound and optimized for weight (Figure 8.38). For example, one design rule is that chemical milling to reduce the thickness of the

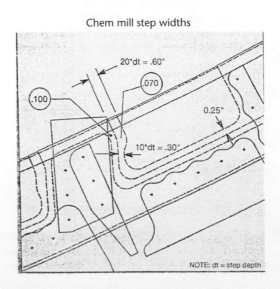

8.38 A manufacturing rule expresses the limits of the manufacturing process, such as chemical milling, that will be used. *(Copyright © Boeing.)*

part must stop .75 inch from a hole in the part. This ensures that the strength of the area around the attachment will not be compromised. Manufacturing rules are derived from the limits of the manufacturing process used to remove excess weight. No part may be designed that cannot be milled. The parameters defining each feature of the part are evaluated against these rules both to optimize the design and to automate much of the part creation.

As you use your CAD database for analysis in downstream applications you should familiarize yourself with the options being developed that can facilitate the exchange of information among packages and the use of analytical tools in design refinement.

Getting Easier Being Green

SolidWorks Sustainability is a module that provides life cycle assessment (LCA) within the SolidWorks user interface. It gives you information about environmentally friendly options for your engineering design decisions. This software allows you to conduct life cycle analyses of parts and assemblies. You can use it to see the environmental impact in terms of air, carbon, energy, and water.

To see if you improve your design's environmental impact as you modify it, you can capture the existing baseline, then track how your new design compares. The **Find Similar Material** dialog box shown lets you search the material database for alternative materials based on their mechanical properties. As you choose them you can visually assess the magnitude and direction of their environmental impact with the charts in the panel at the bottom.

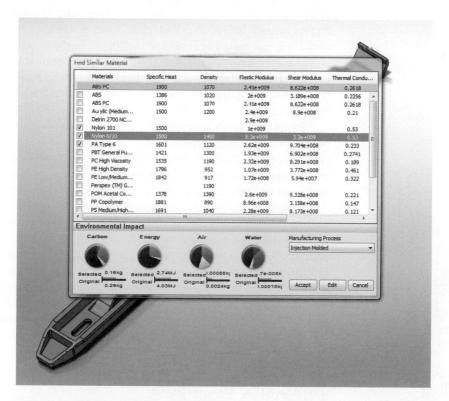

(Courtesy of DS SolidWorks Corp.)

8.39 Complex molded shapes can be prototyped cost effectively with rapid prototyping systems.

8.4 PROTOTYPING YOUR DESIGN

No matter how accurate your CAD model, it is never exactly the same as the manufactured part. Traditionally, building a prototype was the best way to ensure that a design could be manufactured and that it would operate as desired when it was built. Today, the information in the 3D solid modeling database has affected the role of the physical prototype in two key ways: it has made it possible to simulate a prototype with the 3D model, and it has made it faster and cheaper to create a physical prototype than ever before.

A simulated prototype—often referred to as a *virtual* prototype—can serve many of the purposes of a physical model. Throughout this text, you have seen how the 3D solid model can be used to evaluate appearance, customer appeal, fit and clearance for assembled parts, mass properties, kinematics, and other characteristics of the design.

The same information in the CAD database can also be used to direct rapid prototyping processes that generate physical models relatively inexpensively.

Rapid Prototyping

Because the volumes contained in a design are fully defined in a solid modeling system, physical models can be created by nontraditional technologies that translate the data into a physical entity. ***Rapid prototyping*** (RP) systems let you create a prototype directly from a CAD design within minutes or hours instead of the days or weeks it might otherwise take.

What is it worth to the design process to have an actual part that people can hold in their hands? As a visualization tool and a means for checking the fit with other parts, a physical model is a valuable aid in reducing the time it takes a company to develop a product idea from a sketch to a product that is available in the marketplace. Approximately 10% of all manufacturing and design shops spend over $100,000 each year for prototypes. The resulting confidence in the design and improved ability to communicate with the customer about the design in an understandable way is an important advantage of rapid prototyping. Rapid tooling processes use the same 3D CAD information to produce molds and other tooling that can reduce the time to market even further.

Rapid prototypes are especially useful for prototypes of complex molded parts. Molds for fairly simple plastic parts can cost from $20,000 to $50,000, making them prohibitively expensive to create just for checking a design appearance. With rapid prototyping, single parts can be produced in a matter of hours and used to verify the design. In addition, complex shapes can be created as easily as simpler ones.

Despite its advantages over traditional processes, rapid prototyping is not lightning fast. A part that is $2 \times 3 \times 1$ inch may take 3 or more hours to create. However, the time does not usually depend on the complexity of the part, just the size and accuracy built into the prototype file, and the type of process used. Parts that would ordinarily have to be molded or cast can be created in the same amount of time required for a rectangular block of about the same dimensions. The complexity of each slice does not have much effect on the time needed to create the part. Although a rectangular block is easy to manufacture using traditional machining methods, the buckle shown in Figure 8.39 is a good candidate for rapid prototyping because of its complex shape.

Translating the Model

Rapid prototyping systems all work on a similar principle: they slice the CAD model into thin layers, then create the model layer by layer from a material that can be fused to the next layer until the entire part is realized.

To send a CAD file to most rapid prototyping systems, you usually export a file in the STL file format. This file type was developed to export CAD data to an early rapid prototyping system. Since then, it has become the de facto standard for exporting CAD data to RP systems.

8.40 Triangular facets that define the boundaries of the model will be more visible on curved surfaces and when the facet size used is large.

STL files define the boundaries of the CAD model using triangular facets. This format transforms any model into a standardized definition, but it has the disadvantage of generating a very large file when a realistic shape is required. You usually have the option of setting the size of the facets when you export your model. If the facets are small and the model complex, the resulting STL file will be very large. If a larger size for the triangular facets is used, however, the prototyped part will have noticeable facets on its curved surfaces, as shown in Figure 8.40.

Once the CAD file has been exported, thin slices through the model are generated to create the layers (Figure 8.41). Generally speaking, the thinner the slice, the more accurate the part—however, thinner slices also mean that it will take longer to generate the prototype part.

The accuracy of the model's surface is limited by the material used in the process and how small the layers and features are that it can make. A prototype part created by depositing individual layers of material in the X-Y plane and dropping the table down in the Z-plane necessarily produces a jagged edge. The size of the *jaggies*, as these are often called, is dependent on the thickness of the layer of material that is deposited. This thickness is limited by the size of the smallest particles that can be fused together. RP systems that have more than three axes of movement can reduce or eliminate the jagged appearance by filling in material on angled edge surfaces.

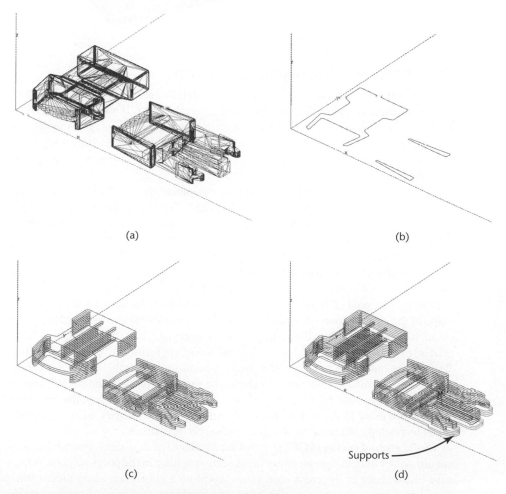

(a)

(b)

(c)

Supports

(d)

8.41   (a) The model is imported as an STL file.
(b & c) Slices through the model are generated (.01 intervals shown here).
(d) The software generates supports (shown here with lighter lines) for segments of the part that do not rest on one of the lower slices.

8.42 SLA uses laser-hardened resin to prototype parts. *(Courtesy of 3D Systems.)*

Current Rapid Prototyping Systems

Rapid prototyping systems vary in the types of materials used, in the size of the model that can be created, and in the time it takes to generate a part. If your company owns a rapid prototyping system, the choice of system may be moot. If not, you should consider the design questions the prototype needs to answer when selecting an appropriate system. The accuracy, size, durability, and time it takes to create a prototype are dependent on the process and material used.

The main categories of rapid prototyping equipment are stereolithography, selective laser sintering, fused deposition modeling, and 3D printing. Most rapid prototyping systems can create parts up to about 10 cubic inches in size. Laminate object manufacturing and topographic shell fabrication are two less common methods that allow you to create prototypes of larger parts.

Stereolithography apparatus (SLA) uses laser-hardened resins to form the model. Figure 8.42 shows an SLA system. The system software controls a focused laser beam in a pool of light-sensitive polymer. The laser hardens each layer in the shape of the cross section or slice of the part. As successive layers are hardened they are submerged slightly into the resin pool (see Figure 8.43) and the next layer is hardened on top of them. Holes and pockets in the model are formed by uncured resin that easily pours out of the resulting part. SLA systems create durable parts that can be painted and finished to look very similar to the finished product. The range of accuracy for SLA parts can be up to ± 0.05 mm. Because of this accuracy, the prototype parts created using SLA can have relatively smooth surface finishes. SLA is an established technology, since it was the first method on the market.

Solid ground curing (SGC) systems are similar to SLA systems except that they use ultraviolet light to cure an entire cross section at once in the polymer pool. A negative of the shape of the cross section is created on a glass plate using electrostatic toner (similar to the process used in a copying machine), then used to mask ultraviolet light in the shape of the cross section. With no lasers to replace, these systems are cost-effective and create accurate, durable parts.

8.43 Cross-sectional slices are hardened in a resin pool by the SLA system. *(Courtesy of 3D Systems.)*

8.44 Selective laser sintering fuses powdered metals and other materials to form prototype parts that can be machined. *(Courtesy of 3D Systems.)*

Selective laser sintering (SLS) uses a focused laser to fuse powdered metals, plastics, or ceramics. (Figure 8.44 shows an SLS-type rapid prototyping system.) The fused layer is covered with additional powder, and the next layer is fused to it. To form a hole in the prototyped piece, the powdered material is simply not fused in that area. The unfused powder still acts as a base for the next layer, but when the part is completed, the unfused portions are simply poured out (see Figure 8.45). This process has the advantage that models created from powdered metal can sometimes be machined for further refinement. The parts can also be strong enough to be used in certain types of assemblies as one-of-a-kind parts. An SLS system can create parts with accuracies of plus or minus 0.127 mm. Other materials, such as a glass-filled nylon, may be used with the sintering process to create parts with varying degrees of flexibility and durability. Figure 8.46 illustrates an elastomeric material with rubberlike characteristics that makes it suitable for prototyping gaskets and athletic equipment.

8.45 This artist's laser sintered design would be very difficult to manufacture using traditional methods. *(Courtesy of Vladimir Bulatov.)*

8.46 A material from DuPont used with the DTM laser sintering process produces a rapid prototype with rubberlike qualities. *(Courtesy of DTM Corporation.)*

Fused deposition modeling (FDM) systems use molten plastic deposited in layers corresponding to cross sections on the part. Because the soft molten plastic cannot be deposited in thin air, to make a hole or an overhang, a second type of plastic is used to create a support structure. Because the two plastics are different materials that do not readily adhere to one another, the support structure can be separated from the actual part. (Figure 8.41(d) shows these supports being formed by the RP software.) Figure 8.47 shows an FDM system from Stratasys Corporation. A part that is about $3 \times 2 \times 1$ inch takes about three hours to create.

Laminated object manufacturing (LOM) produces solid parts from sheets of material, such as paper or vinyl. LOM systems can be used to create larger prototype parts, as shown in Figures 8.48 and 8.49. As in all rapid prototyping processes, software first generates cross-sectional slices through the model. Instead of fusing the slice, however, a computer-controlled laser cuts it from the first sheet of material. Then, a heated roller bonds the next sheet to the previous layer, and the next cross section is cut from this sheet. The material that will later be removed is cut into crosshatched shapes to make removal easier. The Helisys LOM-2030H system can create a part with a maximum size of 32 inches in length, 22 inches in width, and 20 inches in height (813 millimeters in length, 559 millimeters in width, and 508 millimeters in height) and a maximum part weight up to 450 pounds (204 kilograms).

Topographic shell fabrication (TSF) uses layers of high-quality silica sand fused together with wax to build shells that can be used to mold rapid prototypes of large-scale parts. The sand is deposited in layers and then fused with molten wax sprayed from a computer-controlled three-axis nozzle. More sand is deposited and then the next layer is fused. The layers range from .05-inch to .15-inch thick and take about ten minutes per square foot of model to print. Once all the slices have been deposited, the sand/wax shell is smoothed, then lined with plaster or other material. The shell is then used as a temporary mold for creating parts of fiberglass, epoxy, foam, concrete, or other materials. This method is able to handle very large shapes up to $11 \times 6 \times 4$ feet.

8.47 The fused deposition modeling systems require supports for sections of the part that are unsupported by lower slices. *(Courtesy of Stratsys, Inc.)*

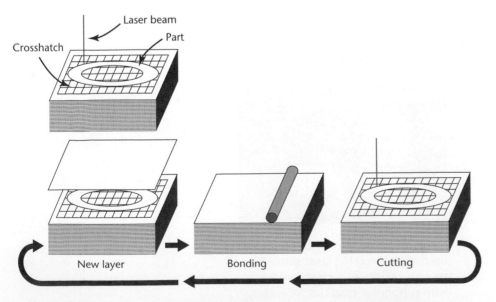

8.48 Laminated object manufacturing forms large-scale prototype parts by bonding sheets of material. The crosshatched material is later removed from the block within which the part was formed. *(Courtesy of Helisys, Inc.)*

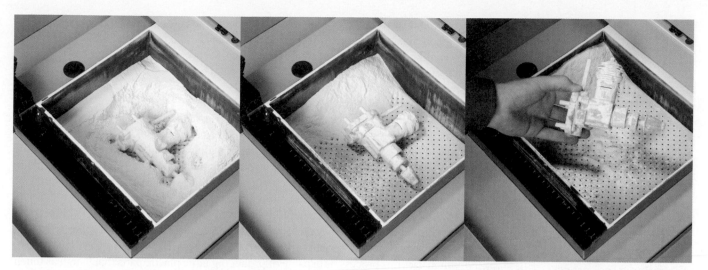

8.49 A 3D Printer *(Courtesy of Z Corporation.)*

3D Printing

Rapid prototyping systems referred to as 3D printing systems "print" layers of molten thermoplastic material. These low-cost machines were designed to enable the use of prototypes early and often in the design cycle. At Kodak, for example, a proposed product idea was to be discussed at a 10:30 meeting. At 8:30, the team was asked to bring a model of the idea to the meeting. Z Corporation's 3D printer, similar to that shown in Figure 8.49 was able to build a prototype from the CAD files in time for the meeting.

The relatively low cost 3D printing systems (currently under $50,000) can be operated safely enough that they can sit next to the regular office printer or copier. Another advantage of 3D printer systems is that those having a three-axis print head can deposit the plastic on the edges of the part in a way that creates a smooth model surface (see Figure 8.50).

Rapid Tooling

Rapid prototyping systems were developed to produce parts without having to create a mold or complete intermediate steps needed to manufacture a part. *Rapid tooling* is a similar process, but one that creates the *tool* (usually a mold for molded plastic or cast metal parts) through a rapid prototyping process, not the part itself. Metal injection molds and molds for cast metal parts are often one of the most expensive and time-consuming parts of the design process. Rapid tooling processes can reduce the amount of time involved in producing these tools. The resulting rapid tool can be used to produce test products and to get products to market early.

Rapid tooling can be accomplished by several different methods. One is direct mold design, in which the tool itself is created using a selective laser sintering–type process. Another method uses rapid prototyping to produce a master part from which a silicone rubber mold is formed. That mold is then used to make other parts. A third method (a more traditional process) uses computer-controlled machining technology to create the cavity in a mold blank to quickly create the mold.

8.50 ZPrinter 450 from Zcorp "printed" the colored part shown in about four hours. *(Courtesy of Z Corporation.)*

SPOTLIGHT

Cores and Cavities

Molded parts are formed by cavities and cores. The *cavity* is the part of the mold that forms the outside shape of the object. The colloquial use of the term *mold* generally refers to the cavity. Holes that will be formed in molded parts are formed by cores. A *core* is a solid shape that fits inside the mold. It will form a hole in the cooled cast metal or molten plastic. Cores for cast metal parts are often made of packed sand. After the cast metal is cooled, the part is pounded to loosen the particles of sand from one another so they can be poured out through a small hole. This is similar to the way nonhardened resin or unfused powdered metal is poured out of holes in rapid prototypes.

8.51 This mold was created by DTM's RapidTool process. The cavity side of the mold is on the left, the core on the right. *(Courtesy of DTM Corporation.)*

Using a CAD file as input, the SLS process melts (or sinters) the powder to form a green mold shape consisting of metal particles bound by smaller areas of polymer. When the green mold is heated in a furnace, the plastic polymer burns off, leaving only the metal mold, as shown in Figure 8.51. The mold is then machined to a tolerance of ±.005 inch to eliminate any defects and can be drilled, tapped, welded, and plated like a conventional mold.

With different materials, cores and molds for sand casting can be created directly using an SLS system, then cured (hardened) in a conventional sand casting oven. Another application is in *investment casting*, in which an original shape called a *master* is used to create the proper-shaped opening in a mold. The master is typically made of a wax material so that it will melt out of the mold when molten metal is poured in. Investment casting masters can be produced using SLS processes.

Silicon rubber molds are another means of producing rapid tooling. This process uses an accurately prototyped part that is then coated with silicon rubber to form a mold from which more parts can be molded. The MCP Vacuum Casting System, which uses this process, is capable of producing large plastic parts that weigh as much as 12 pounds and span 2 feet by 3 feet.

Rapid NC machining of mold inserts from a 3D CAD file also promotes rapid tooling. Even though this is more of a traditional process that creates the mold cavity by removing material, the NC-machined cavity combined with standard mold blanks can often lead to shorter tooling times for injection-molded parts. Metal spraying is another method that can be used in rapid tool production for less complex parts.

Despite the ease with which rapid tooling and rapid prototyping create physical models from CAD data, a strong understanding of traditional and current manufacturing methods will enable you to produce more cost-effective and producible parts. If you would like to learn more about key manufacturing processes and how they can inform your design decisions, read the Implementation chapter available at MyCADKit.com.

SPOTLIGHT

RepRap

RepRap is a desktop 3D printer capable of printing plastic objects. Many of the parts for RepRap are made from plastic parts that RepRap can print (Figure 8.52), making it close to a self-replicating machine. RepRap is meant for the hobbyist to build, given time and materials. Once you get the parts for yours, you become a responsible owner by printing another RepRap for someone else.

Open-source electronics and software are available online for the RepRap project, but if you do not want to build your own, the Makerbot Cupcake rapid prototyper is a version you can buy ready-made.

8.52 A Controller Board for the RepRap. It has an onboard Sanguino prewired to drive the peripherals needed to control a RepRap machine. *(Courtesy of Zach Smith, MakerBot Industries.)*

TESTING THE MODEL: VIBRATION ANALYSIS

Lasers are subjected to some intense conditions. Installed in a wind tunnel, the assembly needs to withstand supersonic wind velocities. In a helicopter, the laser needs to function despite vibrations from the engine and rotors. Each of these generates vibrations at different frequencies. Before Quantel USA builds a prototype for testing, it uses the 3D model and SolidWorks Simulation to "shake up" the laser assembly to see how the parts will respond to vibrations at frequencies typical for the application.

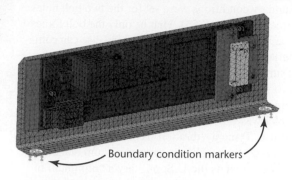

8.54 The Mesh Applied to the Model

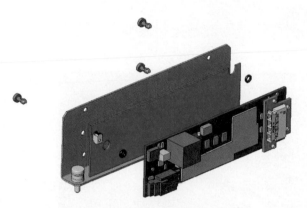

8.53 The circuit board attaches to the sheet metal mount, which attaches via two bolts.

The circuit board shown in Figure 8.53 generates a high-voltage pulse to a laser optic. It is mounted to a cantilevered sheet metal part that attaches to the laser assembly with two bolts. When the circuit board and mounting assembly are subjected to vibrations at different frequencies, will the resultant deflection (bending) result in failure? Laine McNeil, Senior Mechanical Engineer, set up a simulation to help find out.

First, he prepared the model for testing by generating a mesh appropriate for each part. Choosing the right kind of mesh and the right density (number of nodes) is a judgment call driven by the goals of the test. For example, the board components had been tested before, so McNeil could focus on the board and the cantilevered sheet metal part. For the circuit board, McNeil specified a 3D mesh to reflect the thickness of the board (see Figure 8.54). For the sheet metal mount, he chose a shell mesh and selected sheet metal as the material whose properties the mesh would use.

It is common to simplify the 3D model (by removing fillets, perhaps) to reduce the number of nodes in the mesh. Adding nodes improves accuracy, but it also increases the time needed for analysis. For example, the shell mesh chosen for the mount uses one surface of the part and applies the stiffness properties uniformly over the entire part. This generated fewer nodes, but provided enough complexity to adequately represent the sheet metal part.

McNeil next set up the boundary conditions for the assembly: what are the constraints on how the parts can move? First, how are the parts attached to each other? Parts can be bonded or attached in a way that they cannot move or slide (fixed geometry), hinged, connected via a roller/slider, or supported by a spring. The circuit board was mounted in place with three rubber interfacing pads. The rubber was bonded on one end to the sheet metal mount and to the circuit board at the other for this analysis.

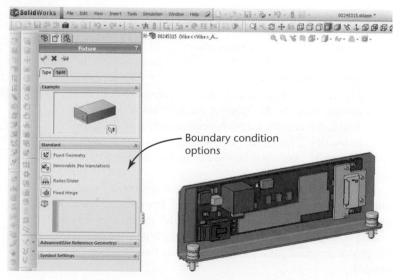

8.55 Choosing Boundary Conditions for the Circuit Board

(continued)

Second, how are the parts connected to the outside world? The sheet metal mount is attached solely by its two bolts, so boundary conditions were set for the two bolt holes (shown in green in Figure 8.55). That is, only the bolt forces and friction are holding the mount; everything else can compress and deflect in the simulation. McNeil could have chosen "on flat surface" for the mount so that the part would not be allowed to deflect downward, only up, but that was not needed for this test.

The model was now analyzed at frequencies expected for its target use. Typical vibration frequencies range from 100 to 2000 hertz. Some clients have minimum standards that must be met, such as the U.S. military's "minimum integrity test." For the circuit board analysis, no deflection was detected in the sheet metal mount until almost 280 hertz. Figure 8.56(c) shows the deflection. The material between the bolt holes is shaded a lighter blue to indicate a small amount of movement, and the upper cantilevered portion of the sheet metal mount is colored red, indicating significantly higher motion. The circuit board had already been tested to conform to the minimum integrity test. Because the circuit board operated reliably despite the first resonant frequency mode shape showing that deformation occurred at as low as 120 hertz, McNeil could be confident that the assembly would be satisfactory as designed.

This result was not always the case. When the circuit board components were initially tested, the cube (representing a power supply) at the top left in Figure 8.56(a) fell off. It had been soldered onto the board, but the solder pads were not strong enough to hold at typical vibrations. Glue was added beneath the power supply to strengthen the bond.

In addition to simple resonant frequency shape generation, analysis can also simulate the energy driving the vibration and the associated displacement. Two examples of vibration-generating devices can readily demonstrate the difference. A cell phone vibrates, but the energy driving the vibration is not enough to shake a room. A hydraulically driven vibrator, used to dislodge material from the bed of a dump truck, has enough energy to shake a building one city block away. For the circuit board assembly, McNeil used the built-in capabilities of SolidWorks, so the energy behind the vibration was assumed to be infinite, and the displacement shown for the circuit board parts was not meaningful. In this instance, this was not the focus of the analysis. If determining actual displacement values is required, the model can be analyzed with a higher-end FEA package, in which a power spectral density curve can be used to define the energy levels causing motion.

At Quantel USA, testing the "virtual prototype" does not replace the need for a physical model. For each design a physical model is built and shaken at the frequencies expected during use. But if the model testing is done properly, the physical model passes the test and only one prototype is needed.

(All images courtesy of Quantel USA.)

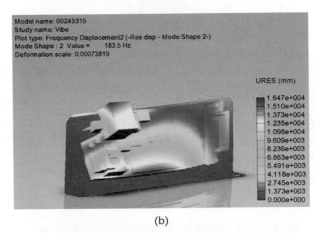

(a)

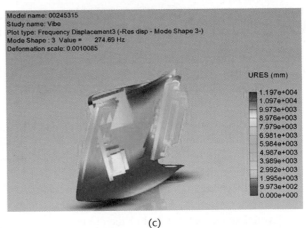

(b)

(c)

8.56 Results of Testing at (a) 121.70, (b) 183.5, and (c) 274.69 Hertz

TELESCOPE PAINT $\boxed{15\ DEC\ 97}$

$$V_{\substack{AMES \\ B.B.}} = 5.35m^2 \cdot 6.0\times10^{-4}m = 0.00321\,m^3$$

$$M_{\substack{AMES \\ B.B.}} = \left(1300^{kg}/m^3 \cdot 0.00321m^3 \cdot 0.8\right) + \left(3200^{kg}/m^3 \cdot 0.00321m^3 \cdot 0.2\right)$$

$$= 3.3384\,kg + 2.0544\,kg = 5.3928\,kg$$

$$V_{\substack{AMES \\ M.T.\&A.T.}} = 1.19m^2 \cdot 6.0\times10^{-4}m = 0.000714\,m^3$$

$$M_{\substack{AMES \\ M.T.\&A.T.}} = \left(1300\,kg/m^3 \cdot 0.000714m^3 \cdot 0.8\right) + \left(3200\,kg/m^3 \cdot 0.000714m^3 \cdot 0.2\right)$$

$$= 0.74256\,kg + 0.45696\,kg = 1.1995\,kg$$

$$V_{\substack{DESOTO \\ B.B.}} = 5.35m^2 \cdot 2.032E\text{-}4m = 0.001087\,m^3$$

$$M_{\substack{DESOTO \\ B.B.}} = \left(1300\,kg/m^3 \cdot 0.001087m^3 \cdot 0.8\right) + \left(3200\,kg/m^3 \cdot 0.001087m^3 \cdot 0.2\right)$$

$$= 1.1306\,kg + 0.6958\,kg = 1.8264\,kg$$

$$V_{\substack{DESOTO \\ M.T.\&A.T.}} = 1.19m^2 \cdot 2.032E\text{-}4m = 0.000242\,m^3$$

$$M_{\substack{DESOTO \\ M.T.\&A.T.}} = \left(1300\,kg/m^3 \cdot 0.000242m^3 \cdot 0.8\right) + \left(3200\,kg/m^3 \cdot 0.000242m^3 \cdot 0.2\right)$$

$$= 0.2515\,kg + 0.1548\,kg = 0.4062\,kg$$

$\boxed{\text{OUTER SHELL PAINT} \quad \text{DUST COVER}}$

STANDARD DESOTO BLACK $\varrho = 1300\,kg/m^3$

DUST COVER -24 MILS

$$A_{D.C.} = 1.56883\,m^2$$

$$V_{D.C.} = 1.56883\,m^2\,(2.032E\text{-}4m) = 0.000319\,m^3$$

OUTER SHELL - BOTTOM

$$A_{O.S.B.} = 8.48329m^2/4 = 2.12082m^2$$

$$V_{O.S.B} = 2.12082\,m^2\,(2.032E\text{-}4m) = 0.000431\,m^3$$

OUTER SHELL - MIDDLE

$$A_{O.S.M.} = 9.24948m^2/2$$
$$= 4.62474m^2$$
$$= 4.41144m^2/2$$
$$= 2.20572\,m^2$$

$$V_{O.S.M} = 2.20572m^2\,(2.032E\text{-}4m)$$
$$= 0.000448m^3$$

8.57 Even the mass of the paint is calculated into the mass properties for space vehicles. For the SIRTF space telescope, the low-emissivity, silicon-grit-based coating was to be 24 mils thick. In the calculations shown here, the surface area data from the model were used to calculate the mass of the various "paints," one of which turned out to be almost 9 kg. This additional mass resulted in the use of a different coating to keep the mass of the telescope within the design constraints for the project. *(Courtesy of Scott Adams, Ball Aerospace & Technologies Corp.)*

HANDS ON TUTORIALS

SolidWorks Tutorials

Simulating Motion

In this tutorial, you will learn to set up your model and use it to conduct a motion study.

- Launch SolidWorks from the icon on your desktop.
- Click to expand the topics listed below the "**?**" help icon.
- Click: **SolidWorks Tutorials.**
- Click: **Working with Models** from the list.
- Click: **Animations Overview** from the list that appears, and complete the tutorial on your own.

If you have SolidWorks Premium, consider doing the Solid-Works Motion Tutorial as well.

- Click: **All SolidWorks Tutorials (Set 2)** from the list.
- Click: **SolidWorks Motion Tutorial** from the list that appears, and complete the tutorial on your own.

Animations Overview

You can use animation motion studies to simulate the motion of assemblies.

In this tutorial, you:

- Set up an animation motion study that uses a motor to move the model.

- Run the motion study while suppressing a mate.

- Save the motor parameters to the Design Library.

SolidWorks Motion Tutorial

SolidWorks Motion uses complete kinematic modeling to compute component motion. You can use SolidWorks Motion to analyze forces in models that include springs, dampers, motors, and friction.

In this tutorial, you learn how to:

- Run a SolidWorks Motion study for a model that includes a spring and a motor.

- Plot the results.

- Duplicate the motion study with modified simulation parameters.

- Use the results to re-design the model.

(Images courtesy of DS SolidWorks Corp.)

REVERSE ENGINEERING PROJECT

The Can Opener Project

In this segment of the project, you will use your can opener assembly for analysis.

1. *Assign materials.* Assign appropriate materials to each of the parts and determine the weight of the can opener from your assembly model. Weigh the actual can opener and compare the two. Are they close?
2. *Use mass properties.* Next, determine the center of gravity (COG) for the can opener assembly. Do you think the can opener will feel balanced when you are operating it based on the location of its center of gravity?
3. *Refine and test again.* Experiment with different options by making changes to your can opener parts and updating the assembly. List two options for changing the design if you are required to reduce the overall weight by 20%.

(Courtesy of Focus Products Group, LLC.)

KEY WORDS

ASCII

Backsolve

Centimeter-Gram-Second

Centroid

Comma-Delimited

Downstream

Drawing Exchange Format

Elasticity

Exporting

Factor of Safety

Finite Elements

Foot-Pound-Second

Graphics Exchange Format

Inch-Pound-Second

Initial Graphics Exchange Specification

Investment Casting

Laminated Object Manufacturing

Mass Density

Material File

Mesh

Moment of Inertia

Native File Format

Neutral Files

Pounds Force

Pounds Mass

Radius of Gyration

Rapid Prototyping

Rapid Tooling

Raster

Right Cylinder

Rule Sheet

Selective Laser Sintering

Simulation

Solid Ground Curing

Space-Delimited

STandard for the Exchange of Product model data

STereo Lithography

Tab-Delimited

Topographic Shell Fabrication

Variable Sheet

Vector

Volume

SKILLS SUMMARY

Now that you have completed this chapter, you should begin practicing using your CAD models for engineering analysis. You should be able to extract mass properties data from your CAD models and use them to solve simple engineering problems. You should also be familiar with exporting CAD data to various packages and selecting the best file format for the export. In addition, you should be able to describe how various rapid prototyping systems work to create physical models from CAD data.

EXERCISES

Exercises 1–6 Mass Properties Evaluation. Use your solids modeler to determine the mass of the following objects. Define a datum, then locate and dimension the centroid. Find the moment of inertia and radii of gyration about a defined major axis for each part.

Assume all objects are milled from 6061-T6 aluminum, with specific gravity of 0.28. Small fillets or chamfers do not significantly change your solution and can be neglected in the model, but do not neglect important machining details such as the internal radii formed by the .25- or .50-inch diameter cutters!

Dimensions shown are in inches.

1. Simple rectangular block

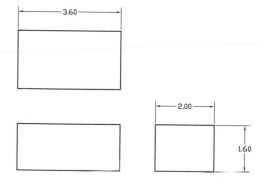

2. Rectangular block with internal recess

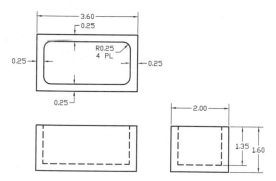

3. Rectangular block with split internal recess

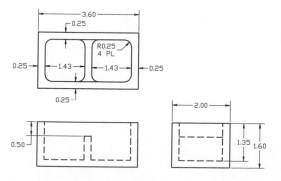

4. Asymmetrically milled block

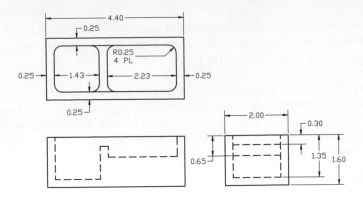

5. Slotted tube

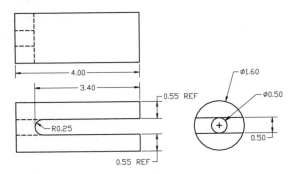

6. Slotted hexagonal hollow rod

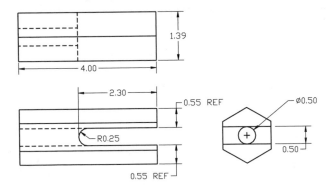

Exercises 7–10 Create a table in Excel or other spreadsheet program comparing the mass properties as calculated by your CAD modeler with hand-calculated exact values, for the following figures. Be sure to define a common datum for each figure, so that proper comparisons can be made. Include a column showing percent difference from the exact value. How does part-axis orientation affect your accuracy?

7.

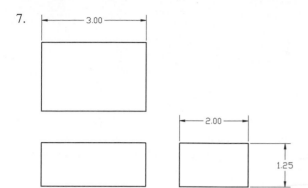

8.

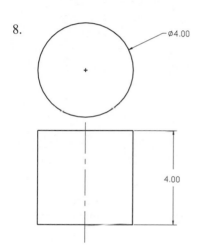

9.

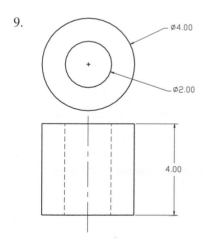

10.

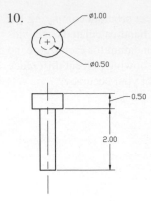

Exercises 11–14 Use the mass properties data for the models in Exercises 7–10 to create a spreadsheet that computes the material, cost, and labor needed to coat the surface area of the parts.

15. Find the center of gravity and total mass for an assembly that you created from the exercises in the previous chapter.

16. Material changes are common in the development phase of engineered parts. Consider a component part modeled first using 6061-T6 Aluminum, and then with 1020 Steel. No dimensional changes are made. Do each of the following properties of a solid change with changing material? How?
    a. Volume
    b. Mass
    c. Density
    d. Centroid
    e. Moment of inertia
    f. Radius of gyration

17. Vector data and raster data are both used in graphics programs. Which type of data format is better suited for CAD systems that have mass properties capability?

18. Define the steps involved in creating a stereolithographic rapid prototype part from a solid model. What are some of the considerations when deciding whether a CAD model is a good candidate for prototyping using stereolithography?

19. Give a brief explanation of the physical process used to create rapid prototype parts by each of the following methods. What are the advantages and disadvantages of each method?
   a. Stereolithography
   b. Fused deposition modeling
   c. Selective laser sintering
   d. 3D printing

20. How does *rapid tooling* differ from *rapid prototyping*? In what situations might either method be used to create actual working components rather than just shape-representative models?

9 DOCUMENTATION DRAWINGS

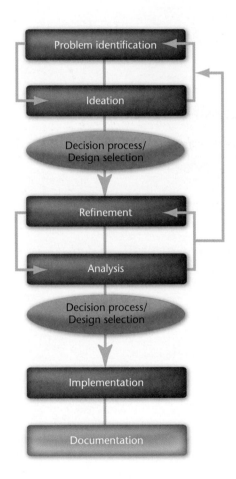

OBJECTIVES

When you have completed this chapter you should be able to:

1. Locate drawing standards that apply to your project.

2. Create drawings from 3D models at an appropriate scale.

3. Create a set of working drawings that fully document a design.

4. Add section, auxiliary, and detail views to a drawing.

5. Dimension a drawing.

6. Add notes and title blocks.

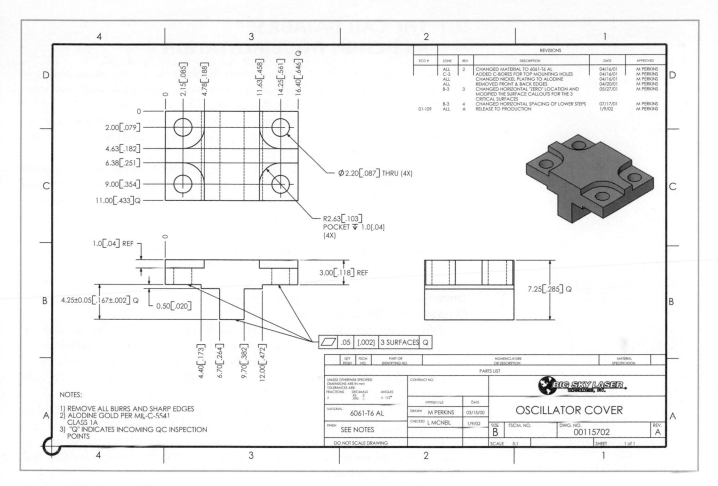

This drawing for a small part shows dimensions in millimeters with the inch values [in brackets] given for reference. *(Courtesy of Quantel USA.)*

WORKING DRAWINGS

Throughout the design process, engineers use a variety of ways to document design ideas. The documentation phase of the design process refers to the production of engineering drawings or controlled, documented models that unambiguously describe the part or process to be made. Documentation locks down a design at a point in time and serves as a contract with a manufacturer, a legal record of the design, and a tool for communicating key aspects of the design's function.

There are standards that govern how to show drawing information. They define symbols and formats that make it possible to express large amounts of information in a small space and to provide the information needed to manufacture the design.

Engineering drawing practices were developed for design communication among people. As more information is conveyed electronically from the CAD database directly to the manufacturing device, drawing practices and standards evolve to support the new processes through methods that increasingly use the electronic database to convey the information.

9.1 THE CAD DATABASE AS DESIGN DOCUMENTATION

An advantage of an accurate CAD database is the ability to use the model as a basis for manufacturing. CAD packages offer tools for incorporating tolerances and manufacturing notes into the 3D digital design database. A good understanding of the type of information that is available in your CAD database—combined with knowing how to show critical dimensions and tolerances—is important to having acceptable parts manufactured.

Companies must consider the legal requirements for maintaining a permanent record of the design. Many industries are required to maintain a permanent record (or snapshot) of the design as used for production. A changeable record on the computer may not be considered a legally acceptable practice, unless a standard of control is met.

Some companies use the 3D model with electronic annotations stored in the file or a related database as the final documentation. Other companies produce 2D drawings from the 3D model to communicate the design for manufacturing and documentation. These companies may store the 3D model but consider the dimensioned 2D drawings as the design record.

Other companies use a combination of the 3D model and 2D drawings to document the design. The 2D drawings communicate information about critical tolerances and other information not easily visible in the 3D file. The CAD file serves as the interface to automated manufacturing processes, and the accompanying drawing calls attention to critical dimensions and tolerances (see Figure 9.1).

Documentation drawings are best provided in a format that can be interpreted by the manufacturer, mold maker, or others who will create or inspect the parts. The manufacturer may not have the software used to create the CAD model. Some common and easy-to-use formats are PDF, SolidWorks eDrawing format, IGES, STEP, DXF, DWF, or a combination of these.

Whether the 2D drawings are printed on paper or stored electronically, correctly shown orthographic views still provide much of the basis for communicating and documenting the design. Correctly shown drawing views are also used to communicate information for user manuals and repair manuals, as well as for manufacturing and inspection.

9.2 STANDARDS

You should be familiar with three broad classes of guidelines for preparing documentation drawings:

- National and international standards applicable to engineering drawings in general
- Discipline-specific standards applicable to a design area
- Individual company drawing standards

The American National Standards Institute (ANSI) describes a series of codes and standards that are commonly used in the United States. The American Society of Mechanical Engineers (ASME) publishes the ANSI/ASME standards, which are organized into groups and referred to by number. Appendix 1 includes a listing of the standards available from ASME and the information for ordering them. Each ANSI standard is broken down into a series of documents covering specific topics. Each of these documents is revised individually as needed. Many of the standards are safety standards or dimensional standards for interchangeable parts, in addition to being drawing standards.

The International Organization for Standardization (ISO) publishes standards that govern international engineering standardization and the creation of metric drawings for international use. Other standards that may apply are those published by the

American Society for Testing and Materials (ASTM) for materials and testing. These are very useful when you wish to specify a particular material, grade, standard stock size, or finish. The Society of Automotive Engineers (SAE) standards also specify material grades. The Industrial Fasteners Institute provides documentation standards for fasteners in addition to those provided by ANSI. The National Institute of Standards and Technology (NIST) is a governmental agency that may publish standards relevant for your design area. For federally and state-funded projects, you should research government contracting regulations that may apply.

There may be other standards and practices for your particular area of interest. Dimensional standards in bridge construction differ greatly from those for manufacturing a laser optical device. Each design area has standards that apply to the processes used in that line of work. The American Congress on Surveying and Mapping (ACSM) provides guidelines for surveying and mapping standards. Civil engineering design and building systems have a number of standards they follow. ASTM also publishes some guidelines for building design and construction. The National Institute for Building Sciences (NIBS) publishes guides for federal construction that can be helpful for these areas. Steel fabrication standards and practices are available from the American Institute of Steel Construction. Appendix 2 lists the preceding organizations and their contact information.

(a)

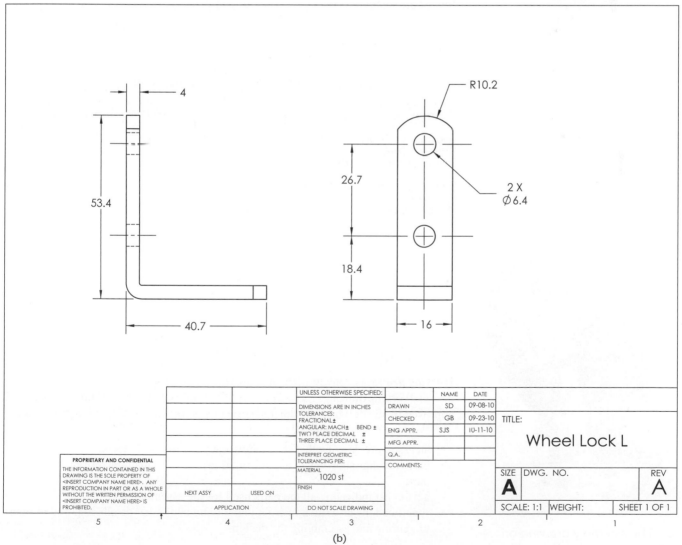

(b)

9.1   (a) A shaded view of the 3D model for the hand brake wheel lock. (b) A dimensioned drawing for the same part. *(Courtesy of Salient Technologies, Inc., www.salient-tech.com.)*

9.2   Standard symbols for pipe fittings, valves, and piping are found in *ANSI Y32.2.3*.

	Flanged	Screwed	Bell & Spigot	Welded	Soldered
1. Joint					
2. Elbow – 90°					
3. Elbow – 45°					
4. Elbow – turned up					
5. Elbow – turned down					

ANSI Drawing Standards

The ANSI standards covered in the Y14, Y1, and Y32 series are pertinent to drawings and terminology, but additional information useful for representing particular types of equipment can be found in other standards as well. Table 9.1 lists some of the drawing topics covered in the Y14 standard. Standards make it easy for you to clearly communicate your design to manufacturers and others. Figure 9.2 shows a few of the standard symbols for piping.

Table 9.1 Selected Topics from the ANSI Y14 Standard.

Name	Designation
Decimal-Inch Drawing Sheet Size and Format	Y14.1
Dimensioning and Tolerancing	Y14.5M
Line Conventions and Lettering	Y14.2M
Metric Drawing Sheet Size and Format	Y14.1M
Multiview and Sectional View Drawings	Y14.3M
Pictorial Drawing	Y14.4M
Screw Thread Representation	Y14.6

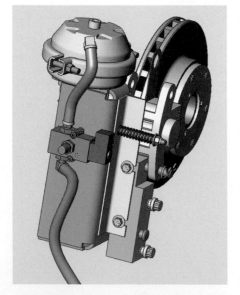

9.3   The assembly drawing on the facing page clearly specifies the locations for all parts of this airbrake. *(Courtesy of Dynojet Research, Inc.)*

9.3 WORKING DRAWINGS

Working drawings convey all the information needed to manufacture and assemble a design. Working drawings may be prepared by drafters and then carefully checked and approved by the designer and other supervisors before being released to production.

Working drawings include

- an assembly drawing that shows how the parts assemble together;
- part drawings showing all the information needed to manufacture each nonstock part;
- a parts list indicating all the parts used for the assembly; and
- specifications or written instructions for any additional details required in manufacturing and assembling the design.

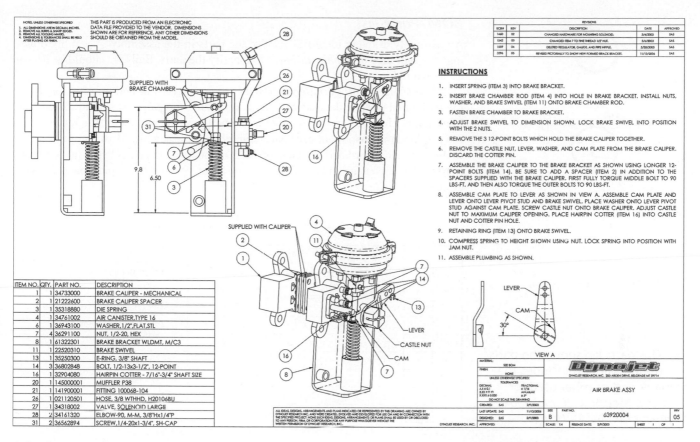

9.3 (*continued*)

Assembly Drawings

Assembly drawings show how the parts in an assembly fit together. A good assembly drawing shows the parts clearly so that a worker who has the parts can assemble them correctly. A single drawing view may be enough. Figure 9.3 shows orthographic and isometric drawing views used for an assembly drawing.

To create an assembly drawing, first create an assembly model by placing all the parts in their assembled locations. Then, use the fully assembled model to create a view that shows the parts clearly.

Section views are often used for assembly drawings because they can show the interior details of how parts fit together (Figure 9.4). To create a section view from an assembly model, use your CAD system's commands to select a cutting plane through the object (typically through the center). Most CAD systems automatically generate correctly shown hatch patterns or hatching boundaries for the section views. Some programs have sophisticated selections for the types of sections you can define for the model.

If the fully assembled model does not show all the parts clearly, another option is to produce an exploded isometric view. To create an exploded isometric view, move the assembled parts along the X-, Y-, or Z-axis directions so that they stay in alignment with their assembled positions. Move each part far enough that it does not overlap other parts. Use this model to generate the assembly drawing. Some CAD system commands generate exploded assembly views, making it even easier to create assembly drawings.

9.4 This rendered assembly model uses a 3D cross section to show the inner workings of the Atomic Aquatics first-stage titanium regulator. An assembly drawing of this model would include section views to show how the interior parts fit together. (*Courtesy of Leo Greene, www.e-Cognition.net.*)

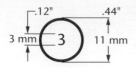

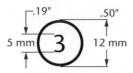

9.5   Dimensions and Sizes for Ball Tags

BOM/Parts List

An assembly drawing identifies each of the parts in the assembly. Assembly drawings use **ball tags** or **bubble numbers** to identify the parts. Ball tags should be drawn as shown in Figure 9.5.

A parts list or **bill of materials** (BOM) provides a list of and information about the individual parts in the assembly. At a minimum, the parts list should include the item number for the part (the number in the ball tag), the part name, the material for the part, and the quantity of each part required in the assembly. Most companies also include their standard part numbers or stock numbers on the parts list to make it easy to track the purchasing and handling of parts. Figure 9.6 shows a typical parts list with company part numbers.

9.6   This parts list shows the parts by item number and includes the company part number used for inventory. *(Courtesy of Implemax Equipment Co., Inc.)*

Item	QTY.:	Part Name:	Mat./Length	Part No.:
1	1	OUTER TONG SIDE PLATE	3/16" MS	6042R.007
2	1	INNER TONG SIDE PLATE	3/16" MS	6042R.008
3	1	TOP TONG PLATE	3/16" MS	6042R.012
4	1	BOTTOM TONG PLATE	3/16" MS	6042R.013
5	1	TONG HYDRAULIC BASE	1" MS	6042R.018
6	2	INTERNAL TONG SPACER	3/16" MS	6042R.031
7	1	3/16" X 1 3/16" B.S.	5"	6042R.458
8	1	3/16" X 1 11/16" B.S.	5"	6042R.033
9	1	2 3/4" x 3/8" WALL DOM	10 11/16	6042R.053
		W/ 1/4-28 ZERK		

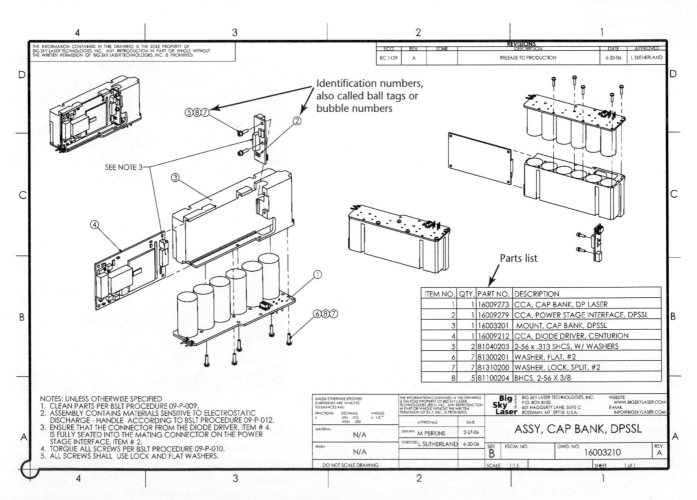

9.7   This exploded assembly drawing show the parts disassembled along major axis lines. *(Courtesy of Quantel USA.)*

CAD software often generates parts lists automatically. If the information describing the part and its material has been stored with the individual part files, this information can be extracted from the assembly model and displayed as the parts list. If a part changes, the assembly drawing and parts list are updated automatically (see Figure 9.7).

For many assemblies, an exterior orthographic view, exploded orthographic view, or exploded isometric view may be sufficient to show how the individual parts assemble (see Figures 9.8–9.11).

9.8 This assembly drawing uses a single orthographic view. Notice that it shows assembly dimensions. *(Copyright 1998 Tektronix, Inc. All rights reserved. Reproduced by permission.)*

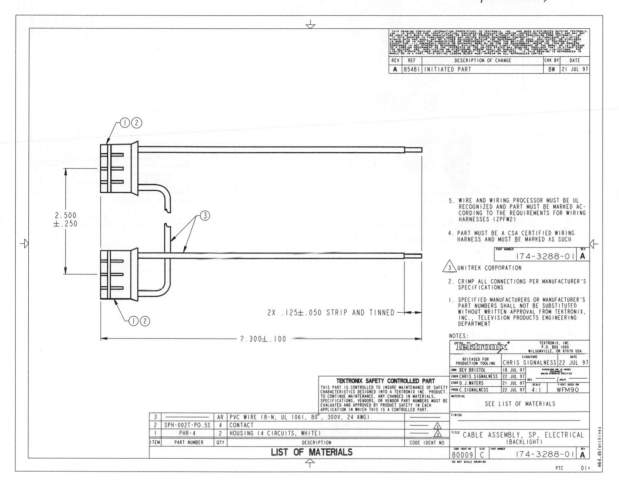

9.9 Exploded assembly drawings move parts along their centerlines, as shown in this drawing that uses a single orthographic view.

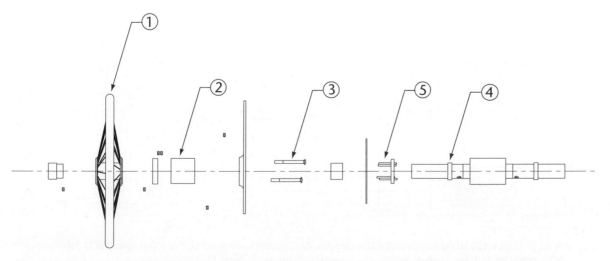

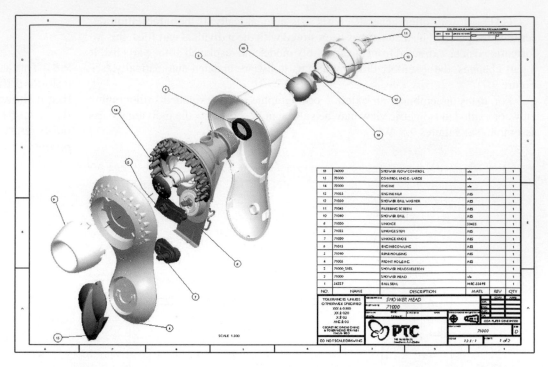

9.10 A Fully Rendered Exploded Isometric View of the Full Assembly *(Courtesy of Parametric Technologies Corporation.)*

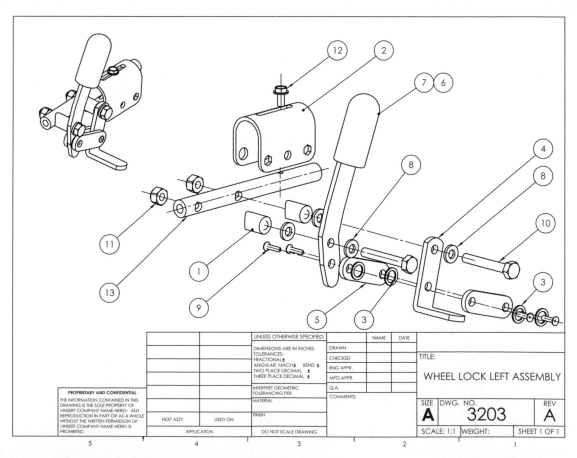

9.11 This exploded assembly includes an assembled view to show how the parts fit. *(Courtesy of Salient Technologies, Inc., www.salient-tech.com.)*

Do Not Show Dimensions

Assembly drawings do not typically show dimensions unless a critical distance needs to be maintained during assembly. For example, at the time of assembly, the wires for the two connectors shown in Figure 9.8 must be made 2.300 long with .125 inch of the end stripped and tinned. These dimensions are shown in the assembly, but not others.

Part Drawings

Part drawings, also called *detail drawings*, are prepared for parts manufactured or modified for the project. They are not necessary for standard or stock parts. Do not waste time creating detailed part drawings for nuts, bolts, bearings, pumps, and other items you will purchase and use as is.

Part drawings usually show one part per drawing sheet, not multiple parts, even if the part is small (Figure 9.12). This way, each part number corresponds to a single drawing that has all the information for that part. If the same part is used in another assembly, the drawing can be used without causing confusion. For very small projects, occasionally more than one part drawing will be shown per sheet, but in general it is not a good practice.

Part drawings include the following:

- drawing views
- dimensions
- tolerances
- material designation
- finishes
- notes
- title and revision block

Figures 9.13–9.15 illustrate part drawings for VKI Technologics' Zuma brewer. Individual part drawings were prepared for 41 of the parts; 7 of the parts were standard parts that did not require part drawings. Only three of the part drawings are shown here; they illustrate detailed drawings for sheet metal, machined, and plastic parts.

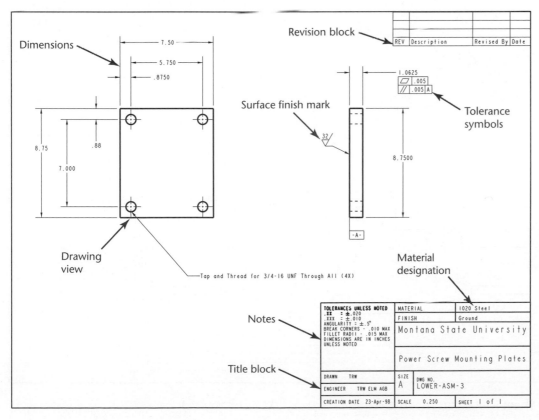

9.12 Part Drawing

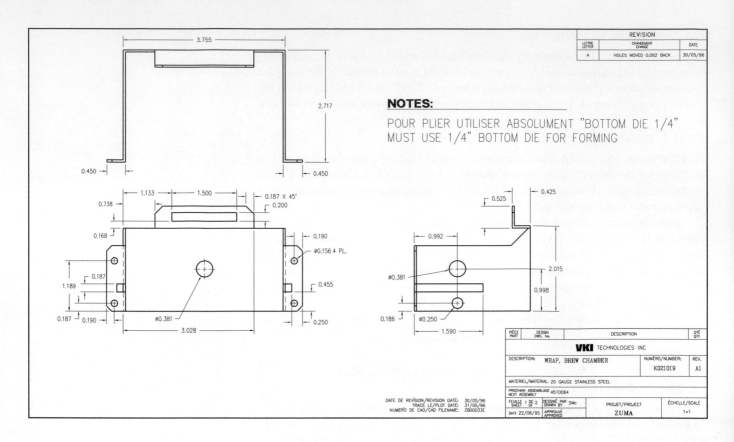

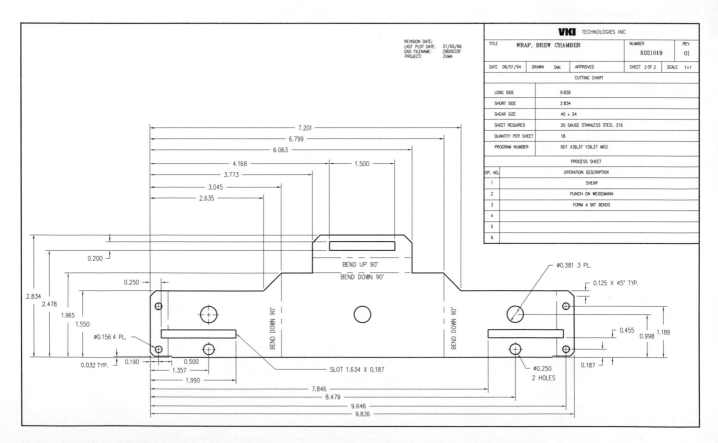

9.13   Part Drawing for a Sheet Metal Part *(Courtesy of Stan McLean, VKI Technologies.)*

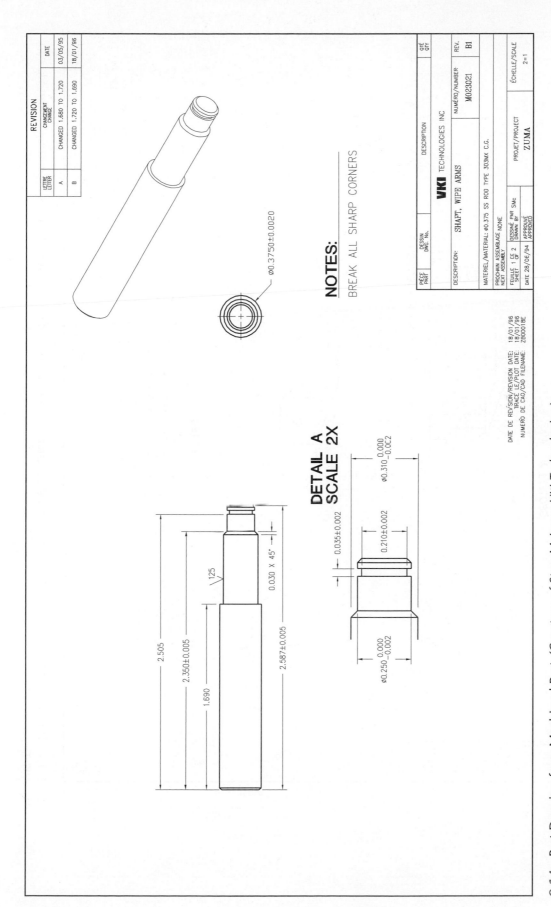

9.14   Part Drawing for a Machined Part (*Courtesy of Stan McLean, VKI Technologies.*)

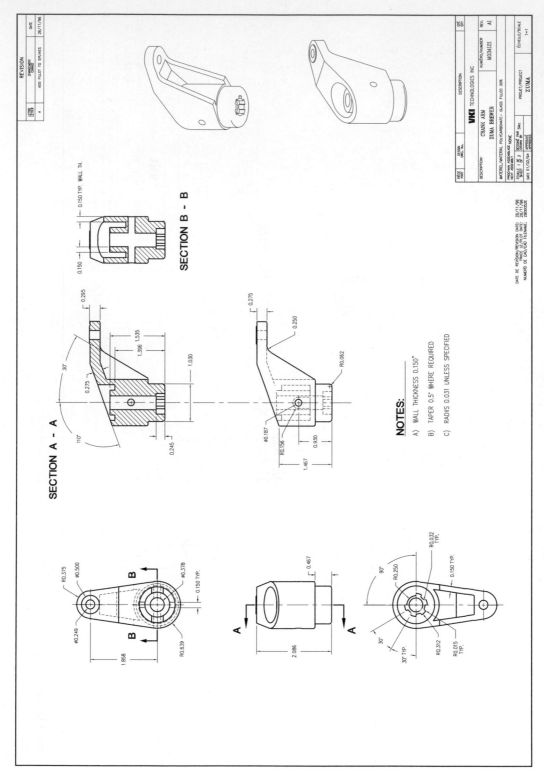

9.15 This part drawing for a molded part includes section views to better define the interior shapes. (*Courtesy of Stan McLean, VKI Technologies.*)

Preparing Working Drawings

Each working drawing is prepared in a similar way:

1. Plan the views needed and determine the drawing size.
2. Determine the scale for the views.
3. Locate the views on the sheet.
4. Add the title, revision blocks, notes, and other standard elements.
5. Dimension and tolerance the views.

If you keep in mind that your intent for the drawing is to communicate, you will be able to judge whether your drawing is good. Ask yourself, Would I be able to make this part from the information shown on the drawing? If not, revise it until the information is clearly and unambiguously presented.

9.4 STANDARD SHEET SIZES

Many companies standardize on a single size sheet for all plotted drawings, or on a few standard sheet sizes. This eliminates the need to stock a large range of paper sizes and makes it easier to work with printed sets of drawings, reducing the risk that smaller sheets will be lost in a set of larger drawing sheets. Tables 9.2 and 9.3 list the ANSI and ISO standards for drawing sheet sizes, which are also illustrated in Figures 9.16 and 9.17.

Table 9.2 ANSI Standard Sheet Sizes (for Inch Sheets).

Designation	Sheet Size in Inches
A	11 × 8.5
B	17 × 11
C	22 × 17
D	34 × 22
E	44 × 34

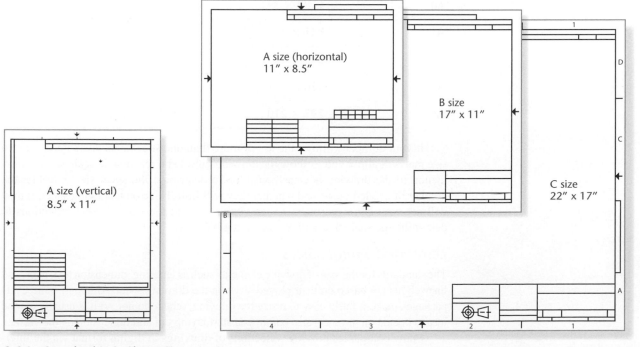

9.16 Standard U.S. Sheet Sizes A–C

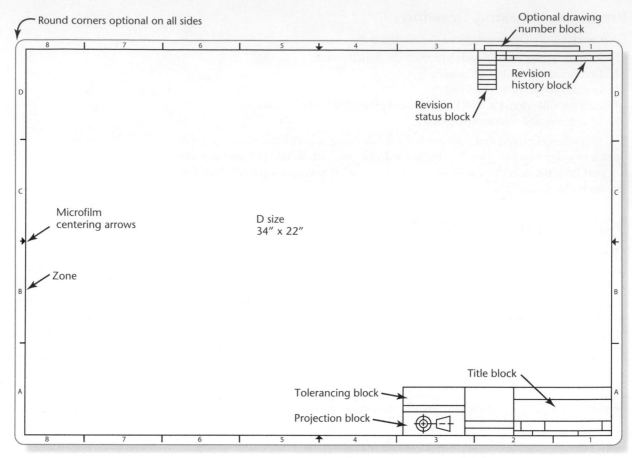

9.17 Standard U.S. Sheet Size D

Table 9.3 ISO Standard Sheet Sizes.

Designation	Metric Sheet Size (in mm)	English Equivalent Sheet (in inches)
A0	1189 × 841	44 × 34(E)
A1	841 × 594	34 × 22(D)
A2	594 × 420	22 × 17(C)
A3	420 × 297	17 × 11(B)
A4	297 × 210	11 × 8.5(A)

If your company allows the use of more than one sheet size, you should select a size that will allow you to show the drawing views at a reasonable scale so the information in the drawing is clearly visible. To determine the sheet size, consider the number of needed views and whether they will fit full size on a drawing sheet. If they will not, you can experiment to see how a view appears at different reductions to determine the scale that will produce a legible view.

Letter and Symbol Sizes

The standards for the size of drawing elements such as lettering, dimension text, gaps, and lineweights are based on their plotted sizes on the drawing sheet. Although some CAD packages make it fairly easy to resize the drawing, others do not, so it is usually a good idea to decide on the sheet size before you start laying out and dimensioning the views.

Many companies provide collections of drawings in digital format via the Internet or on digital media. Printed drawings are becoming less common. Nevertheless, the 2D views in the digital drawing are arranged on standard sheets so they may be viewed and printed easily, with text, dimensions, and drawing details visible at the printed size. Computers with small screens allow you to use zoom commands to enlarge an area of interest to see the details.

9.5 TITLE AND REVISION BLOCKS

Every drawing needs a ***title block***. When changes are made to the released drawing, a ***revision block*** is also necessary. A title block organizes information needed to identify the drawing. Most companies have a standard format for title and revision blocks. Figure 9.18 shows the dimensions and typical information for an ANSI-preferred title block. A variety of sizes and formats of title blocks for standard sheet sizes are shown in Appendix 4.

Companies make the standard title blocks available in a CAD library so they can quickly be inserted into a drawing. This eliminates re-creating the elements of the title block each time and ensures consistency among drawings. Many software packages allow you to define fields in the title block so that the information in those fields, such as part number or designer name, can be extracted into a drawing tracking system or linked to other documents.

A title block generally includes the following information:

- *Name.* Show the originating company or business (and address if desired).
- *Drawing Title.* Briefly describe the item using a noun or noun phrase and if necessary to distinguish it from similar items, adjectives. The title should be clear but brief.
- *Drawing Number.* Give each drawing a unique number, using the company's numbering system.
- *Revision Number.* Track the drawing version using the number of the revision. The original release of the drawing typically shows revision 0.
- *Approval Block.* List the name(s) of the person(s) approving the drawing and the date it was approved. Additional areas of this block can be used for approvals needed at different design stages. For example, a company may use separate areas for structural design or manufacturing engineering approvals.

9.18 Title Block for A-, B-, and C-Size Sheets. Some title block formats may include the revision block; others place it separately in another location on the drawing.

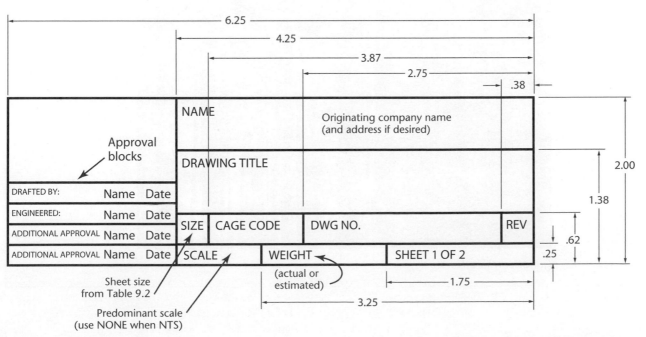

- *Scale.* List the predominant scale for the drawing. If the drawing is not made to a particular scale, note NONE in the scale area. Drawings may include details at other scales, but those scales should be noted below the detail itself.
- *Drawing Size.* List the sheet size used for the drawing. This helps track the original size when the drawing is reproduced at a smaller size.
- *Sheet Number.* List the number of the sheet in the set, using whole numbers starting at 1. A format that lists this sheet out of the total number helps keep track of the entire set, for example, 1 OF 2.
- *CAGE Code.* List the Commercial and Government Entity (CAGE) code if applicable. This is a number assigned to entities that manufacture items for the government. The code is assigned based on the original design activity.
- *Weight.* List the actual or estimated weight of the part if required

Additional information such as the following may also be included in the title block:

- material
- general tolerance
- heat treatment
- finish
- hardness
- superseded drawing numbers

Paper Conservation

According to Worldwatch Institute, 40% of the trees harvested worldwide are used to make paper. The U.S. Environmental Protection Agency (EPA) estimates that paper makes up 38% of municipal solid waste. In 2006 yearly paper production was estimated at 1,250 sheets of paper per inhabitant of the world. The good news is that efforts in green manufacturing have reduced carbon dioxide and sulfur dioxide emissions associated with the production of paper. The bad news is that even with digital data storage, paper consumption has not seen any decreases. Websites related to paper conservation include the following:

http://www2.sims.berkeley.edu/research/projects/
    how-much-info-2003/print.htm
http://www.lesk.com/mlesk/ksg97/ksg.html

(janet carr\Shutterstock.)

ECR#	REV	DESCRIPTION	DATE	APPROVED
		REVISIONS		
1480	02	INCREASED WIDTH OF KEY SLOTS TO EASE KEY INSTALLATION.	4/1/2003	SAS
	02	ADDED GDT TO CENTER KEYWAYS ON THE SHAFT.		
1813	03	ADDED TOLERANCE TO Ø1.497.	1/13/2004	SAS
3560	04	ADDED GTD CONCENTRICITY TOLERANCE TO Ø1.497	2/1/2007	JE
	05	ADDED REQUIREMENT FOR CENTER DRILLING ONE END OF SHAFT.	5/2/2007	SAS

9.19 This typical revision block describes changes and indicates the revision in which they were made. *(Courtesy of Dynojet Research, Inc.)*

A ***revision block*** is used to document changes made to the drawing. A typical revision block is shown in Figure 9.19. To call attention to the revised area, a revision symbol is added to the drawing near the area that has been changed, as shown in Figure 9.20.

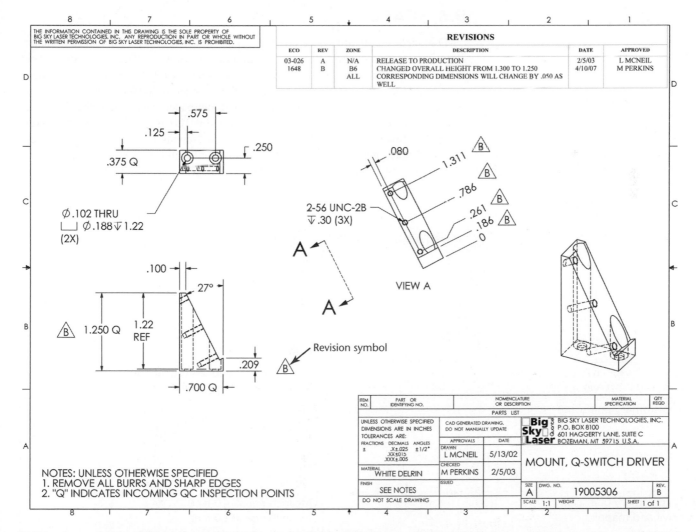

9.20 A symbol matching the item in the revision block indicates the revised feature. *(Courtesy of Quantel USA.)*

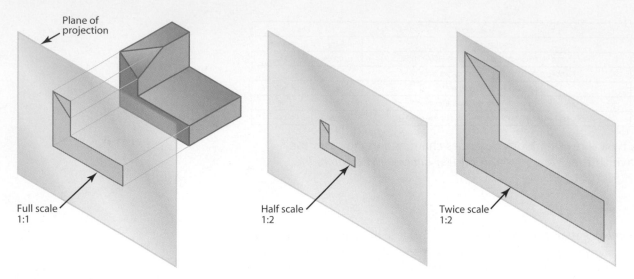

9.21 Full size, half-size, and twice-size projections are illustrated here.

9.6 SCALE

CAD models should always be created full size to be useful for analysis and manufacturing. However, when 2D views are plotted on a drawing sheet, the part may not fit on the paper full size. Drawing views are shown to *scale* on the sheet. The **scale** is the ratio of the full size object to the size of the enlarged or reduced view on the sheet (Figure 9.21). Common scales are used so that the reader can relate to the size of the actual object. It is easier to relate to an object that is 100 or 200 times larger than one that is 157 times larger. The scale for the plotted drawing is noted in the title block. The scale note gives the relationship between plotted units to real world units.

Some typical scales are listed in Table 9.4. Also see Appendix 3.

> Scales are indicated in the following ways:
> - SCALE: 1:2
> - SCALE: 1/2
> - SCALE: .5

Table 9.4 Typical Scales.

Scale	Note Format	Description
Metric scales	1:1 1:2 1:5 1:10 1:20 1:100 2:1	Full Half Fifth Tenth Twentieth Hundredth Double
Metric scales for civil and mapping (*the units are typically meters*)	1:100 1:500	One plotted unit equals one hundred real-world units One plotted unit equals five hundred real-world units
U.S. customary unit scales	1:1 1:2 1:4 1:8 1:10 2:1	Full Half Quarter Eighth Tenth Double
U.S. customary unit scales for civil and mapping	1" = 10' 1" = 20' 1" = 50' 1" = 1000' 1:24000 or 1" = 2000' 1:63360 or 1" = 1 mile	One inch equals 10 feet One inch equals 20 feet One inch equals 50 feet One inch equals 1000 feet One inch equals 2000 feet One inch equals 1 mile

9.7 DEVELOPING VIEWS FROM 3D MODELS

For many constraint-based packages, creating orthographic views from the model is simple. Two basic steps are required: selecting a location on the page where the orthographic view will be placed and selecting the viewing direction. Adding more orthographic views is even easier, as they are automatically placed in relation to the initial view. The software automatically orients and aligns the views around the central view in the drawing. The same basic approach allows the easy creation of section, isometric, and auxiliary views.

Even if your CAD package does not automate the creation of orthographic drawing views to this extent, creating orthographic views directly from the 3D model still offers an advantage. Even people trained in orthographic projection of views have difficulty visualizing complex orthographic views correctly all the time—and the views can be time-consuming to create. Using the 3D object to develop the views helps get the projection right.

When you are selecting which views to show, keep in mind the same practices that make a good sketch:

- Show the shape of the object clearly in the front view, which is placed in a central location on the drawing sheet (see Figure 9.22).

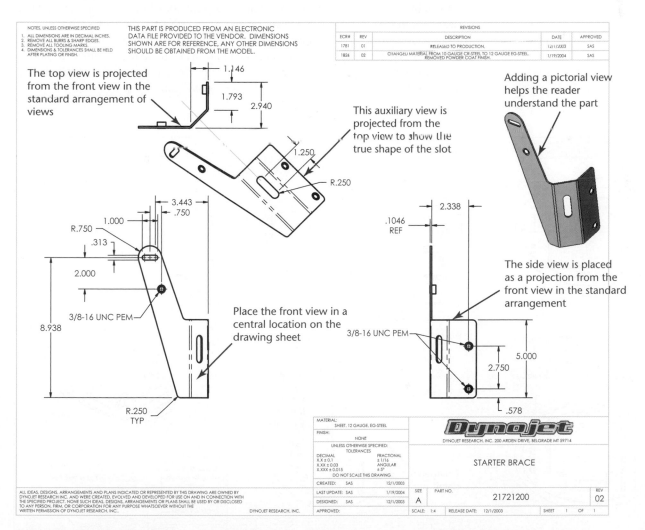

9.22   Orthographic Views and Shaded Pictorial View Generated from the Part Model *(Courtesy of Dynojet Research, Inc.)*

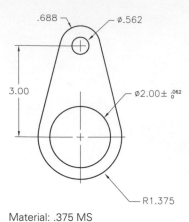

Material: .375 MS

9.23   This object is fully defined with a single view and a note. *(Courtesy of Implemax Equipment Co., Inc.)*

- For objects with uniform thickness, use a single view and specify the thickness in a note (see Figure 9.23).
- For complicated parts, show at least two drawing views in the standard arrangement to make it easy to interpret the views. The reader is not as familiar with the shape of the part as you are. Place additional views on a second drawing sheet if necessary; it is preferred to have all the views on one sheet if they are clear. When multiple sheets are used, indicate this in the title block with a note such as "Sheet 1 of 3."
- Make the drawing clear and easy to interpret. Use different line thicknesses to make the object stand out clearly.
- Show the object at a size so that its features are visible, or include an enlarged detail view at a larger scale (see Figure 9.24).
- Except for enlarged details, show all views at the same scale.
- Clearly note the difference in scale for enlarged detail views.
- Show only as many views as necessary. Do not add unnecessary views. They confuse the reader. You can simplify views by leaving out hidden lines that do not add to the interpretation of the view. If you leave out lines, it is a good idea to label the view as a partial view.

Placing the Views

As you plan the views to be included, allow for plenty of white space on the drawing. Leave room for dimensions between the views, and separate the views from one another with white space. The space between drawing views does not have to be equal, yet the views should appear to be related. Do not space them too far apart. If there is not enough white space in your drawing, or if the views are too small to see the details clearly, consider using a larger sheet size (see Figure 9.25).

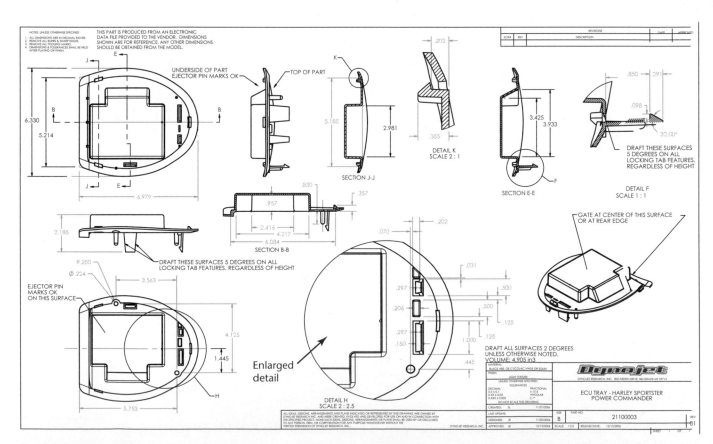

9.24   Drawing of an Injection-Molded Part with Sections and Enlarged Details *(Courtesy of Dynojet Research, Inc.)*

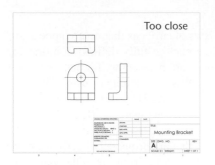
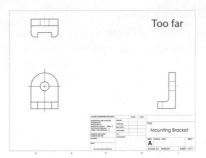
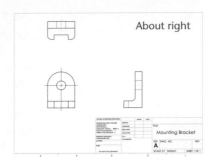

9.25   Spacing Drawing Views

Isometric Views

Including isometric views in detail drawings helps others easily interpret the drawing. Isometric views are often shown in the upper right-hand area of the drawing, as there is often room there. The isometric view does not have to be at the same scale as the other views. Often, a smaller scale is just as clear and fits on the sheet better. It is not necessary to indicate the scale of the isometric view on the drawing. Figure 9.26 shows an example of an isometric view added to a detail drawing. Remember that hidden lines are not usually shown in isometric views. Most modern CAD packages have preset isometric viewing directions that you can select for the model.

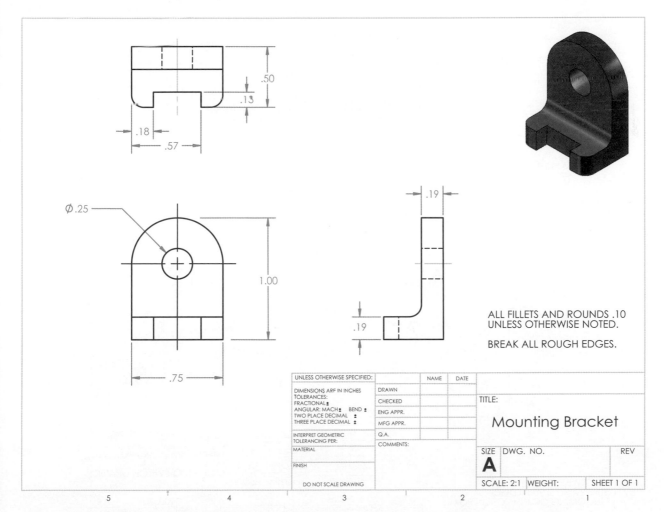

9.26   The isometric view provides an easy visual reference for the part described in the orthographic views.

Auxiliary Views

Auxiliary views are used to create true-size views of surfaces that do not appear true size in a standard orthographic view. A surface may be fully defined in a drawing without appearing true size, but a true-size view is often necessary for pattern layouts and to show features true size for dimensioning. When sketching or using 2D CAD, you project an auxiliary view from existing views. Using 3D CAD, you can generate a true-size view by aligning the viewing plane with the desired surface. This is the same as selecting a direction of sight for the view that is perpendicular to the surface.

Primary auxiliary views are auxiliary views that are projected directly from one of the standard orthographic views, such as the top, front, or side view (see Figure 9.27). ***Secondary auxiliary views*** are auxiliary views that are projected from a primary auxiliary view.

Auxiliary views should project directly from a drawing view when possible. The auxiliary view should be shown in its proper orientation to the view from which it is projected.

Auxiliary views are frequently used to show features true size for dimensioning (see Figure 9.28). If the auxiliary view does not fit on the drawing sheet in projection, then a viewing-plane line (representing the edge view of the auxiliary viewing plane—similar to a cutting-plane line for section views) is used to indicate the viewing direction. The rotation of the auxiliary view should be preserved in its projected orientation even when it is placed elsewhere in the drawing. Figure 9.29(a) shows a viewing-plane line used to indicate the direction of sight for an auxiliary view. The arrows on the viewing-plane line show the direction of sight.

9.27 View A in the drawing below is a primary auxiliary view of an inclined surface. *(Courtesy of Quantel USA.)*

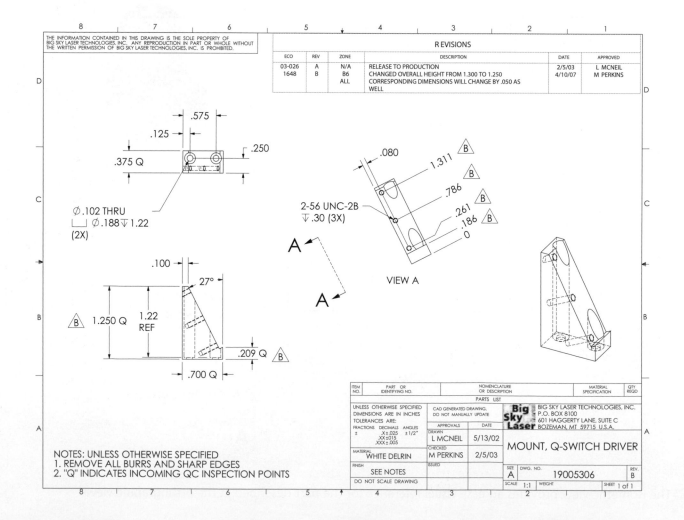

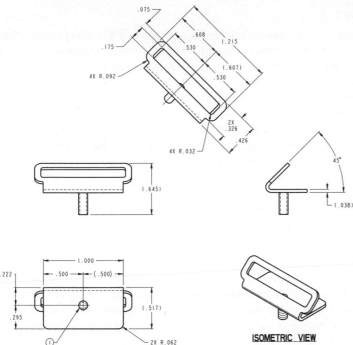

ISOMETRIC VIEW

9.28 An auxiliary view for an inclined surface. *(Copyright 1998 Tektronix, Inc. All rights reserved. Reproduced by permission.)*

Drawing a second auxiliary view in 2D required drawing the primary auxiliary view first. Now, with 3D CAD, secondary auxiliary views can be created and placed directly without the need for a primary auxiliary view. This practice leaves out the view where a viewing-plane line or direct projection could be used to show the relationship. In these cases, draw a line perpendicular to the indicated surface in the orthographic view where the surface is shown foreshortened and label it as the viewing direction for the auxiliary view. A viewing-direction arrow is shown in Figure 9.29(c).

Viewing-plane line

A   B

VIEW A-A   VIEW B-B

(a)

Centerline relates views

(b)

Viewing-direction arrow

A

View A-A

9.29 (a) A viewing plane line shows the direction of sight for the auxiliary view. (b) A centerline can be extended to indicate the view relationship when the auxiliary view is shown in direct projection. (c) A viewing plane arrow is used to show the direction of sight for this removed auxiliary view. A second auxiliary view would be required to show the view in projection.

(c)

Section Views

Any number of section views may be included in a drawing. When many interior features are shown using hidden lines, the view may be difficult to interpret. A section view shows interior parts clearly and makes the drawing easier to read. You can use a section view to dimension interior details where the features show clearly. Section views are useful for parts that have interior shapes.

Each section view is related to one of the orthographic views in the drawing by a cutting-plane line. Each section view is identified by the letters on the cutting-plane line. Sections may either replace standard orthographic views or be located anywhere on the sheet (or a separate sheet) arranged in a logical order (usually alphabetically by the letters identifying the cutting-plane lines).

3D CAD software makes it easy to create section views quickly. Often, you need only to select where to place the view on the sheet and to select a feature that can be used to define the cutting plane.

Hatching

Hatching, also called *cross-hatching* or *section lining*, is used to indicate the solid portions of the object that were cut through when producing the section view. Hatching should be plotted with a thin black line. The typical spacing for hatch lines is 1/16 to 1/8 inch apart. Hatching should always run in the same direction on any single part.

To show different parts in a section view of an assembly, use different hatch angles and/or different hatch patterns for each part. Thin parts, such as gaskets, can be filled in solid when they are too small to show hatching effectively. Do not hatch parts that have no interior detail in assembly section views, such as solid rods, bolts, and screws. Parts like these are shown "in the round," meaning that you show them sitting in the assembly as though they were not cut through by the cutting plane. Figure 9.30 shows an example of hatching practices in an assembly section.

Ribs, webs, gear teeth, and other thin flat parts are not sectioned when the cutting plane passes lengthwise through them because it would give a false impression of the part's thickness. Leave the hatching off these features, as shown in Figure 9.31.

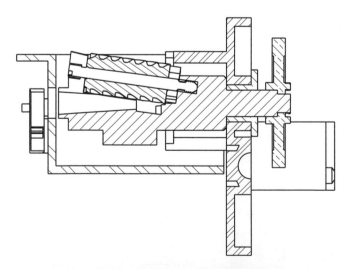

9.30 Use a different angle for the hatch pattern to indicate different parts in an assembly section. Plot hatching with thin, black lines. Thin, flat parts can be filled in solidly. Do not hatch parts that have no interior detail; show them "in the round."

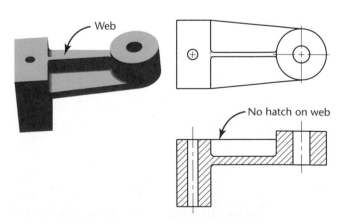

9.31 Leave the hatching off ribs or webs in the section view to show the part's thickness correctly. If the rib were hatched, the object would appear solid.

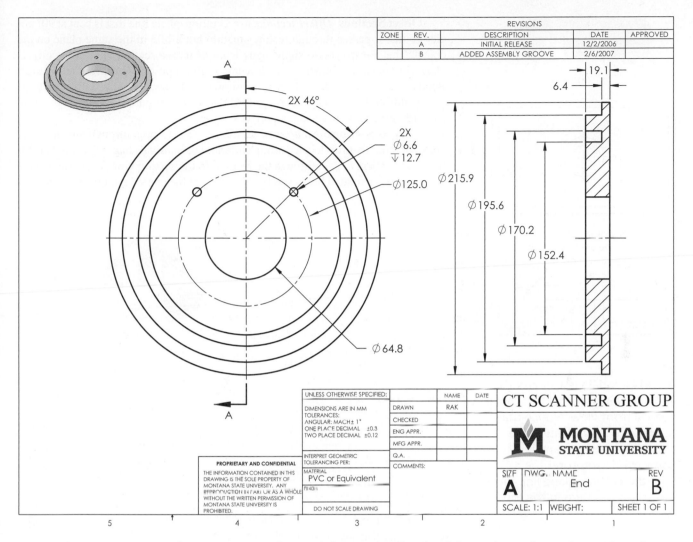

9.32 The section view makes it easy to understand the interior details of the end cap shown here. Note that the isometric view makes it easy to relate the views to the part as a whole. *(Courtesy of Robert Kincaid.)*

Types of Sections

A number of different types of section views can be used to show interior detail.

Full Sections A *full section* shows the part cut entirely through, typically along the center plane (see Figure 9.32).

Half Sections A *half section* shows the object as though a quarter of it were removed, so that half of the resulting view is shown in section and the other half shows the exterior view of the objects. Figure 9.33 shows a pictorial and orthographic half section. The cutting-plane line indicates the direction of sight and the portion of the object that was removed to produce the section. Notice that the two halves of the half-section view are divided by a centerline pattern. Also, hidden lines are typically not shown in section views, including half-section views. Of course, you would use this type of view only when the part being shown in section is symmetrical.

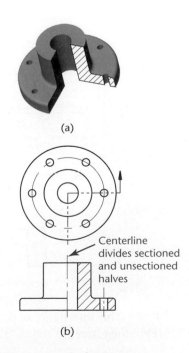

(a)

Centerline
divides sectioned
and unsectioned
halves

(b)

9.33 (a) Pictorial view of a half section; (b) this front view is shown as a half section. Notice the cutting-plane line in the top view.

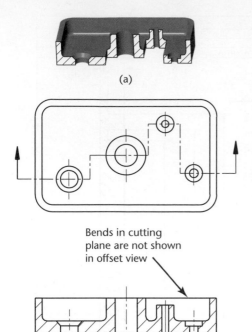

(a)

(b)

9.34 (a) Pictorial view of an offset section; (b) this offset section replaces the typical front view. Notice that there are no lines indicating the bends in the cutting-plane line in the section view. To interpret the offset section, the cutting-plane line must be shown, as it is here in the top view.

Bends in cutting plane are not shown in offset view

Offset Sections *Offset sections* use a cutting-plane line that is bent at 90° angles so that it passes through features that do not all lie in the same plane on the object. This allows you to show many features in one section view. The cutting-plane line must be shown in the drawing so that you can interpret the features. Bends in the cutting-plane line are not shown in the section view. Offset sections can be difficult to create with some CAD packages. Figure 9.34 shows pictorial and orthographic views using an offset section.

Aligned Sections An *aligned section* is used to pass through angled arms, holes, or other features located around a central cylindrical shape. The aligned cutting plane is bent to pass through the feature. In the section view, the cutting plane and section are rotated into a single plane, so that the angled feature shows its true

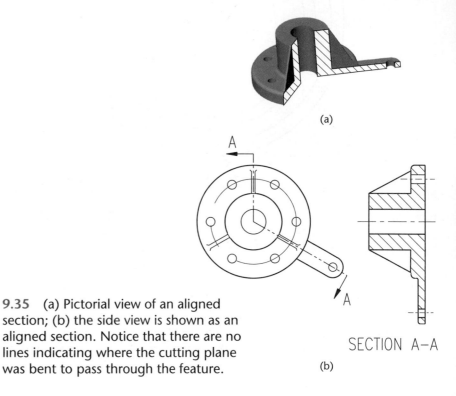

(a)

SECTION A–A

(b)

9.35 (a) Pictorial view of an aligned section; (b) the side view is shown as an aligned section. Notice that there are no lines indicating where the cutting plane was bent to pass through the feature.

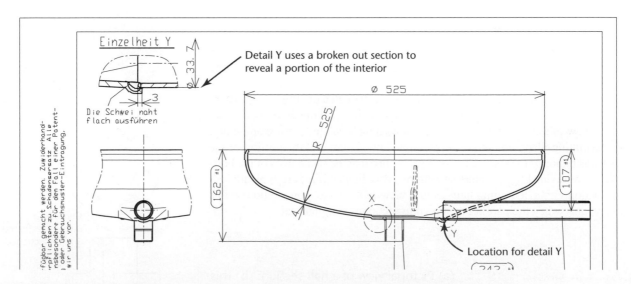

9.36 This international drawing shows enlarged details using broken out sections. (*Einzelheit* means "detail.") (*Viessmann Werke GmbH & Co.*)

size in the section view. Bends in the cutting-plane line are not indicated in the section view. Many CAD packages also have difficulty producing aligned sections. Figure 9.35 shows pictorial and orthographic views of an aligned section.

Broken Out Sections A *broken out section* shows a partial view of the interior of a part or an assembly. A rough-break line in the section view indicates which portions of the surface are broken to reveal the interior. In ANSI standard drawings, the rough break is a thick jagged line. Figure 9.36 shows an international drawing of the bottom base assembly for a boiler. Notice the details X and Y called out in the front view (at the top of the sheet). Detail Y (labeled Einzelheit Y) at left shows a broken out section to reveal details of the interior assembly. Most international drawings use first-angle projection. In first-angle projection, the top view is placed below the front view.

9.8 LINETYPES FOR DRAWING VIEWS

Good use of thick and thin linetypes helps make a readable drawing. The preferred weights for standard engineering linetypes are defined in *ANSI Y14.2M.* (Refer to Figure 3.46 if you need to review standard linetypes.) Object (or visible) lines and cutting-plane, viewing-plane, and short break lines are plotted thick. Hidden lines, centerlines, dimensions, notes, and most other lines are thin. The thin lines do not distract your eye from clearly seeing the shape of the object shown with thick lines. Figure 9.37 shows a good example and a poor example of using different lineweights.

Most CAD systems use templates that control the lineweights for you, but if your system does not do this automatically, check the configuration files and set the lineweights for your drawings.

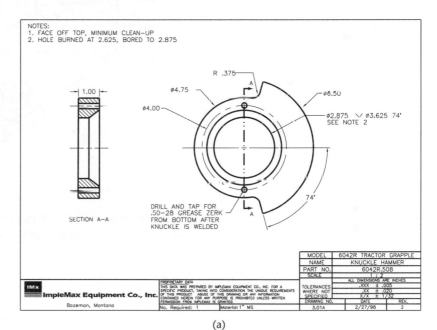

(a)

9.37 (a) Good use of thick and thin lines allows the object lines to stand out. (b) Having all lineweights the same makes it difficult to see the object lines in this drawing. *(Courtesy of Implemax Equipment Co., Inc.)*

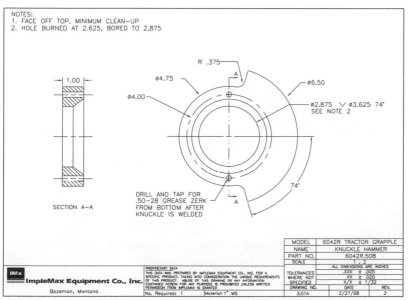

(b)

Phantom Lines

Phantom lines are used to

- show existing equipment,
- indicate portions of a long shape or structure that are not shown, and
- show alternative locations for a part in a single view.

If you are showing how an object will fit with existing equipment, use thin phantom lines for the existing portions and thick object lines for the portion to be built so that it will stand out clearly (see Figure 9.38). Phantom lines can allow you to show part of an object without having to draw it all. In assembly drawings, for example, if an arm can swing to a new location in the assembly, phantom lines can be used to indicate its other extreme location, as shown in Figure 9.39.

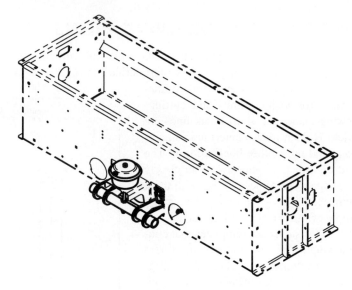

9.38 Phantom lines are used to indicate existing equipment with which the new part needs to work. *(Courtesy of Dynojet Research, Inc.)*

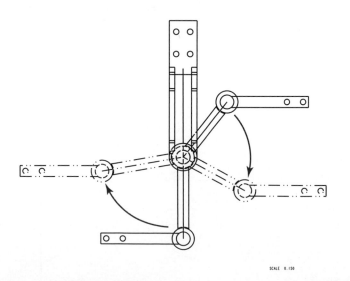

SCALE 0.150

9.39 The phantom lines are used to show the alternate location for the angled arm in the assembly.

9.9 DIMENSIONING

Dimensions are used to indicate the size and location of features on the part. Dimensioning is very important in preparing correct engineering drawings. Once you have clearly shown the drawing views, you must provide the dimensions needed to make the part. Even if a company will manufacture the part directly from the model, it is customary to provide a dimensioned drawing as reference. This drawing may be marked as "reference only" so the manufacturer understands that if there is a discrepancy, he or she is to use the digital model database as the basis for creating the part. In other cases, the drawing may provide only critical dimensions that the manufacturer can use to be sure the digital file has been translated correctly and to resolve any discrepancies between the two. If the part will not be manufactured directly from the model, then it is even more important to provide fully and completely dimensioned drawings, as they are the sole means of communicating the design to the manufacturer.

Dimensioning and tolerancing practices are outlined in *ANSI Y14.5M*. This standard clarifies dimensioning and tolerancing practice so that a drawing cannot be interpreted in more than one way. There are several factors in good dimensioning.

- First is the proper use of dimensioning elements such as extension lines, gaps, text, and arrowheads. Showing them in the proper relationship and at standard sizes is key to their appearance.
- Second is the way the dimensions are placed around the drawing views. Following a few key rules can avoid confusion when the dimensions are read.
- Finally, and perhaps most important, is choosing the dimensions that define the part so they are unambiguous, tolerances can be applied consistently, and the resulting manufactured part will be what the designer intended. To do all these well, you must achieve a balance among the various practices, rules of thumb, and a good understanding of the role the manufacturing technician will play in creating the part.

Units

Dimensioned drawings may use either SI metric standard units or U.S. customary foot/inch units (also called *English units*). Most companies have a standard for the type of drawing units they use. Although the United States is one of a few countries that have not converted to the metric standard, many U.S. companies have adopted metric units for their drawings.

Advantages of Metric Units

Most people agree that the metric system is easier to learn and makes it easier to convert between units. When the Metric Conversion Act was passed by the U.S. Congress in 1975, it encouraged the change to metric units as a voluntary measure. Most companies did not see any advantage to changing their work practices and material sizes, and few companies adopted metric standards for their drawings. When the Omnibus Trade and Competitiveness Act of 1988 designated the SI system as the preferred system for U.S. trade and commerce, it stipulated that federal agencies use the metric standard to the extent economically feasible. It does not require that U.S. industries, states, or other organizations convert to metric, but it may require companies to produce bid documents and plans in metric units for federally funded projects.

Standard Dimension Appearances

Dimensions are drawn between extension lines that relate the dimension to the feature on the part. Dimension lines, dimension values, and extension lines are all thin black lines in the drawing. This way they do not overpower the object outline, they reproduce well, and are easily read. Typical sizes for dimension features are shown in Figure 9.40.

Dimension Lines

Dimension lines are used to show the direction and extent of the dimension. They typically end in arrowheads. The dimension line is usually broken so that the dimension value can be shown in line with the dimension line.

The closest dimension line should be spaced no less than 10 mm (.375 inch) away from the object's outline. Subsequent rows of dimensions should be at least 6 mm (.25 inch) away from the preceding row of dimensions, as shown in Figure 9.41.

Extension Lines

Extension lines extend from the feature that is being dimensioned to 3 mm (.125 inch) beyond the dimension line. The dimension line is drawn perpendicular to the extension lines. Centerlines can be extended beyond the edge of the part to serve as extension lines for dimensions. Extension lines, including extended centerlines, are not broken when they cross the edge of the part.

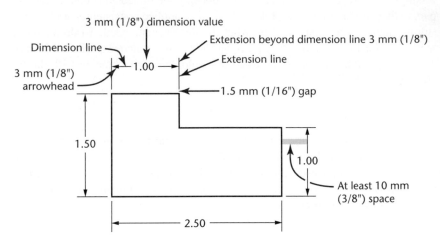

9.40 Typical Sizes for Dimension Features

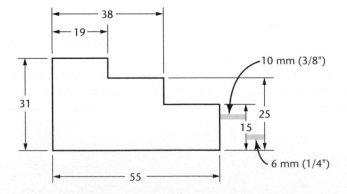

9.41 Dimension lines should be offset consistently from the object outline and from subsequent dimensions.

Dimension Value

The **dimension value** represents the size in units between the extension lines. The dimension value specifies the size of a feature or its distance from another feature on the part.

A simple, single-stroke, Gothic style of lettering should be used to create the dimension values. Most CAD systems provide a font that looks similar to Gothic letter shapes. Many CAD systems allow you to use most of the common font types available in other Windows applications, such as PostScript, TrueType, and OpenType fonts. If you use these, make sure that other people who will open the CAD file have the same fonts available. Otherwise, the software will have to use a substitute font. Substitute fonts are rarely exactly the same size and frequently cause dimensions and drawing notes to overlap other drawing elements. Do not get carried away with the use of exotic fonts. They typically make the drawing difficult to read and can cause problems for other people opening your drawings. You may also find that your ability to control the way the type appears is more limited with exotic fonts than with the fonts built into the CAD package.

Metric Dimension Values The millimeter is the base unit for dimensioning machine drawings in the SI system. Most metric dimensions are given in whole millimeters, unless the precision for the part requires more decimal places to be shown. Whole numbers are not followed by a decimal point and zeros. Dimensions less than 1 mm are written as decimals preceded by a 0. Do not include "mm" or other designations with the dimension values, but make sure that the unit type is clearly labeled in or near the drawing title block. Print the letters "SI" in a larger size than the standard lettering height, for example, or add a note that reads "All measurements in millimeters." Figure 9.42 illustrates millimeter dimensions.

Inch Dimension Values Decimal inches, not fractions, are the base units used for inch dimensioning. Typical inch dimensions are written to two decimal places, unless the precision of the part requires more. The manufacturer will determine what size drill, ream, bore, or other tool is needed to form the hole by reading the value from the drawing along with the stated tolerance. Whole-number dimensions in inch units should typically be followed by a decimal point and two 0s. Values less than 1 are not preceded by a 0. Do not include inch marks (") with the dimension value, but clearly label the units in the title block as inches, or add a note near the title block stating, "All measurements in inches." Figure 9.43 illustrates the preferred form for inch dimensions.

Decimal Points in Values Make sure that the decimal points in dimension values stand out clearly when you plot or print your drawing. They should be dark, uniform, and large enough to stand out and reproduce clearly. In some CAD systems (AutoCAD, for example) you can edit the standard fonts to produce lettering variations. If your decimal points are not easily legible, you may want to edit the font, select a different font, or use a slightly thicker pen width when printing or plotting to make sure that decimals are seen. It is not unheard of for a part or device to be built to 10 times the desired size because a decimal point was not visible on a print.

Orientation

All dimension values and notes should be **unidirectional**; that is, all text is lettered horizontally from left to right so it can be read from the bottom of the sheet, as shown in Figure 9.44(a). Another acceptable, but less common, orientation is **aligned**. When dimensions are aligned, the dimension text is aligned with the dimension line so that it can read from the bottom or right of the sheet, as shown in Figure 9.44(b). Horizontally aligned dimensions are read from the bottom of the sheet, and vertical dimensions from the right side of the sheet (never from the left).

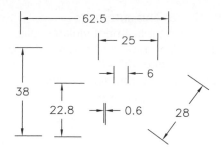

9.42 Dimensions in SI units are typically in whole millimeters or shown to one decimal place. Decimal values less than 1 are preceded by a 0.

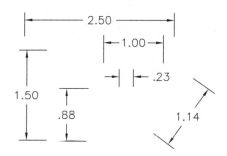

9.43 Inch dimensions are often shown to two decimal places as the default. Decimal values less than 1 are not preceded by a 0.

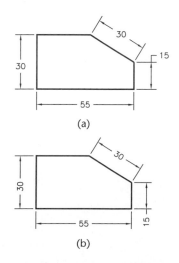

9.44 (a) Unidirectional dimensions are read from the bottom of the sheet. (b) Aligned dimensions are read from the bottom or right side of the sheet.

9.45 (a) This automatic dimension placement violates rules about dimension and extension lines crossing. (b) After repositioning, the dimensions reflect good practice.

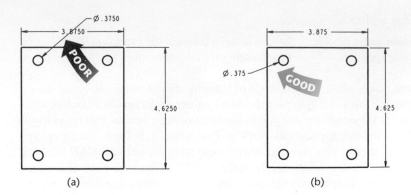

Placement of Dimensions

Placing dimensions around the views requires attention to a few basic rules:

- Dimension lines should not cross one another.
- Extension lines should not be drawn across dimension lines.
- Shorter dimensions should be placed closer to the object to avoid crossing dimension or extension lines with longer dimensions.

These rules of thumb are often poorly observed by the automatic dimensioning algorithms of CAD packages. The drawing in Figure 9.45(a) shows how the dimensions may appear when placed automatically by the software. Tools built into the software allow you to clean up the dimension placement so dimensions can be read clearly and lines do not cross unnecessarily. Figure 9.45(b) shows the same drawing after the dimensions have been adjusted for better placement.

Place dimensions where the feature is shown true size and shows its shape clearly. Because the front view typically shows the shape of the object best, most of the dimensions will be located around the front view. This also makes it easy for the manufacturer to locate the dimensions, as they are mostly grouped together around the central view.

Place dimensions off the view unless placing the dimension on the view makes the drawing easier to read. Placing dimensions off the view makes it easier to see the object lines, hidden lines, and centerlines on the view that define the shape of the features. If placing the dimension off the view has the opposite effect, or if it is in a crowded area, it is better to make an exception. For example, when dimensioning an interior slot on a complex shape where extension lines for the slot cross other features, you may want to place the dimension on the view as shown in Figure 9.46(b).

Show each dimension only one time in the drawing and attached to one view only.

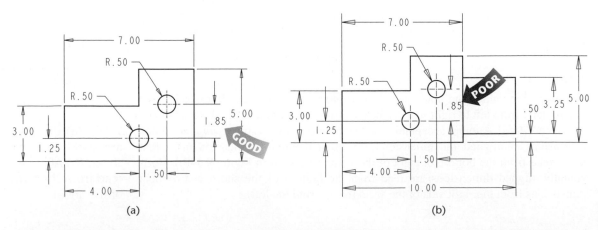

9.46 (a) Place dimensions off the view whenever possible. (b) A dimension may be placed on the view if the alternative would violate other rules of placement.

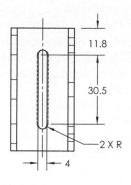

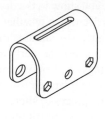

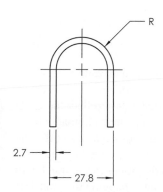

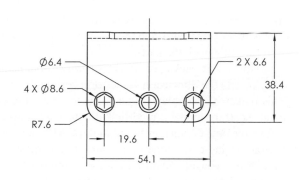

9.47 Show each dimension only once, but choose a location between views that contributes to interpreting the size or location of the feature in both views.

Place dimensions between views when possible to help show that the value is related to the feature as seen in both views. Remember to connect it to only one of the views, as shown in Figure 9.47.

Stagger dimension values so they do not all line up in a neat row, which makes them difficult to read (see Figure 9.48). Avoid using more than three rows of dimensions on one side of a view, which are also difficult to read—but this situation often cannot be avoided.

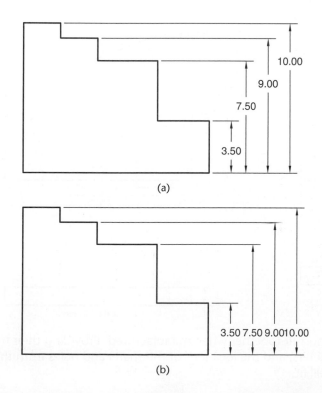

9.48 Staggered dimensions, as shown in (a), are easier to read than those shown in (b).

Choosing the Best Dimensions to Show

The dimensions you show in the drawing are the ones needed to manufacture the part and inspect it to determine whether it meets the specified tolerances for its manufacture. As the designer, you should dimension the drawing so that the manufactured part will perform its function properly in the assembled product.

Some drawings show only critical features and leave the undimensioned features to be created as manufactured from the CAD database, with the tolerances defined as per manufacturing process. Often, these are exterior features, such as the outside of a plastic housing, that do not fit with other parts and are not critical to the function of the assembly.

Mating parts are parts that fit together in the assembly. A ***mating dimension*** is the term for the dimension on mating parts that allows them to fit together when assembled. You must provide mating dimensions or dimension mating features on different parts in a similar way so that the parts fit together when assembled. Figure 9.49 shows two mating parts and their dimensioned drawings. Notice that mating dimensions are shown on the parts.

Most CAD packages offer ***associative dimensioning***, a feature that automatically inserts the dimension value for a selected entity when a dimension is created. This makes it easy to add dimensions to a drawing that accurately reflect the CAD database. If the model is modified, dimension values can be updated automatically to reflect the new dimension stored in the CAD database. Most packages also offer the option to hand-enter dimension values that override the value provided by the database. When you are dimensioning a drawing from the model, it may seem quicker to change a dimension on the drawing instead of going back and changing the model. If you do, however, the CAD database will not reflect the sizes and locations on your drawing and will not be useful for computer-controlled manufacturing.

When constraint-based modeling is the basis for your engineering drawing, choosing the best dimensions to show can be easy. Because parametric design encourages you to give a lot of thought initially to the relationships in the design and how to size the various features, all the part features will already be located in

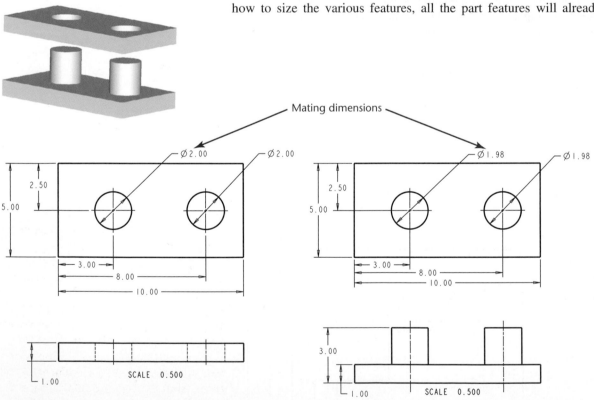

9.49 The two mating parts shown must fit together when manufactured. Providing their mating dimensions—the shaft and hole diameters, and the locations of the shafts and holes from the edge of the part—helps ensure that they will fit.

relation to the existing base feature or to other features on the part. It is a common mistake when dimensioning drawings by hand to remember to give the size of a feature but to forget to locate it on the part. If you captured the design intent in the model, the dimensions you used to create it will generally be the same ones that you should show in the drawing. If they are not, you may want to reconsider your design database and ask if it really reflects your design intent.

The bidirectional associativity of the parametric modeler makes it possible to change dimensions on the model or in the drawing and store the changes in the CAD database. In addition, most parametric modeling software offers the ability to add additional dimensions to a drawing. These dimensions may be updated as an associative dimension will be, but changing them will not change the database. This feature allows you to add dimensions to the drawing that were not used to create the model. Overuse of this option in the CAD package probably means that your parametric model does not adequately reflect your design intent. Traditionally, *reference dimensions* are used to add information to a drawing but are identified with REF to signify that they should not be used to manufacture the part.

Examine Figure 9.50(a) and (b). Notice that the drawing contains dimensions that were not used to create the part model shown in (b). The highlighted dimensions were added in drawing mode but would have been generated automatically if the part had been created with the dimensions shown in (a).

When you are dimensioning a drawing, keep the manufacturing process in mind. Remember that if you can size features to be manufactured with standard tools and stock sizes, the parts will cost less. You do not need to indicate the manufacturing process that will be used to form a hole of the size you have indicated within the specified tolerance range, but you should be aware of standard sizes and the tolerances possible with various methods. The manufacturer will be the best judge of how to make the part cost-effectively.

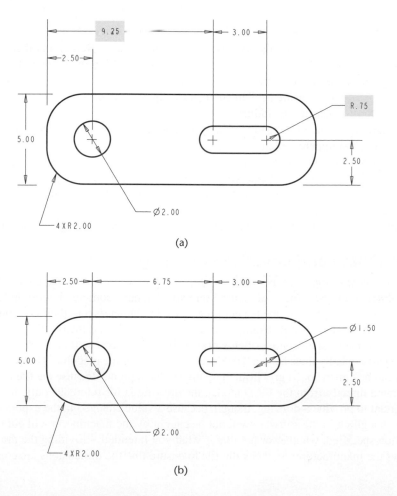

(a)

(b)

9.50 (a) This drawing shows a part that required the addition of placed (or driven) dimensions for the part to be produced as the designer intended. These placed dimensions are highlighted. (b) The driving dimensions for the model show that the slot was dimensioned from the location of the left hole. This dimension had to be overridden in the drawing so that the feature would be located from the left edge of the part. In addition, the R.75 dimension was added to replace the diameter dimension used to model the part. If the model had been created with the dimensions shown in (a), the drawing would not need the placed (driven) dimensions.

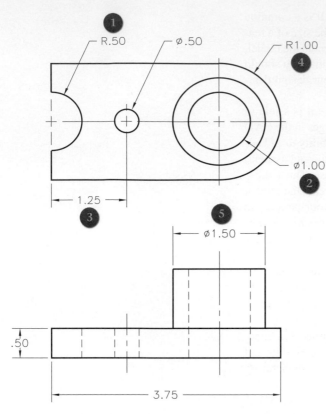

1. Give radius of arcs.
2. Give diameter of circles.
3. Locate holes where you see their circular shapes.
4. Fully rounded ends are self-locating.
5. Dimension the diameters of cylinders where they appear rectangular.

9.51 Dimensioning Circles and Arcs

Dimensioning Arcs and Circles

Specify the diameter of circular shapes and the radius of arcs. Remember that the dimensions you provide are for the manufacture and inspection of the part. Tools for forming holes are sized by their diameter, not radius. Also, once a hole has been drilled, it is easy to measure its diameter using calipers, or even a machinist scale, but it is not easy to measure the radius.

Diameter dimensions are preceded by the diameter symbol: Ø. Radius values are preceded by the uppercase letter R.

Remember to give location dimensions for circular shapes, such as holes, or pins. Most circular shapes are located with a dimension from the center point to another feature. Many radius shapes are self-locating if the entire end of a feature is rounded or a corner is rounded.

Figure 9.51 shows examples of correct practices for dimensioning circles and arcs.

Coordinate and Grid Dimensioning

Grid dimensioning may be used to specify points along an irregular curve. *Coordinate dimensioning* is often used for parts that have complex interior hole patterns. A table may be provided specifying the hole diameters and their coordinates. Figures 9.52 and 9.53 show examples of coordinate and grid dimensioning.

Coordinate dimensioning often works well for parts that will be NC machined, as the part typically must have a 0,0,0 location relative to the machine. The other features can be dimensioned grid fashion relative to the 0 point. Because the file is often generated directly from the CAD model, the drawing is for reference only. It is still important to provide a drawing, though, because a value changed in the exported file through a glitch in the software will not be caught by the machine; it will cut to the location specified, whether or not that is what you intended. Providing the drawing allows the manufacturer to check the file to ensure that the dimensions are correct.

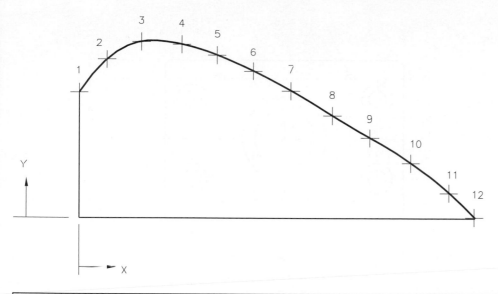

POINT	1	2	3	4	5	6	7	8	9	10	11	12
X	3.2	3.63	4.10	4.66	5.14	5.65	6.18	6.75	7.28	7.85	8.38	8.73
Y	5.0	5.46	5.70	5.67	5.52	5.30	5.00	4.67	4.36	4.00	3.60	3.26

9.52 Grid Dimensioning

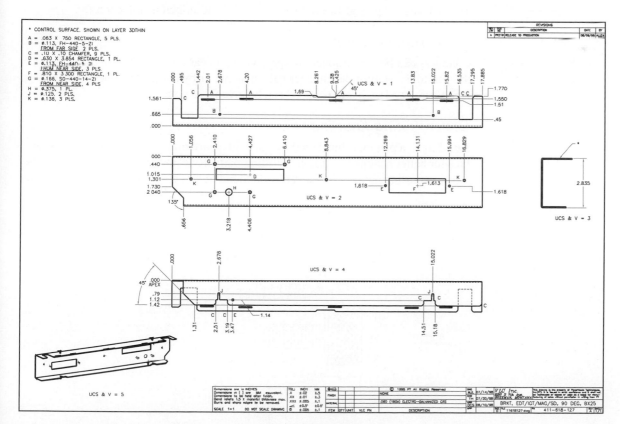

9.53 Coordinate Dimensioning *(Powerhouse Technologies, Inc.)*

9.54 Several dimensioning symbols are used here to dimension common features.

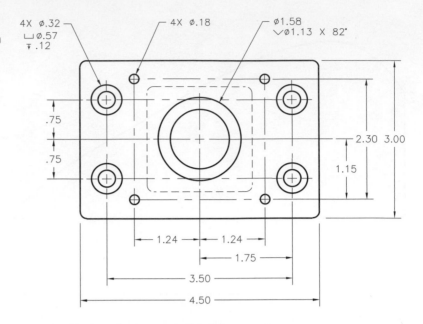

9.55 Methods of Dimensioning Chamfer

Standard Symbols for Dimensioning Common Features

Common features such as countersunk holes, counterbored holes, spotfaced holes, square holes, taper, thread, spherical radius, and spherical diameter all have standardized symbols to help define them clearly on the drawing without a large number of separate dimensions. Figure 9.54 illustrates the use of these symbols in a drawing. The standard shape and size of these symbols is described in Table 9.5.

Dimensioning Chamfers

Chamfers can be dimensioned by specifying a linear dimension from the surface of the part to the start of the chamfer and an angle, or by giving two linear dimensions (see Figure 9.55). Chamfers of 45° can be given as a note.

Dimensioning Knurling

Knurling is a pattern formed on the outside of a part, often to provide a gripping surface for better handling, or for press fitting parts. Knurling is specified by listing the type of knurling pattern (for example, diamond or straight), the pitch (distance from

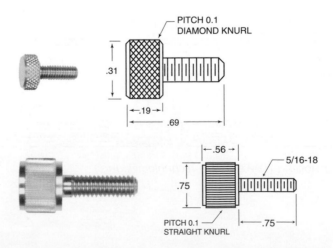

9.56 Dimensioning Knurling (*Reid Tool Supply Company.*)

Table 9.5 Dimensioning Symbols for Common Features.

Feature	Symbol (h = letter/symbol height)
Counterbore	⎣‾‾‾⎦ ⟵2h⟶ ↕h
Spotface	⎣ SF ⎦ ⟵2h⟶ ↕h
Countersink	90° ↕h
Square	▢ ⟵h⟶ ↕h
Conical taper	⟵0.5 h 2 h 30°
Depth	⟵h⟶ 0.6 h ↕h 60°
Diameter	∅
Spherical radius	SR
Spherical diameter	S∅

(a)

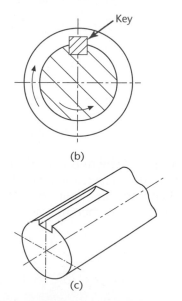

(b)

(c)

9.57   (a) Keyway, (b) Key, and (c) Keyseat. *(Getty Images—IStock Exclusive RF.)*

one groove to the next), and the diameter before and after knurling (see Figure 9.56). If the diameter after knurling is not important, it can be omitted. If the entire feature is not knurled, dimension the length and location of the knurled portion.

Dimensioning Keyseats and Keyways

Keys are rectangular or semicircular shapes that are used to prevent parts, such as gears or wheels, from turning on a shaft. *Keyseats* and *keyways* are the slotted shapes that receive the key. (Keyseats are in the shaft; keyways are in the hub of the part on the shaft; see Figure 9.57) Dimensions for standard key types, such as Woodruff keys, are provided in Appendixes 37 and 38. Keyseats and keyways are dimensioned by specifying the width, depth, and location of the keyseat, as shown in Figure 9.58. If needed, the length of the keyseat is also dimensioned.

9.58 Dimensioning Keyways and Keyseats

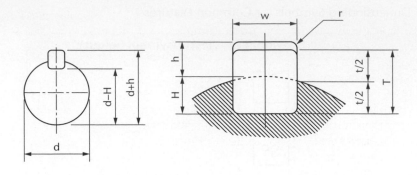

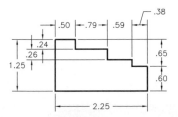

All tolerances ± .02

9.59 Overdimensioned Drawing. Both the length and the height of this part are overdimensioned. Adding the maximum tolerance to each set of dimensions results in a maximum overall length of 2.27 or 2.34, and a maximum height of 1.27 or 1.29.

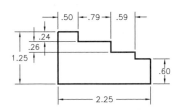

All tolerances ± .02

9.60 Fully Dimensioned Drawing. Removing unnecessary dimensions eliminates the ambiguity about the size of the part when tolerances are applied.

Overdimensioning

Overdimensioning refers to showing the same dimension more than one way in the drawing, or showing the same dimension twice (perhaps in different views). Many constraint-based CAD programs are unable to create a part from an underconstrained or overconstrained sketch because the sketch is ambiguous—it has more than one interpretation. The same thing is true of an overdimensioned drawing. Overdimensioned drawings are difficult to interpret because tolerances (the allowable range of variation for the actual value) must be applied to the dimensions. If the dimension is stated in more than one way, it is not clear how the tolerance is meant to apply.

Figure 9.59 illustrates an overdimensioned drawing. The total width of the part is dimensioned, as is each feature across the top. The tolerance stated in the general note for the part is ±.02 inch, so the overall length of the part can be as much as 2.27 or as small as 2.23. Across the top of the part the overall size is dimensioned as the series .50, .79, .59, and .38. Each of these dimensions can also have a tolerance of ±.02, as stated in the note. This means that the overall width that was just dimensioned as 2.25 ± .02 is dimensioned here as 2.26 (because of the rounded two-digit numbers), and the allowed tolerance is ±.08. The machinist will not be able to interpret this drawing! The vertical dimensions are equally confusing because the height of the drawing is specified two different ways.

Figure 9.60 shows the drawing with extra dimensions removed so that it can be interpreted clearly. If you are specifying the overall size of the part in a dimension, be sure to leave out one dimension in a chain of dimensions. This way it is clear how the tolerance applies. Leave out the dimension for the feature that you can allow to vary in size without affecting the function of the part.

9.10 PATENT DRAWINGS

The same CAD database that is used to produce the design documentation can also be used to create the drawings necessary to apply for patents. Patent drawings were previously produced by hand. You may have seen some of the elaborately hand-drawn representations of devices that were required for patents year ago.

The U.S. Patent office has standards similar to those used to produce good engineering drawings. Drawings may be submitted on one of two sheet sizes:

U.S. size 8.5 × 11 inches
International size 210 × 297 mm (size A4)

No borders may be drawn on the sheets, and minimum margin sizes must be maintained. No writing or drawing lines may extend into the margin area, except for the specific identification required at the top of each sheet. All sheets within a single application must be the same size.

Patent drawings are required to be submitted in black and white. No color plots, drawings, or photographs may be used. Either instrument- or CAD-produced drawings can be used. (Sketches are acceptable during the application process, but since formal drawings are required if the application is accepted, you should probably start with drawings.) Drawing lines must be crisp and sharp, so CAD drawings should be plotted or laser printed on high-quality paper (at least 20# bond). (Jagged dot matrix–style printed lines are not accepted.) Photocopies are acceptable, since three copies of each drawing are required to be submitted. The drawings cannot be returned, even if changes are necessary, so it is not a good idea to send your only original artwork with the patent application.

Unlike most engineering drawings, where it is preferred that views be shown in alignment on the sheet, patent drawings show each separate view as one figure. The figures are numbered consecutively, for example, Fig. 1A, Fig. 1B, Fig. 2. Often, specific dimensions and tolerances are not required to patent the general design or innovation. Instead, exploded isometric or perspective drawings with clear reference numbers identifying the parts, very similar to assembly drawings, are preferred. Centerlines should be used to show how parts in exploded views assemble. The reference number for a part or feature should remain the same throughout the set of drawings. The drawings must show every feature that is listed in the patent claims. If standardized parts (motors, CPUs, etc.) are used, they can be represented symbolically (per an appropriate standard) and do not have to be drawn in detail.

Most rules for creating good-quality engineering drawings still apply. For example, hatching should be thin, black, and uniformly spaced, running in only a single direction on the part; phantom lines may be used to show a secondary position for a part. See Figure 9.61 for a patent drawing.

There is no limit to the number of drawings that may be submitted to fully document the parts, and a variety of views may be used. Border sizes, sheet sizes, how the drawings are signed, and other requirements are spelled out in a guide published by the U.S. patent office that may be ordered from the U.S. Government Printing Office. Another good reference is *Patent It Yourself*, by David Pressman (Nolo Press, 2009).

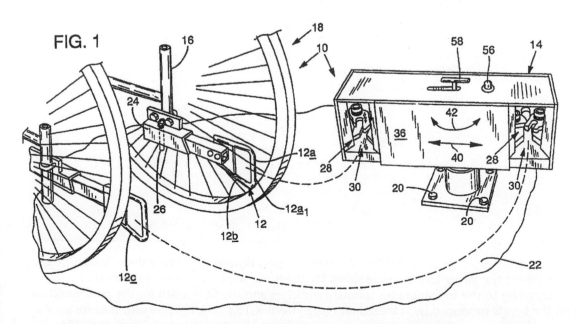

9.61 This patent drawing is taken from the patent for a securement system for a rollable mobility aid that was invented by David Ullman, Katherine Hunter-Zaworski, Derald Herling, and Garrett Clark.

DESIGN DOCUMENTATION: A PRAGMATIC MIX OF 3D AND 2D

Design documentation for mechanical engineer Jae Ellers is a pragmatic mix of constraint-based models and 2D drawings taken from the models—both created in Pro/ENGINEER. The CAD database created for the part conveys the design information to the manufacturer and is the source of the drawing views that define the quality standards for the manufactured part. "The drawing is the legal interface between us and the company doing the manufacturing," explains Ellers. "Until that changes, we have a drawing for every part, and we have to think of the drawing as the actual design documentation."

For Ellers, the process of creating drawings begins after a model has been refined, tested, and prototyped (see Figure 9.62). "There are basically two ways to do drawings:

either build an original drawing and update it as the model changes until it becomes a production drawing, or use the model and sketches as needed to get a working prototype part, then generate the drawings from the part that works. I've seen people do both, but the latter is more efficient for me."

Detail drawings are not required for prototypes because the division has a dedicated, well-equipped machine shop, referred to as the "model shop," on site. Engineers can export an IGES file from the Pro/ENGINEER solid model and send it to the shop to be built. The Mastercam software used in the shop does not need all the solid information in the file, so the engineers can export just a wireframe definition for a simple part, or a surface definition for more complicated parts. An

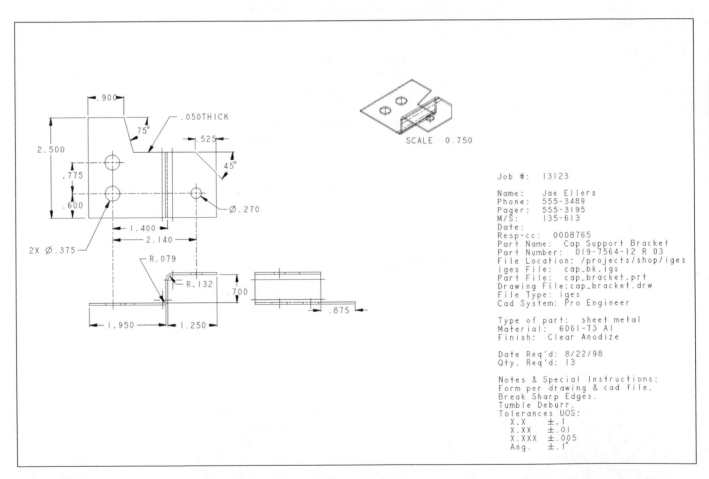

9.62 After conferring with model-shop personnel, Ellers modified his part drawing to reflect their input. The block tolerance was changed to reflect the larger tolerances required for sheet metal parts. Hole sizes were then toleranced individually according to the tolerances for standard punches and the fit needed for the part. Bend radii were changed to match the bends produced by standard dies (e.g., from R.132 to R.125). Finally, hole locations were toleranced to reflect where closer tolerances were required to ensure fit with other parts (e.g., the .775 distance between the two .375-diameter holes is toleranced more closely (±.005) than the standard tolerance in the block (±.015).

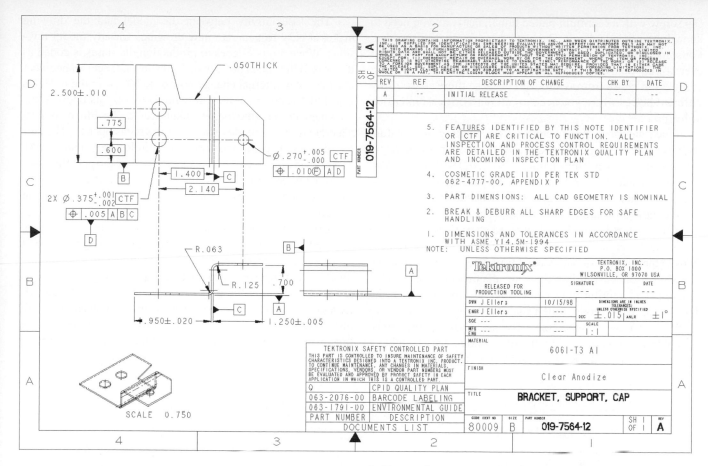

9.63 In the release drawing shown here, Ellers has added the notes, title block, and revision block that conform to company standards. Noncritical dimensions have been removed from the drawing, and critical-to-function dimensions have been identified with the letters CTF.

isometric view of the model with critical tolerances or other manufacturing notes sketched on it is often enough to convey what the shop needs to know to manufacture an acceptable part. Ellers points out that this is simply a time-saving approach that eliminates "spending hours generating drawings that may not be used again if the prototype doesn't work."

Once a design is ready to be produced, however, Ellers begins the drawings for production by identifying what he calls the "critical-to-function" features and dimensions and tolerances. These are the criteria that must be met for the part to operate correctly (see Figure 9.63). From there, he looks at the data transfer capability of the company that will make the part. If the vendor can work with an electronic file—either the Pro/ENGINEER model or an IGES export from the model—the printed drawing will show only dimensions and tolerances for these critical-to-function features. "We only fully dimension a drawing if we need to—for example, if someone is still working on a system where they need to type in the dimensions for a part or if their system can't read the files we write."

Even when a vendor works with an electronic file, there is still room for human error. The drawing provides documentation that will be used to inspect the manufactured parts. While a part is being made, the engineers write up a quality plan (an inspection procedure) that lists the features dimensioned on the drawing. When parts come back, a statistically significant sample (in this case, approximately thirty) of the parts are selected at random and used to measure the features, tolerances, and finishes as specified in the drawing.

"Often, the exact location of a feature isn't as important as its function. A pin that fits in a hole to hold the part together is an example of a part that has some leeway. Even though we model the feature where we want it to be, ultimately it doesn't matter exactly where it comes in as long as the two line up and function as they should." The opposite is true of the critical-to-function features, where dimensions are always provided on the drawing so they can be used to determine if a part has been manufactured acceptably.

Typically, a project will have an assembly drawing at the top that documents how all the parts go together, subassembly drawings for each subassembly, then individual part drawings at the bottom. To document a single part, Ellers uses the tools built into Pro/ENGINEER to create drawing views and add dimensions. "For a simple part, such as a pulley, two views are all that is needed to show the part. If you have a 440 socket-head

(continued)

cap screw with a cadmium-plated finish, that's all you need to know for the standard part. The picture isn't even really needed. For some of the more intricate castings in the print head, however, it may take three orthographic views, an isometric view, and section views from AA to MM because there are so many different places where you need to call out a critical dimension or feature."

Once the views have been created, clicking on a feature will show the driving dimensions used to model it. Ellers refers to these as "shown" dimensions. Pro/ENGINEER will also allow you to add "driven" dimensions by picking on entities and displaying their current length, radius, diameter, etc. "If you capture the design intent in the model, you can show the dimensions in the drawing without having to click lines or points to create the dimensions." Another advantage of "shown" dimensions is their bidirectional associativity. "If you make a change to a shown dimension in a drawing, the model updates. That means manufacturing guys can change the dimension on the drawing and the model will update without their having to know how to use the modeling software." Driven dimensions, in contrast, will update when the model does, but changing a driven dimension on a drawing wiil not update the model.

The associativity between the model and the drawing makes it easy to generate new drawings for any model revisions. When a model goes to production, its status is coded as Rev A and the file is protected. To make a change, an engineer must issue an internal change notice (ICN). Once the model is changed, generating a new drawing is as easy as opening the drawing and changing the revision number. The dimensions regenerate from the model when the drawing is opened. The current drawing is then the revised version, Rev B (Figure 9.64). The drawing database keeps each revision of the part file associated with the appropriate drawing file so that the dimensions shown on a given drawing match the revision of the part that was requested.

Once the dimensions are shown, Ellers uses Pro/ENGINEER to automate their placement—to a degree. He still finds it necessary to make adjustments, however, to align the dimensions properly and display the dimension lines and arrows. "Cleaning up the dimensions is important for readability's sake. If you had two dimensions close to one another and, instead of having the dimension centered between the witness [extension] lines, had to locate the dimension values outside the lines, it might be difficult to interpret them. If this drawing

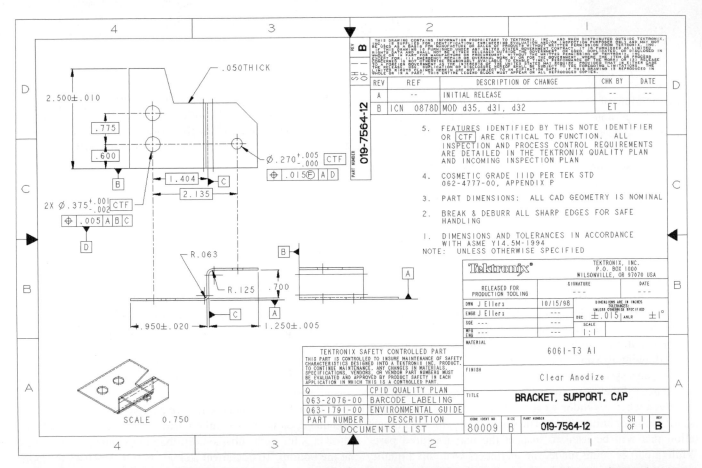

9.64 After the part was made, measurements taken from the actual part were used to modify the dimensions on the drawing (and in the model). Notice that the revision block lists the changed dimensions by the names used in the model.

went to someone who could not use the model but had to enter the data manually, it would be easy to get the numbers in the wrong place."

To finalize the drawings and conform to company standards, Ellers may send the drawing at this point to the in-house drafting group. Four or five draftspersons serve the approximately two hundred fifty engineers in the division and are a resource for preparing accurate drawings with a consistent look. "Pro/ENGINEER is really powerful in that you can communicate in the model the tolerances, the surface finish, and the geometric dimensioning and tolerancing. You can even associate notes with surfaces in the part that can be "shown" in the drawing. You can put all of this in the model itself, generate drawings to show the information you feel is needed, then hand the drawing off to drafting to let them put the notes and tables in the right place, add the bill of materials, and further clean up the drawing." This approach is efficient, in Ellers's view, because it takes advantage of what each group is comfortable with. An engineer could do the entire drawing, but engineers are typically more proficient with the modeling tools. A draftsperson is more proficient with the software's drawing tools, but will not know enough about the design to determine which features are critical to dimension and tolerance.

All drawings are dimensioned according to the *ASME Y14.5M* standard. In addition, company standards for the appearance of the title block, the size and form of the lettering, and the format for legal notices on the drawing are documented and stored centrally. Keeping the standards in a central location discourages individuals from relying on paper copies in their own files that may not be current. (Controlling the standards in this way also meets one of the ISO 9001 documentation requirements.)

Some of the consistency between drawings is achieved by starting all drawings from a template stored in the file. Pro/ENGINEER allows you to store formats that set up the zones on the outside of the drawing and include 2D entities such as the title block that link to data associated with the model file. The engineer's name and the part number of the file are inserted where appropriate in the drawing through these links. Pro/ENGINEER also makes it easy to access boilerplate for notes relating to various processes. If the engineer needs to add a sheet metal note, for example, he or she can bring in a text file that can be modified for that situation. More often, though, the consistency is achieved by the human element. The drafters bring their knowledge and expertise to bear and work together to maintain a consistent look in the drawings.

Ellers feels the biggest challenge in getting from a design of the model to a manufactured part is "unexpected interactions between real parts with real tolerances that don't work together like they did in the prototype. The most important thing to do is to define the functionality you need and get the interface defined. Once you've defined what you need, you can go out and find the process to support it. I've seen a part go from die cast aluminum (at $7 per piece) to injection-molded plastic (at $3 per piece), then to sheet metal (at $1 per piece). There's a whole different set of design considerations for each of these three processes, so you have to define your needs first. If you do, you'll have the information you need to evaluate the trade-offs between processes and select one that will give you the tolerances you need. You can then incorporate these process capabilities into the tolerances specified in your design."

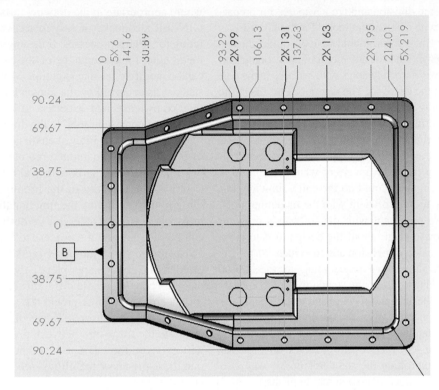

9.65 Many companies transmit models to their vendors to get cost estimates and communicate their designs. This edrawing shows the model with critical dimensions specified. *(Courtesy of Zolo Technologies, Inc.)*

HANDS ON TUTORIALS

SolidWorks Tutorials

Creating Working Drawings

In this tutorial, you will learn more about creating drawings from your parts and assemblies. If you have not already completed Lesson 3 - Drawings, you may choose to complete that tutorial first to get the basics.

In Advanced Drawings, you will create drawing views and dimension them; use tools for adding details, such as annotations, to your drawings; and make an assembly drawing that includes an exploded view and assembly-specific annotations.

- Launch SolidWorks from the icon on your desktop.
- Click to expand the topics listed below the "**?**" help icon.
- Click: **SolidWorks Tutorials.**
- Click: **All SolidWorks Tutorials (Set 1).**
- Click: **Advanced Drawings** from the list that appears, and complete the tutorial on your own.

Advanced Drawings Overview

Lesson 3 introduces drawing basics. This tutorial contains three lessons, wherein four drawing sheets are created. It is recommended that you complete the lessons in this order:

Time	Tutorial
	Creating Drawing Views shows how to create and dimension different drawing views.
	Detailing shows how to use tools to annotate drawings.
	Assembly Drawing Views shows how to create an exploded assembly view and use annotations specifically designed for assemblies.

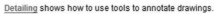
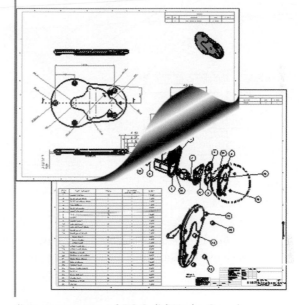

(Image courtesy of DS SolidWorks Corp.)

REVERSE ENGINEERING PROJECT

The Can Opener Project

1. *Document your assembly.* Create an assembly drawing for the can opener. Use an exploded view and a fully assembled view. Add balloon numbers and a parts list. Your exploded view should look similar to the example on the next page.

(Courtesy of Focus Products Group, LLC.)

(continued)

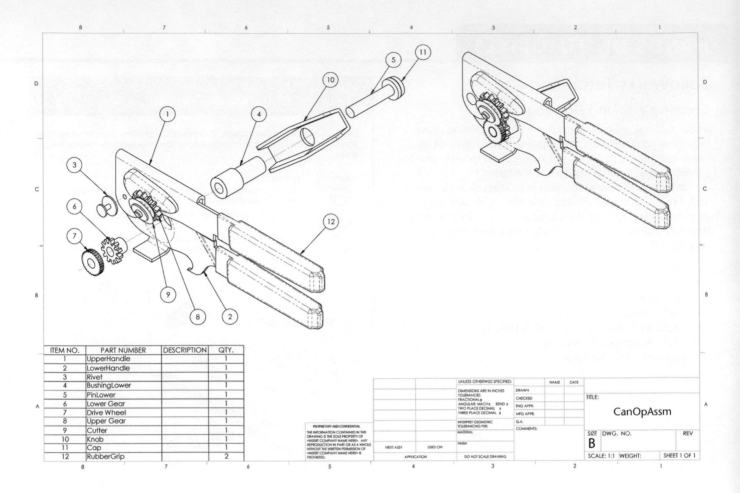

ITEM NO.	PART NUMBER	DESCRIPTION	QTY.
1	UpperHandle		1
2	LowerHandle		1
3	Rivet		1
4	BushingLower		1
5	PinLower		1
6	Lower Gear		1
7	Drive Wheel		1
8	Upper Gear		1
9	Cutter		1
10	Knob		1
11	Cap		1
12	RubberGrip		2

2. *Create a part drawing.* Create a dimensioned part drawing for the upper handle. The upper handle is a sheet metal part that is bent to form its shape. Consider this manufacturing process as you create the dimensioned drawing. An example drawing is shown below.

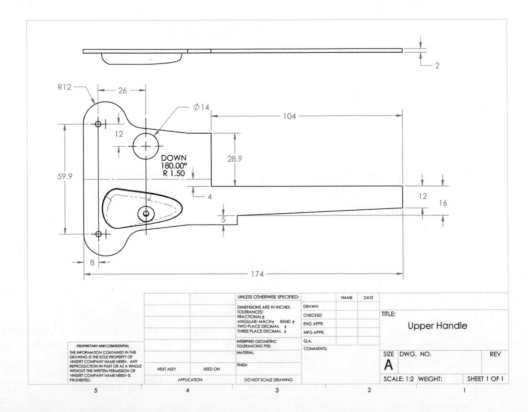

KEY WORDS

Aligned Section
Assembly Drawing
Associative Dimensioning
Ball Tag
Bill of Materials
Broken Out Section
Bubble Number
Coordinate Dimensioning
Cross-Hatching
Detail Drawing
Dimension Line
Dimension Value
Extension Line
Full Section
Grid Dimensioning
Half Section
Hatching
Keyseat
Keyway
Knurling
Mating Dimension
Mating Parts
Offset Section
Overdimensioning
Part Drawing
Primary Auxiliary View
Revision Block
Scale
Secondary Auxiliary View
Section Lining
Title Block
Unidirectional
Working Drawings

SKILLS SUMMARY

Now that you have completed this chapter, you are ready to create dimensioned multiview drawings from CAD models. You should be able to interpret orthographic views and then create the model and a properly dimensioned multiview drawing from it.

EXERCISES

Exercises 1–5: Dimensioning. Using your CAD package, create a solid model of each of the objects shown. From your model, develop orthographic views, and fully dimension all features and locations. Place the views properly so that all views can be included on one sheet. Use size B (11 × 17-inch) sheets for your full-size, dimensioned drawings. Include a border and a title block on all drawings.

1. Dished washer. Assume grid squares are .500 inch.

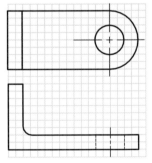

2. Offset bushing. Assume grid squares are .250 inch.

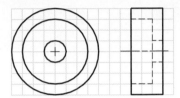

3. L-bracket. Assume grid squares are .250 inch.

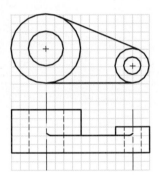

4. Bell crank arm. Assume grid squares are .250 inch, radii all .25 inch.

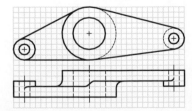

5. Shaft mounting bracket. Grid squares are .250 inch, radii all .25 inch.

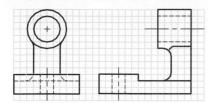

Exercises 6–10: Dimensioning. Create solid models of the objects shown below (some were already encountered in Chapter 2). Perform the following drawing operations on each completed model:

    a. From your solid model, develop the orthographic projection views needed to define the object fully.

    b. Place the required views properly, so that all views can be included on one sheet. Use standard-size drawings (choose size A, B, or C, as appropriate).

    c. Fully dimension all part size elements, features, and locations.

6. Notched block. Each grid square is .50 inch.

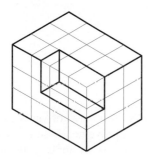

7. Single-rivet nut plate. All holes through. Each grid square is .1250 inch.

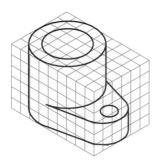

8. Saddle bracket. All holes are through. Assume that each grid square is .50 inch.

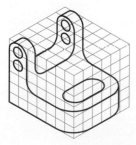

9. Keeper. Large holes pass through; small holes are .375 diameter, drilled and tapped to .50 depth. Assume grid squares are .50 inch. Radii are .50 inch.

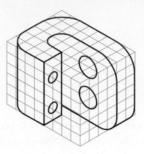

10. Rail mount shaft support. All holes pass through. Rail track on underside of support extends the length of the part. Grid squares are .50 inch apart.

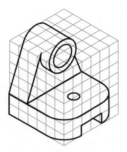

Exercises 11–15: Dimensioning Applied to Real Parts. You may not know exact dimensions of existing parts, or you may have ideas for improvement—so use your knowledge to design your own parts! Here is an opportunity to apply your own creativity and make improvements where you can.

For each object, perform the following operations:
a. Create a new design, including solid models of the objects shown.
b. From your solid model, develop the orthographic projection views needed to define the object fully.
c. Place the required views properly, so that all views can be included on one sheet. Use standard-size drawings (choose size A, B, or C, as appropriate).
d. Fully dimension all part size elements, features, and locations.

11. Bolt hanger for use in rock climbing and mountaineering. Overall size is about 2 inches. A climbing rope is clipped to the hanger, which is bolted to a rock face. This device must stop your fall, so it must be strong! Choose appropriate material and thickness.

12. Climbing carabiner, to be used for general mountaineering purposes. It can clip a rope into the bolt hanger from Exercise 11. Most are made of 6061 aluminum, and have a spring-loaded gate that opens with a light squeeze.

13. A rear cargo rack is a useful addition to a mountain bike. This one is made mostly of extruded aluminum, with welded joints. It is sized to bolt to the chain stays near the rear hub, and to the seat tube near the brake mounting bosses.

14. Campers use lightweight tongs to lift pots off a cookstove. The lighter the better! About 5 inches long.

15. A folding saw for tree pruning. The 1/8-inch-wide stainless steel blade includes extra-sharp teeth. Blade length is around 12 inches. This model has a wood handle, but what material will your handle use?

10 TOLERANCE

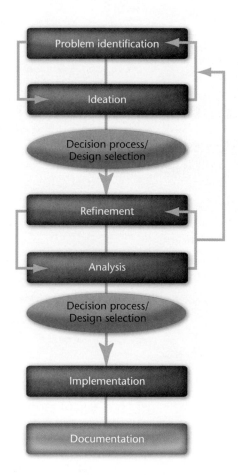

OBJECTIVES

When you have completed this chapter you should be able to:

1. Indicate tolerances on a drawing.

2. Identify six degrees of freedom.

3. Use GD&T symbols.

4. Document surface finish.

5. Indicate tolerances within the model.

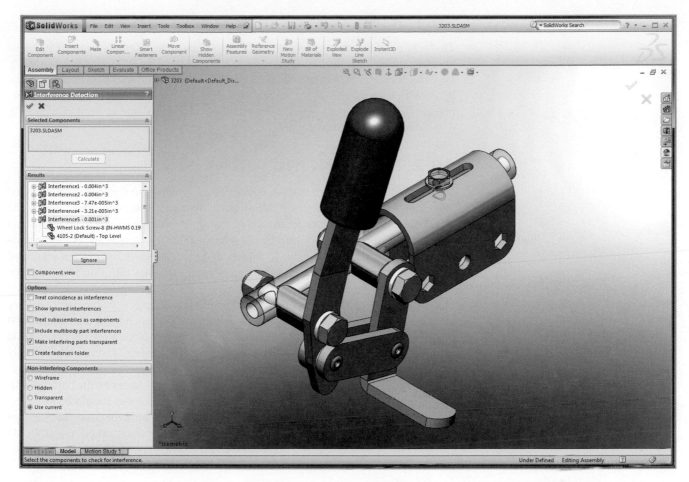

Tolerances are assigned to ensure that parts fit and function properly once they are assembled. *(Courtesy of Salient Technologies, Inc., www.salient-tech.com.)*

MAKING IT FIT

Dimensions and tolerances convey the criteria that the manufactured part must meet. How much variation from the exact measurement is acceptable? In preparing good documentation, the designer specifies the criteria required for the parts to function properly. The tolerances for the manufactured part must be described so that it can be produced as inexpensively as possible and be inspected to see whether it is an acceptable part.

As you create design documentation you are communicating the key aspects of your design for others to interpret. Many CAD programs "automatically" produce dimensioned drawing views. Even though these tools make it much easier to produce documentation drawings, they do not have all the necessary intelligence to produce toleranced drawings. You must add your understanding of the design intent, and desired fits and allowances, to produce good documentation in keeping with drawing standards that clearly communicates the allowable part variations.

10.1 ASME/ANSI Y14.5 STANDARD

This text provides an introduction to the important topic of tolerancing. If you will not be taking a separate course in geometric dimensioning and tolerancing (GD&T), you may want to purchase a dedicated GD&T text, attend training or a course given by ASME, or purchase *ASME/ANSI Y14.5-2009*, the standard for stating and interpreting dimensions and tolerances on drawings. It is available in print or as a PDF download from www.asme.org.

10.2 TOLERANCING

Most of today's goods are mass-produced. High-volume production is one way of keeping the price of goods lower. Many parts are manufactured in places across the world, then brought together for assembly. For mass-produced items to be assembled and repaired later, their parts must be interchangeable. Yet, there is always some variation in the manufactured parts. To ensure that parts are interchangeable, you must consider how the parts fit together in assembly and determine the allowable range for variation of the part features.

Tolerance is the total amount that a measurement on an acceptable part may vary from the specified dimension. Part tolerances are highly related to product quality and reliability. Stating tolerances clearly is an important part of creating documentation drawings. No part can ever be machined to exact dimensions. The more accurately you measure, the smaller the variations in size that can be identified.

A tolerance states a range for the size and location of features on the part that will still allow the part to function properly in the design. Tolerances should be as generous as possible. This allows a wider variety of processes for manufacturing the part at lower cost. Specifying high precisions for parts increases the part price and should be used only if the design requires the added precision.

The primary ways to indicate tolerances in a drawing are as follows:

- a general tolerance note
- a note providing a tolerance for a specific dimension
- a reference on the drawing to another document that specifies the required tolerances
- adding limit tolerances to dimensions
- adding direct plus/minus tolerances to dimensions
- geometric tolerances

Many of these tolerancing methods can be used in combination with one another in the same drawing.

General Tolerance Notes

General notes are usually located in the lower right corner of the drawing sheet near the title block. Often, general tolerance notes are included in the title block itself. For example, a general tolerance note might state, ALL TOLERANCES ±1 MM

SPOTLIGHT

Interchangeable Past

French gunsmith Honoré Blanc was a pioneer in the use of tolerances to produce interchangeable parts. Around 1790, Blanc created 1,000 guns for Napoleon's army. He put the disassembled parts in separate bins and showed how a musket could be assembled using any of the parts. The concept traveled to the United States, and in 1798 Eli Whitney was issued a contract by the U.S. government for guns that were to be based on this system. In the 1970s, David Starbuck led a team of archaeologists to excavate Whitney's Hamden, Connecticut, gun factory. They did not find any evidence of machinery used for mass production of interchangeable parts, such as Whitney claimed to have used. Historians now doubt Whitney's self-proclaimed status as the inventor of interchangeable-part manufacturing. But we can give him credit for a successful marketing ploy.

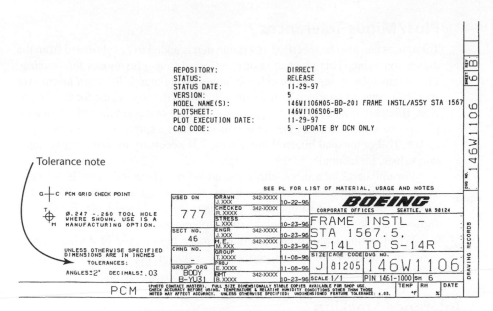

10.1   General Tolerance Note
(Copyright © Boeing.)

UNLESS OTHERWISE NOTED. ANGLES ±1 DEGREE. This indicates that for a dimension value written as 25, for example, any measurement between 24 and 26 on the actual part would be acceptable.

Many companies insert standard CAD title blocks into their drawings. These standard title blocks note general tolerance values acceptable for the type of production common to their industry. Figure 10.1 shows an example of a general tolerance note.

Another way general tolerances are stated is in a table on or near the title block. The following table indicates the tolerance by the number of digits used in the dimension. For example:

Digits	Tolerance
.X	±.2 inch
.XX	±.02 inch
.XXX	±.001 inch
X°	±1°

In this example, a single-place decimal dimension has a tolerance of ±.2. For example, a dimension value written as 3.5 could range anywhere from 3.3 to 3.7 on the actual part and still be acceptable. A dimension written as 3.55 could range from 3.53 to 3.57 on the actual part. And a value written as 3.558 could range from 3.557 to 3.559 and be acceptable. It is uncommon to see more than three decimal places listed for inch drawings, because precisions of ±.0001 are very high precision manufacturing and would be unlikely to be indicated merely by a general tolerance note. Likewise, a range of ±0.001 mm is a small allowable variation, so it is unlikely that millimeter values would be listed to three digits in this type of table.

Limit Tolerances

Limit tolerances state the upper and lower limits for the dimension range in place of the dimension values. Figure 10.2 shows examples of limit tolerances in a drawing. The upper value is always placed above the lower value. If the two values are written horizontally, the lower value is given to the left of the upper value and separated by a dash, as in "29–32."

When stating a pair of limits, make the number of decimal places agree by adding zeros at the end of the value if necessary. For example: 2.500–2.525.

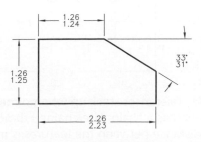

10.2   Limit Tolerances

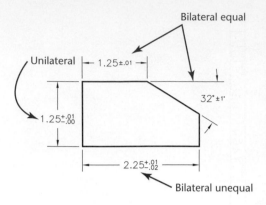

10.3   Unilateral and Bilateral Plus/Minus Tolerances

Plus/Minus Tolerances

Tolerances can also be specified as a range that is added to or subtracted from the dimension value. This practice is often referred to as ***plus/minus tolerancing***. Plus/minus tolerancing can be either bilateral or unilateral. ***Bilateral tolerances*** specify a range to be added to and subtracted from the base value for the dimension. Bilateral tolerances can be equal or unequal. For equal bilateral tolerances, the dimension is written as the value plus or minus a single range, for example, 3.00±.01. For unequal bilateral tolerances, it is necessary to state the two separate values, for example, 3.00+.01/−.02.

For unilateral tolerances, either the added or subtracted value is zero (0). When stating unilateral tolerances, include the zero value, either as just a 0 or preceded by the plus (+) or minus (−) sign. Do not leave it off. Figure 10.3 shows examples of unilateral and bilateral tolerances.

When stating a plus/minus tolerance, make the number of decimal places agree by adding zeros at the end of the value if necessary. For example: 2.500 +/− .002.

Tolerance Stacking

When dimensions are specified as a chain, the tolerances for the part features may accumulate. A ***chained dimension*** uses the end of one dimension as the beginning of the next. ***Tolerance stacking*** occurs when the tolerance for one dimension is added to the next dimension in the chain and so on from one feature to the next, resulting in a large variation in the location of the last feature in the chain.

Figure 10.4 illustrates this effect with a part where the surfaces labeled A and B and the holes labeled C and D are dimensioned in chain fashion. Accumulated tolerances influence the location of surface E relative to the left-hand surface of the part. When surface A is at its maximum size of 19.1 and surface B is at its maximum size of 25.1, surface E is 44.2 away from the left end. Similarly, when the center of hole C is at its maximum distance from the left end, 13.1, the maximum distance for hole D from the left end is 13.1 + 21.1, or 34.2 from the left end of the part. Tolerance stacking is not necessarily bad, if that is the intent for the relative locations of the features. Be aware of the effect that tolerance has on chained dimensions and specify the tolerances this way when tolerance accumulation is useful.

Tolerances also accumulate or "stack" over parts in an assembly, as the size variation for one part changes the alignment of the next.

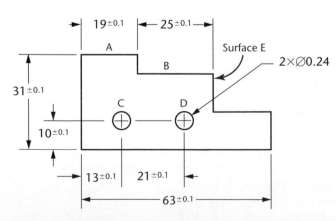

10.4   Tolerances stack up when dimensions are specified as a chain. Use chained dimensions when the distance between the features is more important than their location from a fixed point.

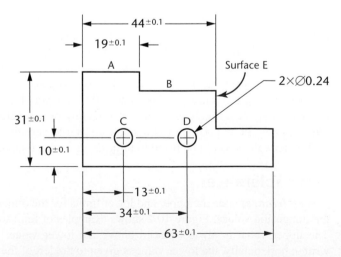

10.5   Use baseline dimensioning when you don't want tolerances to stack up and when your features should be measured from a common base feature.

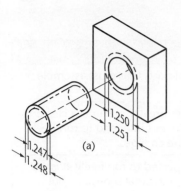

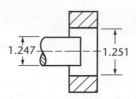

(a)

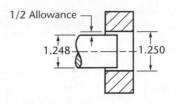

(b) Shaft tolerance = 1.248 – 1.247 = .001
Hole tolerance = 1.251 – 1.250 = .001

(c) Allowance = 1.250 – 1.248 = .002
Max clearance = 1.251 – 1.247 = .004

10.6 The dimensions and tolerances shown here produce a clearance fit because the internal member will always be smaller than the external member.

Baseline Dimensioning

Baseline dimensioning locates a series of features from a common base feature. Tolerances do not stack up because dimensions are not based on other, toleranced dimensions. Figure 10.5 illustrates the part in Figure 10.4 using baseline dimensioning. Surface E is now a maximum of 44.1 from the left surface on an acceptable part, and the location for the center of hole D is a maximum of 34.1 from the left surface.

Baseline dimensioning makes it easy to inspect the part because features are measured from a common base feature. Dimensioning from a zero point as the base feature can also be a useful technique for dimensioning parts for NC machining.

Fit

Fit refers to how tightly or loosely mating parts must fit together when assembled. Names for types of fit were developed through years of manufacturing practice. Before interchangeable parts, the designer and machinist worked closely together. A note or spoken instruction let the machinist know how parts should fit. Sliding fit, running fit, and force fit described to the machinist how the parts should function. The machinist fashioned each part to function as intended.

The categories used to describe types of fit derive from these practices, but now they have been standardized to make them more consistent.

One way to determine the necessary tolerance for a part is by deciding on the type of fit and looking up the required tolerance for that type of fit in a standard fit table. Both U.S. customary inch units and metric unit fit tables are available in Appendixes 6–15.

The following broad categories describe types of fit between mating parts:

A *clearance fit,* as illustrated in Figure 10.6, has an internal member that is always smaller than the external member so that the parts will slide together.

An *interference fit* has an internal member that is always larger than the external member so that the parts must be forced together.

In a *transition fit* either the internal or external member can be larger so that parts either slide together or can be forced together. This sometimes requires parts to be matched during assembly, fitting larger with larger, and smaller with smaller, which can make it difficult to order replacement parts.

A *line fit* indicates that the internal member is sized either so that there is a slight clearance or so that it contacts the external member.

The limit tolerances on the drawing in Figure 10.7 specify an interference fit, and those in Figure 10.8, a transition fit. Within these basic categories are additional designations that more specifically describe the desired fit. These fit classes are listed in Table 10.1.

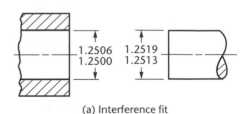

(a) Interference fit

10.7   Interference Fit

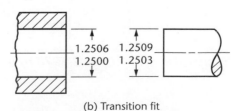

(b) Transition fit

10.8   Transition Fit

Table 10.1 Classes of Fit (ANSI B4.1–1967).

Category	Class	Description
Running or sliding	RC1	Close sliding fit is intended for accurate location of parts that must assemble without perceptible play.
	RC2	Sliding fit is intended for accurate location but with greater maximum clearance than RC1, where parts move and turn easily but are not intended to run freely, and in larger sizes may seize with small temperature changes.
	RC3	Precision running fit, the closest fit that can be expected to run freely. It is intended for precision work at slow speeds and light journal pressures, but is not suitable when appreciable temperature differences will be encountered.
	RC4	Close running fit is intended to run freely on accurate machinery with moderate surface speeds and journal pressures, when accurate location and minimal play are desired.
	RC5, RC6	Medium running fits are intended for higher running speeds or heavy journal pressures, or both.
	RC7	Free running fits are for use when accuracy is not essential, or when large temperature differences are likely.
	RC8, RC9	Loose running fits are for use when wide commercial tolerances are necessary together with an allowance on the external member.
Clearance locational fit	LC1–LC11	Locational clearance fits are intended for parts that are normally stationary but can be freely assembled or disassembled. These fits run from snug parts requiring accuracy of location (LC1) through medium clearance fits for parts such as spigots, to the looser fastener fits, where freedom of assembly is of prime importance.
Transition locational fits	LT1–LT6	Transition fits are a compromise between clearance and interference fits for applications where accuracy of location is important but either a small amount of clearance or interference is permissible.
Interference locational fits	LN1–LN3	Locational interference fits are used where accuracy of location is most important and for parts requiring rigidity and alignment with no special requirements for bore pressure. These fits are not intended for parts designed to transmit frictional loads from one part to another through the tightness of the fit. For these types use force fits.
Force and shrink fits	FN1	Light drive fits require light assembly pressures and produce more or less permanent assemblies. These are suitable for thin sections, long fits, and in cast-iron external members.
	FN2	Medium drive fits are suitable for ordinary steel parts, or for shrink fits on light sections. They are about the tightest fits that can be used with high-grade cast-iron external members.
	FN3	Heavy drive fits are suitable for heavier steel parts or for shrink fits in medium sections.
	FN4, FN5	Force fits are suitable for parts that can be highly stressed, or for shrink fits in which the heavy pressing forces required are impractical.

Using a Fit Table

In this example, the hole to be formed has a nominal 1.00 diameter. The cylindrical shaft that will fit inside the hole is to assemble freely with a loose running fit (RC8).

Based on the nominal size, look up the tolerance for the hole using the fit table (see Table 10.2). For the hole, the values given are +3.5 and −0. For the shaft (the internal member) the values listed are −4.5 and −6.5. The values in the table are given in thousandths of an inch.

For the hole, add the value +.0035 to the 1.0000 nominal size of the hole. The limit tolerance for the hole will be 1.0035–1.0000.

For the shaft, subtract the value .0045 from 1.0000 to find the upper limit, .9955. Subtract the value .0065 from 1.000 to find the lower limit. The limits for the shaft will be .9935–.9955, as shown in Figure 10.9.

The smallest actual hole size is always larger than the largest actual shaft size to produce a clearance fit.

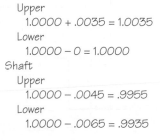

Ø 1.00

Hole
  Upper
    1.0000 + .0035 = 1.0035
  Lower
    1.0000 − 0 = 1.0000
Shaft
  Upper
    1.0000 − .0045 = .9955
  Lower
    1.0000 − .0065 = .9935

Table 10.2 Portion of RC8 Fit Table.

| Nominal Size Range (in.) | Class RC8 | | |
| | Standard Limits (in thousandths) | | |
	Limits of Clearance	Hole H10	Shaft c9
0.71–1.19	4.5	+3.5	−4.5
	10.0	−0	−6.5

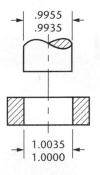

.9955
.9935

1.0035
1.0000

10.9 A 1.00-Nominal Diameter Toleranced for an RC8 Clearance Fit

The Basic Hole System

The basic hole system uses the external member (often a hole) as the size for determining the toleranced dimensions to achieve the fit. The reason for this is that holes are often formed with standard-size tools such as drills and reamers, whereas the internal shaft is machined down to size from standard stock. The shaft can easily be machined to the required size to fit the more standard hole.

The Shaft System

The shaft method for calculating fits takes the size of the shaft as the basic dimension from which the toleranced dimensions are derived. The basic shaft method is not commonly used but may be preferable when multiple parts will fit onto a common shaft size, or when the shaft is formed of a cold-worked material that is difficult to machine to size. (See Figures 10.10 and 10.11, respectively, for the preferred fits for the basic hole system and the preferred fits for the basic shaft system.)

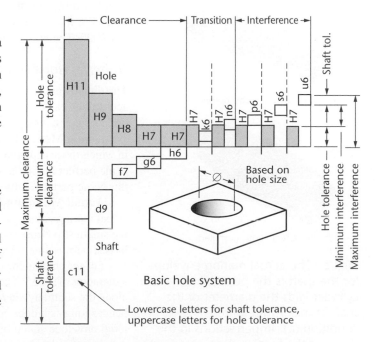

10.10 The Preferred Fits for the Basic Hole System

10.11 The Preferred Fits for the Basic Shaft System

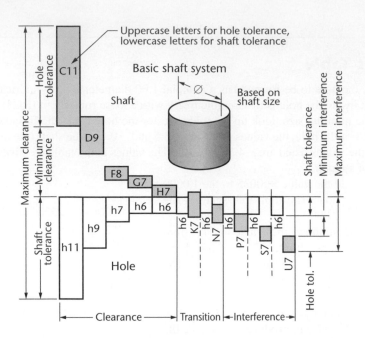

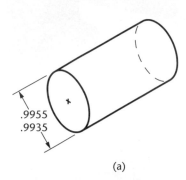

.9955
.9935

(a)

(b) Imperfect form: waisted

(c) Imperfect form: tapered

(d) Imperfect form: bowed

(e) Imperfect form: barrelled

10.12 The actual mating envelope for the shaft is the perfect form—a cylinder with the diameter of the shaft at maximum material condition (a). Imperfections in the shape of the feature shown here are still acceptable as long as they do not exceed the mating envelope.

Allowance

The difference between the smallest hole size and the largest shaft size is the ***allowance.*** Allowance is the minimum clearance or maximum interference that is desired between two members when they are at maximum material condition. The allowance for the parts in Figure 10.9 is .0045.

Maximum Material Condition

The term ***maximum material condition*** is used to eliminate confusion between the largest sizes for holes and shafts. Maximum material condition means the tolerance limit at which the largest amount of material remains on the part. For a hole, or external member, it is the smallest hole. For a shaft, or internal member, it is the largest shaft. For the example in Figure 10.9, the maximum material condition for the hole is 1.0000. The maximum material condition for the shaft is 0.9955.

Imperfections of Form

The dimension and tolerance stated on the drawing define a boundary within which acceptable parts must fit. The following terms are used to describe the tolerance boundary, or envelope, as it is called:

Actual Mating Envelope: This envelope is toward the outside of the material, in which the acceptable actual feature must fit. For external parts, like cylinders, this is the perfect feature at the largest permissible size; for internal features, like holes, this is the perfect feature at the smallest permissible size.

Actual Minimum Material Envelope: This envelope is the counterpart to the actual mating envelope. For an acceptable external feature, it is the perfect feature at the smallest permissible size; for an internal feature this is the perfect feature at the largest permissible size.

For the shaft in the example the actual mating envelope is a perfect cylinder with a diameter of .9955. The actual diameter of an acceptable part can be at any location along the shaft anywhere in the stated tolerance range, and the part can be tapered or bent as long as it does not exceed the actual mating envelope or actual minimum material envelope stated by the tolerance. Figure 10.12(a) illustrates the actual mating envelope for the shaft. When an exterior feature is at its minimum size, the tolerance does not imply that it has to be perfectly formed, but it may not extend beyond the actual mating envelope.

A hole is at maximum material condition when it is at its smallest size limit. The actual mating envelope for an internal feature, like a hole, is an imaginary boundary of the perfect geometry shape at the smallest size limit. If the surface of the actual hole extends inside this actual mating envelope, it is not acceptable. When the hole is at its smallest size, it must be perfectly formed (e.g., it must be perfectly perpendicular if it is shown perpendicular). When the hole is at a larger size, it may be drilled slightly at an angle or slightly misshapen, as long as none of the actual surface forming the hole is inside the actual mating envelope.

Modeling for Tolerance Studies

When you are creating a solid model, you can use the model to perform tolerance studies. Most solid modeling software allows you to check for interferences between different parts in the assembly. Some software has capabilities to apply tolerance ranges to parametric dimensions that can be evaluated at the upper and lower limits for the size range. You can check for interferences when the part is at maximum material condition and least material condition (the smallest shaft and the largest hole). Commands to check interferences usually report the size of the interference. If you are intending a force fit, this will be the allowance.

The type of tolerance designation you use can make it harder or easier to model and change tolerances. Consider the use of limit tolerances in the example in Figure 10.9. The size for the hole is 1.0035–1.0000. The nominal size for the hole is 1.0000, which is the lower limit. A machinist aiming to produce the minimum size hole (a reasonable practice, since it is easy to make holes larger but very difficult to make them smaller!) can easily see the starting dimension, making the drawing easy to interpret. If you want to change the stated tolerance, you can add all the additional tolerance to the upper range (making the 1.0035 dimension larger than 1.0000) and leave the model at the lower-end size. Then, you can change the value (if it is a driving dimension) to the upper range to model the tolerance easily. For the mating shaft part, its upper limit size can remain the same, and the lower limit can be changed to add additional tolerance if desired.

Equal bilateral tolerances can make drawings easy to check, because the value for a dimension includes the nominal size intended for the part (again easily visible to the machinist), with equal upper and lower tolerances. However, changing the tolerance zone for a bilateral tolerance means changing the mean size for either the hole or the shaft. If you add additional plus/minus range to the size of the hole, you will have to change the starting diameter size for the shaft for it still to fit. In general, bilateral tolerances work well for parts where the allowable variation is equally distributed around both sides of the nominal size. Welded assemblies and loose tolerances are an example.

Metric Fit Tables and Designations

The ISO system of preferred metric limits and fits is included in the *ANSI B4.2* standard. Like the basic hole system for inch designations, this system can be used for holes, cylinders, and shafts, and is also adaptable for fits between parallel surfaces including features such as keys and slots. Terms for metric fits are similar to those for decimal-inch fits.

Tolerance: the difference between the permitted minimum and maximum sizes of a part.

Basic size: the size from which limits or deviations are assigned.

Deviation: the difference between the basic size and the hole or shaft size (comparable to the tolerance in the U.S. customary inch system).

Upper deviation: the difference between the basic size and the maximum allowable size of the part, similar to the maximum tolerance in the U.S. customary inch system.

Lower deviation: the difference between the basic size and the minimum allowable size of the part, comparable to minimum tolerance in the U.S. customary inch system.

Table 10.3 Basic Size (in millimeters).

1st Choice	2nd Choice	1st Choice	2nd Choice	1st Choice	2nd Choice	1st Choice	2nd Choice
1	1.1	5	5.5	25	28	120	140
1.2	1.4	6	7	30	35	160	180
1.6	1.8	8	9	40	45	200	220
2	2.2	10	11	50	55	250	280
2.5	2.8	12	14	60	70	300	350
3	3.5	16	18	80	90	400	450
4	4.5	20	22	100	110	500	550

Fundamental deviation: the deviation closest to the basic size, comparable to the minimum allowance in the U.S. customary inch system.

International tolerance grade (IT): a set of tolerances that vary according to the basic size to provide a uniform level of accuracy within the grade. For example, in the dimension 50H8 (for a close-running fit), the IT grade is indicated by the number 8. There are 18 IT grades—IT01, IT0, and IT1 through IT16. Smaller grade numbers indicate smaller tolerance zones.

Tolerance zone: describes the relationship of the tolerance to basic size. In the dimension 50H8, for the close-running fit, the tolerance zone is specified by a letter and a number, H8. An uppercase letter indicates that the tolerance applies to the hole; a lowercase letter indicates that the tolerance applies to the shaft. The number is the IT grade.

Hole-basis system of fits: uses the hole diameter (basic size) as the minimum size. A table of fits will indicate whether the hole basis or shaft basis is being presented.

Shaft-basis system of fits: uses the shaft diameter (basic size) as the maximum size.

Preferred Metric Sizes: The preferred basic sizes for computing tolerances are given in Table 10.3. Choosing basic diameters from the first column yields readily available stock sizes for round, square, and hexagonal products.

10.3 GEOMETRIC DIMENSIONING AND TOLERANCING

Geometric dimensioning and tolerancing (GD&T) uses special drawing symbols to specify geometric characteristics as well as dimensional requirements on engineering drawings. GD&T is a means of more closely controlling how much a feature can deviate from the perfect geometry implied by the drawing. Using GD&T, you can specify tolerance zones that reflect the geometry of the feature you are trying to control. GD&T symbols allow you to control features such as parallelism, perpendicularity, and other feature geometry that may be implied by limit or plus/minus tolerances but not controlled specifically. Use GD&T to control the geometry of features when it is important that they are formed accurately to ensure the fit and function of mating parts. In any particular drawing, not every feature needs to be tightly controlled. The manufacture of some parts or features can be stated with more general tolerances.

ANSI/ASME Y14.5M–2009 is the current GD&T standard. As you build your skill in engineering drawing you may wish to take a separate course in GD&T to learn to apply the basic principles presented here.

Geometric Characteristic Symbols

The geometric characteristic symbols can be divided into those for individual features, such as ***form tolerances*** (which relate to the shape of a feature) and those used

Table 10.4 Geometric Characteristics (*ANSI Y14.5M–2009*).

	Tolerance	Characteristic	Symbol
For individual features	Form	Straightness	—
		Flatness	▱
		Circularity	○
		Cylindricity	⌭
For individual or related features	Profile	Profile of a line	⌒
		Profile of a surface	⌓
For related features	Orientation	Angularity	∠
		Perpendicularity	⊥
		Parallelism	//
	Location	Position	⌖
		Concentricity	◎
		Symmetry	⚌
	Runout	Circular runout	↗
		Total runout	↗↗

for related features such as ***orientation*** or ***location tolerances*** (which control a feature in relation to another feature). Table 10.4 shows the geometric characteristics divided into these categories and the symbols used to indicate them.

The Feature Control Frame

Geometric tolerances are indicated by means of geometric characteristic symbols in a feature control frame. A ***feature control frame*** contains all the symbols and modifiers necessary for specifying the geometric tolerance for a feature. It begins with the ***geometric characteristic symbol,*** which tells the type of geometry being controlled. The next compartment in the frame contains the geometric tolerance value for the feature, which specifies the total width of the tolerance zone. The simplest feature control frame contains just the geometric characteristic symbol and the tolerance, as shown in Figure 10.13. The placement of additional symbols in the feature control frame is governed by the *Y14.5* standard and will be illustrated throughout this section.

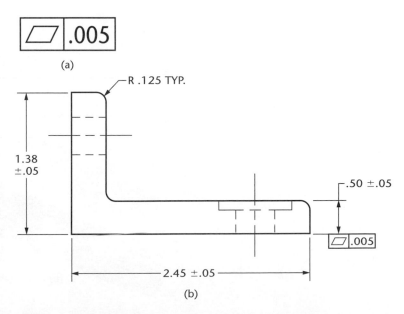

10.13 (a) A Simple Feature Control Frame; (b) Feature Control Frame Attached to an Extension Line Controlling the Flatness of the Bottom Surface

Feature control frames can be related to a feature in the drawing by

- adding the frame below a note or dimension pertaining to the feature;
- extending a leader line from the frame to the feature;
- attaching a side or end of the frame to an extension line from the feature (if it is a plane surface); or
- attaching a side or end of the symbol frame to a dimension line pertaining to a feature of size.

Form Tolerances for Individual Features

Straightness, flatness, circularity (roundness), cylindricity, and sometimes profile are form tolerances that apply to single features. Think back to the discussion of the actual mating envelope implied by a tolerance for the size of a feature. When the feature is at its maximum material condition (largest size for external features like shafts, smallest size for internal features like holes), its form must be perfect to fit inside the actual mating envelope. That is, a shaft would have to be perfectly straight. But if the shaft were at its smallest size, it could be tapered or bent and still meet the specified tolerance. Sometimes you may need to control the tolerance more precisely than this. Specifying the straightness, flatness, roundness, cylindricity, and even the profile tolerance can provide more control over the feature.

Figure 10.14 illustrates a form tolerance using a dimensioned and toleranced drawing of a shaft. The feature control frame indicates that any longitudinal element of the cylinder forming the shaft must be straight within .002 inch. The pictorial drawing in Figure 10.14(b) shows an exaggerated view of two straight parallel lines .002 inch apart running the length of the shaft. Any line element of the shaft must fall within the zone described by the straight lines. The orthographic view in Figure 10.14(c) shows an acceptably straight element of the cylinder as it fits in the tolerance zone.

A *straightness tolerance* specifies a tolerance zone within which an axis or all points of the indicated element must lie. Straightness indicates that the element of a surface or an axis is a straight line (see Figure 10.15).

A *flatness tolerance* specifies a tolerance zone defined by two parallel planes within which the surface must lie. Surface flatness means having all elements in one plane (see Figure 10.16).

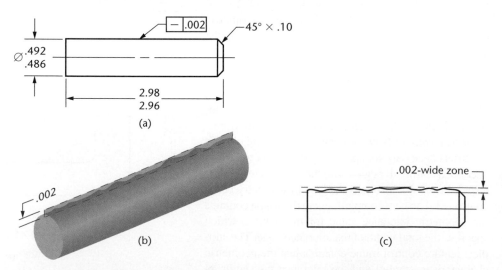

10.14 (a) The feature control frame indicates that any element of the shaft must be straight within .002. (b) To interpret the straightness control, imagine two straight parallel lines set apart by the tolerance amount. Any line element making up the feature must lie entirely between the perfectly straight lines. (c) An acceptably straight element fits within the tolerance zone.

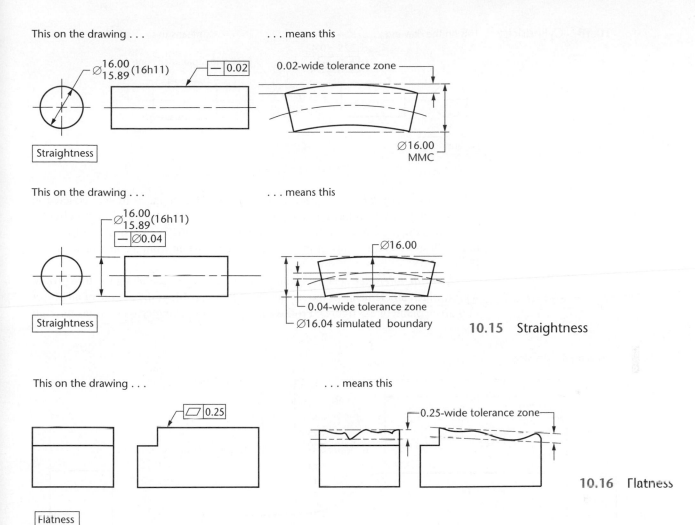

10.15   Straightness

10.16   Flatness

A *circularity (roundness) tolerance* specifies a tolerance zone bounded by two concentric circles within which each circular element of the surface must lie. Circularity of a surface of revolution for a cone or cylinder indicates that wherever it is intersected by a perpendicular plane, its cross section will be circular. For a sphere, all points of the surface intersected by any plane passing through its center are equidistant from that center (form a circle) (see Figure 10.17).

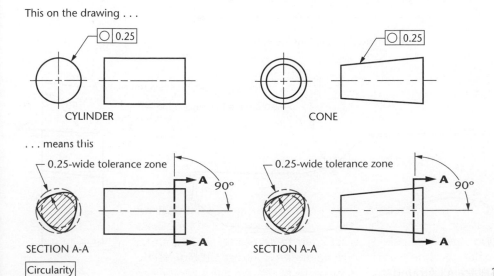

10.17   Circularity (Roundness)

10.18   Cylindricity   This on the drawing . . .   . . . means this

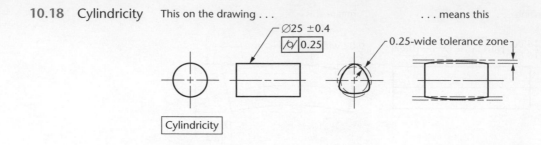

A *cylindricity tolerance* specifies a tolerance zone bounded by two concentric cylinders within which the surface must lie. This tolerance applies to both circular and longitudinal elements of the entire surface. Cylindricity of a surface of revolution indicates that all points of the surface are equidistant from a common axis (see Figure 10.18).

A *profile tolerance* specifies a uniform boundary or zone along the true profile within which all elements of the surface must lie. A profile is the outline of an object in a given plane (a 2D figure). Profiles are formed by taking cross sections through

This on the drawing . . .

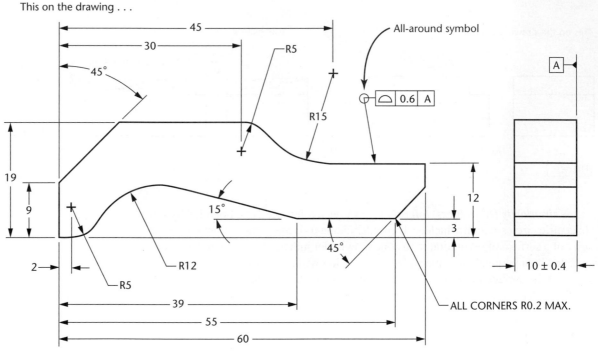

Untoleranced dimensions are basic

. . . means this

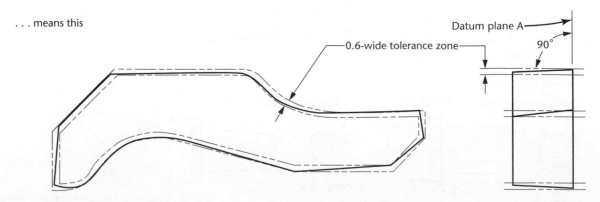

10.19   Profile of a Surface All Around

This on the drawing . . .

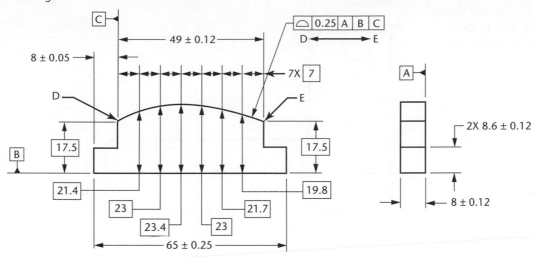

. . . means this

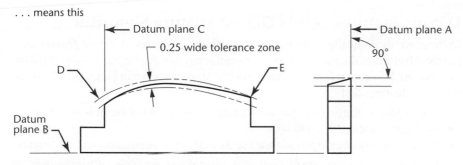

10.20   Profile of a Surface Between Points

the figure. The resulting profile is made up of straight lines, arcs, other curved lines, and 2D elements (see Figures 10.19–10.21.) The profile of a surface is a 3D zone that extends along the length and width (or circumference) of the indicated feature. A profile tolerance also can be applied to parts that have a constant cross section, to a surface, to a revolution, or to parts such as castings by profile tolerances that apply all over (through the use of the all-over symbol) on the feature control frame leader.

This on the drawing . . .

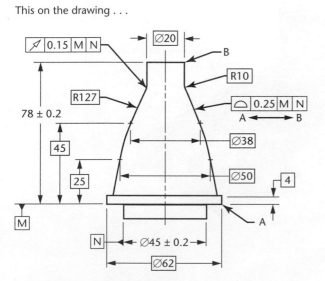

. . . means this

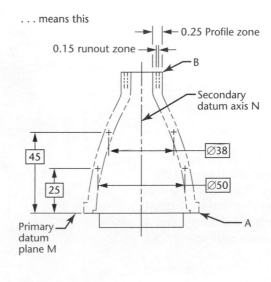

10.21   Profile of a Surface of Revolution

10.22 A datum identifying symbol on the drawing specifies the datum feature. In this case, the bottom surface is identified as datum A.

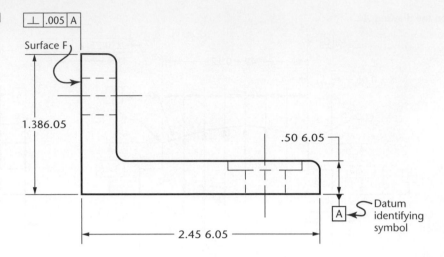

Datum Features Versus Datum Feature Simulator

Datums are theoretically exact points, axes, lines, or planes. A *datum feature* is a physical feature on the part that is being manufactured or inspected. It is identified on the drawing by a datum feature symbol, such as the one shown in Figure 10.22. Specify datum features that

- correspond to the mating feature on another part to which the fit is required.
- are readily accessible on the part.
- are sufficiently large to permit their use for making measurements and alignments.

Datum features, since they are features of the actual part, are inherently imperfect. To inspect a part to determine whether it meets the tolerance requirement specified in relation to a datum, a datum feature simulator is used. The datum feature simulator is a boundary derived from the datum feature. It is typically the inverse of the datum

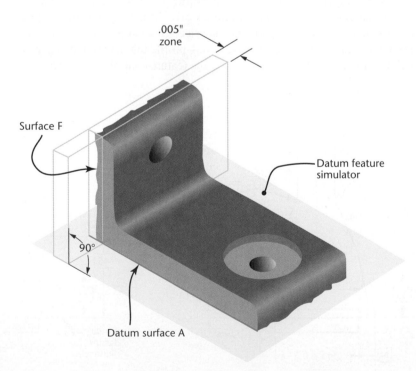

10.23 Surface A on the part is used to establish the theoretically exact datum A. The feature control frame indicates that surface F must be perpendicular to datum plane A within a .005-inch-wide tolerance zone.

feature. For example, if the datum feature is a hole, then the datum feature simulator is a positive cylinder, such as a go/no-go plug gage. A datum feature simulator can be one of two types:

- a theoretically exact datum feature simulator
- a physical datum feature simulator

An example of a theoretically exact datum feature simulator is a plane defined in a CAD system. Figure 10.23 illustrates the actual surfaces of a part and the theoretically exact datum surface A. Datum surfaces are important features in your CAD models. Definition of model datums that reflect how you will inspect your parts is as important as modeling features that reflect your design intent. They go hand in hand.

An example of a physical datum feature simulator is a precisely finished granite slab on which a part to be inspected is placed (Figure 10.24). As physical datum feature simulators themselves cannot be exactly perfect, the tolerances defined for a feature must take into account the imperfections in precision gauges and fixtures. Standards for these are specified in *ASME Y14.43*.

For example, surface F of the short leg of the bracket shown in Figure 10.23 must be perpendicular to surface A, the bottom of the bracket, within a certain tolerance. Not every geometric characteristic requires a datum reference—only those that relate two different features to each other, for example, perpendicularity, parallelism, or concentricity.

Datum Identifying Symbols

A ***datum identifying symbol*** is a capital letter in a square frame with a leader attaching it to the indicated feature via a triangle. The triangle may be filled or not filled. Letters of the alphabet (except I, O, and Q) are used as datum identifying letters (see Figure 10.25).

Datum Reference Frame

Imagine trying to measure the location of a hole in an air hockey puck while it is floating on an air hockey table. That would be difficult! The example exaggerates the concept, but to make precise measurements the thing being measured has to be held still. A datum reference frame provides a framework to eliminate the translations (movement) and rotations of which the free part is capable.

A datum reference frame is often defined by three mutually perpendicular planes, used to immobilize the part to be inspected and provide a way to make accurate and repeatable measurements. The datum feature symbols on the drawing identify the primary datum, secondary datum, and tertiary datum (meaning first, second, and third).

The datum reference frame is a theoretical construct that does not exist on the actual part. It is established from the datum features identified on the part drawing. Often, a fixture is used in which the primary datum surface contacts the measurement fixture at three points, thus defining a plane. Once the primary datum plane is established, only two additional contact points are needed to establish the secondary datum plane. Once the primary and secondary planes are established, a single additional point establishes the tertiary datum plane. The datum reference frame is established from the datum feature simulators, not from the datum features themselves.

Constraining Degrees of Freedom

Datums listed in a feature control frame are applied to constrain degrees of freedom of the part with respect to a datum reference frame. Datums can be identified to constrain the degrees of freedom of the part to immobilize it for inspection.

The free part shown in Figure 10.26 can translate (move) back and forth in the X-, Y-, and Z-directions. It can also have rotations u, v, and w. It is said to have six degrees of freedom owing to these motions.

10.24 Inspecting a Part on a Physical Datum Feature Simulator *(Joerg Reimann/Getty Images— IStock Exclusive RF.)*

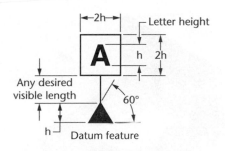

10.25 Datum Identifying Symbol Proportions

10.26 The free part has six degrees of freedom.

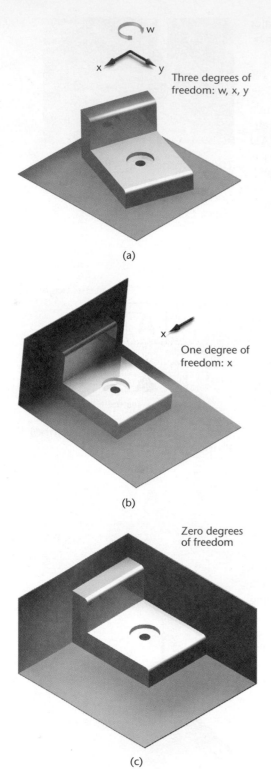

(a)

Three degrees of freedom: w, x, y

(b)

One degree of freedom: x

(c)

Zero degrees of freedom

10.27 (a) The part restricted by datum A has three degrees of freedom. (b) The part restricted by datum surfaces A and B has one degree of freedom. (c) The part restricted by datums A, B, and C has no remaining degrees of freedom.

If the part rests on datum plane A, the part can no longer move in the Z-direction, nor can it rotate in u, or v motions. Now, it has three degrees of freedom, as shown in Figure 10.27(a). It can rotate in w and move in x and y.

Adding datum B further restricts the motion of the part. Now, it can no longer move in the Y-direction, or rotate in w. It has one remaining degree of freedom, as shown in Figure 10.27(b).

Adding datum C restricts the final motion to produce a part that is stationary, as shown in Figure 10.27(c).

The X-, Y-, and Z-axes may be specified on the drawing and in the feature control frame as needed for clarity.

Orientation Tolerances for Related Features

Parallelism, perpendicularity, and angularity are tolerances that relate the orientation of one feature to another. Tolerances for related features require that datum references be specified in the feature control frame.

Orientation tolerances must be related to one or more datums. Only the rotational degrees of freedom are constrained when an orientation tolerance is specified. Orientation tolerances may specify the following:

- A zone defined by two parallel planes at the specified basic angle from, perpendicular to, or parallel to one or more datum planes or a datum axis. The surface or center plane of the feature must lie within this zone.
- A zone defined by two parallel planes at the specified basic angle from, perpendicular to, or parallel to one or more datum planes or a datum axis. The axis of the feature must lie within this zone.
- A cylindrical zone at the specified base angle from, perpendicular to, or parallel to one or more datum planes or a datum axis. The axis of the feature must lie within this zone.
- A zone defined by two parallel lines at the specified basic angle from, perpendicular to, or parallel to one or more datum planes or a datum axis. Any line element of the feature must lie within this zone.

Angularity specifies the condition of a surface, feature's center plane, or feature's axis at a specified angle from a datum plane or datum axis (Figure 10.28).

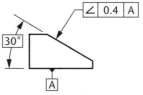

This on the drawing . . .

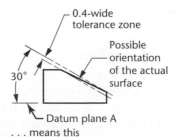

10.28 Angularity   . . . means this

Parallelism specifies the condition of a surface or feature's center plane equidistant from a datum plane, and for a feature's axis, equidistant along its length from one or more datum planes or datum axes (Figures 10.29 and 10.30).

Perpendicularity specifies the condition of a surface, feature's center plane, or feature's axis at a right angle from a datum plane or datum axis. Perpendicularity is the same as angularity when specifying a 90° angle. Either method can be used (Figures 10.31 and 10.32).

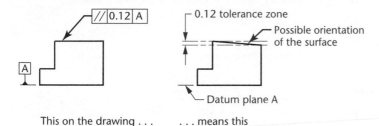

10.29 Parallelism for a Plane Surface

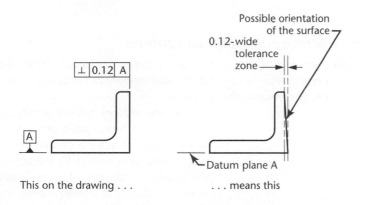

10.30 Parallelism for an Axis

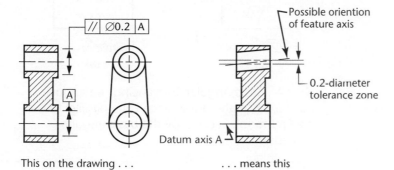

10.31 Perpendicularity for a Plane

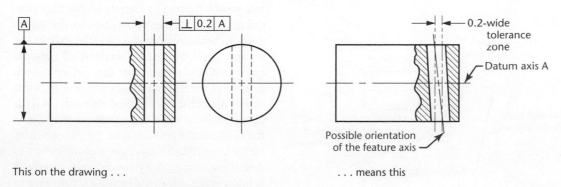

10.32 Perpendicularity for an Axis

Angular Tolerance: GD&T Versus +/− Degrees

Bilateral tolerances have traditionally been used to specify a plus/minus variation for an angular dimension (often of 1°). With bilateral tolerances, the wedge-shaped tolerance zone increases as the distance from the vertex of the angle increases (see Figure 10.33).

To avoid a wider tolerance zone farther away from the vertex of the angle, make the angular measure a basic dimension, as illustrated in Figure 10.34.

The angle is indicated as a basic dimension, so no angular tolerance is specified. The tolerance zone is defined by two parallel planes around the true location of the surface, resulting in improved angular control.

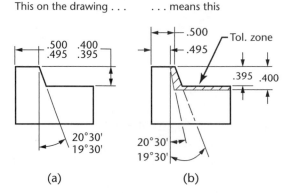

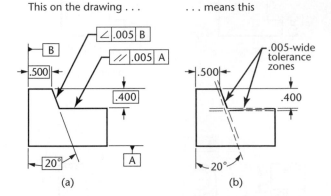

10.33 Bilateral Tolerancing of Angular Dimensions. The effect of the 1° tolerance on the angular measure is magnified as the distance from the vertex increases.

10.34 Basic Angular Tolerancing. An angular tolerance creates a tolerance zone parallel to the perfect form specified by the basic angular dimension.

Location Tolerances for Related Features

Position and concentricity tolerances are geometric characteristics that relate the location of one feature to another. Runout tolerances relate how well a circular or cylindrical part rotates about its axis.

A *concentricity tolerance* indicates that a cylinder, cone, hex, square, or surface of revolution feature shares a common axis with a datum feature. A concentricity tolerance controls the location for the axis of the indicated feature within a cylindrical tolerance zone whose axis coincides with a datum axis. The controlled feature must have its axis entirely within the cylindrical tolerance zone established by the datum axis. To measure concentricity, the axis of the feature is established by analysis of multiple cross-sectional elements of the surface (which calculates the median points between multiple sets of diametrically opposed elements). All the center points for the cross sections must lie in the cylindrical tolerance zone for concentricity (see Figure 10.35). Because measurements to determine concentricity are difficult to make, it is often preferable to use positional tolerance or runout tolerance instead.

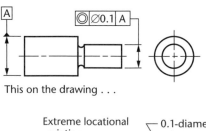

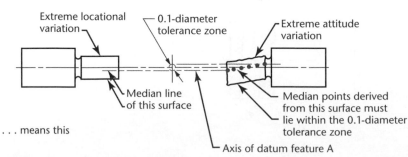

10.35 Concentricity

Runout

Runout is another geometric characteristic that can be specified in a feature control frame. Runout is a measure of how a circular or cylindrical part rotates about its axis. When a perfectly formed wheel is rotated on its axis, the surface of the wheel does not dip above or below the reference plane. To measure runout, an indicator is mounted on a stationary surface and touching the rotating feature so it can determine how much the rotating feature moves above or below the plane. The full indicator movement (FIM) is used to measure the runout. In other words, if the indicator moved from −1 to +1, the FIM would be 2.

Two types of runout can be specified in the feature control frame. *Circular runout* positions the indicator at a single location, where the variation in the circular shape of the rotating object is measured by the indicator (see Figure 10.36.)

This on the drawing . . .

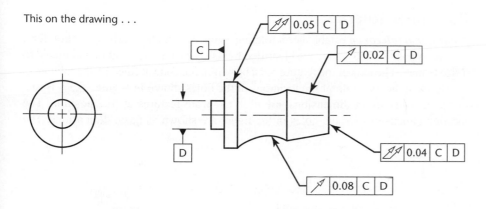

Means this . . .

Datum plane C

90°

At any measuring position, each circular element (for circular runout) and each surface (for total runout) must be within specified runout tolerance when the part is mounted on datum surface C and rotated 360° about datum axis D

Datum axis D

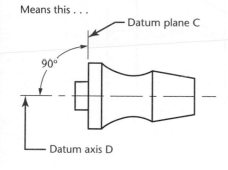

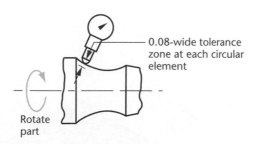

0.05-wide tolerance zone along surface

Rotate part

0.08-wide tolerance zone at each circular element

Rotate part

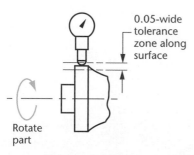

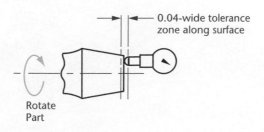

0.02-wide tolerance zone at each circular element

Rotate part

0.04-wide tolerance zone along surface

Rotate Part

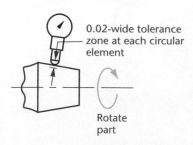

10.36 Circular and Total Runout

To measure **total runout**, the indicator is moved along the rotating surface, and the total allowable variation of the indicator over the surface is specified.

Basic Dimensions

Basic dimensions are the theoretically exact, untoleranced dimensions that specify the perfect location, size, shape, or angle of a feature. A basic dimension is identified by enclosing the dimension in a box, as shown in Figure 10.37(a). Basic dimensions specify theoretically exact values from which variations are defined through feature control frames as well as toleranced features or notes.

Sometimes it is noted on the drawing that GENERAL TOLERANCES DO NOT APPLY TO BASIC DIMENSIONS to prevent general tolerances from being applied to basic dimensions by those not skilled in interpreting GD&T symbols.

Positional Tolerance

A **positional tolerance** establishes a tolerance zone around the perfect location for a feature. A positional tolerance is identified by its characteristic symbol and linked to a size feature. The axis of the feature must be within the zone defined by the tolerance.

Look at the example of the locations of the holes shown in Figure 10.37. Basic, or theoretically exact dimensions, are used to locate features at true position. The location dimensions for the holes in the figure are shown as basic dimensions. The

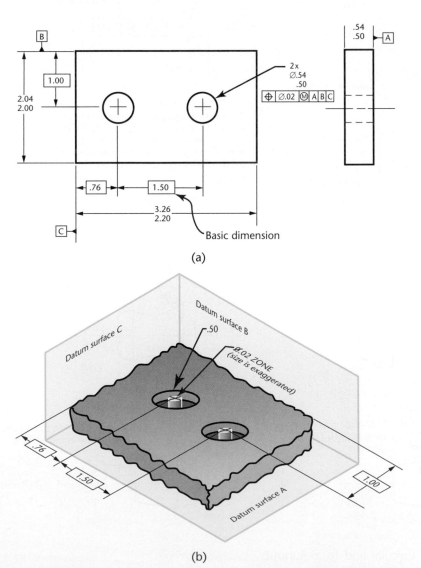

(a)

10.37   (a) The feature control frame in the orthographic view means that when the holes are at their smallest size, their positions can vary from the theoretically exact basic dimension locations within a zone that is .02 in diameter. (b) Pictorially, the small cylinders at the center of the holes represent the tolerance zone for the locations of the hole centers. The centerline of the cylinder forming the hole must remain inside this zone.

(b)

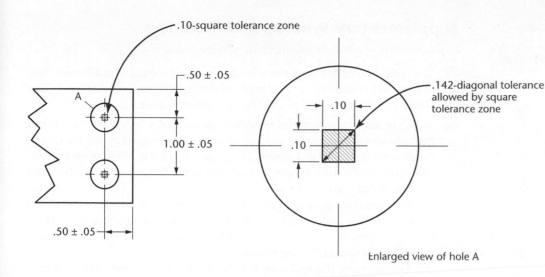

Enlarged view of hole A

positional tolerance specification requires that the location for the center of the hole remain inside a cylindrical zone .02 in diameter. The circled M in the example is a modifier that indicates that the tolerance applies when the hole is at maximum material condition. The envelope for the hole requires all elements on the surface of the hole to be on or outside a cylinder whose diameter is equal to the minimum diameter (actual minimum material envelope). In this case, the hole must have its center in the .02 cylindrical zone but also may not exceed the actual mating envelope.

Using GD&T symbols to specify positional tolerances overcomes the limitations of stating dimensions using only limit tolerances or plus/minus ranges. The general shape of the tolerance zone described by limit or plus/minus tolerances is typically rectangular, regardless of the shape of the feature. This rectangular tolerance zone may not fully convey the tolerance needed for certain features.

Figure 10.38 shows the end of a plate with two circular holes. If the location dimension for the center of a hole is stated using a bilateral tolerance as shown, the center of the actual hole can be anywhere in the shaded area shown and meet the stated tolerance. If the center is located in the upper right corner (along the diagonal) of that square shape, however, the distance from the true position for the center of the circle can be off by .071 when the stated tolerance is only .05. Parts that are toleranced in this way may be accepted even when they may be truly unacceptable. If the tolerance was reduced so that this part would be rejected, a part with a center that varied in a vertical or horizontal direction would be rejected although it would be an acceptable part.

In comparison, GD&T (often called *true position tolerancing*) specifies how the part may vary from the perfect geometry implied on the drawing by defining either the diameter or the width of a tolerance zone.

Figure 10.39 shows a more complex example of positional tolerance for holes.

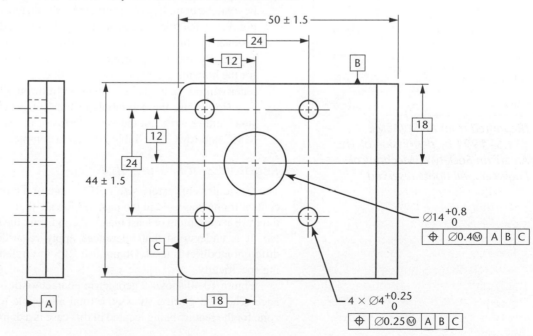

10.39 Basic Dimensions and Positional Tolerance

Table 10.5 Modifying Symbols.

Term	Symbol
At maximum material condition (tolerance value)/At maximum material boundary (datum reference)	Ⓜ
At least material condition (tolerance value)/At least material boundary (datum reference)	Ⓛ
Translation	▷
Projected tolerance zone	Ⓟ
Free state	Ⓕ
Tangent plane	Ⓣ
Unequally disposed profile	Ⓤ
Independency	Ⓘ
Statistical tolerance	⟨ST⟩
Continuous feature	⟨CF⟩
Diameter	∅
Spherical diameter	S∅
Radius	R
Spherical radius	SR
Controlled radius	CR
Square	□
Reference	( )
Arc length	⌒
Dimension origin	⊕→
Between	↔
All around	⟲
All over	⟲

(Reprinted from ANSI /ASME Y14.5–1994 by permission of The American Society of Mechanical Engineers. All rights reserved.)

Supplementary Symbols and Modifiers

Supplementary symbols or modifiers include the following:

∅   The diameter symbol often precedes the specified tolerance in a feature control symbol to describe a circular or cylindrical shaped tolerance zone. This symbol for diameter should precede the tolerance value.

Ⓜ   *MMC,* or *maximum material condition,* means that a feature on the finished part contains the maximum amount of material permitted by the toleranced dimensions shown for that feature. Holes and slots are at MMC when at minimum size. Shafts, tabs, bosses, and other external features are at MMC when they are at their maximum size.

Because all features may vary in size, it is necessary to make clear on the drawing at what limit of size for the feature the characteristic applies. In all but a few cases, when a feature is not at MMC, additional positional tolerance is available without affecting function.

For example, if a tolerance is indicated to apply at maximum material condition, then the full tolerance stated is available when a hole is at its smallest size or a shaft is at its greatest size. As the actual part deviates from maximum material condition toward the minimum material condition a bonus tolerance may be allowable. Consider trying to fit a fixed pin into a hole. As the hole size becomes larger the pin can be off from its perfect location by a larger amount and still fit through the hole. When MMC is indicated, the **bonus tolerance** is the difference between the actual hole size measured on the feature and the maximum material condition. In this case, this extra tolerance can be added to the positional tolerance without affecting the function of the parts.

To avoid possible misinterpretation as to whether maximum material condition applies, it should be clearly stated on the drawing by adding an MMC symbol to each applicable tolerance or by suitable coverage in a document referenced on the drawing.

Ⓛ   *LMC,* or *least material condition,* indicates that the feature contains the least amount of material (as opposed to MMC); LMC indicates the maximum size for an internal feature like a hole or slot and the minimum for an external feature such as a shaft or tab.

Ⓟ   The projected tolerance zone symbol indicates that the tolerance zone applies beyond the boundary of the object by the distance stated to the left of the symbol. Projected tolerance zones are typically used when a threaded or press fit fastener might interfere with a mating part if it is inserted into a hole that is angled. By stating a projected tolerance zone, the true position of the center for the hole must apply beyond the edge of the part, ensuring that a fixed fastener will not be angled too much to fit through the hole in a mating part.

Do not use these symbols in local or general notes in the drawing. Instead, either write out the entire terms or use standard abbreviations such as MMC and LMC. See Table 10.5 for a complete list.

Regardless of Feature Size

The "regardless of feature size" (RFS) note or a circled S modifier in the feature control frame was used in the past to indicate that it does not matter whether features are at maximum or least material condition; the tolerance still applies as written. It is understood that tolerances apply regardless of feature size unless a different modifier is used. Do not use RFS or circled S to indicate it on the drawing specifically.

Figure 10.40 shows a geometric characteristic symbol for a positional tolerance. The next portion shows the total allowable tolerance zone. The tolerance zone for the feature being located in this case is a diametral-shaped zone (indicated

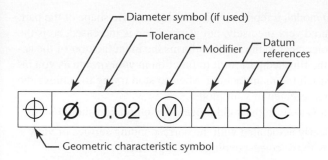

Diameter symbol (if used)
Tolerance
Modifier
Datum references

⊕ | Ø 0.02 | Ⓜ | A | B | C

Geometric characteristic symbol

10.40 This feature control frame uses modifiers and datum references.

by the diameter symbol) .02 inch around the perfect position for the feature. This is followed by any modifiers, in this case a circled M, for maximum material condition. The remaining boxes indicate that the measurements are to be made from datums A, B, and C. The order of the datum symbols indicates which is the primary, secondary, and tertiary datum, respectively.

10.4 TOLERANCES AND DIGITAL PRODUCT DEFINITION

Dimensioning and tolerancing can take place directly in the 3D digital database (Figure 10.41). The electronic file can be transmitted as the digital product definition specifying the shape, size, and finish to the company that will manufacture and/or assemble the parts.

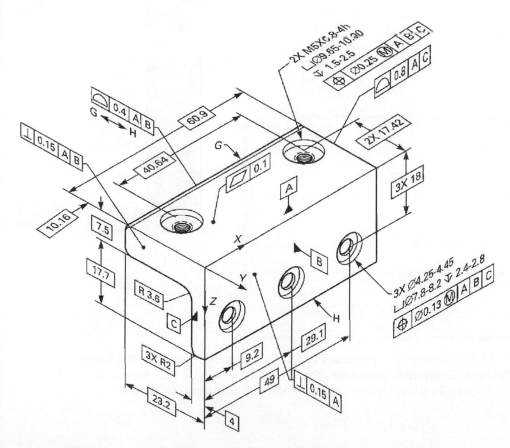

10.41 Tolerances can be added directly to a 3D model so that it can be used as the digital product definition. *(Reprinted from ASME Y14.41-2003 by permission of The American Society of Mechanical Engineers. All rights reserved.)*

When you create a 3D model, it represents the ideal geometric shape of the part. The part can be manufactured very precisely, but as precision is increased, so is the price of the part. Adding tolerances to the model informs the manufacturer of the accuracy that is required on the finished part for it to function in your design as you intend. Essentially, you must tell the manufacturer when to stop trying to achieve the level of perfection that is represented in your model.

When you include annotations in the solid model, the annotation should

- be in a plane that is clearly associated with the corresponding surface or view.
- be clearly associated with the corresponding model geometry.
- be capable of being printed and meet applicable drawing standards.
- be possible to display on the screen or turn off.

In general, tolerances and annotations provided directly in the model need to be interpreted to achieve the result that you intend. A combination of drawings and the model, only drawings, or a fully annotated model are all available methods for documenting a design. Each method has its advantages and disadvantages. You should investigate which method is suitable for your use and understand the applicable standards. Partially dimensioned drawings should note: SEE DATASET FOR FULL GEOMETRIC DESCRIPTION.

Figure 10.42 shows an example of tolerancing in a 3D CAD model. In general, you can choose whether or not to show annotations such as these tolerance symbols. If the model is properly documented, it is not necessary to make a drawing to accompany the model when obtaining bids from vendors. 2D drawings from the model are still often useful as a way of clearly presenting the information in a familiar way.

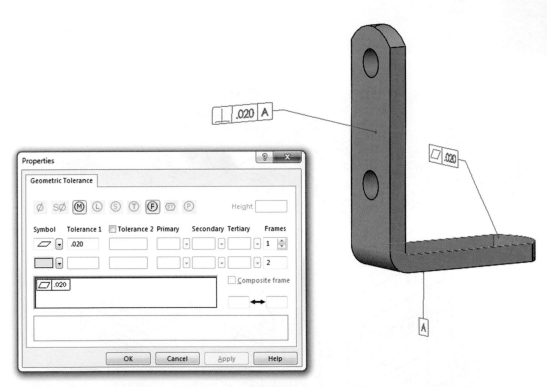

10.42 GD&T symbols can be added to SolidWorks models by selecting **Insert, Annotations, Geometric Tolerance.** The dialog box makes it easy for you to build GD&T symbols. To show the symbols, pick **View, Annotations.**

10.5 SURFACE CONTROL

Surface finish marks may be added to drawings to specify the degree of smoothness for the machined surfaces. Surfaces on a cast or forged part are not smooth when the part is formed. Machining removes material from the surface to provide a smooth finish. *Finish marks* indicate which surfaces are to be machined and the quality of the finish (see Figure 10.43). These marks are not needed for parts to be completely made by machining processes or from sheet metal. If all surfaces of a part are to be finished, "Finish all over" (or FAO) may be added as a note in lieu of finish marks.

Before a part is cast or forged, additional material is added to surfaces that will be machined later. For example, if a slot is to be formed by a core in a casting, its surface will be rough and pebbly. Additional material is added to the casting so the slot can be machined smooth when the part is finished. The determination of how much material to add is often left up to the manufacturer. When it is not, some companies provide two sets of drawings for parts that are to be forged or cast. One drawing shows all the information necessary for the forging or casting, and a second drawing—for the machine shop—shows the finished dimensions for the part.

The finish marks may indicate not only that the surface is to be machined but also that there are special requirements for the finish. The process used to finish the part, such as milling or grinding, and the direction in which a tool is applied to the surface can result in varying degrees of smoothness, called *surface texture*. Controlling the surface texture can be critical to the function of bearings and seals and to reducing the friction between moving parts. If the texture of the surface is important to its function, it should be documented in the drawing by adding additional information to the finish mark.

Finished surfaces have the following qualities:

- roughness
- waviness
- lay

The *roughness* of a surface is a measure of the irregularities in the finish of the surface, as shown in Figure 10.44. Roughness is measured in micrometers (μm), which are millionths of a meter, or in microinches (μin.), which are millionths of an inch.

Figure 10.45 shows the surface finish mark used with a value indicating the average allowable roughness height.

Figure 10.46 shows variations of the surface finish mark. Use these to indicate when additional material for later machining must be added to a casting or forging and when removing material by machining the surface is prohibited.

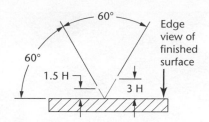

10.43   Surface finish marks are placed in every view where a finished surface appears on edge, even if it is shown as a hidden line. (H = letter/symbol height.)

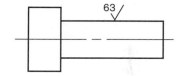

10.45   The finish mark indicates that an arithmetic average roughness of 63 microinches must be met for the surface of the cylinder.

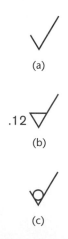

(a)

(b)

(c)

10.46   (a) The standard finish mark indicates that the surface may need to be machined to meet requirements. (b) Machining is required. Additional material must be allowed. A number to the left of the symbol indicates the value for the additional material required. (c) Machining is prohibited. The surface must be produced by a process such as forging or casting.

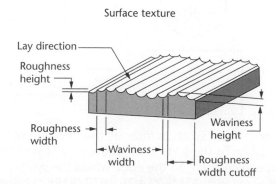

10.44   Roughness and waviness are defined in terms of their height (deviation from the mean plane of the surface) and width (the distance between peaks and valleys). Both measures can be used to define the degree of smoothness of the surface.

10.47 The rectangular block shown has a visible pattern (lay) for the primary surface of the part.

The *waviness* of a surface is the more widely spaced irregularity of the surface due to warping, machine vibrations, heat treating, or other similar factors. Waviness is measured in millimeters or inches.

The *lay* of a surface describes the direction or arrangement of the primary surface pattern, as shown in Figure 10.47. If it is important to the design of the part, the lay symbol is used to indicate the desired arrangement for the surface pattern caused by the machining process (see Figures 10.48 and 10.49). Different degrees of roughness can be expected from different manufacturing processes (see Figure 10.50.)

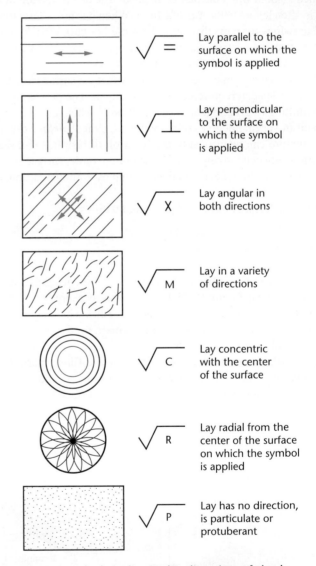

10.48 These symbols indicate the direction of the lay in relation to the surface or, in the case of parallel or perpendicular lay, to the edge where the symbol is located. (*Reprinted from* ASME Y14.36-1978 (R1987) *by permission of The American Society of Mechanical Engineers. All rights reserved.*)

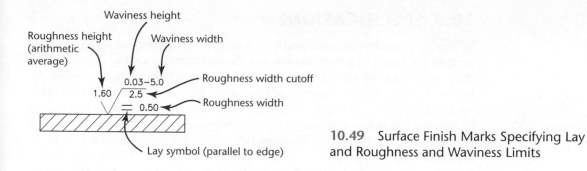

10.49 Surface Finish Marks Specifying Lay and Roughness and Waviness Limits

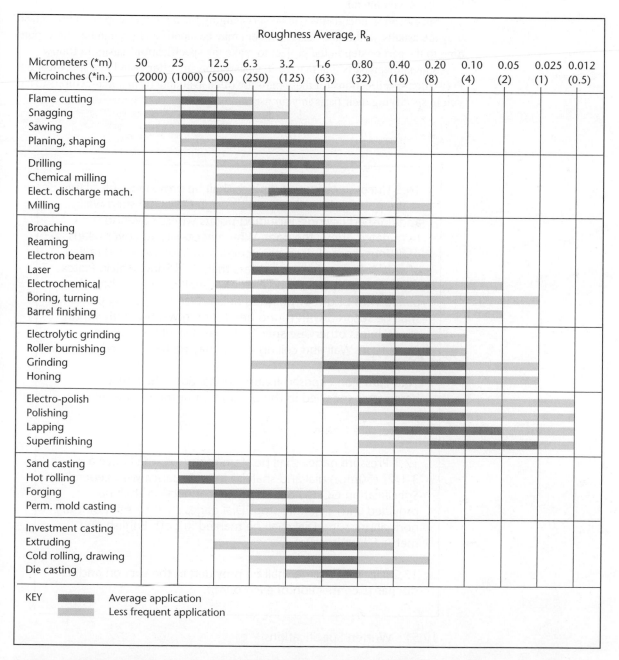

	Roughness Average, R_a												
Micrometers (*m)	50	25	12.5	6.3	3.2	1.6	0.80	0.40	0.20	0.10	0.05	0.025	0.012
Microinches (*in.)	(2000)	(1000)	(500)	(250)	(125)	(63)	(32)	(16)	(8)	(4)	(2)	(1)	(0.5)

KEY
■ Average application
■ Less frequent application

10.50 Different manufacturing processes produce different degrees of smoothness. The range of values for surface roughness shown here is typical of the processes listed under normal condition. (*Reprinted from* ASME B46.1-1985 *by permission of The American Society of Mechanical Engineers. All rights reserved.*)

10.6 SPECIFICATIONS

Written specifications are often used to supplement the requirements for constructing designs. For example, a written document that accompanies a set of drawings for a design might indicate special considerations for material type, strength, textures, appearances, handling, tolerances, finishes, work processes, assembly techniques, safety requirements, packaging, shipping, or any other detailed information that may be necessary to bid or construct the design. An example specification is shown in Figure 10.51. Well-written specifications help ensure that products quoted from various vendors are equal quality. Many organizations, such as ASTM, ASME, UL, among others, publish standards and specifications that can help guide you in writing the detailed specifications required to ensure that the product or design is manufactured as you intend.

Specification documents can be very lengthy owing to the scope of a project, so they are usually well organized. Each item may be identified by a number corresponding to the part or step in the project to make the specifications easier to follow.

Standardized parts that may be in an assembly are often listed in a specification. Prewritten specifications for these parts are often available from manufacturers to aid you in specifying their parts in your assembled product.

16.8 Uncovered, exposed pipes shall be provided with plates at the points where they pass through floors, finished walls, and finished ceilings, including points where covering is terminated above the floor. Where necessary to cover heads of fittings, special deep escutcheons shall be provided in lieu of plates. Plates shall be not less than 0.018-inch thick. Plates on chromium plated pipe or tubing shall be brass, chromium plated with bright or polished finish. Wall and ceiling plates shall be secured with round head set screws, not with spring clips. Unless otherwise specified, plates shall be of the one-piece type. Wall and ceiling plates may be flat, hinged pattern.

16.9 Plates on exposed fixture connections shall be as hereinafter specified in the description of the fixture trim.

17 Pressure Gages

17.1 Pressure gages shall be stem mounted, shall have a 3-1/2" (80mm) dial and shall be in accordance with Federal Specification GG-G-760. The gage connection shall be provided with T-handle stop. Dial gages shall be midpoint of normal pressure. Dial shall be marked in both English and metric S.I. units (kPa).

17.2 A pressure gage shall be provided in the suction and discharge connection of each pump.

10.51 Written Specifications

TOLERANCE ANALYSIS: KEEP THE FOCUS

When something is "focused like a laser beam" you know that it is right on target. For the solid-state lasers designed at Quantel USA, focusing high energy into the smallest beam possible and steering it to the desired target means that designers like Mark Perkins must specify many different kinds of tolerances, including geometric tolerances for parallelism, perpendicularity, and flatness.

A solid-state flashlamp focuses light on a laser rod inside a sealed pump cavity (Figure 10.52). The laser rod emits the laser beam (photons), which must pass through other optics and bounce off mirrors at either end of the laser resonator—an optical cavity containing at a minimum the pump cavity and, commonly, several other optics. The laser operates by having photons bounce back and forth between the two mirrors, with each pass through the pump cavity providing additional photons (Figure 10.53). Some of the light is allowed to leave, as one mirror is usually a partial reflector, sending anywhere from 10%–50% of the photons that hit its surface back toward the laser rod and other mirror while the remaining photons leave the resonator. The laser will emit light as long as the gain from the pump cavity exceeds the losses in the resonator. To generate stronger beams, multiple pump cavities may be used. A

10.54 This assembly model shows the retainers (transparent in part a) that hold the wedge optics and are rotated to steer the beam.

laser beam with a diameter of 1/4 inch may need to center on a rod in the next pump cavity that has a diameter of no more than 5/16 inch.

To align the laser beam, the mirrors in the resonator must be parallel to a very high degree of accuracy (within 0.0006°) so that the reflected beam stays properly aligned (allowing the photons to continue to bounce back and forth between the two mirrors). All the mirrors and optics also have flatness, parallelism, and finish tolerances to enable the beam to resonate and stay aligned in the optical cavity. Optics are commonly manufactured with flatness tolerances of 4×10^{-6} inch over their diameter, to prevent distortion of the laser beam as it propagates through the optic!

Another major goal for the laser designers is efficiency. Most of the light energy going into the laser rod generates heat; only about 1%–2% is transformed to laser light. Any misalignment causes photons to leave the clear aperture. This results in output energy loss, and because the photons are absorbed by the components they strike, may also result in damage to optical mounting components. Ultimately, this could cause contamination of the optical surfaces. Particles generated by ablation from the laser beam could coat the optics and absorb energy—increasing the losses in the resonator.

To keep the beam within its tolerances, a pair of optical wedges are used to steer the laser beam inside the cavity (see Figure 10.54). Each wedge acts like a prism, as it is thicker on one edge than the other, and refracts the beam by an amount proportional to the angle between the two faces, and the index of refraction (material property) of the wedge.

Rotating the wedges steers the laser beam anywhere within the cone angle equal to the refraction angle sum of the

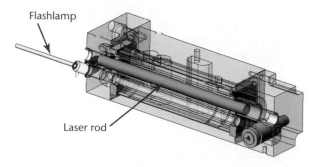

10.52 A 3D Model of a Pump Cavity

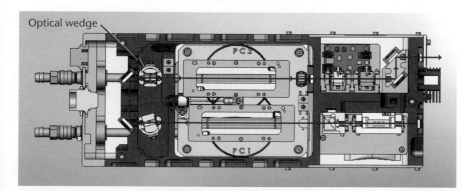

10.53 A Model of the Beam Path (Red) in a Laser

(continued)

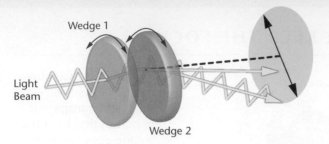

Wedge 1

Light Beam

Wedge 2

10.55 Turning the wedge prisms steers the beam.

10.56 This 3D model of the path that the laser beam takes through an optical system shows the worst-case scenarios (which is why the beam looks conical in some places).

wedge pair. The yellow-green circle in Figure 10.55 indicates the maximum steering possible with this set of wedges.

Every mirror and optic that the beam passes through presents an opportunity for misalignment, as the optical surfaces on each side can never be made perfectly parallel. Selecting the right amount of wedge angle for the wedge pair requires the designer to strike a balance between the accuracy and ease of alignment it provides and the difficulty of achieving the desired tolerances when manufacturing the mechanical parts that make it work.

Selecting the wedge angle also involves a worst-case analysis of the entire system. By combining the variation allowed by the flatness tolerances of the wedges, the perpendicularity and parallelism tolerances for the beam, and tolerances for the mirrors and other optics, the designer can generate a worst-case scenario. That is, if all the components varied by the maximum amount allowed, how much would the beam be misaligned? If the beams can be steered that much by the wedges, then the laser can stay aligned.

A typical analysis involves determining the maximum amount of angular misalignment that could be encountered as a sum of the nonparallelism component due to every optic inside the resonator, plus the inaccuracy of the mounting components for the end mirrors. For example, if the mounting surfaces of the end mirrors are specified to be parallel to each other within .002 inch, with 3/4-inch diameter mirrors, the misalignment due to the mirrors themselves is equal to $\sin^{-1}(.002/.75) = 0.15° [2.6 \times 10^{-3}$ rad].

Mark Perkins uses 3D modeling to similarly assess existing lasers that need to be realigned. Using the maximum tolerances, he revolves a cone shape that represents the maximum spread of the beam (Figure 10.56). Using this model, he can assess whether the maximum spread will cause the beam to clip the aperture. The aperture or steering mechanism can then be adjusted as needed.

Specialized tools for optical engineers work in a similar way. In addition to many other analyses, the software simulates the passage of the beam through an optic to see how much it can vary. It generates a ray tracing of the beam that can then be imported into the 3D model of the laser housing to see whether it will clip any parts of the housing or the aperture.

In addition to the tolerances that affect the alignment of the beam, other tolerances are important to the laser's effectiveness. The laser pump cavity is typically water cooled and must be sealed to prevent leakage of cooling fluids. The laser resonator must be hermetically sealed to prevent dust and other contaminants from entering the cavity. The O-rings used to seal these cavities fit into grooves that are specified to a certain depth, and also to a specified degree of surface finish. This ensures that there will be no gaps in the seal caused by irregularities in the surface finish, which commonly result from machining marks (Figure 10.57).

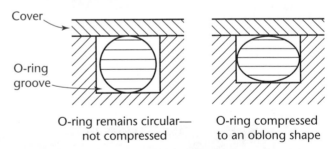

Cover

O-ring groove

O-ring remains circular— not compressed

O-ring compressed to an oblong shape

10.57 If the O-ring groove is sized and toleranced properly, the O-ring will be compressed and create a seal.

(Figures courtesy of Quantel USA.)

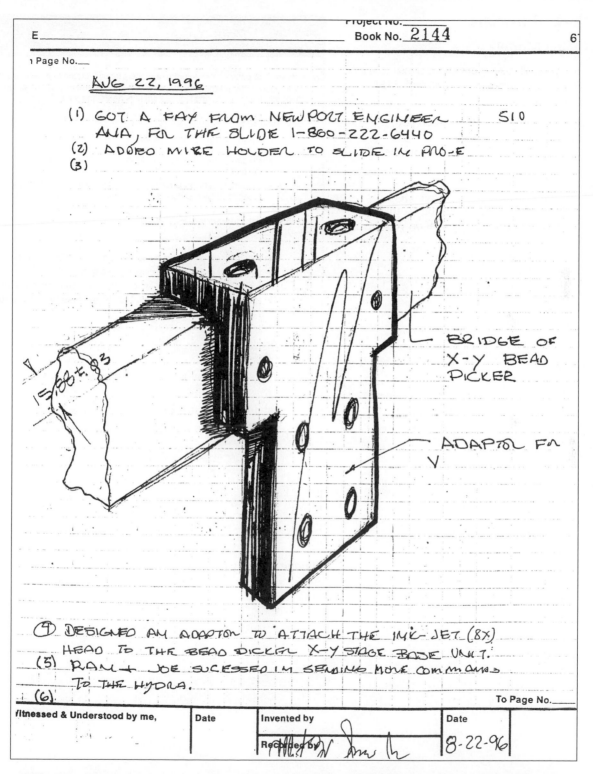

AUG 22, 1996

(1) GOT A FAX FROM NEWPORT ENGINEER   510
    ANA, FOR THE SLIDE 1-860-222-6440
(2) ADDED MIRE HOLDER TO SLIDE IN PRO-E
(3)

15.88 ± .03

BRIDGE OF
X-Y BEAD
PICKER

ADAPTOL FM
V

(4) DESIGNED AN ADAPTOR TO ATTACH THE INK-JET (8X)
    HEAD TO THE BEAD PICKER X-Y STAGE BADE UNIT.
(5) RAN + JOE SUCESSED IN SENDING MORE COMMANDS
    TO THE HYDRA.
(6)

Witnessed & Understood by me,	Date	Invented by	Date
		Recorded by	8-22-96

10.58 The tolerances built into the manufacture of the X-Y bead picker bridge are noted on this sketch of an adapter that fits onto it. The adapter must accommodate a bridge width that can vary from 15.85 to 15.91. (*Courtesy of Albert W. Brown, Jr., Affymax Research Institute.*)

HANDS ON TUTORIALS

SolidWorks Tutorials

Tolerancing a Model

In this tutorial, you will learn more about semiautomatic dimensioning and tolerancing features available for use with your models.

- Launch SolidWorks from the icon on your desktop.
- Click to expand the topics listed below the "**?**" help icon.
- Click: **SolidWorks Tutorials.**
- Click: **All SolidWorks Tutorials (Set 1).**
- Click: **DimXpert Tutorials** from the list that appears, and complete the tutorial on your own.

Overview

DimXpert for parts helps you prepare models for conversion to drawings or for use in TolAnalyst. DimXpert works by inserting dimensions and tolerances, automatically or manually, in manufacturing features such as holes and slots.

Use DimXpert to:

- Prepare a model for conversion to a manufacturing drawing to ensure that the part will be built correctly.
- Prepare several parts for the TolAnalyst add-in. TolAnalyst automatically recognizes tolerances and dimensions created in DimXpert.

In this tutorial, you learn how to:

- Automatically dimension a prismatic part using plus-minus tolerancing.
- Create a drawing from the dimensioned part.
- Automatically dimension a turned part using geometric tolerancing.
- Use manual and automatic dimensioning to prepare a part for TolAnalyst.

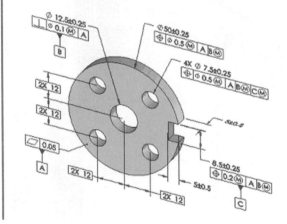

(Image courtesy of DS SolidWorks Corp.)

REVERSE ENGINEERING PROJECT

The Can Opener Project

1. *Practicing tolerance.* Assign tolerances to the lower handle part model of your can opener. Then, add tolerances to your drawing of the lower handle part. Your drawing should look similar to the example below.
2. *How much is enough?* What parts of the can opener have fits that are critical to the can opener function?

    This can opener sells for around $5.00. Consider and describe some of the effects on the price of the can opener if highly precise fits are required.

(Courtesy of Focus Products Group, LLC.)

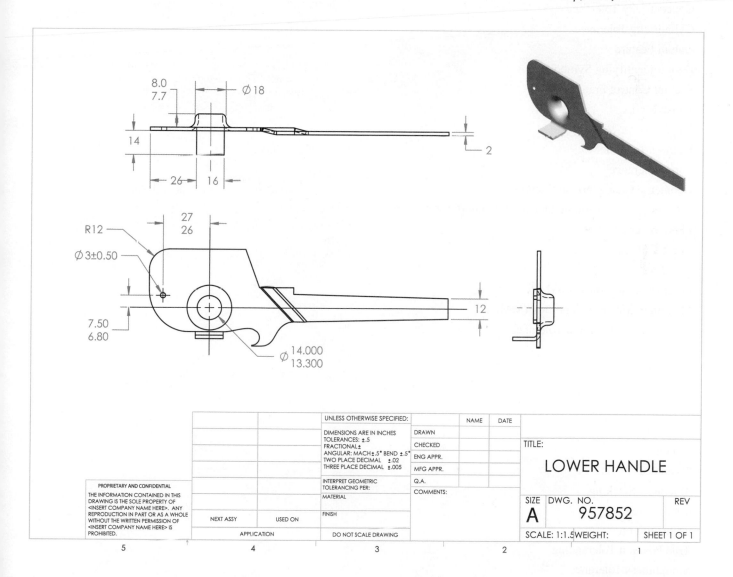

KEY WORDS

Allowance

Angularity

Baseline Dimensioning

Basic Dimension

Bilateral Tolerance

Bonus Tolerance

Chained Dimension

Circularity (Roundness) Tolerance

Circular Runout

Clearance Fit

Concentricity Tolerance

Cylindricity Tolerance

Datum Feature

Datum Identifying Symbol

Feature Control Frame

Finish Mark

Fit

Flatness Tolerance

Form Tolerance

Geometric Characteristic Symbol

Geometric Dimensioning and Tolerancing (GD&T)

Interference Fit

Limit Tolerance

Line Fit

Least Material Condition (LMC)

Maximum Material Condition (MMC)

Orientation or Location Tolerance

Parallelism

Perpendicularity

Plus/Minus Tolerancing

Positional Tolerance

Profile Tolerance

Runout

Tolerance

Tolerance Stacking

Total Runout

Transition Fit

True Position Tolerancing

Straightness Tolerance

SKILLS SUMMARY

Now that you have completed this chapter, you should be able to add tolerances to your drawings to describe the allowable variation that an acceptable part may have from the stated dimensions. You should also be able to interpret orthographic views and model parts to specified tolerances. In addition, you should be able to model a given shape and create properly dimensioned and toleranced drawings to describe it. The following exercises will give you practice in all these skills.

EXERCISES

Exercises 1–5: Tolerances. Using your CAD package, create a solid model of each of the objects shown.

1. Dished washer. Grid squares represent .500 inch. Assume you are the manufacturer of this washer. To what tolerance range will you be able to manufacture your part? Parts that fall outside the range you specify may be rejected by your customer. Specify tolerances in your model for critical features.

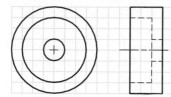

2. Offset bushing. Grid squares represent .250 inch. The central hole at the left of the part must be perpendicular to the bottom surface within .02 inch. Specify this in your model.

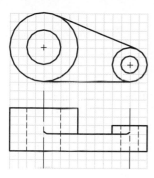

3. L-bracket. Grid squares represent .250 inch. The left end surface must be perpendicular to the bottom surface within .10 inch. The bottom surface of the part must be flat within .002 inch. Specify this in your model.

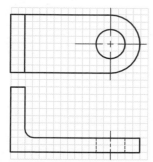

4. Bellcrank arm. Assume that grid squares are .250 inch. Radii are all .25 inch. The two arm surfaces must be parallel within .01 inch. Specify this in your model.

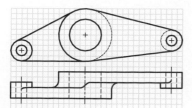

5. Shaft mounting bracket. Grid squares are .250 inch. Radii are all 0.25 inch. The axes of the two holes must be perpendicular. Specify a reasonable tolerance range in your model.

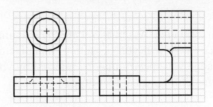

Exercises 6–10: Dimensioned and Toleranced Drawings. Create solid models of the objects shown below (or use the models you created in Chapter 9). Perform the following drawing operations on each completed model:

a. From your solid model, develop the orthographic projection views needed to fully define the object.

b. Place the required views properly, so that all views can be included on one sheet. Use standard-size drawings (choose size A, B, or C, as appropriate).

c. Fully dimension all part size elements, features, and locations.

d. Include tolerances as follows:

Feature	Tolerance
Through-hole diameter	+/−0.005 inch
Hole location from datum	+/−0.030 inch
Overall part dimensions	+/−0.010 inch
Fillets and chamfers	+/−0.010 inch

6. Notched block. Each grid square is .50 inch.

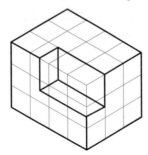

7. Single-rivet nut plate. All holes through. Each grid square is .1250 inch.

8. Saddle bracket. All holes are through. Assume that each grid square is .50 inch.

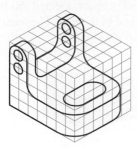

9. Keeper. Large holes pass through; small holes are .375 diameter, drilled and tapped to .50 depth. Assume grid squares are .50 inch. Radii are .50 inch.

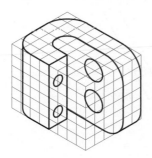

10. Rail mount shaft support. All holes pass through. Rail track on underside of support extends the length of the part. Grid squares are .50 inch apart.

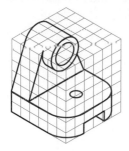

11. Explain "Tolerance Stacking." How can the selection of *chain dimensioning* versus *baseline dimensioning* affect the tolerances on a finished part?

12. Why would you want to leave out one of the chain dimensions in the following part? Redraw the part using correct dimensioning practices. Add surface finish callouts as needed to specify roughness height of 32 microinches on each of the steps.

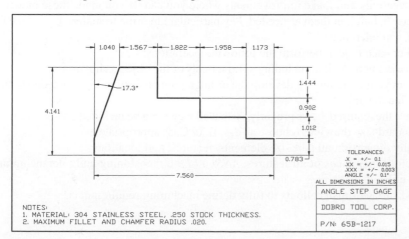

1.040   1.567   1.822   1.958   1.173

17.3°

1.444

0.902

1.012

4.141

0.783

7.560

TOLERANCES:
.X = +/- 0.1
.XX = +/- 0.015
.XXX = +/- 0.003
ANGLE +/- 0.1°
ALL DIMENSIONS IN INCHES

ANGLE STEP GAGE

DOBRO TOOL CORP.

P/N: 65B-1217

NOTES:
1. MATERIAL: 304 STAINLESS STEEL, .250 STOCK THICKNESS.
2. MAXIMUM FILLET AND CHAMFER RADIUS .020.

13. Sketch the figure shown. Use either limit dimensions, bilateral tolerances, or geometric tolerancing to add a hole to the left end of the part, located .50 inch from the bottom surface and 2 inches from the right end of the part. The location should be accurate to ±.005 and its size accurate to within ±.002.

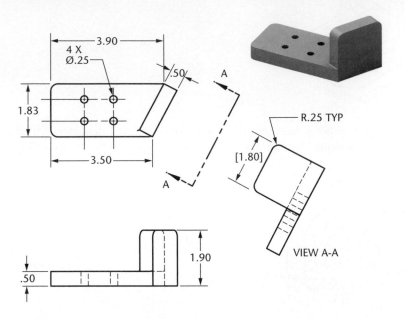

14. Add geometric dimensioning and tolerancing symbols to the drawing to do the following: (a) Control the flatness of the bottom surface to a total tolerance of .001. (b) Control perpendicularity of the left surface and bottom surface to .003. (c) Control the tolerance for the 30° angle to .01.

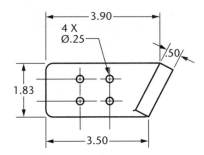

Exercises 15–22: Dimensioning and Tolerancing Applied to Real Parts. You may not know exact dimensions of existing parts, or you may have ideas for improvement—so use your knowledge to design your own parts! Here is an opportunity to apply your own creativity and make improvements where you can. If you made these models in Chapter 9, build on them as needed. Pay particular attention to materials, surface finishes, and tolerances.

For each object, perform the following operations:

a. Create a new design, including solid models of the objects shown.

b. From your solid model, develop the orthographic projection views needed to fully define the object.

c. Place the required views properly, so that all views can be included on one sheet. Use standard-size drawings (choose size A, B, or C, as appropriate).

d. Fully dimension all part size elements, features, and locations.

e. Include tolerances of all features. *ANSI-Y14.5* dimensioning and tolerancing may be utilized.

f. Add surface finish callouts to fully define machining requirements.

15. Bolt hanger for use in rock climbing and mountaineering. Overall size is about 2 inches. A climbing rope is clipped to the hanger, which is bolted to a rock face. This device must stop your fall, so it must be strong! Choose appropriate material and thickness.

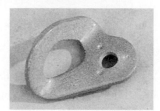

16. Climbing carabiner, to be used for general mountaineering purposes. It can clip a rope into the bolt hanger from Exercise 15. Most are made of 6061 aluminum and have a spring-loaded gate that opens with a light squeeze.

17. A rear cargo rack is a useful addition to a mountain bike. This one is made mostly of extruded aluminum, with welded joints. It is sized to bolt to the chainstays near the rear hub, and to the seat tube near the brake-mounting bosses.

18. Campers use lightweight tongs to lift pots off a cookstove. The lighter the better! About 5 inches long.

19. Socket wrench sets often include a U-joint to permit wrenching at a slight angle. This one is for a 3/8-inch square-drive socket set.

20. A folding saw for tree pruning. The 1/8-inch-wide stainless steel blade includes extra-sharp teeth. The blade length is around 12 inches. The handle on this model is wood, but what material will your handle use?

21. An optic mount is needed for a dielectric mirror as shown. You have been tasked with mounting a dielectric mirror at a 45° angle. You will do this by creating a mount out of 6061-T6 aluminum machined with a 45° face. The face will have a machined counterbore in which to mount the optic and three bond spots for attaching the optic. The specifications for the design are as follows:
   - Reference the drawing shown for a stock 1-inch dielectric mirror.
   - The diameter for the counterbore must be .005 inch to .010 inch larger than the maximum diameter of the optic.
   - The optic must sit at a minimum of .010 inch and maximum of .020 inch above the mounting surface.
   - The optic will be bonded to the mount using ultraviolet curing adhesive. The adhesive bond spots will be .100-inch diameter by .050 inch deep, equally spaced around the perimeter of the optic counterbore.
   - The center point of the outer surface of the optic will be located 1±.05 inch from the bottom of the mount.
   - The mount will be attached to a laser structure using two 6-32 UNC socket head cap screws.
     a. What diameter of counterbore is used to ensure that it is .005 to .010 inch larger than the maximum diameter of the optic?
     b. What depth of counterbore is used to ensure a .010/.020-inch protrusion height of the optic?
     c. Create a full mechanical drawing with appropriate tolerances for the designed mount.

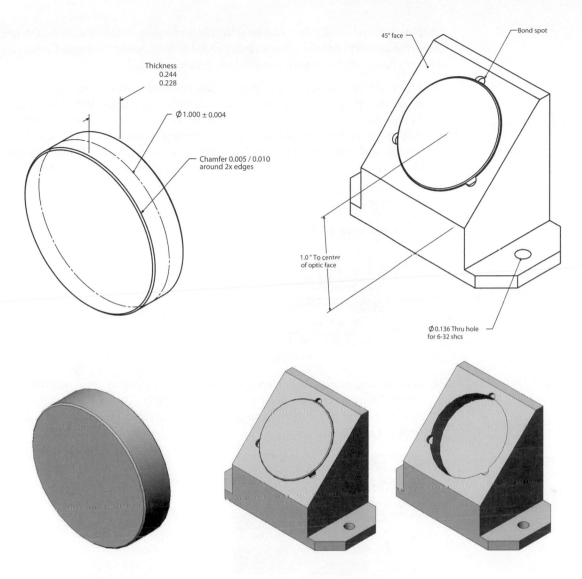

Thickness
0.244
0.228

Ø 1.000 ± 0.004

Chamfer 0.005 / 0.010
around 2x edges

45° face

Bond spot

1.0 " To center
of optic face

Ø 0.136 Thru hole
for 6-32 shcs

22. Design a bike rack using standard tubing products and standard fittings. Provide a means for mounting the rack to concrete. What accuracy is required for your design to function? Research the manufacturing accuracy for the parts you specify. What is the maximum allowance your mounting can be off and still allow the parts to fit?

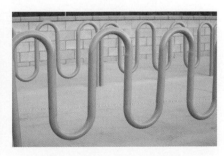

(©Igorsky/Shutterstock.)   *(©Stacy Barnett/Shutterstock.)*   *(©Margo Harrison/Shutterstock.)*

Challenge Exercises 23 & 24: Advanced Dimensioning, Tolerancing and Design.
Create the model(s) and the 2D drawings needed to fully define the parts. Create section views as needed to clarify construction details.

23. This cutaway view shows a gear-type shaft coupling used to transmit power between such elements as a motor output shaft and a pump input shaft. The hubs are fit onto the two shafts—usually with a shrink fit or interference fit. The external ring provides the connection. Precise alignment is mandatory, as is extreme strength and perfect balance. Design yours for mounting on 0.500-inch-diameter shafts.

24. Air-powered actuator cylinders are used in many industries for positioning and lifting. Many sizes and configurations are available. This one uses a stainless steel body and a 1/4-inch-diameter bore, with a captured air-powered piston stroke of about 3 inches.

APPENDIXES

CONTENTS

1 ASME/ANSI CODES AND STANDARDS

Available from www.asme.org.

Name	Designation
Authorized Inspection	QAI
Automotive Lifting Devices	PALD
Boiler and Pressure Vessel Code-2007	
Chains	B29
Compressors	B19
Controls	CSD
Conveyors	B20
Cranes and Hoists	B30, HST
Dimensions	B4, B32, B36
Drawings and Terminology	Y1, Y14, Y32
Elevators and Escalators	A17, QEI
Fasteners	FAP, B18, B27
Flow Management	MFC
Gage Blanks	B47
Gauges	B40
High Pressure Systems	HPS
Industrial Trucks	B56
Keys	B17
Machine Guarding	B15
Manlifts	A90
Measurement	B89, MC88
Metric System	SI
Nuclear	AG, B16, N278, N509, N510, N626, NOG, NQA, OM, QME
Offshore	SPPE
Operator Qualification & Certification	QHO, QRO
Pallets	MH1
Performance Test Codes	PTC
Piping	A13, B31
Plumbing	A112
Pressure Vessels	PVHO
Pumps	B73
Reinforced Thermostat Plastic Corrosion Resistant Equipment	RTP
Screw Threads	B1
Steel Stacks	STS
Storage Tanks	B96
Surface Quality	B46
Tools	B5, B94, B107
Turbines	B133
Valves, Fittings, Flanges, Gaskets	B16

ANSI Y14 STANDARDS

Name	Designation
Associated Lists	Y14.34-2008
Castings and Molded Parts	Y14.8-2009
Certification of Geometric Dimensioning and Tolerancing Professionals	Y14.5.2-2000
Chassis Frames–Passenger Car and Light Truck Ground Vehicle Practices	Y14.32.1M-1994 (R2005)
Decimal Inch Drawing Sheet Size and Format	Y14.1-2006
Digital Product Definition Data Practices	Y14.41-2003 (R2008)
Dimensioning and Tolerancing	Y14.5-2009
Engineering Drawing Practices	Y14.100-2004
Line Conventions and Lettering	Y14.2-2008
Mathematical Definition of Dimensioning and Tolerancing Principles	Y14.5.1M-1994 (R2004)
Metric Drawing Sheet Size and Format	Y14.1M-2006
Multiview and Sectional View Drawings	Y14.3-2003 (R2008)
Pictorial Drawing	Y14.4M-1989 (R2009)
Screw Thread Representation	Y14.6-2001 (R2007)
Surface Texture Symbols	Y14.36M-1996 (R2008)
Types and Applications of Engineering Drawings	Y14.24M-1999 (R2009)
Undimensioned Drawings	Y14.31-2008

ANSI Y32 STANDARDS

Name	Designation
Graphics Symbols for Railroad Maps and Profiles	Y32.7-1972 (R2009)
Symbols for Mechanical and Acoustical Elements as Used in Schematic Diagrams	Y32.18-1972 (R2008)

2 ORGANIZATIONS THAT PUBLISH DRAWING STANDARDS

AAMA	American Architectural Manufacturers Association	AAMA 1827 Walden Office Square Suite 550 Schaumberg, IL 60173-4628	T 847.303.5664	www.aamanet.org
ACSM	American Congress on Surveying and Mapping	ACSM 6 Montgomery Village Avenue Suite 403 Gaithersburg, MD 20879	T 240.632.9716	www.acsm.net
AGA	American Gas Association	AGA 400 N. Capitol Street, NW Suite 450 Washington, DC 20001	T 202.824.7000	www.aga.org
AGMA	American Gear Manufacturers Association	AGMA 500 Montgomery Street Suite 350 Alexandria, VA 22314-1581	T 703.684.0211	www.agma.org
AHAM	Association of Home Appliance Manufacturers	AHAM 1111 19th Street, NW Suite 402 Washington, DC 20036	T 202.872.5955	www.aham.org
AISC	American Institute of Steel Construction	AISC One E. Wacker Drive Suite 700 Chicago, IL 60601-1802	T 312.670.2400	www.aisc.org
AMCA	Air Movement and Control Association International	AMCA 30 W. University Drive Arlington Heights, IL 60004-1893	T 847.394.0150	www.amca.org
ANSI	American National Standards Institute	ANSI 1819 L Street, NW 6th Floor Washington, DC 20036	T 202.293.8020	web.ansi.org
AHRI	Air-conditioning, Heating & Refrigeration Institute	AHRI 2111 Wilson Blvd. Suite 500 Arlington, VA 22201	T 703.524.8800	www.ari.org
ARL	Applied Research Laboratories	Thermo ARL S.A. En Vallaire Ouest C Case postale CH-1024 Ecublens SWITZERLAND	T 41.21.694.71.11	
ASABE	American Society of Agricultural and Biological Engineers	ASABE 2950 Niles Road St. Joseph, MI 49085-9659	T 269.429.0300	www.asabe.org
ASHRAE	American Society of Heating, Refrigeration, and Air Conditioning Engineers	ASHRAE 1791 Tullie Circle, NE Atlanta, GA 30329	T 404.636.8400	www.ashrae.org
ASME	American Society of Mechanical Engineers	ASME 22 Law Drive P.O. Box 2900 Fairfield, NJ 07007-2900	T 800.843.2763 T 973.882.1170	www.asme.org

ASSE	American Society of Sanitary Engineering	ASSE International Office 901 Canterbury Suite A Westlake, OH 44145	T 440.835.3040	ww.asse-plumbing.org
ASTM	American Society for Testing and Materials	ASTM 100 Barr Harbor Drive West Conshohocken, PA 19428-2959	T 610.832.9500	www.astm.org
CGA	Canadian Gas Association	CGA 350 Sparks Street, Suite 809 Ottawa, Ontario K1R 7S8 CANADA	T 613.748.0057	www.cga.ca
CSA	Canadian Standards Association	CSA International 178 Rexdale Blvd. Toronto, ON M9W 1R3 CANADA	T 416.747.4000	www.csa-international .org
DOT	Department of Transportation	US DOT 1200 New Jersey Avenue, SE Washington, DC 20590	T 202.366.4000	www.dot.gov
ETL	ETL Testing Laboratories	ETL Testing Laboratories Intertek Testing Services 3933 US Route 11 Cortland, NY 13045	T 800.967.5352	www.etl.com www.intertek.com
FM	Factory Mutual Engineering	Factory Mutual 1151 Boston-Providence Tpke P.O. Box 9102 Norwood, MA 02062-9102	T 781.278.8805	
HRAI	Heating, Refrigeration, Air Conditioning Institute of Canada	HRAI 2800 Skymark Avenue Suite 201 Mississauga, ON L4W 5A6 CANADA	T 800.267.2231	www.hrai.ca
IAPMO	International Association of Plumbing and Mechanical Officials	IAPMO 4755 E. Philadelphia St. Ontario, CA 91761	T 909.472.4100	www.iapmo.org
IEEE	Institute of Electrical and Electronics Engineers, Inc.	IEEE-USA 2001 L Street, NW, Suite 700 Washington, DC 20036	T 800.678.4333 T 202.785.0017 F 202.785.0835	www.ieee.org
IFI	Industrial Fasteners Institute	IFI 6363 Oak Tree Blvd. Independence, OH 44131	T 216.241.1482	www.industrial-fasteners.org
ISO	International Organization for Standardization	ISO 1, ch. de la Voie-Creuse Case postale 56 CH-1211 Genève 20 SWITZERLAND	T 41.22.749.01.11	www.iso.ch
NEMA	National Electrical Manufacturers Association	NEMA 1300 N. 17th Street Suite 1752 Rosslyn, VA 22209	T 703.841.3200	www.nema.org
NFPA	National Fire Protection Association	NFPA 1 Batterymarch Park Quincy, MA 02169-7471	T 617.770.3000	www.nfpa.org

NIBS	National Institute of Building Sciences	NIBS 1090 Vermont Avenue, NW Suite 700 Washington, DC 20005-4905	T 202.289.7800	www.nibs.org
NIST	National Institute of Standards and Technology	NIST 100 Bureau Drive, Stop 1070 Gaithersburg, MD 20899	T 301.975.NIST	www.nist.gov
NSF	National Sanitation Foundation International	NSF International P.O. Box 130140 789 N. Dixboro Road Ann Arbor, MI 48113-0140	T 800.673.6275	www.nsf.org
OPEI	Outdoor Power Equipment Institute	OPEI 341 S. Patrick Street Alexandria, VA 22314	T 703.549.7600	www.opei.org
OSHA	Occupational Safety and Health Administration	OSHA Public Affairs Office, Room 3647 200 Constitution Avenue Washington, DC 20210	T 800.321.OSHA	www.osha.gov
SAE	Society of Automotive Engineers	SAE World Headquarters 400 Commonwealth Drive Warrendale, PA 15096-0001	T 724.776.4841	www.sae.org
SEV	Electrosuisse SEV Association for Electrical Engineering, Power and Information Technologies	Electrosuisse Luppmenstrasse 1 CH-8320 Fehraltorf SWITZERLAND	T 41.44.956.11.11	www.electrosuisse.ch
SRCC	Solar Rating and Certification Corporation	SRCC 1679 Clearlake Road Cocoa, FL 32922-5703	T 321.638.1537	www.solar-rating.org
SSPMA	Sump and Sewage Pump Manufacturers Association	SSPMA P.O. Box 647 Northbrook, IL 60065-9233	T 847.559.9233	www.sspma.org
UL	Underwriters Laboratories	UL 2600 N.W. Lake Rd. Camas, WA 98607-8542	T 877.854.3577	www.ul.com
USDA	United States Department of Agriculture	USDA 14th & Independence Avenue, SW Washington, DC 20250	T 202.720.2791	www.usda.gov
VDE	Verband der Eletrotechnik Elektronik Informationstechnik	VDE Stresemannallee 15 60596 Frankfurt am Main GERMANY	T 069.6308.284	www.vde.de

3 ABBREVIATIONS FOR USE ON DRAWINGS AND IN TEXT

Selected from ASME/ANSI Y1.14.38, 2007

A

absolute	ABS
accelerate	ACCEL
accessory	ACCESS
account	ACCT
accumulate	ACCUM
actual	ACTL
adapter	ADPTR
addendum	ADD
addition	ADDL
adjacent	ADJ
advance	ADV
after	AFT
aggregate	AGGR
air condition	AIRCOND
aircraft	ACFT
allowance	ALLOW
alloy	ALY
alteration	ALTRN
alternate	ALTN
alternating current	AC
aluminum	AL
American National Standard	AMER NATL STD
American wire gage	AWG
amount	AMT
ampere	AMP
amplifier	AMPL
anneal	ANL
antenna	ANT
apartment	APT
apparatus	APPAR
appendix	APPX
approved	APVD
approximate	APPROX
arc weld	ARCW
armature	ARM
arrange	ARR
artificial	ARTF
asbestos	ASB
asphalt	ASPH
assemble	ASSEM
assembly	ASSY

assistant	ASST
associate	ASSOC
association	ASSN
atomic	AT
audible	AUD
audio frequency	AF
authorized	AUTH
automatic	AUTO
automatic transformer	AXFMR
auxiliary	AUX
avenue	AVE
average	AVG
aviation	AVN
azimuth	AZ

B

Babbitt	BAB
back-feed	BF
back-pressure control	BPC
back to back	BB
backview	BV
balance	BAL
ball bearing	BBRG
barometer	BARO
base line	BL
base pitch	BP
bearing	BRG
bench mark	BM
bent	BT
bessemer	BESS
between	BETW
between centers	BC
between perpendiculars	BP
bevel	BEV
bill of material	BM
Birmingham wire gage	BWG
black	BK
blank	BLK
block	BLK
blueprint	BP
board	BD
boiler	BLR
boiler feed	BF
boiling	BOG

bolt circle	BC
both faces	BF
both sides	BS
both ways	BW
bottom	BOT
bottom chord	BC
bottom face	BF
bracket	BRKT
brake	BK
brass	BRS
brazing	BRZG
break	BRK
Brinell hardness	BH
British Standard	BSI
British thermal unit	BTU
broach	BRCH
bronze	BRZ
Brown & Sharpe Wire Gage	BS
building	BLDG
bulkhead	BHD
burnish	BNSH
bushing	BSHG
button	BTN

C

calibration	CAL
capacitance	CAP
carbide steel	CS
cartridge	CTRG
cast iron	CI
cast steel	CS
casting	CSTG
center	CTR
center line	CL
center of gravity	CG
center of pressure	CP
center to center	C to C
centering	CTRG
chamfer	CHAM
change in design	CID
check	CHK
check valve	CV
chord	CHD

circle	CIR	double	DBL	feeder	FDR
clear	CLR	dovetail	DVTL	feet	FT
clockwise	CW	dowel	DWL	figure	FIG
coated	CTD	down	DN	fillet	FIL
cold drawn copper	CDC	dozen	DOZ	fillister head	FILH
cold-drawn steel	CDS	drafting	DFTG	finish	FNSH
cold-rolled steel	CRS	drawing	DWG	finish all over	FAO
combining	COMB	drill or drill rod	DR	flange	FLG
commercial and government entity code	CAGE	driver	DRVR	flat head	FLH
		drive fit	DF	floor	FL
		drop forge	DF	fluid flow	FDFL
compressor	COMPR	duplicate	DUP	focal	FOC
concentric	CONC	*E*		foot	(') FT
condition	COND	each	EA	for example	EG
connection	CONN	east	E	force	F
contact	CONT	eccentric	ECC	forged steel	FST
continued	CONTD	effective	EFF	forging	FORG
copper	CU	elbow	ELL	forward	FWD
corner	COR	electric	ELEC	foundry	FDRY
counter	CTR	element	ELEM	frequency	FREQ
counterbore	CBORE	elevate	ELEV	front	FR
counter clockwise	CCW	elevation	EL	furnish	FURN
countersink	CSK	engine	ENG	*G*	
cubic	CU	engineer	ENGR	gage or gauge	GA
cubic foot	CUFT	engineering	ENGRG	gallon	GAL
cubic inch	CUIN	entrance	ENTR	galvanize	GALV
current	CUR	equal	EQL	galvanized iron	GALVI
cyanide	CYN	equation	EQ	galvanized steel	GALVS
D		equipment	EQPT	gasket	GSKT
decimal	DEC	equivalent	EQUIV	general	GENL
dedendum	DED	estimate	EST	glass	GL
deflect	DEFL	exchange	EXCH	government	GOVT
degree	DEG	exhaust	EXH	governor	GOV
density	DENS	existing	EXST	grade	GR
department	DEPT	exterior	EXT	gradient	GRAD
design	DSGN	extra fine	EF	graphite	GPH
detail	DET	extra heavy	XHVY	grind	GRD
develop	DVL	extra strong	X STR	groove	GRV
diagonal	DIAG	extrude	EXTR	ground	GND
diagram	DIAG	*F*		*H*	
diameter	DIA	fabricate	FAB	half-round	1/2RD
diametral pitch	DP	face to face	FF	handle	HDL
dimension	DIM	Fahrenheit	F	hanger	HGR
discharge	DISCH	far side	FS	harden	HDN
distance	DIST	federal	FED	hardware	HDW
division	DIV				

header	HDR	length over all	LOA	normal	NORM
headless	HDLS	letter	LTR	north	N
heat treat	HTTR	light	LT	not to scale	NTS
heavy	HVY	line of sight	LOS	number	NO
hexagon	HEX	locate	LCT	numeral	NUM
high pressure steam	HPS	logarithm	LOG	*O*	
high speed	HS	lubricate	LUB	obsolete	OBS
high voltage	HV	lumber	LBR	octagon	OCT
horizontal	HORIZ	*M*		office	OFC
horsepower	HP	machine	MACH	on center	OC
hot rolled	HR	machine steel	MST	opposite	OPP
hot-rolled steel	HRS	maintenance	MAINT	optical	OPT
hour	HR	malleable	MAL	origin	ORIG
housing	HSG	malleable iron	MI	outlet	OUT
hydraulic	HYDR	manual	MNL	outside diameter	OD
I		manufacture	MFR	outside face	OF
illustrate	ILLUS	manufactured	MFD	outside radius	OR
inboard	INBD	manufacturing	MFG	overall	OA
inch	IN	material	MATL	*P*	
inches per second	IPS	maximum	MAX	package	PKG
inclosure	INCLS	mechanical	MECH	paragraph	PARA
include	INCL	mechanism	MECH	part number	PN
inside diameter	ID	median	MDN	patent	PAT
inside radius	IR	metal	MET	pattern	PATT
instrument	INSTR	meter	M	permanent	PERM
interior	INTR	mile	MI	perpendicular	PERP
internal	INTL	miles per hour	MPH	piece	PC
intersect	INTSCT	millimeter	MM	piece mark	PCMK
iron	I	minimum	MIN	pint	PT
irregular	IRREG	miscellaneous	MISC	pitch	P
J		month	MO	pitch circle	PC
joint	JT	morse taper	MORT	pitch diameter	PD
joint army-navy	JAN	motor	MOT	plastic	PLSTC
journal	JNL	mounted	MTD	plated	PLD
junction	JCT	mounting	MTG	plumbing	PLMB
K		multiple	MULT	point	PT
keyseat	KST	music wire gage	MWG	point of curve	PC
keyway	KWY	*N*		point of intersection	PI
L		national	NATL	point of tangent	PT
laboratory	LAB	natural	NAT	polish	POL
laminate	LAM	near face	NF	position	POSN
lateral	LATL	near side	NS	potential	POT
left	L	negative	NEG	pound	LB
left hand side	LHS	neutral	NEUT	pounds per square inch	PSI
length	LG	nominal	NOM	power	PWR

prefabricated	PREFAB	*S*		technical	TECH
preferred	PFD	schedule	SCHED	template	TEMPL
preparation	PREP	schematic	SCHEM	tension	TNSN
pressure	PRESS	scleroscope	SCLER	terminal	TERM
process	PRCS	screw	SCR	thick	THK
production	PROD	second	SEC	thread	THD
profile	PF	section	SECT	through	THRU
propeller	PROP	semi-steel	SS	tolerance	TOL
publication	PUBN	separate	SEP	tongue	TNG
push button	PB	set screw	SSCR	tool steel	TS
Q		shaft	SFT	total	TOT
quadrant	QDRNT	sheet	SH	transfer	XFR
quality	QUAL	shoulder	SHLDR	transparent	TRANS
quarter	QTR	single	SGL	typical	TYP
R		sketch	SK	*U*	
radial	RDL	sleeve	SLV	ultimate	ULT
radius	RAD	sliding	SL	universal	UNIV
railroad	RR	slotted	SLTD	*V*	
reamer	RMR	small	SM	vacuum	VAC
received	RCVD	socket	SKT	valve	V
record	RCD	spacer	SPCR	variable	VAR
rectangle	RECT	special	SPCL	versus	VS
reduce	RDC	specification	SPEC	vertical	VERT
reference line	REFL	spot face	SF	volt	V
reinforce	REINF	spring	SPG	volume	VOL
release	REL	square	SQ	*W*	
relief	RLF	standard	STD	washer	WSHR
remove	RMV	station	STA	watt	W
require	REQ	stationary	STA	week	WK
required	REQD	steel	STL	weight	WT
return	RTN	stock	STK	west	W
reverse	RVS	straight	STR	width	WD
revolution	REV	street	ST	Woodruff	WDF
revolutions per minute	RPM	structural	STRL	working point	WP
right	R	substitute	SUBST	working pressure	WPR
right hand side	RHS	summary	SMY	wrought	WRT
rivet	RVT	support	SPRT	wrought iron	WI
Rockwell hardness	RH	surface	SURF	*Y*	
roller	RLR	symbol	SYM	yard	YD
root diameter	RD	system	SYS	year	YR
root mean square	RMS	*T*			
rough	RGH	tangent	TAN		
round	RND	taper	TPR		

4 STANDARD SHEET SIZES AND TITLE BLOCKS

ANSI Standard Sheet Sizes (for Inch Sheets)

Designation	Sheet Size (inches)
A	11 × 8.5
B	17 × 11
C	22 × 17
D	34 × 22
E	44 × 34

ISO Standard Sheet Sizes

Designation	Metric Sheet Size (mm)	English Equivalent Sheet (inches)
A0	1189 × 841	44 × 34 (E)
A1	841 × 594	34 × 22 (D)
A2	594 × 420	22 × 17 (C)
A3	420 × 297	17 × 11 (B)
A4	297 × 210	11 × 8.5 (A)

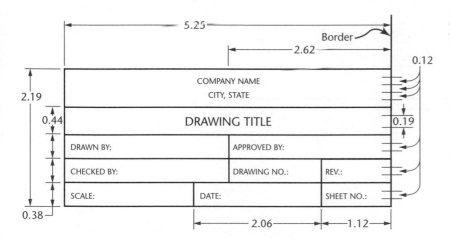

A.1   Typical Title Block for Size A Sheet

A.2 Size B Sheet
(17 × 11 inches)

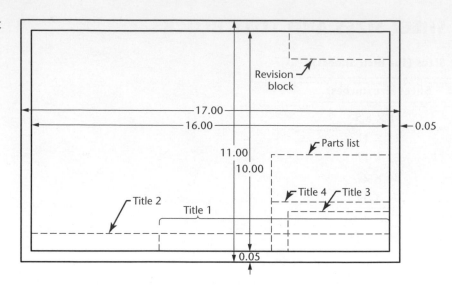

A.3 Size C Sheet
(22 × 17 inches)

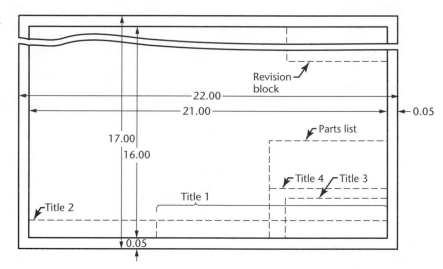

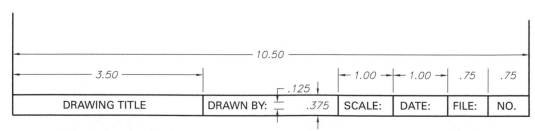

A.4 Title Strip for Size A Sheet

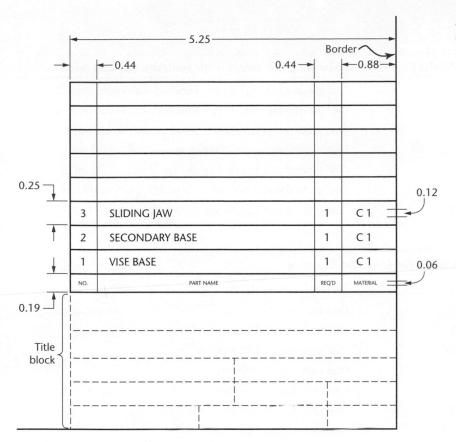

A.5 Typical Parts List or Materials List

NO.	PART NAME	REQ'D	MATERIAL
3	SLIDING JAW	1	C 1
2	SECONDARY BASE	1	C 1
1	VISE BASE	1	C 1

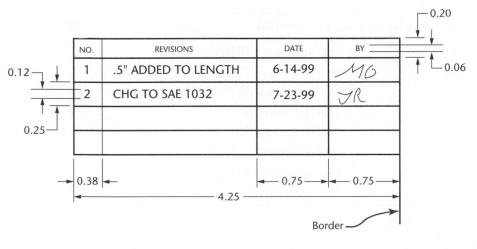

A.6 Typical Revision Block

NO.	REVISIONS	DATE	BY
1	.5" ADDED TO LENGTH	6-14-99	MO
2	CHG TO SAE 1032	7-23-99	JR

5 STANDARD SCALES

Metric Scales

1:1	Full
1:2	Half
1:5	Fifth
1:10	Tenth
1:20	Twentieth
1:100	Hundredth
2:1	Double

Metric Scales for Civil and Mapping (*for mapping the units are typically meters*)

1:100	One plotted unit equals one hundred real-world units
1:200	One plotted unit equals two hundred real-world units
1:500	One plotted unit equals five hundred real-world units
1:1000	One plotted unit equals one thousand real-world units
1:2000	One plotted unit equals two thousand real-world units
1:5000	One plotted unit equals five thousand real-world units
1:10000	One plotted unit equals ten thousand real-world units

English Scales

1 = 1	Full
1 = 2	Half
1 = 4	Quarter
1 = 8	Eighth
1 = 10	Tenth
2 = 1	Double

English Scales for Civil and Mapping

1" = 10'	One inch equals 10 feet
1" = 20'	One inch equals 20 feet
1" = 50'	One inch equals 50 feet
1" = 100'	One inch equals 100 feet
1" = 200'	One inch equals 200 feet
1" = 400'	One inch equals 400 feet
1" = 500'	One inch equals 500 feet
1" = 1000'	One inch equals 1000 feet
1:24000 or 1" = 2000'	One inch equals 2000 feet
1:63360 or 1" = 1 mile	One inch equals 1 mile

English Scales for Building Construction

1" = 10'	One inch equals 10 feet
1" = 20'	One inch equals 20 feet
1" = 50'	One inch equals 50 feet
1" = 100'	One inch equals 100 feet
1" = 200'	One inch equals 200 feet
1" = 400'	One inch equals 400 feet
1" = 500'	One inch equals 500 feet
1" = 1000'	One inch equals 1000 feet

6 AMERICAN STANDARD RUNNING AND SLIDING FITS (HOLE BASIS)

Limits are in thousandths of an inch.
Limits for hole and shaft are applied algebraically to the basic size to obtain the limits of size for the parts.
Data in bold face are in accordance with ABC agreements.
Symbols H5, g5, etc., are Hole and Shaft designations used in ABC System.

Nominal Size Range Inches Over To	Class RC 1 Limits of Clearance	Standard Limits Hole H5	Standard Limits Shaft g4	Class RC 2 Limits of Clearance	Standard Limits Hole H6	Standard Limits Shaft g5	Class RC 3 Limits of Clearance	Standard Limits Hole H7	Standard Limits Shaft f6	Class RC 4 Limits of Clearance	Standard Limits Hole H8	Standard Limits Shaft f7
0–0.12	**0.1**	**+0.2**	**−0.1**	**0.1**	**+0.25**	**−0.1**	**0.3**	**+0.4**	**−0.3**	**0.3**	**+0.6**	**−0.3**
	0.45	0	−0.25	0.55	0	−0.3	0.95	0	−0.55	1.3	0	−0.7
0.12–0.24	0.15	+0.2	−0.15	0.15	+0.3	−0.15	0.4	+0.5	−0.4	0.4	+0.7	−0.4
	0.5	0	−0.3	0.65	0	−0.35	1.12	0	−0.7	1.6	0	−0.9
0.24–0.40	0.2	0.25	−0.2	0.2	+0.4	−0.2	0.5	+0.6	−0.5	0.5	+0.9	−0.5
	0.6	0	−0.35	0.85	0	−0.45	1.5	0	−0.9	2.0	0	−1.1
0.40–0.71	**0.25**	**+0.3**	**−0.25**	**0.25**	**+0.4**	**−0.25**	**0.6**	**+0.7**	**−0.6**	**0.6**	**+1.0**	**−0.6**
	0.75	0	−0.45	0.95	0	−0.55	1.7	0	−1.0	2.3	0	−1.3
0.71–1.19	0.3	+0.4	−0.3	0.3	+0.5	−0.3	0.8	+0.8	−0.8	0.8	+1.2	−0.8
	0.95	0	−0.55	1.2	0	−0.7	2.1	0	−1.3	2.8	0	−1.6
1.19–1.97	0.4	+0.4	−0.4	0.4	+0.6	−0.4	1.0	+1.0	−1.0	1.0	+1.6	−1.0
	1.1	0	−0.7	1.4	0	−0.8	2.6	0	−1.6	3.6	0	−2.0
1.97–3.15	0.4	+0.5	−0.4	0.4	+0.7	−0.4	1.2	+1.2	−1.2	1.2	+1.8	−1.2
	1.2	0	−0.7	1.6	0	−0.9	3.1	0	−1.9	4.2	0	−2.4
3.15–4.73	0.5	+0.6	−0.5	0.5	+0.9	−0.5	1.4	+1.4	−1.4	1.4	+2.2	−1.4
	1.5	0	−0.9	2.0	0	−1.1	3.7	0	−2.3	5.0	0	−2.8
4.73–7.09	0.6	+0.7	−0.6	0.6	+1.0	−0.6	1.6	+1.6	−1.6	1.6	+2.5	−1.6
	1.8	0	−1.1	2.3	0	−1.3	4.2	0	−2.6	5.7	0	−3.2
7.09–9.85	**0.6**	**+0.8**	**−0.6**	**0.6**	**+1.2**	**−0.6**	**2.0**	**+1.8**	**−2.0**	**2.0**	**+2.8**	**−2.0**
	2.0	0	−1.2	2.6	0	−1.4	5.0	0	−3.2	6.6	0	−3.8
9.85–12.41	0.8	+0.9	−0.8	0.8	+1.2	−0.8	2.5	+2.0	−2.5	2.5	+3.0	−2.5
	2.3	0	−1.4	2.9	0	−1.7	5.7	0	−3.7	7.5	0	−4.5
12.41–15.75	1.0	+1.0	−1.0	1.0	+1.4	−1.0	3.0	+	−3.0	3.0	+3.5	−3.0
	2.7	0	1.7	3.4	0	−2.0	6.6	0	−4.4	8.7	0	−5.2
15.75–19.69	1.2	+1.0	−1.2	1.2	+1.6	−1.2	4.0	+1.6	−4.0	4.0	+4.0	−4.0
	3.0	0	−2.0	3.8	0	−2.2	8.1	0	−5.6	10.5	0	−6.5
19.69–30.09	1.6	+1.2	−1.6	1.6	+2.0	−1.6	5.0	+3.0	−5.0	5.0	+5.0	−5.0
	3.7	0	−2.5	4.8	0	−2.8	10.0	0	−7.0	13.0	0	−8.0
30.09–41.49	2.0	+1.6	−2.0	2.0	+2.5	−2.0	6.0	+4.0	−6.0	6.0	+6.0	−6.0
	4.6	0	−3.0	6.1	0	−3.6	12.5	0	−8.5	16.0	0	−10.0
41.49–56.19	2.5	+2.0	−2.5	2.5	+3.0	−2.5	8.0	+5.0	−8.0	8.0	+8.0	−8.0
	5.7	0	−3.7	7.5	0	−4.5	16.0	0	−11.0	21.0	0	−13.0
56.19–76.39	3.0	+2.5	−3.0	3.0	+4.0	−3.0	10.0	+6.0	−10.0	10.0	+10.0	−10.0
	7.1	0	−4.6	9.5	0	−5.5	20.0	0	−14.0	26.0	0	−16.0
76.39–100.9	4.0	+3.0	−4.0	4.0	+5.0	−4.0	12.0	+8.0	−12.0	12.0	+12.0	−12.0
	9.0	0	−6.0	12.0	0	−7.0	25.0	0	−17.0	32.0	0	−20.0
100.9–131.9	5.0	+4.0	−5.0	5.0	+6.0	−5.0	16.0	+10.0	−16.0	16.0	+16.0	−16.0
	11.5	0	−7.5	15.0	0	−9.0	32.0	0	−22.0	36.0	0	−26.0
131.9–171.9	6.0	+5.0	−6.0	6.0	+8.0	−6.0	18.0	+8.0	−18.0	18.0	+20.0	−18.0
	14.0	0	−9.0	19.0	0	−1.0	38.0	0	−26.0	50.0	0	−30.0
171.9–200	8.0	+6.0	−8.0	8.0	+10.0	−8.0	22.0	+16.0	−22.0	22.0	+25.0	−22.0
	18.0	0	−12.0	22.0	0	−12.0	48.0	0	−32.0	63.0	0	−38.0

Class RC 5			Class RC 6			Class RC 7			Class RC 8			Class RC 9			Nominal Size Range Inches
Limits of Clearance	Standard Limits		Limits of Clearance	Standard Limits		Limits of Clearance	Sandard Limits		Limits of Clearance	Standard Limits		Limits of Clearance	Standard Limits		
	Hole H8	Shaft e7		Hole H9	Shaft e8		Hole H9	Shaft d8		Hole H10	Shaft c9		Hole H11	Shaft	Over To
0.6	+0.6	−0.6	0.6	+1.0	−0.6	1.0	+1.0	−1.0	2.5	+1.6	−2.5	4.0	+2.5	−4.0	0−0.12
1.6	−0	−1.0	2.2	−0	−1.2	2.6	0	−1.6	5.1	0	−3.5	8.1	0	−5.6	
0.8	+0.7	−0.8	0.8	+1.2	−0.8	1.2	+1.2	−1.2	2.8	+1.8	−2.8	4.5	+3.0	−4.5	0.12−0.24
2.0	−0	−1.3	2.7	−0	−1.5	3.1	0	−1.9	5.8	0	−4.0	9.0	0	−6.0	
1.0	+0.9	−1.0	1.0	+1.4	−1.0	1.6	+1.4	−1.6	3.0	+2.2	−3.0	5.0	+3.5	−5.0	0.24−0.40
2.5	−0	−1.16	3.3	−0	−1.9	3.9	0	−2.5	6.6	0	−4.4	10.7	0	−7.2	
1.2	+1.0	−1.2	1.2	+1.6	−1.2	2.0	+1.6	−2.0	3.5	+2.8	−3.5	6.0	+4.0	−6.0	0.40−0.71
2.9	−0	−1.9	3.8	−0	−2.2	4.6	0	−3.0	7.9	0	−5.1	12.8	−0	−8.8	
1.6	+1.2	−1.6	1.6	+2.0	−1.6	2.5	+2.0	−2.5	4.5	+3.5	−4.5	7.0	+5.0	−7.0	0.71−1.19
3.6	−0	−2.4	4.8	−0	−2.8	5.7	0	−3.7	10.0	0	−6.5	15.5	0	−10.5	
2.0	+1.6	−2.0	2.0	+2.5	−2.0	3.0	+2.5	−3.0	5.0	+4.0	−5.0	8.0	+6.0	−8.0	1.19−1.97
4.6	−0	−3.0	6.1	−0	−3.6	7.1	0	−4.6	11.5	0	−7.5	18.0	0	−12.0	
2.5	+1.8	−2.5	2.5	+3.0	−2.5	4.0	+3.0	−4.0	6.0	+4.5	−6.0	9.0	+7.0	−9.0	1.97−3.15
5.5	−0	−3.7	7.3	−0	−4.3	8.8	0	−5.8	13.5	0	−9.0	20.5	0	−13.5	
3.0	+2.2	−3.0	3.0	+3.5	−3.0	5.0	+3.5	−5.0	7.0	+5.0	−7.0	10.0	+9.0	−10.0	3.15−4.73
6.6	−0	−4.4	8.7	−0	−5.2	10.7	0	−7.2	15.5	0	−10.5	24.0	0	−15.0	
3.5	+2.5	−3.5	3.5	+4.0	−3.5	6.0	+4.0	−6.0	8.0	+6.0	−8.0	12.0	+10.0	−12.0	4.73−7.09
7.6	−0	−5.1	10.0	−0	−6.0	12.5	0	−8.5	18.0	0	−12.0	28.0	0	−18.0	
4.0	+2.8	−4.0	4.0	+4.5	−4.0	7.0	+4.5	−7.0	10.0	+7.0	−10.0	15.0	+12.0	−15.0	7.09−9.85
8.6	−0	−5.8	11.3	0	−6.8	14.3	0	−9.8	21.5	0	−14.5	34.0	0	−22.0	
5.0	+3.0	−5.0	5.0	+5.0	−5.0	8.0	+5.0	−8.0	12.0	+8.0	−12.0	18.0	+12.0	−18.0	9.85−12.41
10.0	0	−7.0	13.0	0	−8.0	16.0	0	−11.0	25.0	0	−17.0	38.0	0	−26.0	
6.0	+3.5	−6.0	6.0	+6.0	−6.0	10.0	+6.0	−10.0	14.0	+9.0	−14.0	22.0	+14.0	−22.0	12.41−15.75
11.7	0	−8.2	15.5	0	−9.5	19.5	0	13.5	29.0	0	−20.0	45.0	0	−31.0	
8.0	+4.0	−8.0	8.0	+6.0	−8.0	12.0	+6.0	−12.0	16.0	+10.0	−16.0	25.0	+16.0	−25.0	15.75−19.69
14.5	0	−10.5	18.0	0	−12.0	22.0	0	−16.0	32.0	0	−22.0	51.0	0	−35.0	
10.0	+5.0	−10.0	10.0	+8.0	−10.0	16.0	+8.0	−16.0	20.0	+12.0	−20.0	30.0	+20.0	−30.0	19.69−30.09
18.0	0	−13.0	23.0	0	−15.0	29.0	0	−21.0	40.0	0	−28.0	62.0	0	−42.0	
12.0	+6.0	−12.0	12.0	+10.0	−12.0	20.0	+10.0	−20.0	25.0	+16.0	−25.0	40.0	+25.0	−40.0	30.09−41.49
22.0	0	−16.0	28.0	0	−18.0	36.0	0	−26.0	51.0	0	−35.0	81.0	0	−56.0	
16.0	+8.0	−16.0	16.0	+12.0	−16.0	25.0	+12.0	−25.0	30.0	+20.0	−30.0	50.0	+30.0	−50.0	41.49−56.19
29.0	0	−21.0	36.0	0	−24.0	45.0	0	−33.0	62.0	0	−42.0	100	0	−70.0	
20.0	+10.0	−20.0	20.0	+16.0	−20.0	30.0	+16.0	−30.0	40.0	+25.0	−40.0	60.0	+40.0	−60.0	56.19−76.39
36.0	0	−26.0	46.0	0	−30.0	56.0	0	−40.0	81.0	0	−56.0	125	0	−85.0	
25.0	+12.0	−25.0	25.0	+20.0	−25.0	40.0	+20.0	−40.0	50.0	+30.0	−50.0	80.0	+50.0	−80.0	76.39−100.9
45.0	0	−33.0	57.0	0	−37.0	72.0	0	−52.0	100	0	−70.0	160	0	−110	
30.0	+16.0	−30.0	30.0	+35.0	−30.0	50.0	+25.0	−50.0	60.0	+40.0	−60.0	100	+60.0	−100	100.9−131.9
56.0	0	−40.0	71.0	0	−46.0	91.0	0	−66.0	125	0	−85.0	200	0	−140	
35.0	+20.0	−35.0	35.0	+30.0	−35.0	60.0	+30.0	−60.0	80.0	+50.0	−80.0	130	+80.0	−130	131.9−171.9
57.0	0	−47.0	85.0	0	−55.0	110.0	0	−80.0	160	0	−110	260	0	−180	
45.0	+25.0	−45.0	45.0	+40.0	−45.0	80.0	+40.0	−80.0	100	+60.0	−100	150	+100	−150	171.9−200
86.0	0	−61.0	110.0	0	−70.0	145.0	0	−105.0	200	0	−140	310	0	−210	

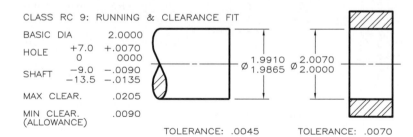

CLASS RC 9: RUNNING & CLEARANCE FIT

BASIC DIA   2.0000

HOLE   +7.0   +.0070
                 0   0000

SHAFT   −9.0   −.0090
            −13.5   −.0135

MAX CLEAR.   .0205

MIN CLEAR.   .0090
(ALLOWANCE)

⌀ 1.9910   ⌀ 2.0070
  1.9865   2.0000

TOLERANCE: .0045   TOLERANCE: .0070

(Reprinted from ANSI B4.1–1967 (R2009). Copyright © American Society of Mechanical Engineers. All rights reserved.)

7 AMERICAN STANDARD CLEARANCE LOCATIONAL FITS (HOLE BASIS)

Limits are in thousandths of an inch.
Limits for hole and shaft are applied algebraically to the basic size to obtain the limits of size for the parts.
Data in bold face are in accordance with ABC agreements.
Symbols H9, f8, etc., are Hole and Shaft designations used in ABC System.

Nominal Size Range Inches Over To	Class LC 1 Limits of Clearance	Class LC 1 Hole H6	Class LC 1 Shaft h5	Class LC 2 Limits of Clearance	Class LC 2 Hole H7	Class LC 2 Shaft h6	Class LC 3 Limits of Clearance	Class LC 3 Hole H8	Class LC 3 Shaft h7	Class LC 4 Limits of Clearance	Class LC 4 Hole H10	Class LC 4 Shaft h9	Class LC 5 Limits of Clearance	Class LC 5 Hole H7	Class LC 5 Shaft g6
0–0.12	0	+0.25	+0	0	+0.4	+0	0	+0.6	+0	0	+1.6	+0	0.1	+0.4	−0.1
	0.45	−0	−0.2	0.65	−0	−0.25	1	−0	−0.4	2.6	−0	−1.0	0.75	−0	−0.35
0.12–0.24	0	+0.3	+0	0	+0.5	+0	0	+0.7	+0	0	+1.8	+0	0.15	+0.5	−0.15
	0.5	−0	−0.2	0.8	−0	−0.3	1.2	−0	−0.5	3.0	−0	−1.2	0.95	−0	−0.45
0.24–0.40	0	+0.4	+0	0	+0.6	+0	0	+0.9	+0	0	+2.2	+0	0.2	+0.6	−0.2
	0.65	−0	−0.25	1.0	−0	−0.4	1.5	−0	−0.6	3.6	−0	−1.4	1.2	−0	−0.6
0.40–0.71	0	+0.4	+0	0	+0.7	+0	0	+1.0	+0	0	+2.8	+0	0.25	+0.7	−0.25
	0.7	−0	−0.3	1.1	−0	−0.4	1.7	−0	−0.7	4.4	−0	−1.6	1.35	−0	−0.65
0.71–1.19	0	+0.5	+0	0	+0.8	+0	0	+1.2	+0	0	+3.5	+0	0.3	+0.8	−0.3
	0.9	−0	−0.4	1.3	−0	−0.5	2	−0	−0.8	5.5	−0	−2.0	1.6	−0	−0.8
1.19–1.97	0	+0.6	+0	0	+1.0	+0	0	+1.6	+0	0	+4.0	+0	0.4	+1.0	−0.4
	1.0	−0	−0.4	1.6	−0	−0.6	2.6	−0	−1	6.5	−0	−2.5	2.0	−0	−1.0
1.97–3.15	0	+0.7	+0	0	+1.2	+0	0	+1.8	+0	0	+4.5	+0	0.4	+1.2	−0.4
	1.2	−0	−0.5	1.9	−0	−0.7	3	−0	−1.2	7.5	−0	−3	2.3	−0	−1.1
3.15–4.73	0	+0.9	+0	0	+1.4	+0	0	+2.2	+0	0	+5.0	+0	0.5	+1.4	−0.5
	1.5	−0	−0.6	2.3	−0	−0.9	3.6	−0	−1.4	8.5	−0	−3.5	2.8	−0	−1.4
4.73–7.09	0	+1.0	+0	0	+1.6	+0	0	+2.5	+0	0	+6.0	+0	0.6	+1.6	−0.6
	1.7	−0	−0.7	2.6	−0	−1.0	4.1	−0	−1.6	10	−0	−4	3.2	−0	1.6
7.09–9.85	0	+1.2	+0	0	+1.8	+0	0	+2.8	+0	0	+7.0	+0	0.6	+1.8	−0.6
	2.0	−0	−0.8	3.0	−0	−1.2	4.6	−0	−1.8	1.5	−0	−4.5	3.6	−0	−1.8
9.85–12.41	0	+1.2	+0	0	+2.0	+0	0	+3.0	+0	0	8.0	+0	0.7	+2.0	−0.7
	2.1	−0	−0.9	3.2	−0	−1.2	5	−0	−2.0	13	−0	−5	3.9	−0	−1.9
12.41–15.75	0	+1.4	+0	0	+2.2	+0	0	+3.5	+0	0	9.0	+0	0.7	+2.2	−0.7
	2.4	−0	−1.0	3.6	−0	1.4	5.7	−0	−2.2	15	−0	−6	4.3	−0	−2.1
15.75–19.69	0	+1.6	+0	0	+2.5	+0	0	+4	+0	0	+10.0	+0	0.8	+2.5	−0.8
	2.6	−0	−1.0	4.1	−0	−1.6	6.5	−0	−2.5	16	−0	−6	4.9	−0	−2.4
19.69–30.09	0	+2.0	+0	0	+3	+0	0	+5	+0	0	+12.0	+0	0.9	+3.0	−0.9
	3.2	−0	−1.2	5.0	−0	−2	8	−0	−3	20	−0	−8	5.9	−0	−2.9
30.09–41.49	0	+2.5	+0	0	+4	+0	0	+6	+0	0	+16.0	+0	1.0	+4.0	−1.0
	4.1	−0	−1.6	6.5	−0	−2.5	10	−0	−4	26	−0	−10	7.5	−0	−3.5
41.49–56.19	0	+3.0	+0	0	+5	+0	0	+8	+0	0	+20.0	+0	1.2	+5.0	−1.2
	5.0	−0	−2.0	8.0	−0	−3	13	−0	−5	32	−0	−12	9.2	−0	−4.2
56.19–76.39	0	+4.0	+0	0	+6	+0	0	+10	+0	0	+25.0	+0	1.2	+6.0	−1.2
	6.5	−0	−2.5	10	−0	−4	16	−0	−6	41	−0	−16	1.2	−0	−5.2
76.39–100.9	0	+5.0	+0	0	+8	+0	0	+12	+0	0	+30.0	+0	1.4	+8.0	−1.4
	8.0	−0	−3.0	13	−0	−5	20	−0	−8	50	−0	−20	14.4	−0	−6.4
100.9–131.9	0	+6.0	+0	0	+10	+0	0	+16	+0	0	+40.0	+0	1.6	+10.0	−1.6
	10.0	−0	−4.0	16	−0	−6	26	−0	−10	65	−0	−25	17.6	−0	−7.6
131.9–171.9	0	+8.0	+0	0	+12	+0	0	+20	+0	0	+50.0	+0	1.8	+12.0	−1.8
	13.0	−0	−5.0	20	−0	−8	32	−0	−12	8	−0	−30	21.8	−0	−9.8
171.9–200	0	+10.0	+0	0	+16	+0	0	+25	+0	0	+60.0	+0	1.8	+16.0	−1.8
	16.0	−0	−6.0	26	−0	−10	41	−0	−16	100	−0	−40	27.8	−0	−11.8

Class LC 6			Class LC 7			Class LC 8			Class LC 9			Class LC 10			Class LC 11			Nominal Size Range Inches
Limits of Clearance	Hole H9	Shaft f8	Limits of Clearance	Hole H10	Shaft e9	Limits of Clearance	Hole H10	Shaft d9	Limits of Clearance	Hole H11	Shaft c10	Limits of Clearance	Hole H12	Shaft	Limits of Clearance	Hole H13	Shaft	Over To
0.3	+1.0	−0.3	0.6	+1.6	−0.6	1.0	+0.6	−1.0	2.5	+2.5	−2.5	4	+4	−4	5	+6	−5	0−0.12
1.9	0	−0.9	3.2	0	−1.6	3.6	−0	−2.0	6.6	−0	−4.1	12	−0	−8	17	−0	−11	
0.4	+1.2	−0.4	0.8	+1.8	−0.8	1.2	+1.8	−1.2	2.8	+3.0	−2.8	4.5	+5	−4.5	6	+7	−6	0.12−0.24
2.3	0	−1.1	3.8	0	−2.0	4.2	−0	−2.4	7.6	−0	−4.6	14.5	−0	−9.5	20	−0	−13	
0.5	+1.4	−0.5	1.0	+2.2	−1.0	1.6	+2.2	−1.6	3.0	+3.5	−3.0	5	+6	−5	7	+9	−7	0.24−0.40
2.8	0	−1.4	4.6	0	−2.4	5.2	−0	−3.0	8.7	−0	−5.2	17	−0	−11	25	−0	−16	
0.6	+1.6	−0.6	1.2	+2.8	−1.2	2.0	+2.8	−2.0	3.5	+4.0	−3.5	6	+7	−6	8	+10	−8	0.40−0.71
3.2	0	−1.6	5.6	0	−2.8	6.4	−0	−3.6	10.3	−0	−6.3	20	−0	−13	28	−0	−18	
0.8	+2.0	−0.8	1.6	+3.5	−1.6	2.5	+3.5	−2.5	4.5	+5.0	−4.5	7	+8	−7	10	+12	−10	0.71−1.19
4.0	0	−2.0	7.1	0	−3.6	8.0	−0	−4.5	13.0	−0	−8.0	23	−0	−15	34	−0	−22	
1.0	+2.5	−1.0	2.0	+4.0	−2.0	3.0	+4.0	−3.0	5	+6	−5	8	+10	−8	12	+16	−12	1.19−1.97
5.1	0	−2.6	8.5	0	−4.5	9.5	−0	−5.5	15	−0	−9	28	−0	−18	44	−0	−28	
1.2	+3.0	−1.2	2.5	+4.5	−2.5	4.0	+4.5	−4.0	6	+7	−6	10	+12	−10	14	+18	−14	1.97−3.15
6.0	0	−3.0	10.0	0	−5.5	11.5	−0	−7.0	17.5	−0	−10.5	34	−0	−22	50	−0	−32	
1.4	+3.5	−1.4	3.0	+5.0	−3.0	5.0	+5.0	−5.0	7	+9	−7	11	+14	−11	16	+22	−16	3.15−4.73
7.1	0	−3.6	13.5	0	−6.5	13.5	−0	−8.5	21	−0	−12	39	−0	−25	60	−0	−38	
1.6	+4.0	−1.6	3.5	+6.0	−3.5	6	+6	−6	8	+10	−8	12	+16	−12	18	+25	−18	4.73−7.09
8.1	0	−4.1	13.5	0	−7.5	16	−0	−10	24	−0	−14	44	−0	−28	68	−0	−43	
2.0	+4.5	−2.0	4.0	+7.0	−4.0	7	+7	−7	10	+12	−10	16	+18	−16	22	+28	−22	7.09−9.85
9.3	0	−4.8	15.5	0	−8.5	18.5	−0	−11.5	29	−0	−17	52	−0	−34	78	−0	−50	
2.2	+5.0	−2.2	4.5	+8.0	−4.5	7	+8	−7	12	+12	−12	20	+20	−20	28	+30	−28	9.85−12.41
10.2	0	−5.2	17.5	0	−9.5	20	−0	−12	32	−0	−20	60	−0	−40	88	−0	−58	
2.5	+6.0	−2.5	5.0	+9.0	−5	8	+9	−8	14	+14	−14	22	+22	−22	30	+35	−30	12.41−15.75
12.0	0	−6.0	20.0	0	−11	23	−0	−14	37	−0	−23	66	−0	−44	100	−0	−65	
2.8	+6.0	−2.8	5.0	+10.0	−5	9	+10	−9	16	+16	−16	25	+25	−25	35	+40	−35	15.75−19.69
12.8	0	−6.8	21.0	0	−11	25	−0	−15	42	−0	−26	75	−0	−50	115	−0	−75	
3.0	+8.0	−3.0	6.0	+12.0	−6	10	+12	−10	18	+20	−18	28	+30	−28	40	+50	−40	19.69−30.09
16.0	0	−8.0	26.0	−0	−14	30	−0	−18	50	−0	−30	88	−0	−58	140	−0	−90	
3.5	+10.0	−3.5	7.0	+16.0	−7	12	+16	−12	20	+25	−20	30	+40	−30	45	+60	−45	30.09−41.49
19.5	0	−9.5	33.0	−0	−17	38	−0	−22	61	−0	−36	110	−0	−70	165	−0	−105	
4.0	+12.0	−4.0	8.0	+20.0	−8	14	+20	−14	25	+30	−25	40	+50	−40	60	+80	−60	41.49−56.19
24.0	0	−12.0	40.0	−0	−20	46	−0	−26	75	−0	−45	140	−0	−90	220	−0	−140	
4.5	+16.0	−4.5	9.0	+25.0	−9	16	+25	−16	30	+40	−30	50	+60	−50	70	+100	−70	56.19−76.39
30.5	0	−14.5	50.0	−0	−25	57	−0	−32	95	−0	−55	170	−0	110	270	−0	−170	
5.0	+20.0	−5	10.0	+30.0	−10	18	+30	−18	35	+50	−35	50	+80	−50	80	+125	−80	76.39−100.9
37.0	0	−17	60.0	−0	−30	68	−0	−38	115	−0	−65	210	−0	−130	330	−0	−205	
6.0	+25.0	−6	12.0	+40.0	−12	20	+40	−20	40	+60	−40	60	+100	−60	90	+160	−90	100.9−131.9
47.0	0	−22	67.0	−0	−27	85	−0	−45	140	−0	−80	260	−0	−160	410	−0	−250	
7.0	+30.0	−7	14.0	+50.0	−14	25	+50	−25	50	+80	−50	80	+125	−80	100	+200	−100	131.9−171.9
57.0	0	−27	94.0	−0	−44	105	−0	−55	180	−0	−100	330	−0	−205	500	−0	−300	
7.0	+40.0	−7	14.0	+60.0	−14	25	+60	−25	50	+100	−50	90	+160	−90	125	+250	−125	171.9−200
72.0	0	−32	114.0	−0	−54	125	−0	−65	210	−0	−110	410	−0	−250	625	−0	−375	

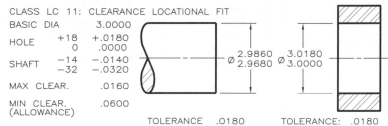

CLASS LC 11: CLEARANCE LOCATIONAL FIT

BASIC DIA		3.0000
HOLE	+18 / 0	+.0180 / .0000
SHAFT	−14 / −32	−.0140 / −.0320
MAX CLEAR.		.0160
MIN CLEAR. (ALLOWANCE)		.0600

Ø 2.9860 / 2.9680   Ø 3.0180 / 3.0000

TOLERANCE .0180   TOLERANCE: .0180

8 AMERICAN STANDARD TRANSITION LOCATIONAL FITS (HOLE BASIS)

Limits are in thousandths of an inch.
Limits for hole and shaft are applied algebraically to the basic size to obtain the limits of size for the mating parts.
Data in bold face are in accordance with ABC agreements.
"Fit" represents the maximum interference (minus values) and the maximum clearance (plus values).
Symbols H7, js6, etc., are Hole and Shaft designations used in ABC System.

Nominal Size Range Inches (Over–To)	Class LT 1 Fit	Class LT 1 Hole H7	Class LT 1 Shaft js6	Class LT 2 Fit	Class LT 2 Hole H8	Class LT 2 Shaft js7	Class LT 3 Fit	Class LT 3 Hole H7	Class LT 3 Shaft k6	Class LT 4 Fit	Class LT 4 Hole H8	Class LT 4 Shaft k7	Class LT 5 Fit	Class LT 5 Hole H7	Class LT 5 Shaft n6	Class LT 6 Fit	Class LT 6 Hole H7	Class LT 6 Shaft n7
0–0.12	−0.10 / +0.50	+0.4 / −0	+0.10 / −0.10	−0.2 / +0.8	+0.6 / −0	+0.2 / −0.2							−0.5 / +0.15	+0.4 / −0	+0.5 / +0.25	−0.65 / +0.15	+0.4 / −0	+0.65 / +0.25
0.12–0.24	−0.15 / +0.65	+0.5 / −0	+0.15 / −0.15	−0.25 / +0.95	+0.7 / −0	+0.25 / −0.25							−0.6 / +0.2	+0.5 / −0	+0.6 / +0.3	−0.8 / +0.2	+0.5 / −0	+0.8 / +0.3
0.24–0.40	−0.2 / +0.8	+0.6 / −0	+0.2 / −0.2	−0.3 / +1.2	+0.9 / −0	+0.3 / −0.3	−0.5 / +0.5	+0.6 / −0	+0.5 / +0.1	−0.7 / +0.8	+0.9 / −0	+0.7 / +0.1	−0.8 / +0.2	+0.6 / −0	+0.8 / +0.4	−1.0 / +0.2	+0.6 / −0	+1.0 / +0.4
0.40–0.71	−0.2 / +0.9	+0.7 / −0	+0.2 / −0.2	−0.35 / +1.35	+1.0 / −0	+0.35 / −0.35	−0.5 / +0.6	+0.7 / −0	+0.5 / +0.1	−0.8 / +0.9	+1.0 / −0	+0.8 / +0.1	−0.9 / +0.2	+0.7 / −0	+0.9 / +0.5	−1.2 / +0.2	+0.7 / −0	+1.2 / +0.5
0.71–1.19	−0.25 / +1.05	+0.8 / −0	+0.25 / −0.25	−0.4 / +1.6	+1.2 / −0	+0.4 / −0.4	−0.5 / +0.7	+0.8 / −0	+0.6 / +0.1	−0.9 / +1.1	+1.2 / −0	+0.9 / +0.1	−1.1 / +0.2	+0.8 / −0	+1.1 / +0.6	−1.4 / +0.2	+0.8 / −0	+1.4 / +0.6
1.19–1.97	−0.3 / +1.3	+1.0 / −0	+0.3 / −0.3	−0.5 / +2.1	+1.6 / −0	+0.5 / −0.5	−0.7 / +0.9	+1.0 / −0	+0.7 / +0.1	−1.1 / +1.5	+1.6 / −0	+1.1 / +0.1	−1.3 / +0.3	+1.0 / −0	+1.3 / +0.7	−1.7 / +0.3	+1.0 / −0	+1.7 / +0.7
1.97–3.15	−0.3 / +1.5	+1.2 / −0	+0.3 / −0.3	−0.6 / +2.4	+1.8 / −0	+0.6 / −0.6	−0.8 / +1.1	+1.2 / −0	+0.8 / +0.1	−1.3 / +1.7	+1.8 / −0	+1.3 / +0.1	−1.5 / +0.4	+1.2 / −0	+1.5 / +0.8	−2.0 / +0.4	+1.2 / −0	+2.0 / +0.8
3.15–4.73	−0.4 / +1.8	+1.4 / −0	+0.4 / −0.4	−0.7 / +2.9	+2.2 / −0	+0.7 / −0.7	−1.0 / +1.3	+1.4 / −0	+1.0 / +0.1	−1.5 / +2.1	+2.2 / −0	+1.5 / +0.1	−1.9 / +0.4	+1.4 / −0	+1.9 / +1.0	−2.4 / +0.4	+1.4 / −0	+2.4 / +1.0
4.73–7.09	−0.5 / +2.1	+1.6 / −0	+0.5 / −0.5	−0.8 / +3.3	+2.5 / −0	+0.8 / −0.8	−1.1 / +1.5	+1.6 / −0	+1.1 / +0.1	−1.7 / +2.4	+2.5 / −0	+1.7 / +0.1	−2.2 / +0.4	+1.6 / −0	+2.2 / +1.2	−2.8 / +0.4	+1.6 / −0	+2.8 / +1.2
7.09–9.85	−0.6 / +2.4	+1.8 / −0	+0.6 / −0.6	−0.9 / +3.7	+2.8 / −0	+0.9 / −0.9	−1.4 / +1.6	+1.8 / −0	+1.4 / +0.2	−2.0 / +2.6	+2.8 / −0	+2.0 / +0.2	−2.6 / +0.4	+1.8 / −0	+2.6 / +1.4	−3.2 / +0.4	+1.8 / −0	+3.2 / +1.4
9.85–12.41	−0.6 / +2.6	+2.0 / −0	+0.6 / −0.6	−1.0 / +4.0	+3.0 / −0	+1.0 / −1.0	−1.4 / +1.8	+2.0 / −0	+1.4 / +0.2	−2.2 / +2.8	+3.0 / −0	+2.2 / +0.2	−2.6 / +0.6	+2.0 / −0	+2.6 / +1.4	−3.4 / +0.6	+2.0 / −0	+3.4 / +1.4
12.41–15.75	−0.7 / +2.9	+2.2 / −0	+0.7 / −0.7	−1.0 / +4.5	+3.5 / −0	+1.0 / −1.0	−1.6 / +2.0	+2.2 / −0	+1.6 / +0.2	−2.4 / +3.3	+3.5 / −0	+2.4 / +0.2	−3.0 / +0.6	+2.2 / −0	+3.0 / +1.6	−3.8 / +0.6	+2.2 / −0	+3.8 / +1.6
15.75–19.69	−0.8 / +3.3	+2.5 / −0	+0.8 / −0.8	−1.2 / +5.2	+4.0 / −0	+1.2 / −1.2	−1.8 / +2.3	+2.5 / −0	+1.8 / +0.2	−2.7 / +3.8	+4.0 / −0	+2.7 / +0.2	−3.4 / +0.7	+2.5 / −0	+3.4 / +1.8	−4.3 / +0.7	+2.5 / −0	+4.3 / +1.8

9 AMERICAN STANDARD INTERFERENCE LOCATIONAL FITS (HOLE BASIS)

Limits are in thousandths of an inch.
Limits for hole and shaft are applied algebraically to the basic size to obtain the limits of size for the parts.
Data in bold face are in accordance with ABC agreements.
Symbols H7, p6, etc., are Hole and Shaft designations used in ABC System.

Nominal Size Range Inches Over / To	Class LN 1 Limits of Interference	Class LN 1 Standard Limits Hole H6	Class LN 1 Standard Limits Shaft n5	Class LN 2 Limits of Interference	Class LN 2 Standard Limits Hole H7	Class LN 2 Standard Limits Shaft p6	Class LN 3 Limits of Interference	Class LN 3 Standard Limits Hole H7	Class LN 3 Standard Limits Shaft r6
0–0.12	0	+0.25	+0.45	0	+0.4	+0.65	0.1	+0.4	+0.75
	0.45	−0	+0.25	0.65	−0	+0.4	0.75	−0	+0.5
0.12–0.24	0	+0.3	+0.5	0	+0.5	+0.8	0.1	+0.5	+0.9
	0.5	−0	+0.3	0.8	−0	+0.5	0.9	−0	+0.6
0.24–0.40	0	+0.4	+0.65	0	+0.6	+1.0	0.2	+0.6	+1.2
	0.65	−0	+0.4	1.0	−0	+0.6	1.2	−0	+0.8
0.40–0.71	0	+0.4	+0.8	0	+0.7	+1.1	0.3	+0.7	+1.4
	0.8	−0	+0.4	1.1	−0	+0.7	1.4	−0	+1.0
0.71–1.19	0	+0.5	+1.0	0	+0.8	+1.3	0.4	+0.8	+1.7
	1.0	−0	+0.5	1.3	−0	+0.8	1.7	−0	+1.2
1.19–1.97	0	+0.6	+1.1	0	+1.0	+1.6	0.4	+1.0	+2.0
	1.1	−0	+0.6	1.6	−0	+1.0	2.0	−0	+1.4
1.97–3.15	0.1	+0.7	+1.3	0.2	+1.2	+2.1	0.4	+1.2	+2.3
	1.3	−0	+0.7	2.1	−0	+1.4	2.3	−0	+1.6
3.15–4.73	0.1	+0.9	+1.6	0.2	+1.4	+2.5	0.6	+1.4	+2.9
	1.6	−0	+1.0	2.5	−0	+1.6	2.9	−0	+2.0
4.73–7.09	0.2	+1.0	+1.9	0.2	+1.6	+2.8	0.9	+1.6	+3.5
	1.9	−0	+1.2	2.8	−0	+1.8	3.5	−0	+2.5
7.09–9.85	0.2	+1.2	+2.2	0.2	+1.8	+3.2	1.2	+1.8	+4.2
	2.2	−0	+1.4	3.2	−0	+2.0	4.2	−0	+3.0
9.85–12.41	0.2	+1.2	+2.3	0.2	+2.0	+3.4	1.5	+2.0	+4.7
	2.3	−0	+1.4	3.4	−0	+2.2	4.7	−0	+3.5
12.41–15.75	0.2	+1.4	+2.6	0.3	+2.2	+3.9	2.3	+2.2	+5.9
	2.6	−0	+1.6	3.9	−0	+2.5	5.9	−0	+4.5
15.75–19.69	0.2	+1.6	+2.8	0.3	+2.5	+4.4	2.5	+2.5	+6.6
	2.8	−0	+1.8	4.4	−0	+2.8	6.6	−0	+5.0
19.69–30.09		+2.0		0.5	+3	+5.5	4	+3	+9
		−0		5.5	−0	+3.5	9	−0	+7
30.09–41.49		+2.5		0.5	+4	+7.0	5	+4	+11.5
		−0		7.0	−0	+4.5	11.5	−0	+9
41.49–56.19		+3.0		1	+5	+9	7	+5	+15
		−0		9	−0	+6	15	−0	+12
56.19–76.39		+4.0		1	+6	+11	10	+6	+20
		−0		11	−0	+7	20	−0	+16
76.39–100.9		+5.0		1	+8	+14	12	+8	+25
		−0		14	−0	+9	25	−0	+20
100.9–131.9		+6.0		2	+10	+18	15	+10	+31
		−0		18	−0	+12	31	−0	+25
131.9–171.9		+8.0		4	+12	+24	18	+12	+38
		−0		24	−0	+16	38	−0	+30
171.9–200		+10.0		4	+16	+30	24	+16	+50
		−0		30	−0	+20	50	−0	+40

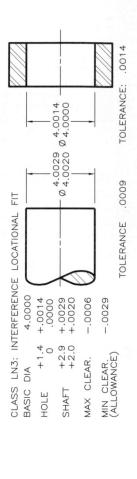

CLASS LN3: INTERFERENCE LOCATIONAL FIT
BASIC DIA   4.0000
HOLE   +1.4   +.0014
         0   .0000
SHAFT   +2.9   +.0029
        +2.0   +.0020
MAX CLEAR.   −.0006
MIN CLEAR. (ALLOWANCE)   −.0029

Ø 4.0014 / Ø 4.0000   TOLERANCE: .0014
Ø 4.0029 / Ø 4.0020   TOLERANCE .0009

10 AMERICAN STANDARD FORCE AND SHRINK FITS (HOLE BASIS)

Limits are in thousandths of an inch.
Limits for hole and shaft are applied algebraically to the basic size to obtain the limits of size for the parts.
Data in bold face are in accordance with ABC agreements.
Symbols H7, s6, etc., are Hole and Shaft designations used in ABC System.

Nominal Size Range Inches Over–To	Class FN 1 Limits of Interference	FN1 Hole H6	FN1 Shaft	Class FN 2 Limits of Interference	FN2 Hole H7	FN2 Shaft s6	Class FN 3 Limits of Interference	FN3 Hole H7	FN3 Shaft t6	Class FN 4 Limits of Interference	FN4 Hole H7	FN4 Shaft u6	Class FN 5 Limits of Interference	FN5 Hole H8	FN5 Shaft x7
0–0.12	0.05	+0.25	+0.5	0.2	+0.4	+0.85				0.3	+0.4	+0.95	0.3	+0.6	+1.3
	0.5	−0	+0.3	0.85	−0	+0.6				0.95	−0	+0.7	1.3	−0	+0.9
0.12–0.24	0.1	+0.3	+0.6	0.2	+0.5	+1.0				0.4	+0.5	+1.2	0.5	+0.7	+1.7
	0.6	−0	+0.4	1.0	−0	+0.7				1.2	−0	+0.9	1.7	−0	+1.2
0.24–0.40	0.1	+0.4	+0.75	0.4	+0.6	+1.4				0.6	+0.6	+1.6	0.5	+0.9	+2.0
	0.75	−0	+0.5	1.4	−0	+1.0				1.6	−0	+1.2	2.0	−0	+1.4
0.40–0.56	0.1	+0.4	+0.8	0.5	+0.7	+1.6				0.7	+0.7	+1.8	0.6	+1.0	+2.3
	0.8	−0	+0.5	1.6	−0	+1.2				1.8	−0	+1.4	2.3	−0	+1.6
0.56–0.71	0.2	+0.4	+0.9	0.5	+0.7	+1.6				0.7	+0.7	+1.8	0.8	+1.0	+2.5
	0.9	−0	+0.6	1.6	−0	+1.2				1.8	−0	+1.4	2.5	−0	+1.8
0.71–0.95	0.2	+0.5	+1.1	0.6	+0.8	+1.9				0.8	+0.8	+2.1	1.0	+1.2	+3.0
	1.1	−0	+0.7	1.9	−0	+1.4				2.1	−0	+1.6	3.0	−0	+2.2
0.95–1.19	0.3	+0.5	+1.2	0.6	+0.8	+1.9	0.8	+0.8	+2.1	1.0	+0.8	+2.3	1.3	+1.2	+3.3
	1.2	−0	+0.8	1.9	−0	+1.4	2.1	−0	+1.6	2.3	−0	+1.8	3.3	−0	+2.5
1.19–1.58	0.3	+0.6	+1.3	0.8	+1.0	+2.4	1.0	+1.0	+2.6	1.5	+1.0	+3.1	1.4	+1.6	+4.0
	1.3	−0	+0.9	2.4	−0	+1.8	2.6	−0	+2.0	3.1	−0	+2.5	4.0	−0	+3.0
1.58–1.97	0.4	+0.6	+1.4	0.8	+1.0	+2.4	1.2	+1.0	+2.8	1.8	+1.0	+3.4	2.4	+1.6	+5.0
	1.4	−0	+1.0	2.4	−0	+1.8	2.8	−0	+2.2	3.4	−0	+2.8	5.0	−0	+4.0
1.97–2.56	0.6	+0.7	+1.8	0.8	+1.2	+2.7	1.3	+1.2	+3.2	2.3	+1.2	+4.2	3.2	+1.8	+6.2
	1.8	−0	+1.3	2.7	−0	+2.0	3.2	−0	+2.5	4.2	−0	+3.5	6.2	−0	+5.0
2.56–3.15	0.7	+0.7	+1.9	1.0	+1.2	+2.9	1.8	+1.2	+3.7	2.8	+1.2	+4.7	4.2	+1.8	+7.2
	1.9	−0	+1.4	2.9	−0	+2.2	3.7	−0	+3.0	4.7	−0	+4.0	7.2	−0	+6.0
3.15–3.94	0.9	+0.9	+2.4	1.4	+1.4	+3.7	2.1	+1.4	+4.4	3.6	+1.4	+5.9	4.8	+2.2	+8.4
	2.4	−0	+1.8	3.7	−0	+2.8	4.4	−0	+3.5	5.9	−0	+5.0	8.4	−0	+7.0
3.94–4.73	1.1	+0.9	+2.6	1.6	+1.4	+3.9	2.6	+1.4	+4.9	4.6	+1.4	+6.9	5.8	+2.2	+9.4
	2.6	−0	+2.0	3.9	−0	+3.0	4.9	−0	+4.0	6.9	−0	+6.0	9.4	−0	+8.0
4.73–5.52	1.2	+1.0	+2.9	1.9	+1.6	+4.5	3.4	+1.6	+6.0	5.4	+1.6	+8.0	7.5	+2.5	+11.6
	2.9	−0	+2.2	4.5	−0	+3.5	6.0	−0	+5.0	8.0	−0	+7.0	11.6	−0	+10.0
5.52–6.30	1.5	+1.0	+3.2	2.4	+1.6	+5.0	3.4	+1.6	+6.0	5.4	+1.6	+8.0	9.5	+2.5	+13.6
	3.2	−0	+2.5	5.0	−0	+4.0	6.0	−0	+5.0	8.0	−0	+7.0	13.6	−0	+12.0
6.30–7.09	1.8	+1.0	+3.5	2.9	+1.6	+5.5	4.4	+1.6	+7.0	6.4	+1.6	+9.0	9.5	+2.5	+13.6
	3.5	−0	+2.8	5.5	−0	+4.5	7.0	−0	+6.0	9.0	−0	+8.0	13.6	−0	+12.0
7.09–7.88	1.8	+1.2	+3.8	3.2	+1.8	+6.2	5.2	+1.8	+8.2	7.2	+1.8	+10.2	11.2	+2.8	+15.8
	3.8	−0	+3.0	6.2	−0	+5.0	8.2	−0	+7.0	10.2	−0	+9.0	15.8	−0	+14.0
7.88–8.86	2.3	+1.2	+4.3	3.2	+1.8	+6.2	5.2	+1.8	+8.2	8.2	+1.8	+11.2	13.2	+2.8	+17.8
	4.3	−0	+3.5	6.2	−0	+5.0	8.2	−0	+7.0	11.2	−0	+10.0	17.8	−0	+16.0
8.86–9.85	2.3	+1.2	+4.3	4.2	+1.8	+7.2	6.2	+1.8	+9.2	10.2	+1.8	+13.2	13.2	+2.8	+17.8
	4.3	−0	+3.5	7.2	−0	+6.0	9.2	−0	+8.0	13.2	−0	+12.0	17.8	−0	+16.0
9.85–11.03	2.8	+1.2	+4.9	4.0	+2.0	+7.2	7.0	+2.0	+10.2	10.0	+2.0	+13.2	15.0	+3.0	+20.0
	4.9	−0	+4.0	7.2	−0	+6.0	10.2	−0	+9.0	13.2	−0	+12.0	20.0	−0	+18.0
11.03–12.41	2.8	+1.2	+4.9	5.0	+2.0	+8.2	7.0	+2.0	+10.2	12.0	+2.0	+15.2	17.0	+3.0	+22.0
	4.9	−0	+4.0	8.2	−0	+7.0	10.2	−0	+9.0	15.2	−0	+14.0	22.0	−0	+20.0
12.41–13.98	3.1	+1.4	+5.5	5.8	+2.2	+9.4	7.8	+2.2	+11.4	13.8	+2.2	+17.4	18.5	+3.5	+24.2
	5.5	−0	+4.5	9.4	−0	+8.0	11.4	−0	+10.0	17.4	−0	+16.0	24.2	+0	+22.0
13.98–15.75	3.6	+1.4	+6.1	5.8	+2.2	+9.4	9.8	+2.2	+13.4	15.8	+2.2	+19.4	21.5	+3.5	+27.2
	6.1	−0	+5.0	9.4	−0	+8.0	13.4	−0	+12.0	19.4	−0	+18.0	27.2	−0	+25.0
15.75–17.72	4.4	+1.6	+7.0	6.5	+2.5	+10.6	9.5	+2.5	+13.6	17.5	+2.5	+21.6	24.0	+4.0	+30.5
	7.0	−0	+6.0	10.6	−0	+9.0	13.6	−0	+12.0	21.6	−0	+20.0	30.5	−0	+28.0
17.72–19.69	4.4	+1.6	+7.0	7.5	+2.5	+11.6	11.5	+2.5	+15.6	19.5	+2.5	+23.6	26.0	+4.0	+32.5
	7.0	−0	+6.0	11.6	−0	+10.0	15.6	−0	+14.0	23.6	−0	+22.0	32.5	−0	+30.0

(Reprinted from ANSI B4.1–1967 (R2009). Copyright © American Society of Mechanical Engineers. All rights reserved.)

11 INTERNATIONAL TOLERANCE GRADES

Dimensions are in mm.

Basic sizes		Tolerance grades[1]																		
Over	Up to and including	ITO1	ITO	IT1	IT2	IT3	IT4	IT5	IT6	IT7	IT8	IT9	IT10	IT11	IT12	IT13	IT14	IT15	IT16
0	3	0.0003	0.0005	0.0008	0.0012	0.002	0.003	0.004	0.006	0.010	0.014	0.025	0.040	0.060	0.100	0.140	0.250	0.400	0.600
3	6	0.0004	0.0006	0.001	0.0015	0.0025	0.004	0.005	0.008	0.012	0.018	0.030	0.048	0.075	0.120	0.180	0.300	0.480	0.750
6	10	0.0004	0.0006	0.001	0.0015	0.0025	0.004	0.006	0.009	0.015	0.022	0.036	0.058	0.090	0.150	0.220	0.360	0.580	0.900
10	18	0.0005	0.0008	0.0012	0.002	0.003	0.005	0.008	0.011	0.018	0.027	0.043	0.070	0.110	0.180	0.270	0.430	0.700	1.100
18	30	0.0006	0.001	0.0015	0.0025	0.004	0.006	0.009	0.013	0.021	0.033	0.052	0.084	0.130	0.210	0.330	0.520	0.840	1.300
30	50	0.0006	0.001	0.0015	0.0025	0.004	0.007	0.011	0.016	0.025	0.039	0.062	0.100	0.160	0.250	0.390	0.620	1.000	1.600
50	80	0.0008	0.0012	0.002	0.003	0.005	0.008	0.013	0.019	0.030	0.046	0.074	0.120	0.190	0.300	0.460	0.740	1.200	1.900
80	120	0.001	0.0015	0.0025	0.004	0.006	0.010	0.015	0.022	0.035	0.054	0.087	0.140	0.220	0.350	0.540	0.870	1.400	2.200
120	180	0.0012	0.002	0.0036	0.005	0.008	0.012	0.018	0.025	0.040	0.063	0.100	0.160	0.250	0.400	0.630	1.000	1.600	2.500
180	250	0.002	0.003	0.0045	0.007	0.010	0.014	0.020	0.029	0.046	0.072	0.115	0.185	0.290	0.460	0.720	1.150	1.850	2.900
250	315	0.0025	0.004	0.006	0.008	0.012	0.016	0.023	0.032	0.052	0.081	0.130	0.210	0.320	0.520	0.810	1.300	2.100	3.200
315	400	0.003	0.005	0.007	0.009	0.013	0.018	0.025	0.036	0.057	0.089	0.140	0.230	0.360	0.570	0.890	1.400	2.300	3.600
400	500	0.004	0.006	0.008	0.010	0.015	0.020	0.027	0.040	0.063	0.097	0.156	0.250	0.400	0.630	0.970	1.550	2.500	4.000
500	630	0.0045	0.006	0.009	0.011	0.016	0.022	0.030	0.044	0.070	0.110	0.175	0.280	0.440	0.700	1.100	1.750	2.800	4.400
630	800	0.005	0.007	0.010	0.013	0.018	0.025	0.035	0.050	0.080	0.125	0.200	0.320	0.500	0.800	1.250	2.000	3.200	5.000
800	1000	0.0055	0.008	0.011	0.015	0.021	0.029	0.040	0.056	0.090	0.140	0.230	0.360	0.560	0.900	1.400	2.300	3.600	5.600
1000	1250	0.0065	0.009	0.013	0.018	0.024	0.034	0.046	0.066	0.105	0.165	0.260	0.420	0.660	1.050	1.650	2.600	4.200	6.600
1250	1600	0.008	0.011	0.015	0.021	0.029	0.040	0.054	0.078	0.125	0.195	0.310	0.500	0.780	1.250	1.950	3.100	5.000	7.800
1600	2000	0.009	0.013	0.018	0.025	0.035	0.048	0.065	0.092	0.150	0.230	0.370	0.600	0.920	1.500	2.300	3.700	6.000	9.200
2000	2500	0.011	0.015	0.022	0.030	0.041	0.057	0.077	0.110	0.175	0.280	0.440	0.700	1.100	1.750	2.800	4.400	7.000	11.000
2500	3150	0.013	0.018	0.026	0.036	0.050	0.069	0.093	0.135	0.210	0.330	0.540	0.860	1.350	2.100	3.300	5.400	8.600	13.500

[1]IT Values for tolerance grades larger than IT16 can be calculated by using the following formulas:
IT17 = IT12 × 10; IT18 = IT13 × 10; etc.

12 PREFERRED HOLE BASIS CLEARANCE FITS—CYLINDRICAL FITS

Dimensions are in mm.

BASIC SIZE		LOOSE RUNNING			FREE RUNNING			CLOSE RUNNING			SLIDING			LOCATIONAL CLEARANCE		
		Hole H11	Shaft c11	Fit	Hole H9	Shaft d9	Fit	Hole H8	Shaft f7	Fit	Hole H7	Shaft g6	Fit	Hole H7	Shaft h6	Fit
1	Max	1.060	0.940	0.180	1.025	0.980	0.070	1.014	0.994	0.030	1.010	0.998	0.018	1.010	1.000	0.016
	Min	1.000	0.880	0.060	1.000	0.955	0.020	1.000	0.984	0.006	1.000	0.992	0.002	1.000	0.994	0.000
1.2	Max	1.260	1.140	0.180	1.225	1.180	0.070	1.214	1.194	0.030	1.210	1.198	0.018	1.210	1.200	0.016
	Min	1.200	1.080	0.060	1.200	1.155	0.020	1.200	1.184	0.006	1.200	1.192	0.002	1.200	1.194	0.000
1.6	Max	1.660	1.540	0.180	1.625	1.583	0.070	1.614	1.594	0.030	1.610	1.598	0.018	1.610	1.600	0.016
	Min	1.600	1.480	0.060	1.600	1.555	0.020	1.600	1.584	0.006	1.600	1.592	0.002	1.600	1.594	0.000
2	Max	2.060	1.940	0.180	2.025	1.980	0.070	2.014	1.994	0.030	2.010	1.998	0.018	2.010	2.000	0.016
	Min	2.000	1.880	0.060	2.000	1.955	0.020	2.000	1.984	0.006	2.000	1.992	0.002	2.000	1.994	0.000
2.5	Max	2.560	2.440	0.180	2.525	0.480	0.070	2.514	2.494	0.030	2.510	2.498	0.018	2.510	2.500	0.016
	Min	2.500	2.380	0.060	2.500	2.455	0.020	2.500	2.484	0.006	2.500	2.492	0.002	2.500	2.494	0.000
3	Max	3.060	2.940	0.180	3.025	2.980	0.070	3.014	2.994	0.030	3.010	2.998	0.018	3.010	3.000	0.016
	Min	3.000	2.880	0.060	3.000	2.955	0.020	3.000	2.984	0.006	3.000	2.992	0.002	3.000	2.994	0.000
4	Max	4.075	3.930	0.220	4.030	3.970	0.090	4.018	3.990	0.040	4.012	3.996	0.024	4.012	4.000	0.020
	Min	4.000	3.855	0.070	4.000	3.940	0.030	4.000	3.978	0.010	4.000	3.988	0.004	4.000	3.992	0.000
5	Max	5.075	4.930	0.220	5.030	4.970	0.090	5.018	4.990	0.040	5.012	4.996	0.024	5.012	5.000	0.020
	Min	5.000	4.855	0.070	5.000	4.940	0.030	5.000	4.978	0.010	5.000	4.988	0.004	5.000	4.992	0.000
6	Max	6.075	5.930	0.220	6.030	5.970	0.090	6.018	5.990	0.040	6.012	5.996	0.024	6.012	6.000	0.020
	Min	6.000	5.855	0.070	6.000	5.940	0.030	6.000	5.978	0.010	6.000	5.988	0.004	6.000	5.992	0.000
8	Max	8.090	7.920	0.260	8.036	7.960	0.112	8.022	7.987	0.050	8.015	7.995	0.029	8.015	8.000	0.024
	Min	8.000	7.830	0.080	8.000	7.924	0.040	8.000	7.972	0.013	8.000	7.986	0.005	8.000	7.991	0.000
10	Max	10.090	9.920	0.260	10.036	9.960	0.112	10.022	9.987	0.050	10.015	9.995	0.029	10.015	10.000	0.024
	Min	10.000	9.830	0.080	10.000	9.924	0.040	10.000	9.972	0.013	10.000	9.986	0.005	10.000	9.991	0.000
12	Max	12.110	11.905	0.315	12.043	11.950	0.136	12.027	11.984	0.061	12.018	11.994	0.035	12.018	12.000	0.029
	Min	12.000	11.795	0.095	12.000	11.907	0.050	12.000	11.966	0.016	12.000	11.983	0.006	12.000	11.989	0.000
16	Max	16.110	15.905	0.315	16.043	15.950	0.136	16.027	15.984	0.061	16.018	15.994	0.035	16.018	16.000	0.029
	Min	16.000	15.795	0.095	16.000	15.907	0.050	16.000	15.966	0.016	16.000	15.983	0.006	16.000	15.989	0.000
20	Max	20.130	19.890	0.370	20.052	19.935	0.169	20.033	19.980	0.074	20.021	19.993	0.041	20.021	20.000	0.034
	Min	20.000	19.760	0.110	20.000	19.883	0.065	20.000	19.959	0.020	20.000	19.980	0.007	20.000	19.987	0.000
25	Max	25.130	24.890	0.370	25.052	24.935	0.169	25.033	24.980	0.074	25.021	24.993	0.041	25.021	25.000	0.034
	Min	25.000	24.760	0.110	25.000	24.883	0.065	25.000	24.959	0.020	25.000	24.980	0.007	25.000	24.987	0.000
30	Max	30.130	29.890	0.370	30.052	29.935	0.169	30.033	29.980	0.074	30.021	29.993	0.041	30.021	30.000	0.034
	Min	30.000	29.760	0.110	30.000	29.883	0.065	30.000	29.959	0.020	30.000	29.980	0.007	30.000	29.987	0.000

BASIC SIZE		LOOSE RUNNING			FREE RUNNING			CLOSE RUNNING			SLIDING			LOCATIONAL CLEARANCE		
		Hole H11	Shaft c11	Fit	Hole H9	Shaft d9	Fit	Hole H8	Shaft f7	Fit	Hole H7	Shaft g6	Fit	Hole H7	Shaft h6	Fit
40	Max	40.160	39.880	0.440	40.062	39.920	0.204	40.039	39.975	0.089	40.025	39.991	0.050	40.025	40.000	0.041
	Min	40.000	39.720	0.120	40.000	39.858	0.080	40.000	39.950	0.025	40.000	39.975	0.009	40.000	39.984	0.000
50	Max	50.160	49.870	0.450	50.062	49.920	0.204	50.039	49.975	0.089	50.025	49.991	0.050	50.025	50.000	0.041
	Min	50.000	49.710	0.130	50.000	49.858	0.080	50.000	49.950	0.025	50.000	49.975	0.009	50.000	49.984	0.000
60	Max	60.190	59.860	0.520	60.074	59.900	0.248	60.046	59.970	0.106	60.030	59.990	0.059	60.030	60.000	0.049
	Min	60.000	59.670	0.140	60.000	59.826	0.100	60.000	59.940	0.030	60.000	59.971	0.010	60.000	59.981	0.000
80	Max	80.190	79.850	0.530	80.074	79.900	0.248	80.046	79.970	0.106	80.030	79.990	0.059	80.030	80.000	0.049
	Min	80.000	79.660	0.150	80.000	79.826	0.100	80.000	79.940	0.030	80.000	79.971	0.010	80.000	79.981	0.000
100	Max	100.220	99.830	0.610	100.087	99.880	0.294	100.054	99.964	0.125	100.035	99.988	0.069	100.035	100.000	0.057
	Min	100.000	99.610	0.170	100.000	99.793	0.120	100.000	99.929	0.036	100.000	99.966	0.012	100.000	99.978	0.000
120	Max	120.220	119.820	0.620	120.087	119.880	0.294	120.054	119.964	0.125	120.035	119.988	0.069	120.035	120.000	0.057
	Min	120.000	119.600	0.180	120.000	119.793	0.120	120.000	119.929	0.036	120.000	119.966	0.012	120.000	119.978	0.000
160	Max	160.250	159.790	0.710	160.100	159.855	0.345	160.063	159.957	0.146	160.040	159.986	0.079	160.040	160.000	0.065
	Min	160.000	159.540	0.210	160.000	159.755	0.145	160.000	159.917	0.043	160.000	159.961	0.014	160.000	159.975	0.000
200	Max	200.290	199.760	0.820	200.115	199.830	0.400	200.072	199.950	0.168	200.046	199.985	0.090	200.046	200.000	0.075
	Min	200.000	199.470	0.240	200.000	199.715	0.170	200.000	199.904	0.050	200.000	199.956	0.105	200.000	199.971	0.000
250	Max	250.290	249.720	0.860	250.115	249.830	0.400	250.072	249.950	0.168	250.046	249.985	0.090	250.046	250.000	0.075
	Min	250.000	249.430	0.280	250.000	249.715	0.170	250.000	249.904	0.050	250.000	249.956	0.015	250.000	249.971	0.000
300	Max	300.320	299.670	0.970	300.130	299.810	0.450	300.081	299.944	0.189	300.052	299.983	0.101	300.052	300.000	0.084
	Min	300.000	299.350	0.330	300.000	299.680	0.190	300.000	299.892	0.056	300.000	299.951	0.017	300.000	299.968	0.000
400	Max	400.360	399.600	1.120	400.140	399.790	0.490	400.089	399.938	0.208	400.057	399.982	0.111	400.057	400.000	0.093
	Min	400.000	399.240	0.400	400.000	399.650	0.210	400.000	399.881	0.062	400.000	399.946	0.018	400.000	399.964	0.000
500	Max	500.400	499.520	1.280	500.155	499.770	0.540	500.097	499.932	0.228	500.063	499.980	0.123	500.063	500.000	0.103
	Min	500.000	499.120	0.480	500.000	499.615	0.230	500.000	499.869	0.068	500.000	499.940	0.020	500.000	499.960	0.000

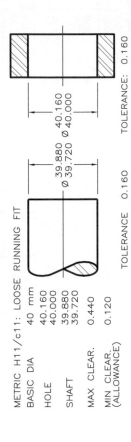

METRIC H11/c11: LOOSE RUNNING FIT

BASIC DIA	40 mm	
HOLE	40.160	
	40.000	
SHAFT	39.880	
	39.720	
MAX CLEAR.	0.440	
MIN CLEAR. (ALLOWANCE)	0.120	

40.160 Ø 40.000

39.880 40.160

Ø 39.880 39.720

TOLERANCE 0.160   TOLERANCE 0.160   TOLERANCE: 0.160

13 PREFERRED HOLE BASIS TRANSITION AND INTERFERENCE FITS—CYLINDRICAL FITS

Dimensions are in mm.

BASIC SIZE		LOCATIONAL TRANSN.			LOCATIONAL TRANSN.			LOCATIONAL INTERF.			MEDIUM DRIVE			FORCE		
		Hole H7	Shaft k6	Fit	Hole H7	Shaft n6	Fit	Hole H7	Shaft p6	Fit	Hole H7	Shaft s6	Fit	Hole H7	Shaft u6	Fit
1	Max	1.010	1.006	0.010	1.010	1.010	0.006	1.010	1.012	0.004	1.010	1.020	−0.004	1.010	1.024	−0.008
	Min	1.000	1.000	−0.006	1.000	1.004	−0.010	1.000	1.006	−0.012	1.000	1.014	−0.020	1.000	1.018	−0.024
1.2	Max	1.210	1.206	0.010	1.210	1.210	0.006	1.210	1.212	0.004	1.210	1.220	−0.004	1.210	1.224	−0.008
	Min	1.200	1.200	−0.006	1.200	1.204	−0.010	1.200	1.206	−0.012	1.200	1.214	−0.020	1.200	1.218	−0.024
1.6	Max	1.610	1.606	0.010	1.610	1.610	0.006	1.610	1.612	0.004	1.610	1.620	−0.004	1.610	1.624	−0.008
	Min	1.600	1.600	−0.006	1.600	1.604	−0.010	1.600	1.606	−0.012	1.600	1.614	−0.020	1.600	1.618	−0.024
2	Max	2.010	2.006	0.010	2.010	2.010	0.006	2.010	2.010	0.004	2.010	2.020	−0.004	2.010	2.024	−0.008
	Min	2.000	2.000	−0.006	2.000	2.004	−0.010	2.000	2.006	−0.012	2.000	2.014	−0.020	2.000	2.018	−0.024
2.5	Max	2.510	2.506	0.010	2.510	2.510	0.006	2.510	2.512	0.004	2.510	2.520	−0.004	2.510	2.524	−0.008
	Min	2.500	2.500	−0.006	2.500	2.504	−0.010	2.500	2.506	−0.012	2.500	2.514	−0.020	2.500	2.518	−0.024
3	Max	3.010	3.006	0.010	3.010	3.010	0.006	3.010	3.012	0.004	3.010	3.020	−0.004	3.010	3.024	−0.008
	Min	3.000	3.000	−0.006	3.000	3.004	−0.010	3.000	3.006	−0.012	3.000	3.014	−0.020	3.000	3.018	−0.024
4	Max	4.012	4.009	0.011	4.012	4.016	0.004	4.012	4.020	0.000	4.012	4.027	−0.007	4.012	4.031	−0.011
	Min	4.000	4.001	−0.009	4.000	4.008	−0.016	4.000	4.012	−0.020	4.000	4.019	−0.027	4.000	4.023	−0.031
5	Max	5.012	5.009	0.011	5.012	5.016	0.004	5.012	5.020	0.000	5.012	5.027	−0.007	5.012	5.031	−0.011
	Min	5.000	5.001	−0.009	5.000	5.008	−0.016	5.000	5.012	−0.020	5.000	5.019	−0.027	5.000	5.023	−0.031
6	Max	6.012	6.009	0.011	6.012	6.016	0.004	6.012	6.020	0.000	6.012	6.027	−0.007	6.012	6.031	−0.011
	Min	6.000	6.001	−0.009	6.000	6.008	−0.016	6.000	6.012	−0.020	6.000	6.019	−0.027	6.000	6.023	−0.031
8	Max	8.015	8.010	0.014	8.015	8.019	0.005	8.015	8.024	0.000	8.015	8.032	−0.008	8.015	8.037	−0.013
	Min	8.000	8.001	−0.010	8.000	8.010	−0.019	8.000	8.015	−0.024	8.000	8.023	−0.032	8.000	8.028	−0.037
10	Max	10.015	10.010	0.014	10.015	10.019	0.005	10.015	10.024	0.000	10.015	10.032	−0.008	10.015	10.037	−0.013
	Min	10.000	10.001	−0.010	10.000	10.010	−0.019	10.000	10.015	−0.024	10.000	10.023	−0.032	10.000	10.028	−0.037
12	Max	12.018	12.012	0.017	12.018	12.023	0.006	12.018	12.029	0.000	12.018	12.039	−0.010	12.018	12.044	−0.015
	Min	12.000	12.001	−0.012	12.000	12.012	−0.023	12.000	12.018	−0.029	12.000	12.028	−0.039	12.000	12.033	−0.044
16	Max	16.018	16.012	0.017	16.018	16.023	0.006	16.018	16.029	0.000	16.018	16.039	−0.010	16.018	16.044	−0.015
	Min	16.000	16.001	−0.012	16.000	16.012	−0.023	16.000	16.018	−0.029	16.000	16.028	−0.039	16.000	16.033	−0.044
20	Max	20.021	20.015	0.019	20.021	20.028	0.006	20.021	20.035	−0.001	20.021	20.048	−0.014	20.021	20.054	−0.020
	Min	20.000	20.002	−0.015	20.000	20.015	−0.028	20.000	20.022	−0.035	20.000	20.035	−0.048	20.000	20.041	−0.054
25	Max	25.021	25.015	0.019	25.021	25.028	0.006	25.021	25.035	−0.001	25.021	25.048	−0.014	25.021	25.061	−0.027
	Min	25.000	25.002	−0.015	25.000	25.015	−0.028	25.000	25.022	−0.035	25.000	25.035	−0.048	25.000	25.048	−0.061
30	Max	30.021	30.015	0.019	30.021	30.028	0.006	30.021	30.035	−0.001	30.021	30.048	−0.014	30.021	30.061	−0.027
	Min	30.000	30.002	−0.015	30.000	30.015	−0.028	30.000	30.022	−0.035	30.000	30.035	−0.048	30.000	30.048	−0.061

BASIC SIZE		LOCATIONAL TRANSN.			LOCATIONAL TRANSN.			LOCATIONAL INTERF.			MEDIUM DRIVE			FORCE		
		Hole H7	Shaft k6	Fit	Hole H7	Shaft n6	Fit	Hole H7	Shaft p6	Fit	Hole H7	Shaft s6	Fit	Hole H7	Shaft u6	Fit
0	Max	40.025	40.018	0.023	40.025	40.033	0.008	40.025	40.042	-0.001	40.025	40.059	-0.018	40.025	40.076	-0.035
	Min	40.000	40.002	-0.018	40.000	40.017	-0.033	40.000	40.026	-0.042	40.000	40.043	-0.059	40.000	40.060	-0.076
50	Max	50.025	50.018	0.023	50.025	50.033	0.008	50.025	50.042	-0.001	50.025	50.059	-0.018	50.025	50.086	-0.045
	Min	50.000	50.002	-0.018	50.000	50.017	-0.033	50.000	50.026	-0.042	50.000	50.043	-0.059	50.000	50.070	-0.086
60	Max	60.030	60.021	0.028	60.030	60.039	0.010	60.030	60.051	-0.002	60.030	60.072	-0.023	60.030	60.106	-0.057
	Min	60.000	60.002	-0.021	60.000	60.020	-0.039	60.000	60.032	-0.051	60.000	60.053	-0.072	60.000	60.087	-0.106
80	Max	80.030	80.021	0.028	80.030	80.039	0.010	80.030	80.051	-0.002	80.030	80.078	-0.029	80.030	80.121	-0.072
	Min	80.000	80.002	-0.021	80.000	80.020	-0.039	80.000	80.032	-0.051	80.000	80.059	-0.078	80.000	80.102	-0.121
100	Max	100.035	100.025	0.032	100.035	100.045	0.012	100.035	100.059	-0.002	100.035	100.093	-0.036	100.035	100.146	-0.089
	Min	100.000	100.003	-0.025	100.000	100.023	-0.045	100.000	100.037	-0.059	100.000	100.071	-0.093	100.000	100.124	-0.146
120	Max	120.035	120.025	0.032	120.035	120.045	0.012	120.035	120.059	-0.002	120.035	120.101	-0.044	120.035	120.166	-0.109
	Min	120.000	120.003	-0.025	120.000	120.023	-0.045	120.000	120.037	-0.059	120.000	120.079	-0.101	120.000	120.144	-0.166
160	Max	160.040	160.028	0.037	160.040	160.052	0.013	160.040	160.068	-0.003	160.040	160.125	-0.060	160.040	160.215	-0.150
	Min	160.000	160.003	-0.028	160.000	160.027	-0.052	160.000	160.043	-0.068	160.000	160.100	-0.125	160.000	160.190	-0.215
200	Max	200.046	200.033	0.042	200.046	200.060	0.015	200.046	200.079	-0.004	200.046	200.151	-0.076	200.046	200.265	-0.190
	Min	200.000	200.004	-0.033	200.000	200.031	-0.060	200.000	200.050	-0.079	200.000	200.122	-0.151	200.000	200.236	-0.265
250	Max	250.046	250.033	0.042	250.046	250.060	0.015	250.046	250.079	-0.004	250.046	250.169	-0.094	250.046	250.313	-0.238
	Min	250.000	250.004	-0.033	250.000	250.031	-0.060	250.000	250.050	-0.079	250.000	250.140	-0.169	250.000	250.284	-0.313
300	Max	300.052	300.036	0.048	300.052	300.066	0.018	300.052	300.088	-0.004	300.052	300.202	-0.118	300.052	300.382	-0.298
	Min	300.000	300.004	-0.036	300.000	300.034	-0.066	300.000	300.056	-0.088	300.000	300.170	-0.202	300.000	300.350	-0.382
400	Max	400.057	400.040	0.053	400.057	400.073	0.020	400.057	400.098	-0.005	400.057	400.244	-0.151	400.057	400.471	-0.378
	Min	400.000	400.004	-0.040	400.000	400.037	-0.073	400.000	400.062	-0.098	400.000	400.208	-0.244	400.000	400.435	-0.471
500	Max	500.063	500.045	0.058	500.063	500.080	0.023	500.063	500.108	-0.005	500.063	500.292	-0.189	500.063	500.580	-0.477
	Min	500.000	500.005	-0.045	500.000	500.040	-0.080	500.000	500.068	-0.108	500.000	500.252	-0.292	500.000	500.540	-0.580

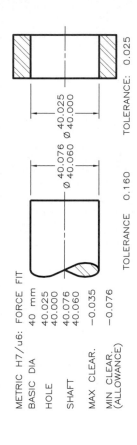

METRIC H7/u6: FORCE FIT

BASIC DIA	40 mm	
HOLE	40.025	40.000
SHAFT	40.076	40.060
MAX CLEAR.	-0.035	
MIN CLEAR. (ALLOWANCE)	-0.076	

Ø 40.076 Ø 40.060
Ø 40.025 Ø 40.000

TOLERANCE 0.160   TOLERANCE: 0.025

(Reprinted from ANSI B4.2–1978 (R2009). Copyright © American Society of Mechanical Engineers. All rights reserved).

14 PREFERRED SHAFT BASIS CLEARANCE FITS—CYLINDRICAL FITS

Dimensions are in mm.

BASIC SIZE		LOOSE RUNNING			FREE RUNNING			CLOSE RUNNING			SLIDING			LOCATIONAL CLEARANCE		
		Hole H11	Shaft c11	Fit	Hole H9	Shaft d9	Fit	Hole H8	Shaft f7	Fit	Hole H7	Shaft g6	Fit	Hole H7	Shaft h6	Fit
1	MAX	1.060	0.940	0.180	1.025	0.980	0.070	1.014	0.994	0.030	1.010	0.998	0.018	1.010	1.000	0.016
	MIN	1.000	0.880	0.060	1.000	0.955	0.020	1.000	0.984	0.006	1.000	0.992	0.002	1.000	0.994	0.000
1.2	MAX	1.260	1.140	0.180	1.225	1.180	0.070	1.214	1.194	0.030	1.210	1.198	0.018	1.210	1.200	0.016
	MIN	1.200	1.080	0.060	1.200	1.155	0.020	1.200	1.184	0.006	1.200	1.192	0.002	1.200	1.194	0.000
1.6	MAX	1.660	1.540	0.180	1.625	1.580	0.070	1.614	1.594	0.030	1.610	1.598	0.018	1.610	1.600	0.016
	MIN	1.600	1.480	0.060	1.600	1.555	0.020	1.600	1.584	0.006	1.600	1.592	0.002	1.600	1.594	0.000
2	MAX	2.060	1.940	0.180	2.025	1.980	0.070	2.014	1.994	0.030	2.010	1.998	0.018	2.010	2.000	0.016
	MIN	2.000	1.880	0.060	2.000	1.955	0.020	2.000	1.984	0.006	2.000	1.992	0.002	2.000	1.994	0.000
2.5	MAX	2.560	2.440	0.180	2.525	2.480	0.070	2.514	2.494	0.030	2.510	2.498	0.018	2.510	2.500	0.016
	MIN	2.500	2.380	0.060	2.500	2.455	0.020	2.500	2.484	0.006	2.500	2.492	0.002	2.500	2.494	0.000
3	MAX	3.060	2.940	0.180	3.025	2.980	0.070	3.014	2.994	0.030	3.010	2.998	0.018	3.010	3.000	0.016
	MIN	3.000	2.880	0.060	3.000	2.955	0.020	3.000	2.984	0.006	3.000	2.992	0.002	3.000	2.994	0.000
4	MAX	4.075	3.930	0.220	4.030	3.970	0.090	4.018	3.990	0.040	4.012	3.996	0.024	4.012	4.000	0.020
	MIN	4.000	3.855	0.070	4.000	3.940	0.030	4.000	3.978	0.010	4.000	3.988	0.004	4.000	3.992	0.000
5	MAX	5.075	4.930	0.220	5.030	4.970	0.090	5.018	4.990	0.040	5.012	4.996	0.024	5.012	5.000	0.020
	MIN	5.000	4.855	0.070	5.000	4.940	0.030	5.000	4.978	0.010	5.000	4.988	0.004	5.000	4.992	0.000
6	MAX	6.075	5.930	0.220	6.030	5.970	0.090	6.018	5.990	0.040	6.012	5.996	0.024	6.012	6.000	0.020
	MIN	6.000	5.855	0.070	6.000	5.940	0.030	6.000	5.978	0.010	6.000	5.988	0.004	6.000	5.992	0.000
8	MAX	8.090	7.920	0.260	8.036	7.960	0.112	8.022	7.987	0.050	8.015	7.995	0.029	8.015	8.000	0.024
	MIN	8.000	7.830	0.080	8.000	7.924	0.040	8.000	7.972	0.013	8.000	7.986	0.005	8.000	7.991	0.000
10	MAX	10.090	9.920	0.260	10.036	9.960	0.112	10.022	9.987	0.050	10.015	9.995	0.029	10.015	10.000	0.024
	MIN	10.000	9.830	0.080	10.000	9.924	0.040	10.000	9.972	0.013	10.000	9.986	0.005	10.000	9.991	0.000
12	MAX	12.110	11.905	0.315	12.043	11.950	0.136	12.027	11.984	0.061	12.018	11.994	0.035	12.018	12.000	0.029
	MIN	12.000	11.795	0.095	12.000	11.907	0.050	12.000	11.966	0.016	12.000	11.983	0.006	12.000	11.989	0.000
16	MAX	16.110	15.905	0.315	16.043	15.950	0.136	16.027	15.984	0.061	16.018	15.994	0.035	16.018	16.000	0.029
	MIN	16.000	15.795	0.095	16.000	15.907	0.050	16.000	15.966	0.016	16.000	15.983	0.006	16.000	15.989	0.000
20	MAX	20.130	19.890	0.370	20.052	19.935	0.169	20.033	19.980	0.074	20.021	19.993	0.041	20.021	20.000	0.034
	MIN	20.000	19.760	0.110	20.000	19.883	0.065	20.000	19.959	0.020	20.000	19.980	0.007	20.000	19.987	0.000
25	MAX	25.130	24.890	0.370	25.052	24.935	0.169	25.033	24.980	0.074	25.021	24.993	0.041	25.021	25.000	0.034
	MIN	25.000	24.760	0.110	25.000	24.883	0.065	25.000	24.959	0.020	25.000	24.980	0.007	25.000	24.987	0.000
30	MAX	30.130	29.890	0.370	30.052	29.935	0.169	30.033	29.980	0.074	30.021	29.993	0.041	30.021	30.000	0.034
	MIN	30.000	29.760	0.110	30.000	29.883	0.065	30.000	29.959	0.020	30.000	29.980	0.007	30.000	29.987	0.000

BASIC SIZE		LOOSE RUNNING			FREE RUNNING			CLOSE RUNNING			SLIDING			LOCATIONAL CLEARANCE		
		Hole C11	Shaft h11	Fit	Hole D9	Shaft h9	Fit	Hole F8	Shaft h7	Fit	Hole G7	Shaft h6	Fit	Hole H7	Shaft h6	Fit
40	Max	40.280	40.000	0.440	40.142	40.000	0.204	40.064	40.000	0.089	40.034	40.000	0.050	40.025	40.000	0.041
	Min	40.120	39.840	0.120	40.080	39.938	0.080	40.025	39.975	0.025	40.009	39.984	0.009	40.000	39.984	0.000
50	Max	50.290	50.000	0.450	50.142	50.000	0.204	50.064	50.000	0.089	50.034	50.000	0.050	50.025	50.000	0.041
	Min	50.130	49.840	0.130	50.080	49.938	0.080	50.025	49.975	0.025	50.009	49.984	0.009	50.000	49.984	0.000
60	Max	60.330	60.000	0.520	60.174	60.000	0.248	60.076	60.000	0.106	60.040	60.000	0.059	60.030	60.000	0.049
	Min	60.140	59.810	0.140	60.100	59.926	0.100	60.030	59.970	0.030	60.010	59.981	0.010	60.000	59.981	0.000
80	Max	80.340	80.000	0.530	80.174	80.000	0.248	80.076	80.000	0.106	80.040	80.000	0.059	80.030	80.000	0.049
	Min	80.150	79.810	0.150	80.100	79.926	0.100	80.030	79.970	0.030	80.010	79.981	0.010	80.000	79.981	0.000
100	Max	100.390	100.000	0.610	100.207	100.000	0.294	100.090	100.000	0.125	100.047	100.000	0.069	100.035	100.000	0.057
	Min	100.170	99.780	0.170	100.120	99.913	0.120	100.036	99.965	0.036	100.012	99.979	0.012	100.000	99.979	0.000
120	Max	120.400	120.000	0.620	120.207	120.000	0.294	120.090	120.000	0.125	120.047	120.000	0.069	120.035	120.000	0.057
	Min	120.180	119.780	0.180	120.120	119.913	0.120	120.036	119.965	0.036	120.012	119.978	0.012	120.000	119.978	0.000
160	Max	160.460	160.000	0.710	160.245	160.000	0.345	160.106	160.000	0.146	160.054	160.000	0.079	160.040	160.000	0.065
	Min	160.210	159.750	0.210	160.145	159.900	0.145	160.043	159.960	0.043	160.014	159.975	0.014	160.000	159.975	0.000
200	Max	200.530	200.000	0.820	200.285	200.000	0.400	200.122	200.000	0.168	200.061	200.000	0.090	200.046	200.000	0.075
	Min	200.240	199.710	0.240	200.170	199.885	0.170	200.050	199.954	0.050	200.015	199.971	0.015	200.000	199.971	0.000
250	Max	250.570	250.000	0.860	250.285	250.000	0.400	250.122	250.000	0.168	250.061	250.000	0.090	250.046	250.000	0.075
	Min	250.280	249.710	0.280	250.170	249.885	0.170	250.050	249.954	0.050	250.015	249.971	0.015	250.000	249.971	0.000
300	Max	300.650	300.000	0.970	300.320	300.000	0.450	300.137	300.000	0.189	300.069	300.000	0.101	300.052	300.000	0.084
	Min	300.330	299.680	0.330	300.190	299.870	0.190	300.056	299.948	0.056	300.017	299.968	0.017	300.000	299.968	0.000
400	Max	400.760	400.000	1.120	400.350	400.000	0.490	400.151	400.000	0.208	400.075	400.000	0.111	400.057	400.000	0.983
	Min	400.400	399.640	0.400	400.210	399.860	0.210	400.062	399.943	0.062	400.018	399.964	0.018	400.000	399.964	0.000
500	Max	500.880	500.000	1.280	500.385	500.000	0.540	500.165	500.000	0.228	500.083	500.000	0.123	500.063	500.000	0.103
	Min	500.480	499.600	0.480	500.230	499.845	0.230	500.068	499.937	0.068	500.020	499.960	0.020	500.000	499.960	0.000

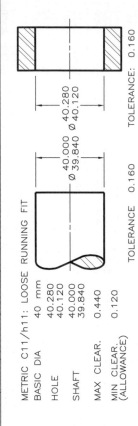

METRIC C11/h11: LOOSE RUNNING FIT

BASIC DIA	40 mm	
HOLE	40.280	40.120
SHAFT	40.000	39.840
MAX CLEAR.	0.440	
MIN CLEAR. (ALLOWANCE)	0.120	

Ø 40.000 Ø 40.280
Ø 39.840 Ø 40.120

TOLERANCE   0.160   TOLERANCE: 0.160

15 PREFERRED SHAFT BASIS TRANSITION AND INTERFERENCE FITS—CYLINDRICAL FITS

Dimensions are in mm.

BASIC SIZE		LOCATIONAL TRANSN. Hole K7	LOCATIONAL TRANSN. Shaft h6	LOCATIONAL TRANSN. Fit	LOCATIONAL TRANSN. Hole N7	LOCATIONAL TRANSN. Shaft h6	LOCATIONAL TRANSN. Fit	LOCATIONAL INTERF. Hole P7	LOCATIONAL INTERF. Shaft h6	LOCATIONAL INTERF. Fit	MEDIUM DRIVE Hole S7	MEDIUM DRIVE Shaft h6	MEDIUM DRIVE Fit	FORCE Hole U7	FORCE Shaft h6	FORCE Fit
1	Max	1.000	1.000	0.006	0.996	1.000	0.002	0.994	1.000	0.000	0.986	1.000	-0.008	0.982	1.000	-0.012
	Min	0.990	0.994	-0.010	0.986	0.994	-0.014	0.984	0.994	-0.016	0.976	0.994	-0.024	0.972	0.994	-0.028
1.2	Max	1.200	1.200	0.006	1.196	1.200	0.002	1.194	1.200	0.000	1.186	1.200	-0.008	1.182	1.200	-0.012
	Min	1.190	1.194	-0.010	1.186	1.194	-0.014	1.184	1.194	-0.016	1.176	1.194	-0.024	1.172	1.194	-0.028
1.6	Max	1.600	1.600	0.006	1.596	1.600	0.002	1.594	1.600	0.000	1.586	1.600	-0.008	1.582	1.600	-0.012
	Min	1.590	1.594	-0.010	1.586	1.594	-0.014	1.584	1.594	-0.016	1.576	1.594	-0.024	1.572	1.594	-0.028
2	Max	2.000	2.000	0.006	1.996	2.000	0.002	1.994	2.000	0.008	1.986	2.000	-0.008	1.982	2.000	-0.012
	Min	1.990	1.994	-0.010	1.986	1.994	-0.014	1.984	1.994	-0.016	1.976	1.994	-0.024	1.972	1.994	-0.028
2.5	Max	2.500	2.500	0.006	2.496	2.500	0.002	2.494	2.500	0.000	2.486	2.500	-0.008	2.482	2.500	-0.012
	Min	2.490	2.494	-0.010	2.486	2.494	-0.014	2.484	2.494	-0.016	2.476	2.494	-0.024	2.472	2.494	-0.028
3	Max	3.000	3.000	0.006	2.996	3.000	0.002	2.994	3.000	0.000	2.986	3.000	-0.008	2.982	3.000	-0.012
	Min	2.990	2.994	-.010	2.986	2.994	-0.014	2.984	2.994	-0.016	2.976	2.994	-0.024	2.972	2.994	-0.028
4	Max	4.003	4.000	0.011	3.996	4.000	0.004	3.992	4.000	0.000	3.985	4.000	-0.007	3.981	4.000	-0.011
	Min	3.991	3.992	-0.009	3.984	3.992	-0.016	3.980	3.992	-0.020	3.973	3.992	-0.027	3.969	3.992	-0.031
5	Max	5.003	5.000	0.011	4.996	5.000	0.004	4.992	5.000	0.000	4.985	5.000	-0.007	4.981	5.000	-0.011
	Min	4.991	4.992	-0.009	4.984	4.992	-0.016	4.980	4.992	-0.020	4.973	4.992	-0.027	4.969	4.992	-0.031
6	Max	6.003	6.000	0.011	5.996	6.000	0.004	5.992	6.000	0.000	5.985	6.000	-0.007	5.981	6.000	-0.011
	Min	5.991	5.992	-0.009	5.984	5.992	-0.016	5.980	5.992	-0.020	5.973	5.992	-0.027	5.969	5.992	-0.031
8	Max	8.005	8.000	0.014	7.986	8.000	0.005	7.991	8.000	0.000	7.983	8.000	-0.008	7.978	8.000	-0.013
	Min	7.990	7.991	-0.010	7.981	7.991	-0.019	7.976	7.991	-0.024	7.968	7.991	-0.032	7.963	7.991	-0.037
10	Max	10.005	10.000	0.014	9.996	10.000	0.005	9.991	10.000	0.0000	9.983	10.000	-0.008	9.978	10.000	-0.013
	Min	9.990	9.991	-0.010	9.981	9.991	-0.019	9.976	9.991	-0.024	9.968	9.991	-0.032	9.963	9.991	-0.037
12	Max	12.006	12.000	0.017	11.995	12.000	0.006	11.989	12.000	0.000	11.979	12.000	-0.010	11.974	12.000	-0.015
	Min	11.988	11.989	-0.012	11.977	11.989	-0.023	11.971	11.989	-0.029	11.961	11.989	-0.039	11.956	11.989	-0.044
16	Max	16.006	16.000	0.017	15.995	16.000	0.006	15.989	16.000	0.000	15.979	16.000	-0.010	15.974	16.000	-0.015
	Min	15.988	15.989	-0.012	15.977	15.989	-0.023	15.971	15.989	-0.029	15.961	15.989	-0.039	15.956	15.989	-0.044
20	Max	20.006	20.000	0.019	19.993	20.000	0.006	19.986	20.000	-0.001	19.973	20.000	-0.014	19.967	20.000	-0.020
	Min	19.985	19.987	-0.015	19.972	19.987	-0.028	19.965	19.987	-0.035	19.952	19.987	-0.048	19.946	19.987	-0.054
25	Max	25.006	25.000	0.019	24.993	25.000	0.006	24.986	25.000	-0.001	24.973	25.000	-0.014	24.960	25.000	-0.027
	Min	24.985	24.987	-0.015	24.972	24.987	-0.028	24.965	24.987	-0.035	24.952	24.987	-0.048	24.939	24.987	-0.061
30	Max	30.006	30.000	0.019	29.993	30.000	0.006	29.986	30.000	-0.001	29.973	30.000	-0.014	29.960	30.000	-0.027
	Min	29.985	29.987	-0.015	29.972	29.987	-0.028	29.965	29.987	-0.035	29.952	29.987	-0.048	29.939	29.987	-0.061

Dimensions are in mm.

BASIC SIZE		LOCATIONAL TRANSN. Hole K7	Shaft h6	Fit	LOCATIONAL TRANSN. Hole N7	Shaft h6	Fit	LOCATIONAL TRANSN. Hole P7	Shaft h6	Fit	MEDIUM DRIVE Hole S7	Shaft h6	Fit	FORCE Hole U7	Shaft h6	Fit
40	Max	40.007	40.000	0.023	39.992	40.000	0.008	39.983	40.000	-0.001	39.966	40.000	-0.018	39.949	40.000	-0.035
	Min	39.982	39.984	-0.018	39.967	39.984	-0.033	39.958	39.984	-0.042	39.941	39.984	-0.059	39.924	39.984	-0.076
50	Max	50.007	50.000	0.023	49.992	50.000	0.008	49.983	50.000	-0.001	49.966	50.000	-0.018	49.939	50.000	-0.045
	Min	49.982	49.984	-0.018	49.967	49.984	-0.033	49.958	49.984	-0.042	49.941	49.984	-0.059	49.914	49.984	-0.086
60	Max	60.009	60.000	0.028	59.991	60.000	0.010	59.979	60.000	-0.002	59.958	60.000	-0.023	59.924	60.000	-0.057
	Min	59.979	59.981	-0.021	59.961	59.981	-0.039	59.949	59.981	-0.051	59.928	59.981	-0.072	59.894	59.981	-0.106
80	Max	80.009	80.000	0.028	79.991	80.000	0.010	79.979	80.000	-0.002	79.952	80.000	-0.029	79.909	80.000	-0.072
	Min	79.979	79.981	-0.021	79.961	79.981	-0.039	79.949	79.981	-0.051	79.922	79.981	-0.078	79.879	79.981	-0.121
100	Max	100.010	100.000	0.032	99.990	100.000	0.012	99.976	100.000	-0.002	99.942	100.000	-0.036	99.889	100.000	-0.089
	Min	99.975	99.978	-0.025	99.955	99.978	-0.045	99.941	99.978	-0.059	99.907	99.978	-0.093	99.854	99.978	-0.146
120	Max	120.010	120.000	0.032	119.990	120.000	0.012	119.976	120.000	-0.002	119.934	120.000	-0.044	119.869	120.000	-0.109
	Min	119.975	119.978	-0.025	119.955	119.978	-0.045	119.941	119.978	-0.059	119.899	119.978	-0.101	119.834	119.978	-0.166
160	Max	160.012	160.000	0.037	159.988	160.000	0.013	159.972	160.000	-0.003	159.915	160.000	-0.060	159.825	160.000	-0.150
	Min	159.972	159.975	-0.028	159.948	159.975	-0.052	159.932	159.975	-0.068	159.875	159.975	-0.125	159.785	159.975	-0.215
200	Max	200.013	200.000	0.042	199.986	200.000	0.015	199.967	200.000	-0.004	199.895	200.000	-0.076	199.781	200.000	-0.190
	Min	199.967	199.971	-0.033	199.940	199.971	-0.060	199.921	199.971	-0.079	199.849	199.971	-0.151	199.735	199.971	-0.265
250	Max	250.013	250.000	0.042	249.986	250.000	0.015	249.967	250.000	-0.004	249.877	250.000	-0.094	249.733	250.000	-0.238
	Min	249.967	249.971	-0.033	249.940	249.971	-0.060	249.921	249.971	-0.079	249.831	249.971	-0.169	249.687	249.971	-0.313
300	Max	300.016	300.000	0.048	299.986	300.000	0.018	299.964	300.000	-0.004	299.850	300.000	-0.188	299.670	300.000	-0.298
	Min	299.964	299.968	-0.036	299.934	299.968	-0.066	299.912	299.968	-0.088	299.798	299.968	-0.202	299.618	299.968	-0.382
400	Max	400.017	400.000	0.053	399.984	400.000	0.020	399.959	400.000	-0.005	399.813	400.000	-0.151	399.586	400.000	-0.378
	Min	399.960	399.964	-0.040	399.927	399.964	-0.073	399.902	399.964	-0.08	399.756	399.964	-0.244	399.529	399.964	-0.471
500	Max	500.018	500.000	0.058	499.983	500.000	0.023	499.955	500.000	-0.005	499.771	500.000	-0.189	499.483	500.000	-0.477
	Min	499.955	499.960	-0.045	499.920	499.960	-0.080	499.892	499.960	-0.1808	499.708	499.960	-0.292	499.420	499.960	-0.580

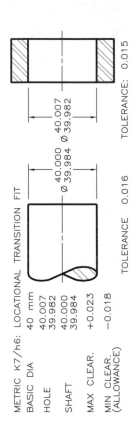

METRIC K7/h6: LOCATIONAL TRANSITION FIT

BASIC DIA	40 mm
HOLE	40.007
	39.982
SHAFT	40.000
	39.984
MAX. CLEAR.	+0.023
MIN CLEAR. (ALLOWANCE)	-0.018

TOLERANCE 0.016   TOLERANCE: 0.015

(Reprinted from ANSI B4.2–1978 (R2009). Copyright © American Society of Mechanical Engineers. All rights reserved.)

16 HOLE SIZES FOR NONPREFERRED DIAMETERS (MILLIMETERS)

Basic Size		C11	D9	F8	G7	H7	H8	H9	H11	K7	N7	P7	S7	U7
Over	0	+0.120	+0.045	+0.020	+0.012	+0.010	+0.014	+0.025	+0.060	0.000	-0.004	-0.006	-0.014	-0.018
To	3	+0.060	+0.020	+0.006	+0.002	0.000	0.000	0.000	0.000	-0.010	-0.014	-0.016	-0.024	-0.028
Over	3	+0.145	+0.060	+0.028	+0.016	+0.012	+0.018	+0.030	+0.075	+0.003	-0.004	-0.008	-0.015	-0.019
To	6	+0.070	+0.030	+0.010	+0.004	0.000	0.000	0.000	0.000	-0.009	-0.016	-0.020	-0.027	-0.031
Over	6	+0.170	+0.076	+0.035	+0.020	+0.015	+0.022	+0.036	+0.090	+0.005	-0.004	-0.009	-0.017	-0.022
To	10	+0.080	+0.040	+0.013	+0.005	0.000	0.000	0.000	0.000	-0.010	-0.019	-0.024	-0.032	-0.037
Over	10	+0.205	+0.093	+0.043	+0.024	+0.018	+0.027	+0.043	+0.110	+0.006	-0.005	-0.011	-0.021	-0.026
To	14	+0.095	+0.050	+0.016	+0.006	0.000	0.000	0.000	0.000	-0.012	-0.023	-0.029	-0.039	-0.044
Over	14	+0.205	+0.093	+0.043	+0.024	+0.018	+0.027	+0.043	+0.110	+0.006	-0.005	-0.011	-0.021	-0.026
To	18	+0.095	+0.050	+0.016	+0.006	0.000	0.000	0.000	0.000	-0.012	-0.023	-0.029	-0.039	-0.044
Over	18	+0.240	+0.117	+0.053	+0.028	+0.021	+0.033	+0.052	+0.130	+0.006	-0.007	-0.014	-0.027	-0.033
To	24	+0.110	+0.065	+0.020	+0.007	0.000	0.000	0.000	0.000	-0.015	-0.028	-0.035	-0.048	-0.054
Over	24	+0.240	+0.117	+0.053	+0.028	+0.021	+0.033	+0.052	+0.130	+0.006	-0.007	-0.014	-0.027	-0.040
To	30	+0.110	+0.065	+0.020	+0.007	0.000	0.000	0.000	0.000	-0.015	-0.028	-0.035	-0.048	-0.061
Over	30	+0.280	+0.142	+0.064	+0.034	+0.025	+0.039	+0.062	+0.160	+0.007	-0.008	-0.017	-0.034	-0.051
To	40	+0.120	+0.080	+0.025	+0.009	0.000	0.000	0.000	0.000	-0.018	-0.033	-0.042	-0.059	-0.076
Over	40	+0.290	+0.142	+0.064	+0.034	+0.025	+0.039	+0.062	+0.160	+0.007	-0.008	-0.017	-0.034	-0.061
To	50	+0.130	+0.080	+0.025	+0.009	0.000	0.000	0.000	0.000	-0.018	-0.033	-0.042	-0.059	-0.086
Over	50	+0.330	+0.174	+0.076	+0.040	+0.030	+0.046	+0.074	+0.190	+0.009	-0.009	-0.021	-0.042	-0.076
To	65	+0.140	+0.100	+0.030	+0.010	0.000	0.000	0.000	0.000	-0.021	-0.039	-0.051	-0.072	-0.106
Over	65	+0.340	+0.174	+0.076	+0.040	+0.030	+0.046	+0.074	+0.190	+0.009	-0.009	-0.021	-0.048	-0.091
To	80	+0.150	+0.100	+0.030	+0.010	0.000	0.000	0.000	0.000	-0.021	-0.039	-0.051	-0.078	-0.121
Over	80	+0.390	+0.207	+0.090	+0.047	+0.035	+0.054	+0.087	+0.220	+0.010	-0.010	-0.024	-0.058	-0.111
To	100	+0.170	+0.120	+0.036	+0.012	0.000	0.000	0.000	0.000	-0.025	-0.045	-0.059	-0.093	-0.146

Basic Size	c11	d9	f8	g7	h7	h8	h9	h11	k7	n7	p7	s7	u7
Over 100 / To 120	+0.400 / +0.180	+0.207 / +0.120	+0.090 / +0.036	+0.047 / +0.012	+0.035 / 0.000	+0.054 / 0.000	+0.087 / 0.000	+0.220 / 0.000	+0.010 / -0.025	-0.010 / -0.045	-0.024 / -0.059	-0.066 / -0.101	-0.131 / -0.166
Over 120 / To 140	+0.450 / +0.200	+0.245 / +0.145	+0.106 / +0.043	+0.054 / +0.014	+0.040 / 0.000	+0.063 / 0.000	+0.100 / 0.000	+0.250 / 0.000	+0.012 / -0.028	-0.012 / -0.052	-0.028 / -0.068	-0.077 / -0.117	-0.155 / -0.195
Over 140 / To 160	+0.460 / +0.210	+0.245 / +0.145	+0.106 / +0.043	+0.054 / +0.014	+0.040 / 0.000	+0.063 / 0.000	+0.100 / 0.000	+0.250 / 0.000	+0.012 / -0.028	-0.012 / -0.052	-0.028 / -0.068	-0.085 / -0.125	-0.175 / -0.215
Over 160 / To 180	+0.480 / +0.230	+0.245 / +0.145	+0.106 / +0.043	+0.054 / +0.014	+0.040 / 0.000	+0.063 / 0.000	+0.100 / 0.000	+0.250 / 0.000	+0.012 / -0.028	-0.012 / -0.052	-0.028 / -0.068	-0.093 / -0.133	-0.195 / -0.235
Over 180 / To 200	+0.530 / +0.240	+0.285 / +0.170	+0.122 / +0.050	+0.061 / +0.015	+0.046 / 0.000	+0.072 / 0.000	+0.115 / 0.000	+0.290 / 0.000	-0.013 / -0.033	-0.014 / -0.060	-0.033 / -0.079	-0.105 / -0.151	-0.219 / -0.265
Over 200 / To 225	+0.550 / +0.260	+0.285 / +0.170	+0.122 / +0.050	+0.061 / +0.015	+0.046 / 0.000	+0.072 / 0.000	+0.115 / 0.000	+0.290 / 0.000	+0.013 / -0.033	-0.014 / -0.060	-0.033 / -0.079	-0.113 / -0.159	-0.241 / -0.287
Over 225 / To 250	+0.570 / +0.280	+0.285 / +0.170	+0.122 / +0.050	+0.061 / +0.015	+0.046 / 0.000	+0.072 / 0.000	+0.115 / 0.000	+0.290 / 0.000	+0.013 / -0.033	-0.014 / -0.060	-0.033 / -0.079	-0.123 / -0.169	-0.267 / -0.313
Over 250 / To 280	+0.620 / +0.300	+0.320 / +0.190	+0.137 / +0.056	+0.069 / +0.017	+0.052 / 0.000	+0.081 / 0.000	+0.130 / 0.000	+0.320 / 0.000	+0.016 / -0.036	-0.014 / -0.066	-0.036 / -0.088	-0.138 / -0.190	-0.295 / -0.347
Over 280 / To 315	+0.650 / +0.330	+0.320 / +0.190	+0.137 / +0.056	+0.069 / 0.017	+0.052 / 0.000	+0.081 / 0.000	+0.130 / 0.000	+0.320 / 0.000	+0.016 / -0.036	-0.014 / -0.066	-0.036 / -0.088	-0.150 / -0.202	-0.330 / -0.382
Over 315 / To 355	+0.720 / +0.360	+0.350 / +0.210	+0.151 / +0.062	+0.075 / +0.018	+0.057 / 0.000	+0.089 / 0.000	+0.140 / 0.000	+0.360 / 0.000	+0.017 / -0.040	-0.016 / -0.073	-0.041 / -0.058	-0.169 / -0.226	-0.369 / -0.426
Over 355 / To 400	+0.760 / +0.400	+0.350 / +0.210	+0.151 / +0.062	+0.075 / +0.018	+0.057 / 0.000	+0.089 / 0.000	+0.140 / 0.000	+0.360 / 0.000	+0.017 / -0.040	-0.016 / -0.073	-0.041 / -0.058	-0.187 / -0.244	-0.414 / -0.471
Over 400 / To 450	+0.840 / +0.440	+0.385 / +0.230	+0.165 / +0.068	+0.083 / +0.020	+0.063 / 0.000	+0.097 / 0.000	+0.155 / 0.000	+0.400 / 0.000	+0.018 / -0.045	-0.017 / -0.080	-0.045 / -0.108	-0.209 / -0.272	-0.467 / -0.530
Over 450 / To 500	+0.880 / +0.480	+0.385 / +0.230	+0.165 / +0.068	+0.083 / +0.020	+0.063 / 0.000	+0.097 / 0.000	+0.155 / 0.000	+0.400 / 0.000	+0.018 / -0.045	-0.017 / -0.080	-0.045 / -0.108	-0.229 / -0.292	-0.517 / -0.580

17 SHAFT SIZES FOR NONPREFERRED DIAMETERS (MILLIMETERS)

Basic Size		c11	d9	f7	g6	h6	h7	h9	h11	k6	n6	p6	s6	u6
Over 0	To 3	−0.060	−0.020	−0.006	−0.002	0.000	0.000	0.000	0.000	+0.006	+0.010	+0.012	+0.020	+0.024
		−0.120	−0.045	−0.016	−0.008	−0.006	−0.010	−0.025	−0.060	0.000	+0.004	+0.006	+0.014	+0.018
Over 3	To 6	−0.070	−0.030	−0.010	−0.004	0.000	0.000	0.000	0.000	+0.009	+0.016	+0.020	+0.027	+0.031
		−0.145	−0.060	−0.022	−0.012	−0.008	−0.012	−0.030	−0.075	+0.001	+0.008	+0.012	+0.019	+0.023
Over 6	To 10	−0.080	−0.040	−0.013	−0.005	0.000	0.000	0.000	0.000	+0.010	+0.019	+0.024	+0.032	+0.037
		−0.170	−0.076	−0.028	−0.014	−0.009	−0.015	−0.036	−0.090	+0.001	+0.010	+0.024	+0.023	+0.028
Over 10	To 14	−0.095	−0.050	−0.016	−0.006	0.000	0.000	0.000	0.000	+0.012	+0.023	+0.029	+0.039	+0.044
		−0.205	−0.093	−0.034	−0.017	−0.011	−0.018	−0.043	−0.110	+0.001	+0.012	+0.018	+0.028	+0.033
Over 14	To 18	−0.095	−0.050	−0.016	−0.006	0.000	0.000	0.000	0.000	+0.012	+0.023	+0.029	+0.039	+0.044
		−0.205	−0.093	−0.034	−0.017	−0.011	−0.018	−0.043	−0.110	+0.001	+0.012	+0.018	+0.028	+0.033
Over 18	To 24	−0.110	−0.065	−0.020	−0.007	0.000	0.000	0.000	0.000	+0.015	+0.028	+0.035	+0.048	+0.054
		−0.240	−0.117	−0.041	−0.020	−0.013	−0.021	−0.052	−0.130	+0.002	+0.015	+0.022	+0.035	+0.041
Over 24	To 30	−0.110	−0.065	−0.020	−0.007	0.000	0.000	0.000	0.000	+0.015	+0.028	+0.035	+0.048	+0.061
		−0.240	−0.117	−0.041	−0.020	−0.013	−0.021	−0.052	−0.130	+0.002	+0.015	+0.022	+0.035	+0.048
Over 30	To 40	−0.120	−0.080	−0.025	−0.009	0.000	0.000	0.000	0.000	+0.018	+0.033	+0.042	+0.059	+0.076
		−0.280	−0.142	−0.050	−0.025	−0.016	−0.025	−0.062	−0.160	+0.002	+0.017	+0.026	+0.043	+0.060
Over 40	To 50	−0.130	−0.080	−0.025	−0.009	0.000	0.000	0.000	0.000	+0.018	+0.033	+0.042	+0.059	+0.086
		−0.290	−0.142	−0.050	−0.025	−0.016	−0.025	−0.062	−0.160	+0.002	+0.017	+0.026	+0.043	+0.070
Over 50	To 65	−0.140	−0.100	−0.030	−0.010	0.000	0.000	0.000	0.000	+0.021	+0.039	+0.051	+0.072	+0.106
		−0.330	−0.174	−0.060	−0.029	−0.019	−0.030	−0.074	−0.190	+0.002	+0.020	−0.032	+0.053	+0.087
Over 65	To 80	−0.150	−0.100	−0.030	−0.010	0.000	0.000	0.000	0.000	+0.021	+0.039	v0.051	+0.078	+0.121
		−0.340	−0.174	−0.060	−0.029	−0.019	−0.030	−0.074	−0.190	+0.002	+0.020	+0.032	+0.059	+0.102
Over 80	To 100	−0.170	−0.120	−0.036	−0.012	0.000	0.000	0.000	0.000	+0.025	+0.045	+0.059	+0.093	+0.146
		−0.390	−0.207	−0.071	−0.034	−0.022	−0.035	−0.087	−0.220	+0.003	+0.023	+0.037	+0.071	+0.124

Basic Size		c11	d9	f7	g6	h6	h7	h9	h11	k6	n6	p6	s6	u6
Over	100	−0.180	−0.120	−0.036	−0.012	0.000	0.000	0.000	0.000	+0.025	+0.045	+0.059	+0.101	+0.166
To	120	−0.400	−0.207	−0.071	−0.034	−0.022	−0.035	−0.087	−0.220	+0.003	+0.023	+0.037	+0.079	+0.144
Over	120	−0.200	−0.145	−0.043	−0.014	0.000	0.000	0.000	0.000	+0.028	+0.052	+0.068	+0.117	+0.195
To	140	−0.450	−0.245	−0.083	−0.039	−0.025	−0.040	−0.100	−0.250	+0.003	+0.027	+0.043	+0.092	+0.170
Over	140	−0.210	−0.145	−0.043	−0.014	0.000	0.000	0.000	0.000	+0.028	+0.052	+0.068	+0.125	+0.215
To	160	−0.460	−0.245	−0.083	−0.039	−0.025	−0.040	−0.100	−0.250	+0.003	+0.027	+0.043	+0.100	+0.190
Over	160	−0.230	−0.145	−0.043	−0.014	0.000	0.000	0.000	0.000	+0.028	+0.052	+0.068	+0.133	+0.235
To	180	−0.480	−0.245	−0.083	−0.039	−0.025	−0.040	−0.100	−0.250	+0.003	+0.027	+0.043	+0.108	+0.210
Over	180	−0.240	−0.170	−0.050	−0.015	0.000	0.000	0.000	0.000	+0.033	+0.060	+0.079	+0.151	+0.265
To	200	−0.530	−0.285	−0.096	−0.044	−0.029	−0.046	−0.115	−0.290	+0.004	+0.031	+0.050	+0.122	+0.236
Over	200	−0.260	−0.170	−0.050	−0.015	0.000	0.000	0.000	0.000	+0.033	+0.060	+0.079	+0.159	+0.287
To	225	−0.550	−0.285	−0.096	−0.044	−0.029	−0.046	−0.115	−0.290	+0.004	+0.031	+0.050	+0.130	+0.258
Over	225	−0.280	−0.170	−0.050	−0.015	0.000	0.000	0.000	0.000	+0.033	+0.060	+0.079	+0.169	+0.313
To	250	−0.570	−0.285	−0.096	−0.044	−0.029	−0.046	−0.115	−0.290	+0.004	+0.031	+0.050	+0.140	+0.284
Over	250	−0.300	−0.190	−0.056	−0.017	0.000	0.000	0.000	0.000	+0.036	+0.066	+0.088	+0.190	+0.347
To	280	−0.620	−0.320	−0.108	−0.049	−0.032	−0.052	−0.130	−0.320	+0.004	+0.034	+0.056	+0.158	+0.315
Over	280	−0.330	−0.190	−0.056	−0.017	0.000	0.000	0.000	0.000	+0.036	+0.066	+0.088	+0.202	+0.382
To	315	−0.650	−0.320	−0.108	−0.049	−0.032	−0.052	−0.130	−0.320	+0.004	+0.034	+0.056	+0.170	+0.350
Over	315	−0.360	−0.210	−0.062	−0.018	0.000	0.000	0.000	0.000	+0.040	+0.073	+0.098	+0.226	+0.426
To	355	−0.720	−0.350	−0.119	−0.054	−0.036	−0.057	−0.140	−0.360	+0.004	+0.037	+0.062	+0.190	+0.390
Over	355	−0.400	−0.210	−0.062	−0.018	0.000	0.000	0.000	0.000	+0.040	+0.073	+0.098	+0.244	+0.471
To	400	−0.760	−0.350	−0.119	−0.054	−0.036	−0.057	−0.140	−0.360	+0.004	+0.037	+0.062	+0.208	+0.435
Over	400	−0.440	−0.230	−0.068	−0.020	0.000	0.000	0.000	0.000	+0.045	+0.080	+0.108	+0.272	+0.530
To	450	−0.840	−0.385	−0.131	−0.060	−0.040	−0.063	−0.155	−0.400	+0.005	+0.040	+0.068	+0.232	+0.490
Over	450	−0.480	−0.230	−0.068	−0.020	0.000	0.000	0.000	0.000	+0.045	+0.080	+0.108	+0.292	+0.580
To	500	−0.880	−0.385	−0.131	−0.060	−0.040	−0.063	−0.155	−0.400	+0.005	+0.040	+0.068	+0.252	+0.540

18 SCREW THREADS: AMERICAN NATIONAL AND UNIFIED (INCHES)

Note: UNJ thread form is similar to UN, but includes a 0.15011P minimum mandatory radius root on external thread.

UNR thread form is similar to UN, but includes a 0.10825P minimum mandatory radius root on external thread.

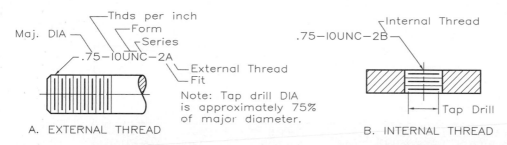

A. EXTERNAL THREAD

Note: Tap drill DIA is approximately 75% of major diameter.

B. INTERNAL THREAD

Nominal Diameter	Basic Diameter	Coarse NC & UNC		Fine NF & UNF		Extra Fine NEF/UNEF		Nominal Diameter	Basic Diameter	Coarse NC & UNC		Fine NF & UNF		Extra Fine NEF/UNEF	
		Thds per in.	Tap Drill DIA	Thds per in.	Tap Drill DIA	Thds per in.	Tap Drill DIA			Thds per in.	Tap Drill DIA	Thds per in.	Tap Drill DIA	Thds per in.	Tap Drill DIA
0	.060			80	.0469			1	1.000	8	.875	12	.922	20	.953
1	.073	64	No. 53	72	No. 53			1-1/16	1.063	...	...	...	...	18	1.000
2	.086	56	No. 50	64	No. 50			1-1/8	1.125	7	.904	12	1.046	18	1.070
3	.099	48	No. 47	56	No. 45			1-3/16	1.188	...	...	...	...	18	1.141
4	.112	40	No. 43	48	No. 42			1-1/4	1.250	7	1.109	12	1.172	18	1.188
5	.125	40	No. 38	44	No. 37			1-5/16	1.313	...	...	...	...	18	1.266
6	.138	32	No. 36	40	No. 33			1-3/8	1.375	6	1.219	12	1.297	18	1.313
8	.164	32	No. 29	36	No. 29			1-7/16	1.438	...	...	...	...	18	1.375
10	.190	24	No. 25	32	No. 21			1-1/2	1.500	6	1.344	12	1.422	18	1.438
12	.216	24	No. 16	28	No. 14	32	No. 13	1-9/16	1.563	...	...	...	...	18	1.500
1/4	.250	20	No. 7	28	No. 3	32	.2189	1-5/8	1.625	...	...	...	...	18	1.563
5/16	.3125	18	F	24	I	32	.2813	1-11/16	1.688	...	...	...	...	18	1.625
3/8	.375	16	.3125	24	Q	32	.3438	1-3/4	1.750	5	1.563	...	...	...	...
7/16	.4375	14	U	20	.3906	28	.4062	2	2.000	4.5	1.781	...	...	...	...
1/2	.500	13	.4219	20	.4531	28	.4688	2-1/4	2.250	4.5	2.031	...	...	...	...
9/16	.5625	12	.4844	18	.5156	24	.5156	2-1/2	2.500	4	2.250	...	...	...	...
5/8	.625	11	.5313	18	.5781	24	.5781	2-3/4	2.750	4	2.500	...	...	...	...
11/16	.6875	...	...	...	...	24	.6406	3	3.000	4	2.750	...	...	...	...
3/4	.750	10	.6563	16	.6875	20	.7031	3-1/4	3.250	4	...	...	...	...	...
13/16	.8125	...	...	...	...	20	.7656	3-1/2	3.500	4	...	...	...	...	...
7/8	.875	9	.7656	14	.8125	20	.8281	3-3/4	3.750	4	...	...	...	...	...
15/16	.9375	...	...	...	...	20	.8906	4	4.000	4	...	...	...	...	...

19 SCREW THREADS: AMERICAN NATIONAL AND UNIFIED (INCHES)

Constant-Pitch Threads

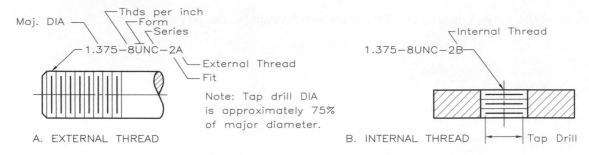

A. EXTERNAL THREAD

Note: Tap drill DIA is approximately 75% of major diameter.

B. INTERNAL THREAD

Nominal Diameter	8 Pitch 8N & 8UN		12 Pitch 12N & 12UN		16 Pitch 16N & 16UN		Nominal Diameter	8 Pitch 8N & 8UN		12 Pitch 12N & 12UN		16 Pitch 16N & 16UN	
	Thds per in.	Tap Drill DIA	Thds per in.	Tap Drill DIA	Thds per in.	Tap Drill DIA		Thds per in.	Tap Drill DIA	Thds per in.	Tap Drill DIA	Thds per in.	Tap Drill DIA
.500	...	...	12	.422	...	...	2.063	...	...	...	...	16	2.000
.563	...	...	12	.484	...	...	2.125	...	...	12	2.047	16	2.063
.625	...	...	12	.547	...	...	2.188	...	...	...	...	16	2.125
.688	...	...	12	.609	...	...	2.250	8	2.125	12	2.172	16	2.188
.750	...	...	12	.672	16	.688	2.313	...	...	...	...	16	2.250
.813	...	...	12	.734	16	.750	2.375	...	...	12	2.297	16	2.313
.875	...	...	12	.797	16	.813	2.438	...	...	...	...	16	2.375
.934	...	...	12	.859	16	.875	2.500	8	2.375	12	2.422	16	2.438
1.000	8	.875	12	.922	16	.938	2.625	...	...	12	2.547	16	2.563
1.063	...	...	12	.984	16	1.000	2.750	8	2.625	12	2.717	16	2.688
1.125	8	1.000	12	1.047	16	1.063	2.875	...	...	12	...	16	...
1.188	...	...	12	1.109	16	1.125	3.000	8	2.875	12	...	16	...
1.250	8	1.125	12	1.172	16	1.188	3.125	...	...	12	...	16	...
1.313	...	...	12	1.234	16	1.250	3.250	8	...	12	...	16	...
1.375	8	1.250	12	1.297	16	1.313	3.375	...	...	12	...	16	...
1.434	...	...	12	1.359	16	1.375	3.500	8	...	12	...	16	...
1.500	8	1.375	12	1.422	16	1.438	3.625	...	...	12	...	16	...
1.563	...	...	...	...	16	1.500	3.750	8	...	12	...	16	...
1.625	8	1.500	12	1.547	16	1.563	3.875	...	...	12	...	16	...
1.688	...	...	...	...	16	1.625	4.000	8	...	12	...	16	...
1.750	8	1.625	12	1.672	16	1.688	4.250	8	...	12	...	16	...
1.813	...	...	...	...	16	1.750	4.500	8	...	12	...	16	...
1.875	8	1.750	12	1.797	16	1.813	4.750	8	...	12	...	16	...
1.934	...	...	...	...	16	1.875	5.000	8	...	12	...	16	...
2.000	8	1.875	12	1.922	16	1.938	5.250	8	...	12	...	16	...

20 SCREW THREADS: AMERICAN NATIONAL AND UNIFIED (METRIC)

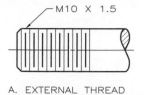

A. EXTERNAL THREAD

Note: Tap drill DIA is approximately 75% of major diameter.

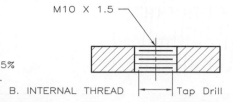

B. INTERNAL THREAD   ⊢ Tap Drill

COARSE		FINE		COARSE		FINE	
MAJ. DIA & THD PITCH	TAP DRILL	MAJ. DIA & THD PITCH	TAP DRILL	MAJ. DIA & THD PITCH	TAP DRILL	MAJ. DIA & THD PITCH	TAP DRILL
M1.6 × 0.35	1.25			M20 × 2.5	17.5	M20 × 1.5	18.5
M1.8 × 0.35	1.45			M22 × 2.5	19.5	M22 × 1.5	20.5
M2 × 0.4	1.6			M24 × 3	21.0	M24 × 2	22.0
M2.2 × 0.45	1.75			M27 × 3	24.0	M27 × 2	25.0
M2.5 × 0.45	2.05			M30 × 3.5	26.5	M30 × 2	28.0
M3 × 0.5	2.5			M33 × 3.5	29.5	M33 × 2	31.0
M3.5 × 0.6	2.9			M36 × 4	32.0	M36 × 3	33.0
M4 × 0.7	3.3			M39 × 4	35.0	M39 × 3	36.0
M4.5 × 0.75	3.75			M42 × 4.5	37.5	M42 × 3	39.0
M5 × 0.8	4.2			M45 × 4.5	40.5	M45 × 3	42.0
M6 × 1	5.0			M48 × 5	43.0	M48 × 3	45.0
M7 × 1	6.0			M52 × 5	47.0	M52 × 3	49.0
M8 × 1.25	6.8	M8 × 1	7.0	M56 × 5.5	50.5	M56 × 4	52.0
M9 × 1.25	7.75			M60 × 5.5	54.5	M60 × 4	56.0
M10 × 1.5	8.5	M10 × 1.25	8.75	M64 × 6	58.0	M64 × 4	60.0
M11 × 1.5	9.5			M68 × 6	62.0	M68 × 4	64.0
M12 × 1.75	10.3	M12 × 1.25	10.5	M72 × 6	66.0	M72 × 4	68.0
M14 × 2	12.0	M14 × 1.5	12.5	M80 × 6	74.0	M80 × 4	76.0
M16 × 2	14.0	M16 × 1.5	14.5	M90 × 6	84.0	M90 × 4	86.0
M18 × 2.5	15.5	M18 × 1.5	16.5	M100 × 6	94.0	M100 × 4	96.0

21 SQUARE AND ACME THREADS

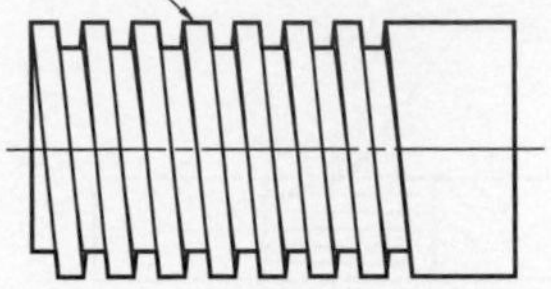

2.00—2.5 SQUARE

Typical
thread
note

Dimensions are in inches.

Size	Size	Thds per inch	Size	Size	Thds per inch	Size	Size	Thds per inch
3/8	.375	12	1-1/8	1.125	4	3	3.000	1-1/2
7/16	.438	10	1-1/4	1.250	4	3-1/4	3.125	1-1/2
1/2	.500	10	1-1/2	1.500	3	3-1/2	3.500	1-1/3
9/16	.563	8	1-3/4	1.750	2-1/2	3-3/4	3.750	1-1/3
5/8	.625	8	2	2.000	2-1/2	4	4.000	1-1/3
3/4	.75	6	2-1/4	2.250	2	4-1/4	4.250	1-1/3
7/8	.875	5	2-1/2	2.500	2	4-1/2	4.500	1
1	1.000	5	2-3/4	2.750	2	Larger		1

(Reprinted from ANSI/ASME B1.5–1997 (R2009). Copyright © American Society of Mechanical Engineers. All rights reserved.)

22 AMERICAN STANDARD TAPER PIPE THREADS (NPT)

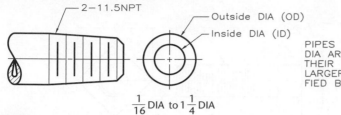

2–11.5NPT

Outside DIA (OD)

Inside DIA (ID)

$\frac{1}{16}$ DIA to $1\frac{1}{4}$ DIA

PIPES THRU 12 INCHES IN DIA ARE SPECIFIED BY THEIR INSIDE DIAMETERS. LARGER PIPES ARE SPECI—FIED BY THEIR OD.

Dimensions are in inches.

Nominal ID	$\frac{1}{16}$	$\frac{1}{8}$	$\frac{1}{4}$	$\frac{3}{8}$	$\frac{1}{2}$	$\frac{3}{4}$	1	$1\frac{1}{4}$
Outside DIA	0.313	0.405	0.540	0.675	0.840	1.050	1.315	1.660
Thds/Inch	27	27	18	18	14	14	$11\frac{1}{2}$	$11\frac{1}{2}$

$1\frac{1}{2}$ DIA to 6 DIA

Nominal ID	$1\frac{1}{2}$	2	$2\frac{1}{2}$	3	$3\frac{1}{2}$	4	5	6
Outside DIA	1.900	2.375	2.875	3.500	4.000	4.500	5.563	6.625
Thds/Inch	$11\frac{1}{2}$	$11\frac{1}{2}$	8	8	8	8	8	8

8 DIA to 24 DIA

Nominal ID	8	10	12	14 OD	16 OD	18 OD	20 OD	24 OD
Outside DIA	8.625	10.750	12.750	14.000	16.000	18.000	20.000	24.000
Thds/Inch	8	8	8	8	8	8	8	8

(Reprinted from ANSI/ASME B1.20.1–1983 (R2006). Copyright © American Society of Mechanical Engineers. All rights reserved.)

23 SQUARE BOLTS (INCHES)

DIA	E Max.	F Max.	G Avg.	H Max.	R Max.
1/4	.250	.375	.530	.188	.031
5/16	.313	.500	.707	.220	.031
3/8	.375	.563	.795	.268	.031
7/16	.438	.625	.884	.316	.031
1/2	.500	.750	1.061	.348	.031
5/8	.625	.938	1.326	.444	.062
3/4	.750	1.125	1.591	.524	.062
7/8	.875	1.313	1.856	.620	.062
1	1.000	1.500	2.121	.684	.093
1-1/8	1.125	1.688	2.386	.780	.093
1-1/4	1.250	1.875	2.652	.876	.093
1-3/8	1.375	2.625	2.917	.940	.093
1-1/2	1.500	2.250	3.182	1.036	.093

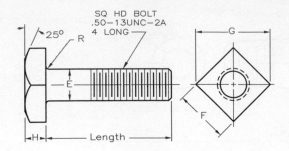

*14 MEANS THAT LENGTHS ARE AVAILABLE
AT 1 INCH INCREMENTS UP TO 10 INCHES.

(Reprinted from ANSI/ASME B18.2.1–1996 (R2005). Copyright © American Society of Mechanical Engineers. All rights reserved.)

24 SQUARE NUTS (INCHES)

Dimensions are in inches.

DIA	DIA	F Max.	G Avg.	H Max.
1/4	.250	.438	.619	.235
5/16	.313	.563	.795	.283
3/8	.375	.625	.884	.346
7/16	.438	.750	1.061	.394
1/2	.500	.813	1.149	.458
5/8	.625	1.000	1.414	.569
3/4	.750	1.125	1.591	.680
7/8	.875	1.313	1.856	.792
1	1.000	1.500	2.121	.903
1-1/8	1.125	1.688	2.386	1.030
1-1/4	1.250	1.875	2.652	1.126
1-3/8	1.375	1.063	2.917	1.237
1-1/2	1.500	2.250	3.182	1.348

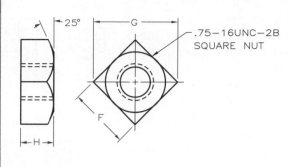

(Reprinted from ANSI/ASME B18.2.2–1987 (R2005). Copyright © American Society of Mechanical Engineers. All rights reserved.)

25 HEXAGON HEAD BOLTS (INCHES)

Dimensions are in inches.

DIA	E Max.	F Max.	G Avg.	H Max.	R Max.
1/4	.250	.438	.505	.163	.025
5/16	.313	.500	.577	.211	.025
3/8	.375	.563	.650	.243	.025
7/16	.438	.625	.722	.291	.025
1/2	.500	.750	.866	.323	.025
9/16	.563	.812	.938	.371	.045
5/8	.625	.938	1.083	.403	.045
3/4	.750	1.125	1.299	.483	.045
7/8	.875	1.313	1.516	.563	.065
1	1.000	1.500	1.732	.627	.095
1-1/8	1.125	1.688	1.949	.718	.095
1-1/4	1.250	1.875	2.165	.813	.095
1-3/8	1.375	2.063	2.382	.878	.095
1-1/2	1.500	2.250	2.598	.974	.095
1-3/4	1.750	2.625	3.031	1.134	.095
2	2.000	3.000	3.464	1.263	.095
2-1/4	2.250	3.375	3.897	1.423	.095
2-1/2	2.500	3.750	4.330	1.583	.095
2-3/4	2.750	4.125	4.763	1.744	.095
3	3.000	4.500	5.196	1.935	.095

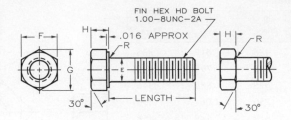

(Reprinted from ANSI/ASME B18.2.1–1996 (R2005). Copyright © American Society of Mechanical Engineers. All rights reserved.)

26 HEX NUTS AND JAM NUTS

MAJOR DIA		F Max.	G Avg.	H1 Max.	H2 Max.
1/4	.250	.438	.505	.226	.163
5/16	.313	.500	.577	.273	.195
3/8	.375	.563	.650	.337	.227
7/16	.438	.688	.794	.385	.260
1/2	.500	.750	.866	.448	.323
9/16	.563	.875	1.010	.496	.324
5/8	.625	.938	1.083	.559	.387
3/4	.750	1.125	1.299	.665	.446
7/8	.875	1.313	1.516	.776	.510
1	1.000	1.500	1.732	.887	.575
1-1/8	1.125	1.688	1.949	.899	.639
1-1/4	1.250	1.875	2.165	1.094	.751
1-3/8	1.375	2.063	2.382	1.206	.815
1-1/2	1.500	2.250	2.589	1.317	.880

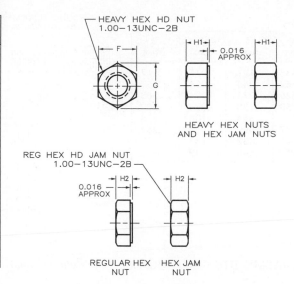

HEAVY HEX HD NUT
1.00–13UNC–2B

HEAVY HEX NUTS
AND HEX JAM NUTS

REG HEX HD JAM NUT
1.00–13UNC–2B

REGULAR HEX NUT   HEX JAM NUT

27 ROUND HEAD CAP SCREWS

Dimensions are in inches.

DIA	D Max.	A Max.	H Avg.	J Max.	T Max.
1/4	.250	.437	.191	.075	.117
5/16	.313	.562	.245	.084	.151
3/8	.375	.625	.273	.094	.168
7/16	.438	.750	.328	.094	.202
1/2	.500	.812	.354	.106	.218
9/16	.563	.937	.409	.118	.252
5/8	.625	1.000	.437	.133	.270
3/4	.750	1.250	.546	.149	.338

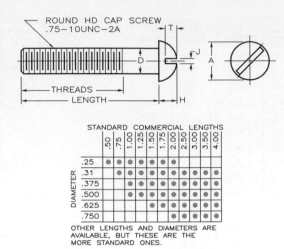

STANDARD COMMERCIAL LENGTHS

OTHER LENGTHS AND DIAMETERS ARE AVAILABLE, BUT THESE ARE THE MORE STANDARD ONES.

(Reprinted from ANSI/ASME B18.3–2003 (R2008). Copyright © American Society of Mechanical Engineers. All rights reserved.)

28 FLAT HEAD CAP SCREWS

Dimensions are in inches.

DIA	D Max.	A Max.	H Avg.	J Max.	T Max.
1/4	.250	.500	.140	.075	.068
5/16	.313	.625	.177	.084	.086
3/8	.375	.750	.210	.094	.103
7/16	.438	.813	.210	.094	.103
1/2	.500	.875	.210	.106	.103
9/16	.563	1.000	.244	.118	.120
5/8	.625	1.125	.281	.133	.137
3/4	.750	1.375	.352	.149	.171
7/8	.875	1.625	.423	.167	.206
1	1.000	1.875	.494	.188	.240
1-1/8	1.125	2.062	.529	.196	.257
1-1/4	1.250	2.312	.600	.211	.291
1-3/8	1.375	2.562	.665	.226	.326
1-1/2	1.500	2.812	.742	.258	.360

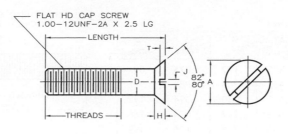

FLAT HD CAP SCREW 1.00—12UNF—2A X 2.5 LG

STANDARD COMMERCIAL LENGTHS

DIAMETER	.50	.75	1.00	1.25	1.50	1.75	2.00	2.50	3.00	3.50	4.00
.25	●	●	●	●	●	●	●				
.31		●	●	●	●	●	●	●	●	●	
.375			●	●	●	●	●	●	●	●	●
.500			●	●	●	●	●	●	●	●	●
.625					●	●	●	●	●	●	●
.750						●	●	●	●	●	●
.875								●	●	●	●
1.00									●	●	●
1.50									●	●	●

OTHER LENGTHS AND DIAMETERS ARE AVAILABLE, BUT THESE ARE THE MORE STANDARD ONES.

(Reprinted from ANSI/ASME B18.6.2–1998 (R2005). Copyright © American Society of Mechanical Engineers. All rights reserved.)

29 FILLISTER HEAD CAP SCREWS

Dimensions are in inches.

DIA	D Max.	A Max.	H Avg.	J Max.	T Max.
1/4	.250	.375	.172	.075	.097
5/16	.313	.437	.203	.084	.115
3/8	.375	.562	.250	.094	.142
7/16	.438	.625	.297	.094	.168
1/2	.500	.750	.328	.106	.193
9/16	.563	.812	.375	.118	.213
5/8	.625	.875	.422	.133	.239
3/4	.750	1.000	.500	.149	.283
7/8	.875	1.125	.594	.167	.334
1	1.000	1.312	.656	.188	.371

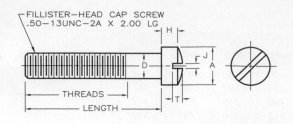

FILLISTER—HEAD CAP SCREW
.50—13UNC—2A X 2.00 LG

THREADS

LENGTH

STANDARD LENGTHS

DIAMETER	.50	.75	1.00	1.25	1.50	1.75	2.00	2.50
.25	●	●	●	●	●	●	●	●
.313	●		●	●			●	
.375		●	●	●			●	
.500		●	●				●	●
.625			●		●		●	●
.750			●			●	●	●
.875					●	●	●	●
1.00				●	●	●	●	●

(Reprinted from ANSI/ASME B18.6.2–1998 (R2005). Copyright © American Society of Mechanical Engineers. All rights reserved.)

30 FLAT SOCKET HEAD CAP SCREWS

| Diameter | | Pitch | A | Ang. | W |
mm	inches				
M3	.118	.5	6	90	2
M4	.157	.7	8	90	2.5
M5	.197	.8	10	90	3
M6	.236	1	12	90	4
M8	.315	1.25	16	90	5
M10	.394	1.5	20	90	6
M12	.472	1.75	24	90	8
M14	.551	2	27	90	10
M16	.630	2	30	90	10
M20	.787	2.5	36	90	12

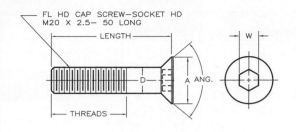

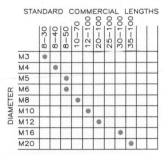

STANDARD COMMERCIAL LENGTHS

DIA 8–16: LENGTHS AT INTERVALS OF 2 MM
DIA 20–100: LENGTHS AT INTERVALS OF 5 MM

(Reprinted from ANSI/ASME B18.3–2003 (R2008) and ANSI/ASME B18.3.5M–1986 (R2008). Copyright © American Society of Mechanical Engineers. All rights reserved.)

31 SOCKET HEAD CAP SCREWS

Diameter		Pitch	A	H	W
mm	inches				
M3	.118	.5	6	3	2
M4	.157	.7	8	4	3
M5	.187	.8	10	5	4
M6	.236	1	12	6	6
M8	.315	1.25	16	8	6
M10	.394	1.5	20	10	8
M12	.472	1.75	24	12	10
M14	.551	2	27	14	12
M16	.630	2	30	16	14
M20	.787	2.5	36	20	17

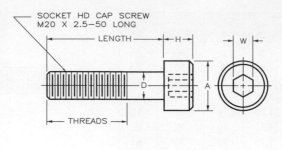

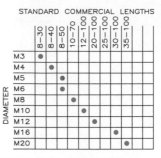

DIA 8–16: LENGTHS AT INTERVALS OF 2 MM
DIA 20–100: LENGTHS AT INTERVALS OF 5 MM

(Reprinted from ANSI/ASME B18.3–2003 (R2008) and ANSI/ASME B18.3.5M–1986 (R2008). Copyright © American Society of Mechanical Engineers. All rights reserved.)

32 ROUND HEAD MACHINE SCREWS

Dimensions are in inches.

DIA	D Max.	A Max.	H Avg.	J Max.	T Max.
0	.060	.113	.053	.023	.039
1	.073	.138	.061	.026	.044
2	.086	.162	.069	.031	.048
3	.099	.187	.078	.035	.053
4	112	.211	.086	.039	.058
5	.125	.236	.095	.043	.063
6	.138	.260	.103	.048	.068
8	.164	.309	.120	.054	.077
10	.190	.359	.137	.060	.087
12	.216	.408	.153	.067	.096
1/4	.250	.472	.175	.075	.109
5/16	.313	.590	.216	.084	.132
3/8	.375	.708	.256	.094	.155
7/16	.438	.750	.328	.094	.196
1/2	.500	.813	.355	.106	.211
9/16	.563	.938	.410	118	.242
5/8	.625	1.000	.438	.133	.258
3/4	.750	1.250	.547	.149	.320

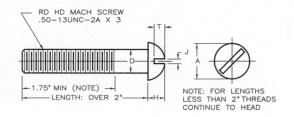

RD HD MACH SCREW
.50–13UNC–2A X 3

1.75" MIN (NOTE)
LENGTH: OVER 2"

NOTE: FOR LENGTHS LESS THAN 2" THREADS CONTINUE TO HEAD

STANDARD LENGTHS

DIAMETER	.25	.50	.75	1.00	1.25	1.50	1.75	2.00	2.50	3.00
0	●	●								
1	●	●								
2	●	●	●	●						
4	●	●	●	●	●	●				
6	●	●	●	●	●	●	●	●	●	●
.125	●	●	●	●	●	●	●	●	●	●
.250	●	●	●	●	●	●	●	●	●	●
.375		●	●	●	●	●	●	●	●	●
.500		●	●	●	●	●	●	●	●	●
.625		●	●	●	●	●	●	●	●	●
.750			●	●	●	●	●	●	●	●

OTHER LENGTHS AND DIAMETERS ARE AVAILABLE; THESE ARE THE MORE STANDARD ONES.

33 SET SCREWS

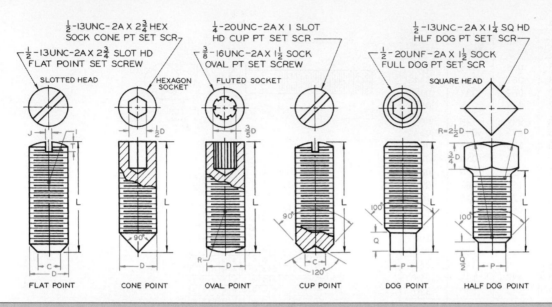

Dimensions for the set screws shown above (dimensions in inches)										
D	**I**	**J**	**T**	**R**	**C**		**P**		**Q**	**q**
Nominal Size	**Radius of Headless Crown**	**Width of Slot**	**Depth of Slot**	**Oval Point Radius**	**Diameter of Cup and Flat Points**		**Diameter of Dog Point**		**Length of Dog Point**	
					Max	**Min**	**Max**	**Min**	**Full**	**Half**
5 0.125	0.125	0.023	0.031	0.094	0.067	0.057	0.083	0.078	0.060	0.030
6 0.138	0.138	0.025	0.035	0.109	0.047	0.064	0.092	0.087	0.070	0.035
8 0.164	0.164	0.029	0.041	0.125	0.087	0.076	0.109	0.103	0.080	0.040
10 0.190	0.190	0.032	0.048	0.141	0.102	0.088	0.127	0.120	0.090	0.045
12 0.216	0.216	0.036	0.054	0.156	0.115	0.101	0.144	0.137	0.110	0.055
$\frac{1}{4}$ 0.250	0.250	0.045	0.063	0.188	0.132	0.118	0.156	0.149	0.125	0.063
$\frac{5}{16}$ 0.3125	0.313	0.051	0.076	0.234	0.172	0.156	0.203	0.195	0.156	0.078
$\frac{3}{8}$ 0.375	0.375	0.064	0.094	0.281	0.212	0.194	0.250	0.241	0.188	0.094
$\frac{7}{16}$ 0.4375	0.438	0.072	0.190	0.328	0.252	0.232	0.297	0.287	0.219	0.109
$\frac{1}{2}$ 0.500	0.500	0.081	0.125	0.375	0.291	0.270	0.344	0.344	0.250	0.125
$\frac{9}{16}$ 0.5625	0.563	0.091	0.141	0.422	0.332	0.309	0.391	0.379	0.281	0.140
$\frac{5}{8}$ 0.625	0.625	0.102	0.156	0.469	0.371	0.347	0.469	0.456	0.313	0.156
$\frac{3}{4}$ 0.750	0.750	0.129	0.188	0.563	0.450	0.425	0.563	0.549	0.375	0.188

(Reprinted from ANSI/ASME B18.6.2–1998 (R2005). Copyright © American Society of Mechanical Engineers. All rights reserved.)

34 TWIST DRILL SIZES

Letter Size Drills

Size	Drill Diameter inches	mm	Size	Drill Diameter inches	mm	Size	Drill Diameter inches	mm	Size	Drill Diameter inches	mm
A	0.234	5.944	H	0.266	6.756	O	0.316	8.026	V	0.377	9.576
B	0.238	6.045	I	0.272	6.909	P	0.323	8.204	W	0.386	9.804
C	0.242	6.147	J	0.277	7.036	Q	0.332	8.433	X	0.397	10.084
D	0.246	6.248	K	0.281	7.137	R	0.339	8.611	Y	0.404	10.262
E	0.250	6.350	L	0.290	7.366	S	0.348	8.839	Z	0.413	10.490
F	0.257	6.528	M	0.295	7.493	T	0.358	9.093			
G	0.261	6.629	N	0.302	7.601	U	0.368	9.347			

(Courtesy of General Motors Corporation.)

Number Size Drills

Size	Drill Diameter inches	mm	Size	Drill Diameter inches	mm	Size	Drill Diameter inches	mm	Size	Drill Diameter inches	mm
1	0.2280	5.7912	21	0.1590	4.0386	41	0.0960	2.4384	61	0.0390	0.9906
2	0.2210	5.6134	22	0.1570	3.9878	42	0.0935	2.3622	62	0.0380	0.9652
3	0.2130	5.4102	23	0.1540	3.9116	43	0.0890	2.2606	63	0.0370	0.9398
4	0.2090	5.3086	24	0.1520	3.8608	44	0.0860	2.1844	64	0.0360	0.9144
5	0.2055	5.2197	25	0.1495	3.7973	45	0.0820	2.0828	65	0.0350	0.8890
6	0.2040	5.1816	26	0.1470	3.7338	46	0.0810	2.0574	66	0.0330	0.8382
7	0.2010	5.1054	27	0.1440	3.6576	47	0.0785	19.812	67	0.0320	0.8128
8	0.1990	5.0800	28	0.1405	3.5560	48	0.0760	1.9304	68	0.0310	0.7874
9	0.1960	4.9784	29	0.1360	3.4544	49	0.0730	1.8542	69	0.0292	0.7417
10	0.1935	4.9149	30	0.1285	3.2639	50	0.0700	1.7780	70	0.0280	0.7112
11	0.1910	4.8514	31	0.1200	3.0480	51	0.0670	1.7018	71	0.0260	0.6604
12	0.1890	4.8006	32	0.1160	2.9464	52	0.0635	1.6129	72	0.0250	0.6350
13	0.1850	4.6990	33	0.1130	2.8702	53	0.0595	1.5113	73	0.0240	0.6096
14	0.1820	4.6228	34	0.1110	2.8194	54	0.0550	1.3970	74	0.0225	0.5715
15	0.1800	4.5720	35	0.1100	2.7940	55	0.0520	1.3208	75	0.0210	0.5334
16	0.1770	4.4958	36	0.1065	0.7051	56	0.0465	1.1684	76	0.0200	0.5080
17	0.1730	4.3942	37	0.1040	2.6416	57	0.0430	1.0922	77	0.0180	0.4572
18	0.1695	4.3053	38	0.1015	2.5781	58	0.0420	1.0668	78	0.0160	0.4064
19	0.1660	4.2164	39	0.0995	2.5273	59	0.0410	1.0414	79	0.0145	0.3638
20	0.1610	4.0894	40	0.0980	2.4892	60	0.0400	1.0160	80	0.0135	0.3428

Metric Size Drills

Decimal-inch equivalents are for reference only.

Drill Diameter		Drill Diameter		Drill Diameter		Drill Diameter		Drill Diameter		Drill Diameter		Drill Diameter	
mm	in.	mm	in.	mm	in.	mm	in.	mm	in.	mm	in.	mm	in.
.40	.0157	1.03	.0406	2.20	.0866	5.00	.1969	10.00	.3937	21.50	.8465	48.00	1.8898
.42	.0165	1.05	.0413	2.30	.0906	5.20	.2047	10.30	.4055	22.00	.8661	50.00	1.9685
.45	.0177	1.08	.0425	2.40	.0945	5.30	.2087	10.50	.4134	23.00	.9055	51.50	2.0276
.48	.0189	1.10	.0433	2.50	.0984	5.40	.2126	10.80	.4252	24.00	.9449	53.00	2.0866
.50	.0197	1.15	.0453	2.60	.1024	5.60	.2205	11.00	.4331	25.00	.9843	54.00	2.1260
.52	.0205	1.20	.0472	2.70	.1063	5.80	.2283	11.50	.4528	26.00	1.0236	56.00	2.2047
.55	.0217	1.25	.0492	2.80	.1102	6.00	.2362	12.00	.4724	27.00	1.0630	58.00	2.2835
.58	.0228	1.30	.0512	2.90	.1142	6.20	.2441	12.50	.4921	28.00	1.1024	60.00	2.3622
.60	.0236	1.35	.0531	3.00	.1181	6.30	.2480	13.00	.5118	29.00	1.1417		
.62	.0244	1.40	.0551	3.10	.1220	6.50	.2559	13.50	.5315	30.00	1.1811		
.65	.0256	1.45	.0571	3.20	.1260	6.70	.2638	14.00	.5512	31.00	1.2205		
.68	.0268	1.50	.0591	3.30	.1299	6.80	.2677	14.50	.5709	32.00	1.2598		
.70	.0276	1.55	.0610	3.40	.1339	6.90	.2717	15.00	.5906	33.00	1.2992		
.72	.0283	1.60	.0630	3.50	.1378	7.10	.2795	15.50	.6102	34.00	1.3386		
.75	.0295	1.65	.0650	3.60	.1417	7.30	.2874	16.00	.6299	35.00	1.3780		
.78	.0307	1.70	.0669	3.70	.1457	7.50	.2953	16.50	.6496	36.00	1.4173		
.80	.0315	1.75	.0689	3.80	.1496	7.80	.3071	17.00	.6693	37.00	1.4567		
.82	.0323	1.80	.0709	3.90	.1535	8.00	.3150	17.50	.6890	38.00	1.4961		
.85	.0335	1.85	.0728	4.00	.1575	8.20	.3228	18.00	.7087	39.00	1.5354		
.88	.0346	1.90	.0748	4.10	.1614	8.50	.3346	18.50	.7283	40.00	1.5748		
.90	.0354	1.95	.0768	4.20	.1654	8.80	.3465	19.00	.7480	41.00	1.6142		
.92	.0362	2.00	.0787	4.40	.1732	9.00	.3543	19.50	.7677	42.00	1.6535		
.95	.0374	2.05	.0807	4.50	.1772	9.20	.3622	20.00	.7874	43.50	1.7126		
.98	.0386	2.10	.0827	4.60	.1811	9.50	.3740	20.50	.0871	45.00	1.7717		
1.00	.0394	2.15	.0846	4.80	.1890	9.80	.3858	21.00	.8268	46.50	1.8307		

35 COTTER PINS: AMERICAN NATIONAL STANDARD

Nominal Diameter	Maximum DIA A	Minimum DIA B	Hole Size
0.031	0.032	0.063	0.047
0.047	0.048	0.094	0.063
0.062	0.060	0.125	0.078
0.078	0.076	0.156	0.094
0.094	0.090	0.188	0.109
0.109	0.104	0.219	0.125
0.125	0.120	0.250	0.141
0.141	0.176	0.281	0.156
0.156	0.207	0.313	0.172
0.188	0.176	0.375	0.203
0.219	0.207	0.438	0.234
0.250	0.225	0.500	0.266
0.312	0.280	0.625	0.313
0.375	0.335	0.750	0.375
0.438	0.406	0.875	0.438
0.500	0.473	1.000	0.500
0.625	0.598	1.250	0.625
0.750	0.723	1.500	0.750

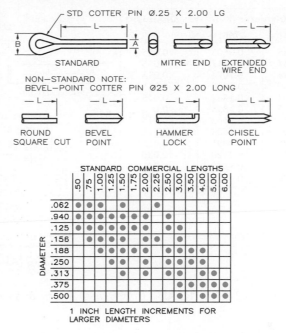

(Reprinted from ANSI/ASME B18.8.1–1994 (R2005). Copyright © American Society of Mechanical Engineers. All rights reserved.)

36 STRAIGHT PINS

Nominal DIA	Diameter A		Chamfer B
	Max	Min	
0.062	0.0625	0.0605	0.015
0.094	0.0937	0.0917	0.015
0.109	0.1094	0.1074	0.015
0.125	0.1250	0.1230	0.015
0.156	0.1562	0.1542	0.015
0.188	0.1875	0.1855	0.015
0.219	0.2187	0.2167	0.015
0.250	0.2500	0.2480	0.015
0.312	0.3125	0.3095	0.015
0.375	0.3750	0.3720	0.030
0.438	0.4345	0.4345	0.030
0.500	0.4970	0.4970	0.030

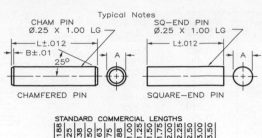

37 WOODRUFF KEYS

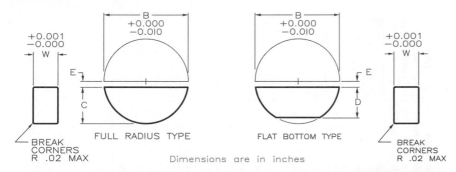

FULL RADIUS TYPE   FLAT BOTTOM TYPE

BREAK CORNERS R .02 MAX

Dimensions are in inches

BREAK CORNERS R .02 MAX

Key No.	W × B	C Max.	D Max.	E	Key No.	W × B	C Max.	D Max.	E
204	1/16 × 1/2	.203	.194	.047	506	5/32 × 3/4	.313	.303	.063
304	3/32 × 1/2	.203	.194	.047	606	3/16 × 3/4	.313	.303	.063
404	1/8 × 1/2	.203	.194	.047	507	5/32 × 7/8	.375	.365	.063
305	3/32 × 5/8	.250	.240	.063	607	3/16 × 7/8	.375	.365	.063
405	1/8 × 5/8	.250	.240	.063	807	1/4 × 7/8	.375	.365	.063
505	5/32 × 5/8	.250	.240	.063	608	3/16 × 1	.438	.428	.063
406	1/8 × 3/4	.313	.303	.063	609	3/16 × 1-1/8	.484	.475	.078

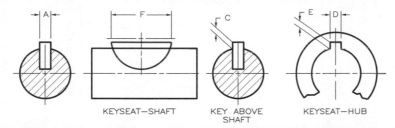

KEYSEAT—SHAFT   KEY ABOVE SHAFT   KEYSEAT—HUB

Key No.	A Min.	C +.005 −.000	F	D +.005 −.000	E +.005 −.000	Key No.	A Min.	C +.005 −.000	F	D +.005 −.000	E +.005 −.000
204	.0615	.0312	.500	.0635	.0372	506	.1553	.0781	.750	.1573	.0841
304	.0928	.0469	.500	.0948	.0529	606	.1863	.0937	.750	.1885	.0997
404	.1240	.0625	.500	.1260	.0685	507	.1553	.0781	.875	.1573	.0841
305	.0928	.0625	.625	.0948	.0529	607	.1863	.0937	.875	.1885	.0997
405	.1240	.0469	.625	.1260	.0685	807	.2487	.1250	.875	.2510	.1310
505	.1553	.0625	.625	.1573	.0841	608	.1863	.3393	1.000	.1885	.0997
406	.1240	.0781	.750	.1260	.0685	609	.1863	.3853	1.125	.1885	.0997

Key sizes vs. Shaft sizes

Shaft DIA	to .375	to .500	to .750	to 1.313	to 1.188	to 1.448	to 1.750	to 2.125	to 2.500
Key Nos.	204	304 305	404 405 406	505 506 507	606 607 608 609	807 808 809	810 811 812	1011 1012	1211 1212

38 STANDARD KEYS AND KEYSEATS

Diagrams: A. PARALLEL KEY, B. TAPER KEY (TAPER 1/8 PER 12 IN.), C. GIB-HEAD TAPER KEY (TAPER 1/8 PER 12 IN.), and shaft cross-section showing KEY.

Sprocket Bore (= Shaft Diam.) Inches D	Keyway Dimensions – Inches				Key Dimensions – Inches						Gib Head Dimensions – Inches				Key Tolerances Taper and Gib Head	
	For Square Key		For Flat Key		Square		Flat		Tolerance on W and T (–)		Square Key		Flat Key		W (–)	T (–)
	Width W	Depth T/2	Width W	Depth T/2	Width W	Height T	Width W	Height T			H	G	H	G		
$\frac{1}{2}-\frac{9}{16}$	$\frac{1}{8}\times\frac{1}{16}$		$\frac{1}{8}\times\frac{3}{64}$		$\frac{1}{8}\times\frac{1}{8}$		$\frac{1}{8}\times\frac{3}{32}$		0.002		$\frac{1}{4}$	$\frac{7}{32}$	$\frac{3}{16}$	$\frac{1}{8}$	0.002	0.002
$\frac{5}{8}-\frac{7}{8}$	$\frac{3}{16}\times\frac{3}{32}$		$\frac{3}{16}\times\frac{1}{16}$		$\frac{3}{16}\times\frac{3}{16}$		$\frac{3}{16}\times\frac{1}{8}$		0.002		$\frac{5}{16}$	$\frac{9}{32}$	$\frac{1}{4}$	$\frac{3}{16}$	0.002	0.002
$\frac{13}{16}-1\frac{1}{4}$	$\frac{1}{4}\times\frac{1}{8}$		$\frac{1}{4}\times\frac{3}{32}$		$\frac{1}{4}\times\frac{1}{4}$		$\frac{1}{4}\times\frac{3}{16}$		0.002		$\frac{7}{16}$	$\frac{11}{32}$	$\frac{5}{16}$	$\frac{1}{4}$	0.002	0.002
$1\frac{5}{16}-1\frac{3}{8}$	$\frac{5}{16}\times\frac{5}{32}$		$\frac{5}{16}\times\frac{1}{8}$		$\frac{5}{16}\times\frac{5}{16}$		$\frac{5}{16}\times\frac{1}{4}$		0.002		$\frac{9}{16}$	$\frac{13}{32}$	$\frac{3}{8}$	$\frac{5}{16}$	0.002	0.002
$1\frac{7}{16}-1\frac{3}{4}$	$\frac{3}{8}\times\frac{3}{16}$		$\frac{3}{8}\times\frac{1}{8}$		$\frac{3}{8}\times\frac{3}{8}$		$\frac{3}{8}\times\frac{1}{4}$		0.002		$\frac{11}{16}$	$\frac{15}{32}$	$\frac{7}{16}$	$\frac{3}{8}$	0.002	0.002
$1\frac{13}{16}-2\frac{1}{4}$	$\frac{1}{2}\times\frac{1}{4}$		$\frac{1}{2}\times\frac{3}{16}$		$\frac{1}{2}\times\frac{1}{2}$		$\frac{1}{2}\times\frac{3}{8}$		0.0025		$\frac{7}{8}$	$\frac{19}{32}$	$\frac{5}{8}$	$\frac{1}{2}$	0.0025	0.0025
$2\frac{5}{16}-2\frac{3}{4}$	$\frac{5}{8}\times\frac{5}{16}$		$\frac{5}{8}\times\frac{7}{32}$		$\frac{5}{8}\times\frac{5}{8}$		$\frac{5}{8}\times\frac{7}{16}$		0.0025		$1\frac{1}{16}$	$\frac{23}{32}$	$\frac{3}{4}$	$\frac{5}{8}$	0.0025	0.0025
$2\frac{7}{16}-3\frac{1}{4}$	$\frac{3}{4}\times\frac{3}{8}$		$\frac{3}{4}\times\frac{1}{4}$		$\frac{3}{4}\times\frac{3}{4}$		$\frac{3}{4}\times\frac{1}{2}$		0.0025		$1\frac{1}{4}$	$\frac{7}{8}$	$\frac{7}{8}$	$\frac{3}{4}$	0.0025	0.0025
$3\frac{3}{8}-3\frac{3}{4}$	$\frac{7}{8}\times\frac{7}{16}$		$\frac{7}{8}\times\frac{5}{16}$		$\frac{7}{8}\times\frac{7}{8}$		$\frac{7}{8}\times\frac{5}{8}$		0.003		$1\frac{1}{2}$	1	$1\frac{1}{16}$	$\frac{7}{8}$	0.003	0.003
$3\frac{7}{8}-4\frac{1}{2}$	$1\times\frac{1}{2}$		$1\times\frac{3}{8}$		1×1		$1\times\frac{3}{4}$		0.003		$1\frac{3}{4}$	$1\frac{3}{16}$	$1\frac{1}{4}$	1	0.003	0.003
$4\frac{3}{4}-5\frac{1}{2}$	$1\frac{1}{4}\times\frac{5}{8}$		$1\frac{1}{4}\times\frac{7}{16}$		$1\frac{1}{4}\times1\frac{1}{4}$		$1\frac{1}{4}\times\frac{7}{8}$		0.003		2	$1\frac{7}{16}$	$1\frac{1}{2}$	$1\frac{1}{4}$	0.003	0.003
$5\frac{3}{4}-7\frac{3}{8}$	$1\frac{1}{2}\times\frac{3}{4}$		$1\frac{1}{2}\times\frac{1}{2}$		$1\frac{1}{2}\times1\frac{1}{2}$		$1\frac{1}{2}\times1$		0.003		$2\frac{1}{2}$	$1\frac{3}{4}$	$1\frac{3}{4}$	$1\frac{1}{2}$	0.003	0.003
$7\frac{1}{2}-9\frac{7}{8}$	$1\frac{3}{4}\times\frac{7}{8}$		..		$1\frac{3}{4}\times1\frac{3}{4}$		..		0.004		3	2	..	..	0.004	0.004
$10-12\frac{1}{2}$	2×1		..		2×2		..		0.004		$3\frac{1}{2}$	$2\frac{3}{8}$	..	..	0.004	0.004

Standard Keyway Tolerances: Straight Keyway – Width (W) +.005   Depth (T/2) +.010
                                                   –.000   –.000

Taper Keyway – Width (W) +.005   Depth (T/2) +.000
                              –.000   –.010

(Reprinted from ANSI B17.1–1967 (R2008). Copyright © American Society of Mechanical Engineers. All rights reserved.)

39 PLAIN WASHERS (INCHES)

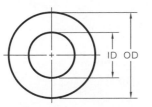

.938 X 2.25 X .165
TYPE A PLAIN WASHER

Dimensioned
Washer

In Screw Size Column
N= Narrow washer
W= Wide washer

Narrow Washer (N)
TYPE A PLAIN WASHERS

WIDE WASHER (W)

SCREW SIZE	ID SIZE	OD SIZE	THICK-NESS	SCREW SIZE	ID SIZE	OD SIZE	THICK-NESS
0.138	0.156	0.375	0.049	0.875N	0.938	1.750	0.134
0.164	0.188	0.438	0.049	0.875W	0.938	2.250	0.165
0.190	0.219	0.500	0.049	1.000N	1.062	2.000	0.134
0.188	0.250	0.562	0.049	1.000W	1.062	2.500	0.165
0.216	0.250	0.562	0.065	1.125N	1.250	2.250	0.134
0.250N	0.281	0.625	0.065	1.125W	1.250	2.750	0.165
0.250W	0.312	0.734	0.065	1.250N	1.375	2.500	0.165
0.312N	0.344	0.688	0.065	1.250W	1.375	3.000	0.165
0.312W	0.375	0.875	0.083	1.375N	1.500	2.750	0.165
0.375N	0.406	0.812	0.065	1.375W	1.500	3.250	0.180
0.375W	0.438	1.000	0.083	1.500N	1.625	3.000	0.165
0.438N	0.469	0.922	0.065	1.500W	1.625	3.500	0.180
0.438W	0.500	1.250	0.083	1.625	1.750	3.750	0.180
0.500N	0.531	1.062	0.095	1.750	1.875	4.000	0.180
0.500W	0.562	1.375	0.109	1.875	2.000	4.250	0.180
0.562N	0.594	1.156	0.095	2.000	2.125	4.500	0.180
0.562W	0.594	1.469	0.190	2.250	2.375	4.750	0.220
0.625N	0.625	1.312	0.095	2.500	2.625	5.000	0.238
0.625N	0.625	1.750	0.134	2.750	2.875	5.250	0.259
0.750W	0.812	1.469	0.134	3.000	3.125	5.500	0.284
0.750W	0.812	2.000	0.148				

40 METRIC WASHERS (MILLIMETERS)

Flat Washers

SCREW SIZE	ID SIZE	OD SIZE	THICK-NESS
3	3.2	9	0.8
4	4.3	12	1
5	5.3	15	1.5
6	6.4	18	1.5
8	8.4	25	2
10	10.5	30	2.5
12	13	40	3
14	15	45	3
16	17	50	3
18	19	56	4
20	21	60	4

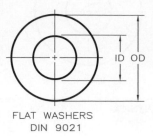

ID OD

FLAT WASHERS
DIN 9021

Wrought Washers

SCREW SIZE	ID SIZE	OD SIZE	THICK-NESS
2.6	2.8	5.5	0.5
3	3.2	6	0.5
4	4.3	8	0.5
5	5.3	10	1.0
6	6.4	1	1.5
8	8.4	15	1.5
10	10.5	18	1.5
12	13	20	2.0
14	15	25	2.0
16	17	27	2.0
18	19	30	2.5
20	21	33	2.5

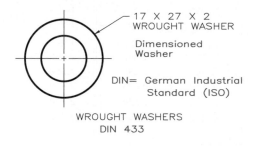

17 X 27 X 2
WROUGHT WASHER

Dimensioned
Washer

DIN= German Industrial
Standard (ISO)

WROUGHT WASHERS
DIN 433

41 REGULAR HELICAL SPRING LOCK WASHERS (INCHES AND MILLIMETERS)

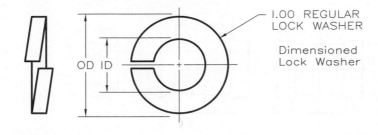

LOCK WASHERS—inches

SCREW SIZE	ID SIZE	OD SIZE	THICK-NESS
0.164	0.168	0.175	0.040
0.190	0.194	0.202	0.047
0.216	0.221	0.229	0.056
0.250	0.255	0.263	0.062
0.312	0.318	0.328	0.078
0.375	0.382	0.393	0.094
0.438	0.446	0.459	0.109
0.500	0.509	0.523	0.125
0.562	0.572	0.587	0.141
0.625	0.636	0.653	0.156
0.688	0.700	0.718	0.172
0.750	0.763	0.783	0.188
0.812	0.826	1.367	0.203
0.875	0.890	1.464	0.219
0.938	0.954	1.560	0.234
1.000	1.017	1.661	0.250
1.062	1.080	1.756	0.266
1.125	1.144	1.853	0.281
1.188	1.208	1.950	0.297
1.250	1.271	2.045	0.312
1.312	1.334	2.141	0.328
1.375	1.398	2.239	0.344
1.438	1.462	2.334	0.359
1.500	1.525	2.430	0.375

METRIC LOCK WASHERS—DIN 127 (Millimeters)

SCREW SIZE	ID SIZE	OD SIZE	THICK-NESS
4	4.1	7.1	0.9
5	5.1	8.7	1.2
6	6.1	1.1	1.6
8	8.2	12.1	1.6
10	10.2	14.2	2
12	12.1	17.2	2.2
14	14.2	20.2	2.5
16	16.2	23.2	3
18	18.2	26.2	3.5
20	20.2	28.2	3.5
22	22.5	34.5	4
24	24.5	38.5	5
27	27.5	41.5	5
30	30.5	46.5	6
33	33.5	53.5	6
36	36.5	56.5	6
39	39.5	59.5	6
42	42.5	66.5	7
45	45.5	69.5	7
48	49	73	7

(Reprinted from ANSI/ASME B18.21.1–2009. Copyright © American Society of Mechanical Engineers. All rights reserved.)

42 AMERICAN NATIONAL STANDARD 125-LB CAST IRON SCREWED FITTINGS (INCHES)

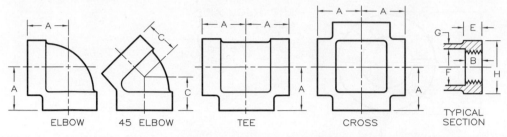

ELBOW   45 ELBOW   TEE   CROSS   TYPICAL SECTION

Nominal Pipe Size	A	C	B	E	F		G	H
			Min	Min	Min	Max	Min	Min
$\frac{1}{4}$	0.81	0.73	0.32	0.38	0.540	0.584	0.110	0.93
$\frac{3}{8}$	0.95	0.80	0.36	0.44	0.675	0.719	0.120	1.12
$\frac{1}{2}$	1.12	0.88	0.43	0.50	0.840	0.897	0.130	1.34
$\frac{3}{4}$	1.31	0.98	0.50	0.56	1.050	1.107	0.155	1.63
1	1.50	1.12	0.58	0.62	1.315	1.385	0.170	1.95
$1\frac{1}{4}$	1.75	1.29	0.67	0.69	1.660	1.730	0.185	2.39
$1\frac{1}{2}$	1.94	1.43	0.70	0.75	1.900	1.970	0.200	2.68
2	2.25	1.68	0.75	0.84	2.375	2.445	0.220	3.28
$2\frac{1}{2}$	2.70	1.95	0.92	0.94	2.875	2.975	0.240	3.86
3	3.08	2.17	0.98	1.00	3.500	3.600	0.260	4.62
$3\frac{1}{2}$	3.42	2.39	1.03	1.06	4.000	4.100	0.280	5.20
4	3.79	2.61	1.08	1.12	4.500	4.600	0.310	5.79
5	4.50	3.05	1.18	1.18	5.563	5.663	0.380	7.05
6	5.13	3.46	1.28	1.28	6.625	0.725	0.430	8.28
8	6.56	4.28	1.47	1.47	8.625	8.725	0.550	10.63
10	8.08	5.16	1.68	1.68	10.750	10.850	0.690	13.12
12	9.50	5.97	1.88	1.88	12.750	12.850	0.800	15.47
14 O.D.	10.40	—	2.00	2.00	14.000	14.100	0.880	16.94
16 O.D.	11.82	—	2.20	2.20	16.000	16.100	1.000	19.30

43 AMERICAN NATIONAL STANDARD 250-LB CAST IRON FLANGED FITTINGS (INCHES)

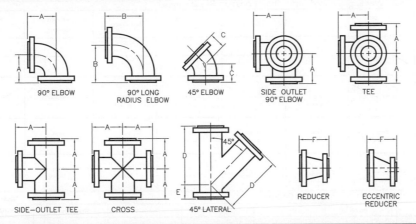

Nominal Pipe Size	Flanges			Fittings		Straight					
	Dia of Flange	Thickness of Flange (Min)	Dia of Raised Face	Inside Dia of Fittings (Min)	Wall Thickness	Center to Face 90 Deg Elbow Tees, Crosses and True "Y"	Center to Face 90 Deg Long Radius Elbow	Center to Face 45 Deg Elbow	Center to Face Lateral	Short Center to Face True "Y" and Lateral	Face to Face Reducer
						A	B	C	D	E	F
1	$4\frac{7}{8}$	$\frac{11}{16}$	$2\frac{11}{16}$	1	$\frac{7}{16}$	4	5	2	$6\frac{1}{2}$	2	....
$1\frac{1}{4}$	$5\frac{1}{4}$	$\frac{3}{4}$	$3\frac{1}{16}$	$1\frac{1}{4}$	$\frac{7}{16}$	$4\frac{1}{4}$	$5\frac{1}{2}$	$2\frac{1}{2}$	$7\frac{1}{4}$	$2\frac{1}{4}$	....
$1\frac{1}{2}$	$6\frac{1}{8}$	$\frac{13}{16}$	$3\frac{9}{16}$	$1\frac{1}{2}$	$\frac{7}{16}$	$4\frac{1}{2}$	6	$2\frac{3}{4}$	$8\frac{1}{2}$	$2\frac{1}{2}$	....
2	$6\frac{1}{2}$	$\frac{7}{8}$	$4\frac{3}{16}$	2	$\frac{7}{16}$	5	$6\frac{1}{2}$	3	9	$2\frac{1}{2}$	5
$2\frac{1}{2}$	$7\frac{1}{2}$	1	$4\frac{15}{16}$	$2\frac{1}{2}$	$\frac{1}{2}$	$5\frac{1}{2}$	7	$3\frac{1}{2}$	$10\frac{1}{2}$	$2\frac{1}{2}$	$5\frac{1}{2}$
3	$8\frac{1}{4}$	$1\frac{1}{8}$	$5\frac{11}{16}$	3	$\frac{9}{16}$	6	$7\frac{3}{4}$	$3\frac{1}{2}$	11	3	6
$3\frac{1}{2}$	9	$1\frac{3}{16}$	$6\frac{5}{16}$	$3\frac{1}{2}$	$\frac{9}{16}$	$6\frac{1}{2}$	8	4	$12\frac{1}{2}$	3	$6\frac{1}{2}$
4	10	$1\frac{1}{4}$	$6\frac{15}{16}$	4	$\frac{5}{8}$	7	9	$4\frac{1}{2}$	$13\frac{1}{2}$	3	7
5	11	$1\frac{3}{8}$	$8\frac{5}{16}$	5	$\frac{11}{16}$	8	$10\frac{1}{4}$	5	15	$3\frac{1}{2}$	8
6	$12\frac{1}{2}$	$1\frac{7}{16}$	$9\frac{11}{16}$	6	$\frac{3}{4}$	$8\frac{1}{2}$	$11\frac{1}{2}$	$5\frac{1}{2}$	$17\frac{1}{2}$	4	9
8	15	$1\frac{5}{8}$	$11\frac{15}{16}$	8	$\frac{13}{16}$	10	14	6	$20\frac{1}{2}$	5	11
10	$17\frac{1}{2}$	$1\frac{7}{8}$	$14\frac{1}{16}$	10	$\frac{15}{16}$	$11\frac{1}{2}$	$16\frac{1}{2}$	7	24	$5\frac{1}{2}$	12
12	$20\frac{1}{2}$	2	$16\frac{7}{16}$	12	1	13	19	8	$27\frac{1}{2}$	6	14
14	23	$2\frac{1}{8}$	$18\frac{15}{16}$	$13\frac{1}{4}$	$1\frac{1}{8}$	15	$21\frac{1}{2}$	$8\frac{1}{2}$	31	$6\frac{1}{2}$	16
16	$25\frac{1}{2}$	$2\frac{1}{4}$	$21\frac{1}{36}$	$15\frac{1}{4}$	$1\frac{1}{4}$	$16\frac{1}{2}$	24	$9\frac{1}{2}$	$34\frac{1}{2}$	$7\frac{1}{2}$	18
18	28	$2\frac{3}{8}$	$23\frac{5}{16}$	17	$1\frac{3}{8}$	18	$26\frac{1}{2}$	10	$37\frac{1}{2}$	8	19
20	$30\frac{1}{2}$	$2\frac{1}{2}$	$25\frac{9}{16}$	19	$1\frac{1}{2}$	$19\frac{1}{2}$	29	$10\frac{1}{2}$	$40\frac{1}{2}$	$8\frac{1}{2}$	20
24	36	$2\frac{3}{4}$	$30\frac{5}{16}$	23	$1\frac{5}{8}$	$22\frac{1}{2}$	34	12	$47\frac{1}{2}$	10	24
30	43	3	$37\frac{3}{16}$	29	2	$27\frac{1}{2}$	$41\frac{1}{2}$	15	....	....	30

(Reprinted from ANSI B16.1–2005. Copyright © American Society of Mechanical Engineers. All rights reserved.)

44 WELDING SYMBOLS

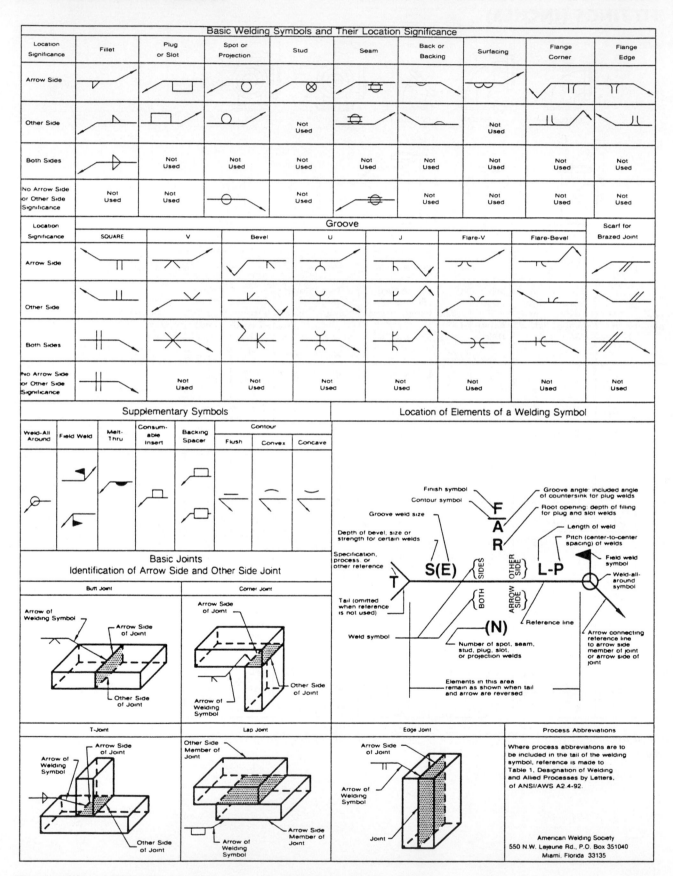

(Reprinted from ANSI/AWS 3.0–2007. *Copyright © American Welding Society.)*

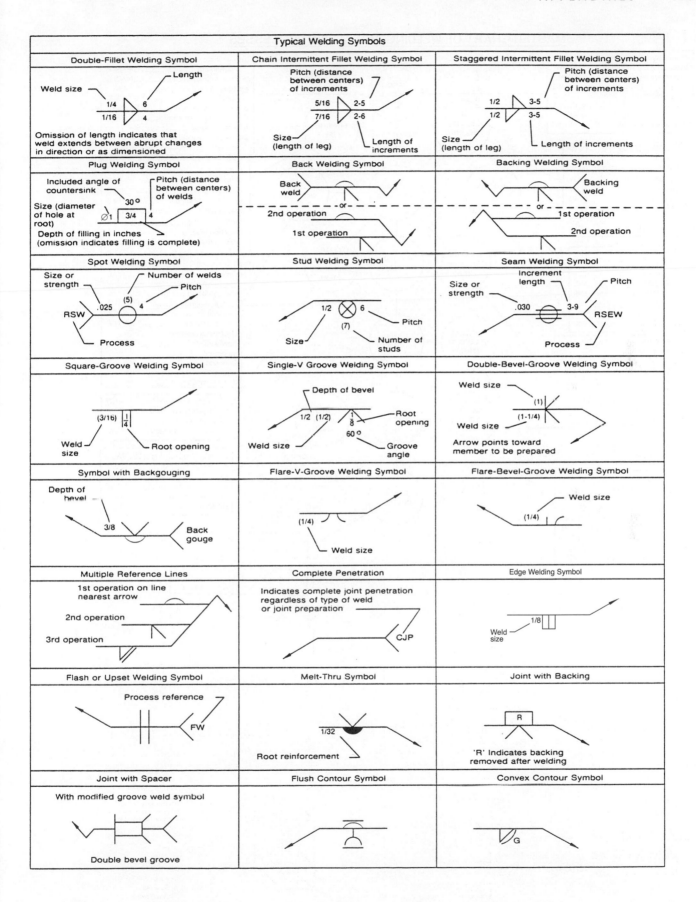

Typical Welding Symbols

Double-Fillet Welding Symbol	Chain Intermittent Fillet Welding Symbol	Staggered Intermittent Fillet Welding Symbol
Plug Welding Symbol	Back Welding Symbol	Backing Welding Symbol
Spot Welding Symbol	Stud Welding Symbol	Seam Welding Symbol
Square-Groove Welding Symbol	Single-V Groove Welding Symbol	Double-Bevel-Groove Welding Symbol
Symbol with Backgouging	Flare-V-Groove Welding Symbol	Flare-Bevel-Groove Welding Symbol
Multiple Reference Lines	Complete Penetration	Edge Welding Symbol
Flash or Upset Welding Symbol	Melt-Thru Symbol	Joint with Backing
Joint with Spacer	Flush Contour Symbol	Convex Contour Symbol

MASTER CHART OF WELDING AND ALLIED PROCESSES

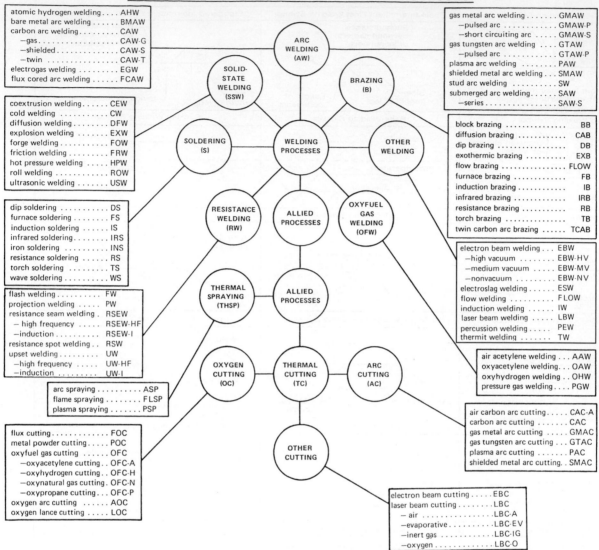

atomic hydrogen welding . . . . AHW
bare metal arc welding . . . . . . BMAW
carbon arc welding . . . . CAW
   —gas . . . . CAW-G
   —shielded . . . . CAW-S
   —twin . . . . CAW-T
electrogas welding . . . . EGW
flux cored arc welding . . . . . . FCAW

coextrusion welding . . . . . . CEW
cold welding . . . . CW
diffusion welding . . . . DFW
explosion welding . . . . EXW
forge welding . . . . FOW
friction welding . . . . FRW
hot pressure welding . . . . . HPW
roll welding . . . . ROW
ultrasonic welding . . . . USW

dip soldering . . . . DS
furnace soldering . . . . FS
induction soldering . . . . . . IS
infrared soldering . . . . IRS
iron soldering . . . . INS
resistance soldering . . . . . . RS
torch soldering . . . . TS
wave soldering . . . . WS

flash welding . . . . FW
projection welding . . . . . PW
resistance seam welding . RSEW
  — high frequency . . . . . RSEW-HF
  —induction . . . . RSEW-I
resistance spot welding . RSW
upset welding . . . . UW
  —high frequency . . . . . UW-HF
  —induction . . . . UW-I

arc spraying . . . . ASP
flame spraying . . . . FLSP
plasma spraying . . . . PSP

flux cutting . . . . FOC
metal powder cutting . . . . . POC
oxyfuel gas cutting . . . . . . OFC
  —oxyacetylene cutting . . OFC-A
  —oxyhydrogen cutting . . OFC-H
  —oxynatural gas cutting . OFC-N
  —oxypropane cutting . . . OFC-P
oxygen arc cutting . . . . . . AOC
oxygen lance cutting . . . . . LOC

ARC WELDING (AW)
SOLID-STATE WELDING (SSW)
BRAZING (B)
SOLDERING (S)
WELDING PROCESSES
OTHER WELDING
RESISTANCE WELDING (RW)
ALLIED PROCESSES
OXYFUEL GAS WELDING (OFW)
THERMAL SPRAYING (THSP)
ALLIED PROCESSES
OXYGEN CUTTING (OC)
THERMAL CUTTING (TC)
ARC CUTTING (AC)
OTHER CUTTING

gas metal arc welding . . . . GMAW
  —pulsed arc . . . . GMAW-P
  —short circuiting arc . . . . . GMAW-S
gas tungsten arc welding . . . . GTAW
  —pulsed arc . . . . GTAW-P
plasma arc welding . . . . PAW
shielded metal arc welding . . . SMAW
stud arc welding . . . . SW
submerged arc welding . . . . . . SAW
  —series . . . . SAW-S

block brazing . . . . BB
diffusion brazing . . . . CAB
dip brazing . . . . DB
exothermic brazing . . . . EXB
flow brazing . . . . FLOW
furnace brazing . . . . FB
induction brazing . . . . IB
infrared brazing . . . . IRB
resistance brazing . . . . RB
torch brazing . . . . TB
twin carbon arc brazing . . . . . . TCAB

electron beam welding . . . EBW
  —high vacuum . . . . EBW-HV
  —medium vacuum . . . . . EBW-MV
  —nonvacuum . . . . EBW-NV
electroslag welding . . . . ESW
flow welding . . . . FLOW
induction welding . . . . . . IW
laser beam welding . . . . . LBW
percussion welding . . . . . PEW
thermit welding . . . . TW

air acetylene welding . . . AAW
oxyacetylene welding . . . OAW
oxyhydrogen welding . . OHW
pressure gas welding . . . . PGW

air carbon arc cutting . . . . . CAC-A
carbon arc cutting . . . . CAC
gas metal arc cutting . . . . . GMAC
gas tungsten arc cutting . . . GTAC
plasma arc cutting . . . . PAC
shielded metal arc cutting . . SMAC

electron beam cutting . . . . . EBC
laser beam cutting . . . . LBC
  — air . . . . LBC-A
  —evaporative . . . . LBC-EV
  —inert gas . . . . LBC-IG
  —oxygen . . . . LBC-O

45 PIPING SYMBOLS, ANSI STANDARD

FLANGED	SCREWED	BELL & SPIGOT	WELDED	SOLDERED	
					Joint
					Elbow—90°
					Elbow—45°
					Elbow—Turned Up
					Elbow—Turned Down
					Elbow—Long Radius
					Reducing Elbow
					Tee
					Tee—Outlet Up
					Tee—Outlet Down
					Side Outlet Tee—Outlet Up
					Cross
					Reducer—Concentric
					Reducer—Eccentric
					Lateral
					Gate Valve—Elev.
					Globe Valve—Elev.
					Check Valve
					Stop Cock
					Safety Valve
					Expansion Joint
					Union
					Sleeve
					Bushing

(Reprinted from ANSI/ASME Y32.2.3–1949 (R1999). Copyright © American Society of Mechanical Engineers. All rights reserved.)

46 HEATING, VENTILATION, AND DUCTWORK SYMBOLS, ANSI STANDARD

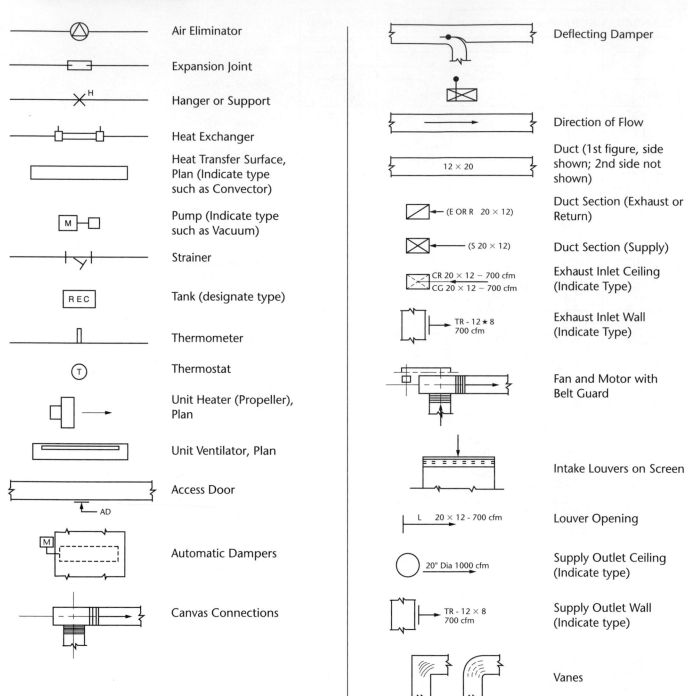

Air Eliminator

Expansion Joint

Hanger or Support

Heat Exchanger

Heat Transfer Surface, Plan (Indicate type such as Convector)

Pump (Indicate type such as Vacuum)

Strainer

Tank (designate type)

Thermometer

Thermostat

Unit Heater (Propeller), Plan

Unit Ventilator, Plan

Access Door

Automatic Dampers

Canvas Connections

Deflecting Damper

Direction of Flow

Duct (1st figure, side shown; 2nd side not shown)

Duct Section (Exhaust or Return)

Duct Section (Supply)

Exhaust Inlet Ceiling (Indicate Type)

Exhaust Inlet Wall (Indicate Type)

Fan and Motor with Belt Guard

Intake Louvers on Screen

Louver Opening

Supply Outlet Ceiling (Indicate type)

Supply Outlet Wall (Indicate type)

Vanes

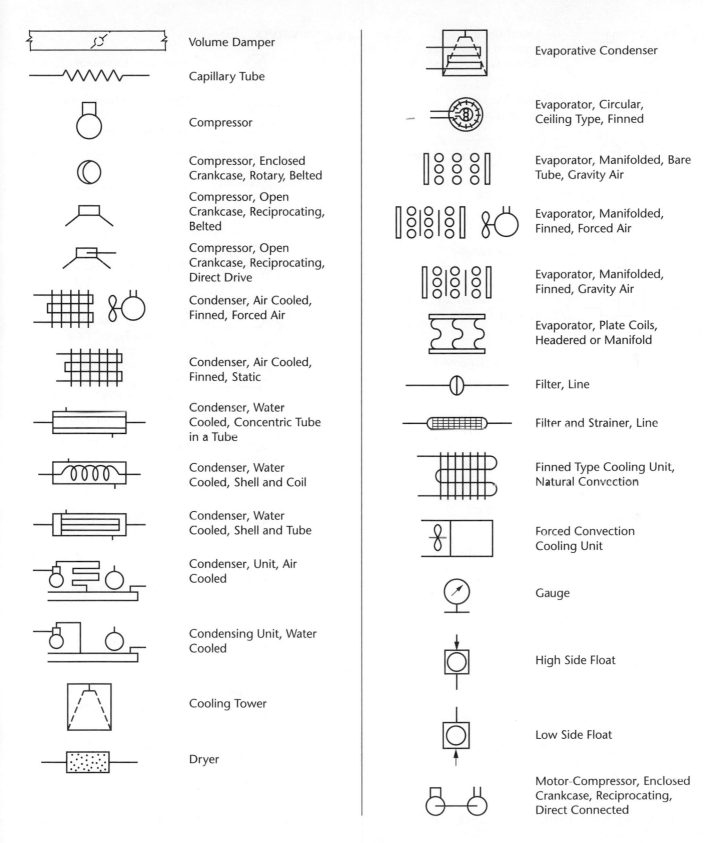

Volume Damper

Capillary Tube

Compressor

Compressor, Enclosed Crankcase, Rotary, Belted

Compressor, Open Crankcase, Reciprocating, Belted

Compressor, Open Crankcase, Reciprocating, Direct Drive

Condenser, Air Cooled, Finned, Forced Air

Condenser, Air Cooled, Finned, Static

Condenser, Water Cooled, Concentric Tube in a Tube

Condenser, Water Cooled, Shell and Coil

Condenser, Water Cooled, Shell and Tube

Condenser, Unit, Air Cooled

Condensing Unit, Water Cooled

Cooling Tower

Dryer

Evaporative Condenser

Evaporator, Circular, Ceiling Type, Finned

Evaporator, Manifolded, Bare Tube, Gravity Air

Evaporator, Manifolded, Finned, Forced Air

Evaporator, Manifolded, Finned, Gravity Air

Evaporator, Plate Coils, Headered or Manifold

Filter, Line

Filter and Strainer, Line

Finned Type Cooling Unit, Natural Convection

Forced Convection Cooling Unit

Gauge

High Side Float

Low Side Float

Motor-Compressor, Enclosed Crankcase, Reciprocating, Direct Connected

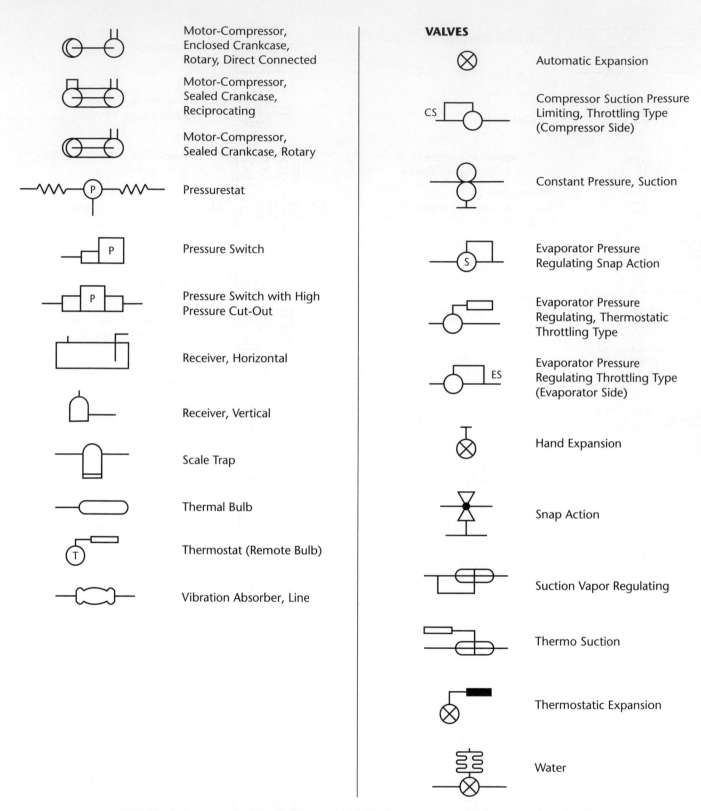

VALVES

	Motor-Compressor, Enclosed Crankcase, Rotary, Direct Connected
	Motor-Compressor, Sealed Crankcase, Reciprocating
	Motor-Compressor, Sealed Crankcase, Rotary
	Pressurestat
	Pressure Switch
	Pressure Switch with High Pressure Cut-Out
	Receiver, Horizontal
	Receiver, Vertical
	Scale Trap
	Thermal Bulb
	Thermostat (Remote Bulb)
	Vibration Absorber, Line

	Automatic Expansion
	Compressor Suction Pressure Limiting, Throttling Type (Compressor Side)
	Constant Pressure, Suction
	Evaporator Pressure Regulating Snap Action
	Evaporator Pressure Regulating, Thermostatic Throttling Type
	Evaporator Pressure Regulating Throttling Type (Evaporator Side)
	Hand Expansion
	Snap Action
	Suction Vapor Regulating
	Thermo Suction
	Thermostatic Expansion
	Water

47 ELECTRONICS SYMBOLS, ANSI STANDARD

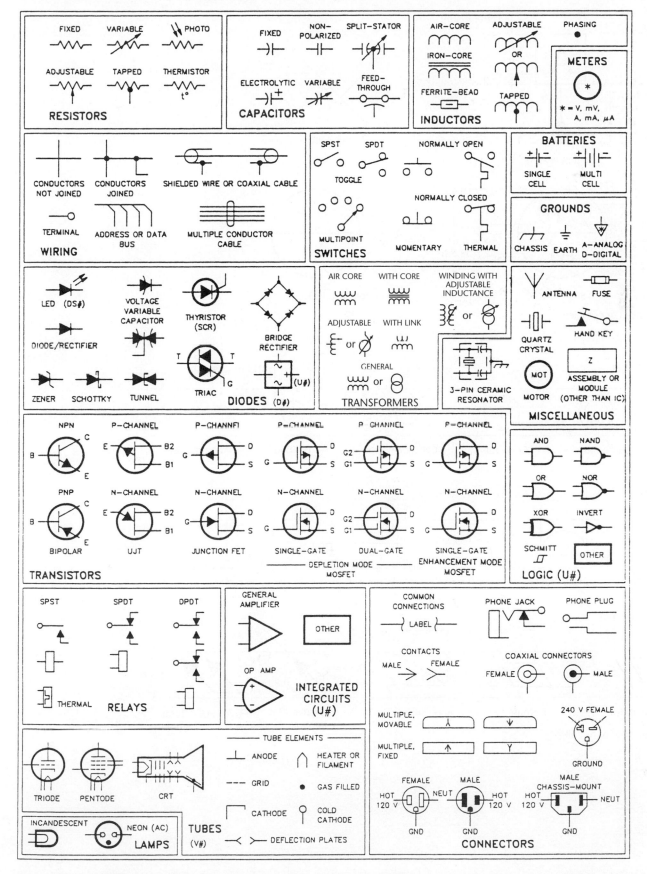

(*Reprinted from* IEEE 315A–1993/ANSI Y32.2–1989. *Copyright © IEEE.*)

48 GEOMETRIC DIMENSIONING AND TOLERANCING SYMBOLS

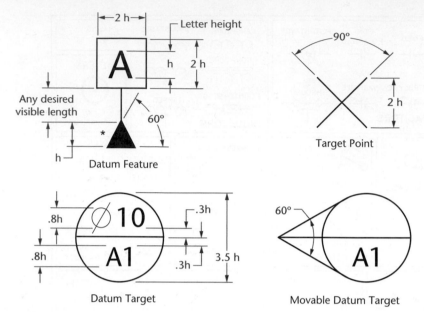

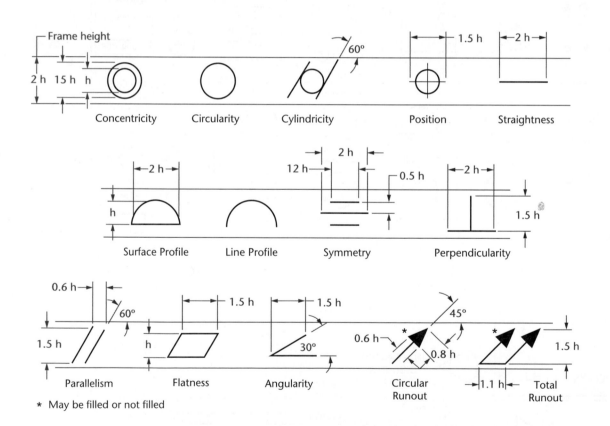

* May be filled or not filled

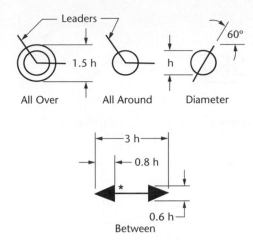

All Over   All Around   Diameter

Between

* May be filled or not filled

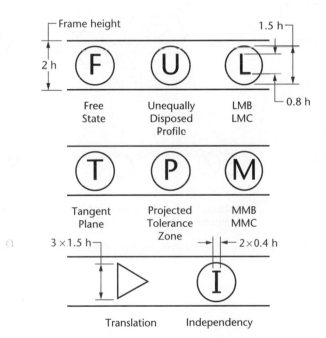

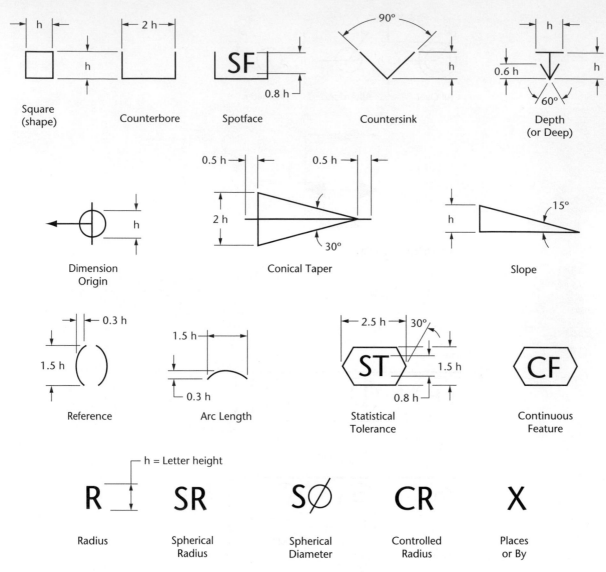

Square
(shape)

Counterbore

Spotface

Countersink

Depth
(or Deep)

Dimension
Origin

Conical Taper

Slope

Reference

Arc Length

Statistical
Tolerance

Continuous
Feature

h = Letter height

R   SR   SØ   CR   X

Radius

Spherical
Radius

Spherical
Diameter

Controlled
Radius

Places
or By

49 USEFUL FORMULAS FOR GEOMETRIC ENTITIES

Formulas for Circles

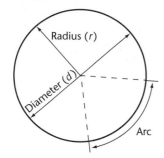

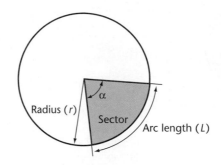

Circle*

Area	$A = \pi r^2$
	$A = 3.141r^2$
	$A = 0.7854d^2$
Radius	$r = d/2$
Diameter	$d = 2r$
Circumference	$C = 2\pi r$
	$C = \pi d$
	$C = 3.141d$

*Note: 22/7 and 3.141 are different approximations for π.

Sector of a Circle

Area	$A = \dfrac{3.141r^2\alpha}{360}$
Arc (length)	$L = \dfrac{2\pi r}{360}\alpha$
	$L = 0.01745r\alpha$
Angle	$\alpha = \dfrac{L}{0.01745r}$
Radius	$r = \dfrac{L}{0.01745\alpha}$

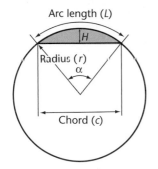

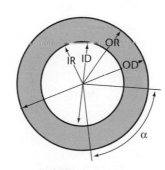

Segment of a Circle

Area	$A = \dfrac{1}{2}(r \bullet L - c\,(r - h))$
Arc (length)	$L = 0.01745r\alpha$
Angle	$\alpha = \dfrac{57.296L}{r}$
Height	$H = r - \dfrac{1}{2}\sqrt{4r^2 - c^2}$
Chord	$c = 2r\sin\alpha$

Circular Ring

Ring area	$A = 0.7854\,(OD^2 - ID^2)$
Ring sector area	$a = 0.00873\,\alpha(OR^2 - IR^2)$
	$a = 0.00218\,\alpha(OD^2 - ID^2)$

OD = outside diameter
ID = inside diameter
α = ring sector angle
OR = outside radius
IR = inside radius

Formulas for Triangles

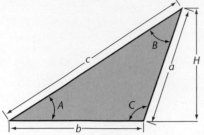

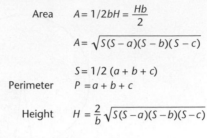

Obtuse angle triangle

Any Triangle

Area $A = 1/2bH = \dfrac{Hb}{2}$

$$A = \sqrt{S(S-a)(S-b)(S-c)}$$

$$S = 1/2\,(a+b+c)$$

Perimeter $P = a + b + c$

Height $H = \dfrac{2}{b}\sqrt{S(S-a)(S-b)(S-c)}$

Sum of angles $180° = A + B + C$

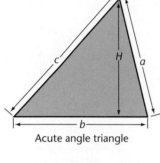

Acute angle triangle

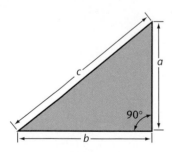

Equilateral Triangle

Area $A = a^2\dfrac{\sqrt{3}}{4} = 0.433\,a^2$

$$A = 0.577H^2$$

$$A = \dfrac{a^2}{2}\ \text{or}\ \dfrac{aH}{2}$$

Perimeter $P = 3a$

Height $H = \dfrac{a}{2}\sqrt{3} = 0.866a$

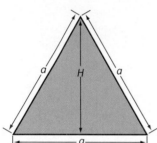

Right Triangle

Area $A = \dfrac{ba}{2}$

Perimeter $P = a + b + c$

Height $a = \sqrt{c^2 - b^2}$

Base $b = \sqrt{c^2 - a^2}$

Hypotenuse $c = \sqrt{a^2 - b^2}$

Formulas for Four-Sided Polygons

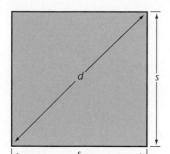

Square

Area $A = s^2$
          $A = 0.5d^2$
Side $s = 0.707d$
Diagonal $d = 1.414s$
Perimeter $P = 4s$

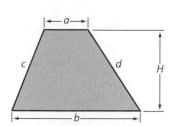

Rectangle

Area $A = ab$

Side a $a = \sqrt{d^2 - b^2}$

Side b $b = \sqrt{d^2 - a^2}$

Diagonal $d = \sqrt{a^2 - b^2}$

Perimeter $P = 2(a + b)$

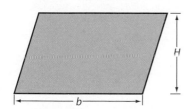

Parallelogram

Area $A = Hb$
Height $H = A/b$
Base $b = A/H$

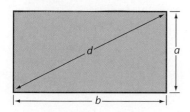

Trapezoid

Area $A = 1/2\,(a + b) \bullet H$
Perimeter $P = a + b + c + d$

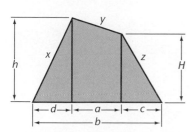

Trapezium

Area $A = \dfrac{a(H + h) + cH + dh}{2}$

Area   Another method is to divide
          the area into two triangles,
          find the area of each, and add
          the areas together.

Perimeter $P = b + x + y + z$

Formulas for Ellipses and Parabolas

Ellipse

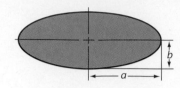

Area $A = \pi ab$
$A = 3.142ab$

Perimeter $P = 6.283 \cdot \dfrac{\sqrt{a^2 + b^2}}{2}$

Parabola

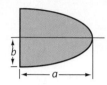

Area $A = 2/3\ ab$

Formulas for Regular Polygons

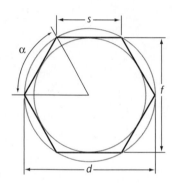

Multisided

Area $A = n\dfrac{s \cdot \frac{1}{2}f}{2}$

n = number of sides

Side $s = 2\sqrt{\frac{1}{2}d^2 - \frac{1}{2}f^2}$

Flats f = distance across flats; diameter of inscribed circle
Diagonal d = diameter of circumscribed circle
Perimeter P = sum of the sides
Angle $\alpha = 360/n$

Hexagon

Area $A = 0.866f^2$
$A = 0.650d^2$
$A = 2.598s^2$
Side $s = 0.577f$
$s = 0.5d$
Flats $f = 1.732s$
$f = 0.866d$
Diagonal $d = 2s$
$d = 1.155f$
Perimeter $P = 6s$
Angle $\alpha = 60°$

Octagon

Area $A = 0.828f^2$
$A = 0.707d^2$
$A = 4.828s^2$
Side $s = 0.414f$
$s = 0.383d$
Flats $f = 2.414s$
$f = 0.924d$
Diagonal $d = 2.613s$
$d = 1.083f$
Perimeter $P = 8s$
Angle $\alpha = 45°$

Formulas for 3D Shapes

Cube

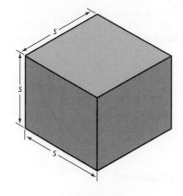

Volume	$V = s^3$
Surface area	$SA = 6s^2$
Side	$s = \sqrt[3]{V}$

Rectangular Prism

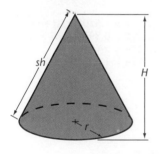

Volume	$V = lwh$
Surface area	$SA = 2(lw + lh + wh)$
Length	$l = V/hw$
Width	$w = V/lh$
Height	$h = V/lw$

Cone (Right Circular)

Volume $V = 1/3\ AH*$
A = area of base
$V = 1/3\pi r^2 H$
r = radius of base

Slant height $sh = \sqrt{r^2 + H^2}$

Surface area $SA = (1/2\ \text{perimeter of base} \bullet sh) + \pi r^2$
$SA = \pi r(sh) + \pi r^2$
Lateral surface area $LSA = \pi r(sh)$

*Note: True for any cone or pyramid

Pyramid

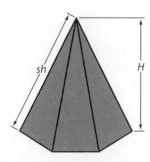

Volume $V = 1/3\ AH$
A = area of base
Surface area $SA = (1/2\ \text{perimeter of base} \bullet sh) + A$

Slant height $sh = \sqrt{r^2 + h^2}$

r = radius of circle circumscribed around base

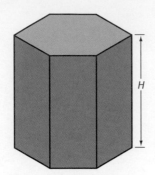

Prism

Volume $V = AH*$
            A = area of base
                 (see multisided polygon)
Surface area   SA = (area of each panel) + $2A$

*Note: True for any prism or cylinder with parallel bases.

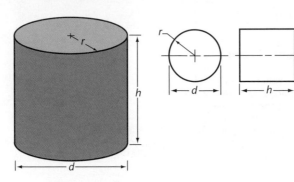

Cylinder (Right Circular)

Volume $V = Ah$
           $V = \pi r^2 h$
           $V = 0.7854d^2 h$
Surface area $SA = \pi dh + 2\pi r^2$
               $SA = 2\pi rh + 2\pi r^2$
               $SA = 6.283rh + 6.283r^2$
Lateral surface area $LSA = 2\pi rh$

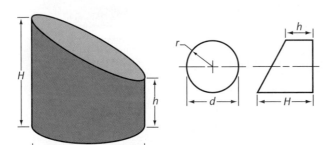

Frustum of a Cylinder

Volume $V = \pi r^2 \dfrac{H + h}{2}$

           $V = 1.5708r^2 (H + h)$
           $V = 0.3927d^2 (H + h)$
Lateral surface area $LSA = \pi r(H + h)$
                      $LSA = 1.5708d(H + h)$

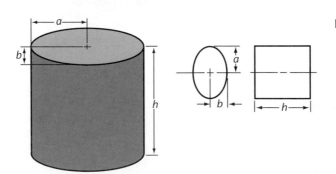

Elliptical Cylinder

Volume $V = \pi abh$

Lateral surface area $LSA = \pi h \sqrt{a^2 + b^2}$

INDEX